Approximate Computing Techniques

Alberto Bosio • Daniel Ménard • Olivier Sentieys
Editors

Approximate Computing Techniques

From Component- to Application-Level

Editors
Alberto Bosio
University of Lyon, ECL, INSA Lyon,
CNRS, UCBL, CPE Lyon, INL, UMR5270
Écully, France

Daniel Ménard
Univ. Rennes, INSA Rennes, IETR,
Rennes, France

Olivier Sentieys
University of Rennes, INRIA/IRISA
Lannion, France

ISBN 978-3-030-94707-1 ISBN 978-3-030-94705-7 (eBook)
https://doi.org/10.1007/978-3-030-94705-7

This Springer imprint is published by the registered company Springer Nature Switzerland AG
The registered company address is: Gewerbestrasse 11, 6330 Cham, Switzerland

Contents

1 **General Introduction** 1
Alberto Bosio, Daniel Ménard, and Olivier Sentieys

Part I Techniques for Approximate Computing

2 **Customizing Number Representation and Precision** 11
Olivier Sentieys and Daniel Ménard

3 **Hardware Level Approximations** 43
Ioannis Tsiokanos, George Papadimitriou, Dimitris Gizopoulos, and Georgios Karakonstantis

4 **Inexact Arithmetic Operators** 81
Lukas Sekanina, Zdenek Vasicek, and Vojtech Mrazek

5 **Approximate Computing at the Algorithmic Level** 109
Justine Bonnot, Alexandre Mercat, Erwan Nogues, and Daniel Ménard

Part II Methods and Tools for Approximate Computing

6 **Analysis of the Impact of Approximate Computing on the Application Quality** 145
Justine Bonnot, Daniel Ménard, and Karol Desnos

7 **Accuracy-Aware Compilers** 177
Sasa Misailovic

8 **Design Space Exploration Tools** 215
Mario Barbareschi, Salvatore Barone, Nicola Mazzocca, and Alberto Moriconi

9 **Wordlength Optimization of Fixed-Point Algorithms** 261
Gabriel Caffarena

Part III Approximate Computing Applied to Real-Life Applications

10 Exploiting Approximations in Real-Time Scheduling 287
Kamyar Mirzazad Barijough, Lin Huang, I-Hong Hou, Sachin S. Sapatnekar, Jiang Hu, and Andreas Gerstlauer

11 Security in an Approximated World: New Threats and Opportunities in the Approximate Computing Paradigm 323
Paolo Palmieri, Ilia Polian, and Francesco Regazzoni

12 Design, Verification, Test, and In-Field Implications of Approximate Digital Integrated Circuits 349
Alberto Bosio, Stefano Di Carlo, Patrick Girard, Annachiara Ruospo, Ernesto Sanchez, Alessandro Savino, Lukas Sekanina, Marcello Traiola, Zdenek Vasicek, and Arnaud Virazel

13 Approximate Computing for Fault Tolerance Mechanisms for Safety-Critical Applications 387
Gennaro S. Rodrigues, Fernanda L. Kastensmidt, and Alberto Bosio

14 Approximate Computing for Scientific Applications 415
Hartwig Anzt, Marc Casas, A. Cristiano I. Malossi, Enrique S. Quintana-Ortí, Florian Scheidegger, and Sicong Zhuang

15 Approximations in Deep Learning 467
Etienne Dupuis, Silviu Filip, Olivier Sentieys, David Novo, Ian O'Connor, and Alberto Bosio

Index 513

List of Figures

Fig. 1.1	Energy consumption trends in computing versus the world energy production. Source: SIA/SRC [3]	2
Fig. 1.2	Energy cost in a processor. Source: [4]	2
Fig. 1.3	Perceptual limitation and removal of instructions. Source: courtesy of Lukas Sekanina	4
Fig. 1.4	AxC design space exploration: accuracy vs. cost	5
Fig. 1.5	Computing system layers	6
Fig. 2.1	Fixed-point specification	13
Fig. 2.2	Example of a sequence of operations: data d is the output of the operation O_j and the input of operation O_k	19
Fig. 2.3	Floating-point representation	21
Fig. 2.4	Relative area, delay, and energy per operation between fixed-point and floating-point of adders for different bit-widths	31
Fig. 2.5	Relative area, delay, and energy per operation between fixed-point and floating-point of multipliers for different bit-widths	32
Fig. 2.6	2D K-means clustering golden output example, obtained using double-precision (64-bit) floating-point	33
Fig. 2.7	K-means clustering outputs for 8- and 16-bit floating-point and fixed-point, with accuracy target of 10^{-4}. (a) ct_float$_8$(5). (b) ac_fixed$_8$(3). (c) ct_float$_{16}$(5). (d) ac_fixed$_{16}$(3)	37
Fig. 2.8	Energy vs. classification error rate for K-Means clustering with stopping conditions of 10^{-4} (top), 10^{-3} (centre) and 10^{-2} (bottom)	38
Fig. 2.9	Fixed-point and floating-point energy per operation (pJ) vs MSE for FFT-16 for different bit-widths	39
Fig. 3.1	Data-paths in a 4-bit ripple carry adder, highlighting the longest latency path (*LLP1*) and two short latency paths (*SLP1* and *SLP2*)	45

Fig. 3.2 X-Gene 2 characterization results for one benchmark on three different chips (TTT, TFF, and TSS). Blue colour in bars represents the Safe region, grey represents the Unsafe region, and black represents the Crash region 52
Fig. 3.3 Impact of voltage down-scaling on gate delay variation [88] 54
Fig. 3.4 Carry propagation in a 6-bit ripple carry adder (RCA) 58
Fig. 3.5 Block diagram of the long latency paths prediction unit (LLPPU) 59
Fig. 3.6 Instruction flow and delay requirements across different paths and stages in a pipelined core 60
Fig. 3.7 Conventional path distribution of a pipelined design 61
Fig. 3.8 (a) Conventional path distribution and (b) path distribution after applying path shaping 62
Fig. 3.9 Average relative error (RE) under different delay increase levels for the original FPU with ($Strat1$) and without statically truncating the 44 LSBs (Orig), the FPU when path shaping and static truncation of 44 LSB are enabled ($Strat1 + Strat3$), and the FPU when path shaping, timing error prediction, and dynamic truncation of 44 LSBs are jointly considered ($Strat1 + Strat2 + Strat3$). (a) Raytrace. (b) CFD. (c) K-means. (d) Heartwall 64
Fig. 3.10 X-Gene 2 with the custom thermal adapters 68
Fig. 3.11 WER for DRAM operating under 0.618 s and 1.173 s at 50 °C 69
Fig. 3.12 WER for DRAM operating under 1.727 s and 2.283 s at 50 °C 69
Fig. 3.13 Temperature controller board [147] 70
Fig. 4.1 The 2-bit approximate multiplier proposed by Kulkarni et al. in [13] and its specification 84
Fig. 4.2 Quality configurable 2-bit multiplier according to [17] 86
Fig. 4.3 Miter for equivalence checking (a) and arithmetic error analysis (b). For the equivalence checking, the output E corresponds with $E(x) = [\![f(x) \neq \hat{f}(x)]\!]$. For arithmetic error analysis, the output E equals the error magnitude $E(x) = \text{nat}(f(x)) - \text{nat}(\hat{f}(x))$ 89
Fig. 4.4 Overview of formal error analysis approaches 90
Fig. 4.5 Example of candidate designs in the objective space (power, MAE). Four types of designs are distinguished: non-dominated solutions (circles), dominated solutions (crosses), infeasible solutions (squares), and original solution (diamond) 93
Fig. 4.6 Employing Cartesian genetic programming for automated design of approximate circuits 98

Fig. 4.7 The 8-bit approximate multipliers (black points) that were selected to EvoApprox8b-Lite from all the evolved approximate multipliers (gray points) and compared to the former version of EvoApprox8b library (red points), broken array multipliers (green points) and truncated multipliers (blue points) ... 100
Fig. 4.8 The computational requirements of the WCEGT procedure proving that $e_{wce} > \mathscr{T}$ of 8-bit approximate multipliers taken from EvoApprox library ... 101
Fig. 4.9 Relative difference between exact and estimated errors for 8-bit approximate multipliers. The whiskers show the 2nd percentile and the 98th percentile. Triangles indicate that there is even higher RD than shown ... 102
Fig. 4.10 Relative difference between exact and estimated errors for 16-bit approximate multipliers. The whiskers show the 2nd percentile and the 98th percentile. Triangles indicate that there is even higher RD than shown ... 102
Fig. 4.11 Relative difference between exact and estimated errors for 16-bit approximate adders. The whiskers show the 2nd percentile and the 98th percentile. Triangles indicate that there is even higher RD than shown ... 102
Fig. 5.1 Exploitation of slack time to reduce the energy consumption for real-time processing. (a) Initial processing. (b) DPM approach. (c) DVFS approach ... 110
Fig. 5.2 Illustration of the loop perforation concept. (a) Initial *for-loop*. (b) Loop perforation with the execution of one iteration loop after n_1. (c) Loop perforation with the skip of one iteration after n_2 ... 112
Fig. 5.3 Illustration of the loop perforation concept at run-time. (a) Dynamic perforation: the number of skipped iterations is decided at run-time. (b) Selective perforation ... 114
Fig. 5.4 Illustration of the difference between loop perforation and early termination for an iterative refinement algorithm. (a) Initial iterative refinement algorithm. (b) Loop perforation. (c) Early termination. The process is stopped when the condition criterion is reached ... 115
Fig. 5.5 Three techniques to handle OSSE algorithms. (a) Exhaustive search: each branch represents a full solution computation. (b) Early termination: the exploration of branches which cannot lead to the best solution is stopped before the end. (c) SSSR technique: coarse estimation is carried-out and a refinement is applied to the best candidates ... 116
Fig. 5.6 Algorithm classification of an HEVC decoder ... 125

Fig. 5.7 Average relative complexity of data processing blocks in an optimized HEVC decoder on a general purpose processor. Input sequences: Kimono RA 1920x1080 with Quantization Parameter from 22 to 32 by steps of 2 126
Fig. 5.8 Interpolation filter of the MC block approximation 127
Fig. 5.9 In-loop filter skip 128
Fig. 5.10 Power consumption of *computation approximation* decoder vs SSIM distortion 130
Fig. 5.11 Power consumption of *computation skip* decoder vs SSIM distortion 130
Fig. 5.12 Block diagram of HEVC intra encoder composed by several blocks: Intra Picture Process (IPP), Intra Picture Estimation (IPE), Transform (T), Quantization (Q), Inverse Quantization (Q^{-1}), Inverse Transform (T_{-1}), Current Picture Buffer (CPB), Deblocking Filter (DF), Sample-Adaptive Offset (SAO) and Entropy Coding (EC) 131
Fig. 5.13 Quad-tree structure of a Coding Tree Block (CTB) divided into Coding Block (CB) 132
Fig. 5.14 Neighboring samples used for intra-prediction in an $N \times N$ PB with $N = 8$ and intra-prediction modes 133
Fig. 5.15 Intra-prediction steps 133
Fig. 5.16 Pareto in Rate-Energy space from the set of parameters F and N_d 136
Fig. 5.17 Pareto in Rate-Energy space generated from all $3 \times 3 \times 2 \times 2 = 36$ combinations of parameter values 137
Fig. 5.18 Pareto in Rate-Energy space from the set of Θ defined in the table 138
Fig. 5.19 Pareto in Rate-Energy space from the CT OSSE, the IM OSSE and the combination of the two OSSE: CT & IM 138
Fig. 6.1 Different steps to analyze the impact of approximation on quality.. 146
Fig. 6.2 Evolution of maximal error amplitude according to its occurrence probability 147
Fig. 6.3 Probability mass function of almost correct adder, $N = 16, C = 4$, and approximate array multiplier, $N = 16$ 150
Fig. 6.4 Comparison of the simulation time for the BALL and floating-point simulation of two inexact arithmetic operators..... 153
Fig. 6.5 The different abstraction levels and times for emulation 155
Fig. 6.6 Widrow model for fixed-point quantization noise 157
Fig. 6.7 Illustration of the simulation-based technique to determine the error metric 161
Fig. 6.8 Model for the computation of output quantization error power based on noise sources e_i and gains α_i 168

Fig. 7.1 Examples of accuracy-aware optimizations (the approximation code is marked in red). (a) Data structure optimization. (b) Loop perforation. (c) Reduction sampling. (d) Approximate tiling. (e) Function substitution. (f) Dropping tasks. (g) Remove locks 181
Fig. 7.2 The approaches for accuracy-aware optimization. We start with the original program and produce the approximate program. The analysis-based compiler takes into consideration the annotations on the kernels, while the profiling-based compiler treats the whole program as one entity 183
Fig. 7.3 The impact of errors on end-to-end acceptability: the kernel computation calculates the correct pixel value with only specified frequency (left) or computes it with the specified absolute error in each pixel component (right) 184
Fig. 7.4 Sensitivity profiling finds the set of perforatable loops S in application A given representative inputs I, accuracy metric Q, and accuracy goal b 188
Fig. 7.5 Results of sensitivity profiling of x264 191
Fig. 7.6 x264 Tradeoff space and Pareto set 192
Fig. 7.7 Chisel overview 194
Fig. 7.8 Model of approximate hardware, with exact (blue) and approximate (orange) components 195
Fig. 7.9 Chisel's intermediate language 197
Fig. 8.1 Quality constraint circuit [25] 220
Fig. 8.2 Sequential quality constraint circuit [27] 223
Fig. 8.3 Hardware model used for simulation in [44] 232
Fig. 8.4 The delta-debugging algorithm used in [18] 235
Fig. 8.5 The PRECIMONIOUS flow [18] 235
Fig. 8.6 The ACCEPT compiler flow [46] 237
Fig. 8.7 Example of hypercubes (a) and CDFs (b) in the ASAC tool [48]. 239
Fig. 8.8 The $\mathbb{IDEA}$ flow, which includes Clang-Chimera and Bellerophon tools... 242
Fig. 8.9 AST representation of a *for* loop 243
Fig. 8.10 Definition of a new mutator class.................................. 249
Fig. 8.11 Defining and registering of operators.............................. 252
Fig. 8.12 Pareto front estimation provided by Bellerophon................. 255
Fig. 9.1 Fixed-point representation. Example a shows a number with no integer part. Example b shows a number with both integer and fractional parts 262
Fig. 9.2 Fixed-point refinement diagram. The wordlength optimization process is detailed 264
Fig. 9.3 WLO techniques 265
Fig. 9.4 WLO with fast estimation of the error 271

Fig. 10.1 Case 1: service preserving interval for an overrun job 297
Fig. 10.2 Case 2: service preserving interval for an active high-criticality job without overrun 297
Fig. 10.3 Case 3: service preserving interval for an immediate newly coming high-criticality job 298
Fig. 10.4 Active low-criticality job with deadline after $t^* + P$ 299
Fig. 10.5 Active low-criticality job with deadline before $t^* + P$ 299
Fig. 10.6 Acceptance ratio vs. normalized utilization of 4 processors ($K_L = 0.1$, $K_H = 5$) 306
Fig. 10.7 Acceptance ratio versus normalized utilization of 4 processors with consideration of overhead 307
Fig. 10.8 Acceptance ratio vs normalized utilization of 8 processors 308
Fig. 10.9 Mean error (with standard deviation) vs. normalized utilization of 8 processors ($K_L = 0.1$, $K_H = 5$) 308
Fig. 10.10 Examples of distributed task graphs. (a) A graph with linear task chain. (b) A graph with multiple source tasks 310
Fig. 10.11 Comparison of different schedules for graph of Fig. 10.10b. (a) A pure data-driven schedule. (b) A schedule with uniform latency budget distribution. (c) A schedule with optimized latency budget distribution........ 312
Fig. 10.12 A linear task graph with N tasks mapped to N hosts 313
Fig. 10.13 A task graph with two parallel task chains 316
Fig. 11.1 Scope of AxC techniques discussed in this chapter 324
Fig. 11.2 Overview of this chapter 326
Fig. 11.3 Adversarial attack example: Image classified as `trombone` with confidence 24.67% (a); perturbed image classified as `mousetrap` with confidence 55.07% (b); perturbations (differences) between the two images (c). Original image from https://commons.wikimedia.org/wiki/Trombone 337
Fig. 11.4 Adversarial attack example: Perturbations calculated by the method from [60] for image from Fig. 11.3a, and the results of adding these perturbations weighted by $\varepsilon = 0.01$ and 0.02 339
Fig. 11.5 Adversarial attack example cont'd: Results of adding perturbations from Fig. 11.4 to image from Fig. 11.3a, weighted by $\varepsilon = 0.05$ and 0.10 340
Fig. 11.6 4-input neuron based on stochastic computing with stanh FSM as activation function (a) and example calculation using its multiply-accumulate part (b)............................. 342

Fig. 11.7 Hybrid binary-stochastic VGG19-SC network used for analysis of adversarial attacks. The network structure corresponds to VGG-19 [67]. The stochastic layer and the randomness-injection circuits (RICs) are shown in black ... 343
Fig. 11.8 Randomness-injection circuit (RIC) [59] ... 343
Fig. 12.1 AxIC design and manufacturing flow ... 350
Fig. 12.2 Digital testing [5] ... 352
Fig. 12.3 Two-bit multiplier example ... 358
Fig. 12.4 Fault Injection based functional approximation ... 359
Fig. 12.5 Approximate two-bit multiplier ... 359
Fig. 12.6 (b) Example of an approximate 1-bit full adder, obtained from the accurate 1-bit full adder in (a). The subfigure (c) shows the truth tables of the two circuits: for each input, the output bit values are reported (S and C_{out}), as well as their unsigned integer representation, calculated as $S * 2^0 + C_{out} * 2^1$. Subfigure (d) reports the error thresholds for the approximate circuit, for different error metrics 362
Fig. 12.7 Approximation-Aware (AxA) fault classification concept ... 365
Fig. 12.8 Approximation-Aware (AxA) test generation methodology ... 368
Fig. 12.9 Approximation-Aware (AxA) test application methodology ... 371
Fig. 12.10 Custom data type ... 372
Fig. 12.11 Fault injection scenario ... 373
Fig. 12.12 Two input neuron Bayesian Network example. (a) Neuron with sigmoid activation function. (b) Neuron's Bayesian network model ... 379
Fig. 12.13 Bayesian conditional probability tables meaning example ... 381
Fig. 13.1 Fault tolerance techniques classification ... 393
Fig. 13.2 Inaccuracy for each ATMR by data precision design applied to a 2×2 matrix multiplication. Source: [40] ... 399
Fig. 13.3 Reliability for each ATMR configuration for an acceptance threshold of 0.01. Source: [40] ... 402
Fig. 13.4 Reliability for each ATMR configuration for an acceptance threshold of 1. Source: [40] ... 402
Fig. 13.5 Diagram of the proposed ATMR method. Source: [42] ... 405
Fig. 13.6 Number of ATMR tasks with errors for a $\approx$0% difference threshold between the tasks outputs and golden value, on the single-precision version of the Newton-Raphson algorithm. Source: [41] ... 407
Fig. 13.7 Number of ATMR tasks with errors for a 2% difference threshold between the tasks outputs and golden value, on the single-precision version of the Newton-Raphson algorithm. Source: [41] ... 407

Fig. 13.8 Number of ATMR tasks with errors for a 5% difference threshold between the tasks outputs and golden value, on the single-precision version of the Newton-Raphson algorithm. Source: [41] ... 408
Fig. 13.9 Number of ATMR tasks with errors for a ≈0% difference threshold between the tasks outputs and golden value, on the double-precision version of the Newton-Raphson algorithm. Source: [41] ... 409
Fig. 13.10 Number of ATMR tasks with errors for a 2% difference threshold between the tasks outputs and golden value, on the double-precision version of the Newton-Raphson algorithm. Source: [41] ... 409
Fig. 13.11 Number of ATMR tasks with errors for a 5% difference threshold between the tasks outputs and golden value, on the double-precision version of the Newton-Raphson algorithm. Source: [41] ... 410
Fig. 14.1 Acceleration of the adaptive precision Jacobi in a 2-segment realization for a sparse matrix stored in CSR format (left) and ELLPACK format (right), respectively ... 425
Fig. 14.2 Benefits of a BiCGSTAB solver enhanced with an adaptive precision block-Jacobi preconditioner relative to a realization that includes a conventional full precision preconditioner in the same method ... 429
Fig. 14.3 The ADt on a 2-GPU system. Variables include: weights which go through the ADt procedure and biases which are sent directly to the GPUs to build the network model together with the unpacked weights ... 437
Fig. 14.4 Bitpack implemented with AVX2, RoundTo=3 ... 440
Fig. 14.5 Alexnet training considering 32 and 16 batch sizes. The two upper plots show the top-5 validation error evolution of *baseline*, *oracle* and A^2DTWP. The two bottom plots provide information on the performance improvement of *oracle* and A^2DTWP against *baseline* during the training process. Experiments run on the x86 system ... 445
Fig. 14.6 VGG training considering 64 and 32 batch sizes. The two upper plots show the top-5 validation error evolution of *baseline*, *oracle* and A^2DTWP. The two bottom figures provide information on the performance improvement of *oracle* and A^2DTWP against *baseline* during the training process. Experiments run on the x86 system ... 446

Fig. 14.7 Normalized execution times of the A^2DTWP and the *oracle* policies with respect to the baseline. Results obtained on the x86 system appear in the upper plot while the evaluation on the POWER system appears at the bottom 447
Fig. 14.8 Normalized execution time of A^2DTWP with respect to *baseline* considering the Imagenet1000 data set. Training for Alexnet, VGG, and Resnet considers up to 20, 8, and 16 epochs, respectively 451
Fig. 14.9 Data type configurations for different quality constraints of the considered reference models. Too low and too high exponent field widths are not required. Allowing for larger quality margins against the reference allows to further reduce the mantissa field. (a) $\Delta Q = 0.000\%$. (b) $\Delta Q < 0.01\%$. (c) $\Delta Q < 0.1\%$. (d) $\Delta Q < 1\%$. (e) $\Delta Q < 2\%$. (f) $\Delta Q < 5\%$ 455
Fig. 14.10 Results of our architecture search compared with reference models. Each dot represents a model according to its size and the obtained accuracy on the CIFAR-10 validation set. Our search finds results over five orders of magnitude and, in particular, finds various models that are much smaller than out-of-the box models. In the restricted IoT domain, our search delivers models that outperform the reference with a wide margin for fixed constraints 457
Fig. 15.1 A basic DNN example and the associated terminology (adapted from [11, Figure 1.3]). (**a**) Artificial neuron. (**b**) Simple neural network example 470
Fig. 15.2 A backpropagation example through a neural network (adapted from [11, Figure 1.6]). (a) Compute the gradient of the loss relative to the layer inputs ($\frac{\partial \ell}{\partial x_i} = \sum_j w_{ij} \frac{\partial \ell}{\partial y_j}$). (b) Compute the gradient of the loss relative to the weights ($\frac{\partial \ell}{\partial w_{ij}} = \frac{\partial \ell}{\partial y_j} x_i$) 471
Fig. 15.3 Expanded view of a typical 2D CONV layer inside a CNN 473
Fig. 15.4 A visual representation of LeNet-5 (adapted from [14, Fig. 2]), an early example of a CNN that promoted the subsequent development of Deep Learning. It contains the main layers that are usually found in CNNs: convolutional, pooling, and fully connected 474
Fig. 15.5 Different types of approximation techniques for DNN inference 476

Fig. 15.6 Diagrams for bit representations of various numerical formats discussed in the context of DL quantization in this chapter. Red, green, and blue shading are used to represent mantissa (M), exponent (E), and sign (S) bits, respectively. In (a), the 16-bit IEEE 754 `float16` *floating-point* format is shown (corresponding to $(-1)^S \times 2^{E-15} \times 1.M_2$ for normalized values), with 1 sign bit, 5 exponent bits, and 10 mantissa bits. (b) illustrates a 16-bit *signed integer* format. By choosing a *fixed* splitting point for integer (I) and fractional (F) parts in the mantissa ($M := I.F$), it can also serve as a representation for a *fixed-point* format (namely to $(-1)^S \times I_2.F_2$). Additionally, (b) can represent a form of *logarithmic number system* (see, for instance, [88]), with the encoded value being $(-1)^S \times 2^M = (-1)^S \times 2^{I.F}$. Part (c) exemplifies a *block floating-point* format, namely the `flex16+5` format [89] with a 15-bit mantissa and 5-bit shared exponent ... 481
Fig. 15.7 Weight sharing techniques allow network compression by storing indices instead of values ... 482
Fig. 15.8 Distribution of the weights composing the first layer of a trained ResNet50V2 [26], original (top), with only 8 (middle) and 16 (bottom) shared values ... 483
Fig. 15.9 The various scopes of applying weight sharing ... 485
Fig. 15.10 Different granularities of pruning in a 4-dimensional weight tensor for DNN inference (adapted from [110, Figure 1]) ... 486
Fig. 15.11 Mixed-precision training iteration for a network layer (adapted from [147, Fig. 1]) ... 491
Fig. 15.12 The loss scaling procedure for updating the master weights in mixed-precision training ... 492
Fig. 15.13 Summary of the precision settings for (a) the GEMM operations during the forward and backward passes in backpropagation and (b) the AXPY operations during a standard SGD weight update process (adapted from [152, Fig. 2]) ... 495
Fig. 15.14 The low-precision training flow with the S2FP8 format, where the truncation function T corresponds to $T(X) = \left[2^{-\beta}\left\{\text{round}_{\text{FP8}}(2^{\beta}|X|^{\alpha})\right\}\right]^{1/\alpha}$. The forward and backward GEMM operations use only S2FP8 values, whereas the weight update step uses FP32 master weights (adapted from [155, Fig. 4]) ... 496

Fig. 15.15 Comparing the systolic arrays based architecture of the ASIC Google TPU [164] (top) and the Processing Element (PE) of the FPGA grid-based Eyeriss [163] architecture (bottom) 499

List of Tables

Table 2.1 IEEE 754 normalized floating-point representation 22
Table 2.2 Cost, delay, and power of floating-point addition vs. integer addition 23
Table 2.3 Cost, delay, and power of FlP multiplication vs. integer multiplication.. 24
Table 2.4 Main properties of the custom floating-point libraries AC_FLOAT, CT_FLOAT, and FLOPOCO 28
Table 2.5 Comparative results for 16-bit FlP addition/subtraction with $F_{\mathtt{clk}} = 200\text{MHz}$ 28
Table 2.6 Comparative results for 16-bit FlP multiplication with $F_{\mathtt{clk}} = 200\text{MHz}$ 29
Table 2.7 Comparative results for 32-bit FlP addition/subtraction with $F_{\mathtt{clk}} = 200\text{MHz}$ 29
Table 2.8 Comparative results for 32-bit FlP multiplication with $F_{\mathtt{clk}} = 200\text{MHz}$ 29
Table 2.9 8- and 16-bit area, energy, and accuracy for K-means clustering experiment 36
Table 3.1 Quality degradation in terms of Relative Error (RE) under "random bit errors" induced by timing errors and "deterministic errors" induced by deliberately setting the last 32 bits of each operand to 0 in a floating-point addition 57
Table 4.1 Parameters of 2-bit and 8-bit exact and configurable approximate multipliers according to [16, 17]. e_{mae} and e_{wce} denote the mean absolute error and the worst-case error 86
Table 4.2 Selected automated approximation methods, benchmark problems used to evaluate them, and the error evaluation approaches 98

Table 4.3 The number of approximate implementations of arithmetic circuits in extended EvoApprox library (December 2019) 99
Table 5.1 Filter size per configuration 128
Table 5.2 Percentage of block skipping per configuration 129
Table 5.3 Energy reduction opportunities (in J) [32] 134
Table 6.1 Various application quality metrics depending on the nature of the application 152
Table 6.2 Ratio r between BALL simulation and simulation of accurate floating-point operation times for 32-bit operators 154
Table 6.3 Comparison between IA and AA 167
Table 8.1 Summary of tools for digital circuits approximation 219
Table 8.2 Summary of tools for software approximation 231
Table 9.1 All compared techniques. Checkmarks indicate if a technique is considered in one comparison work 276
Table 11.1 A list of current libraries supporting multiple FHE schemes 335
Table 12.1 Error magnitude example 360
Table 12.2 Error metric values in presence of different faults affecting the approximate circuit in Fig. 12.6b, and fault coverage report for the exhaustive test set 364
Table 12.3 Values of the AxIC in Fig. 12.6b when the input vectors are applied in presence of the faults 369
Table 12.4 LeNet-5 data type accuracy loss [%] 375
Table 12.5 LeNet-5 fault injection outcomes w.r.t. Golden Std. 376
Table 12.6 LeNet-5 fault list for injection campaigns 378
Table 12.7 Prediction average error 382
Table 13.1 Area usage and performance latency of the ATMR by data reduction designs for 2×2 and 3×3 matrix multiplications 400
Table 13.2 Exhaustive onboard fault injection emulation results for a 2×2 matrix multiplication 403
Table 13.3 Execution time overheads of ATMR configurations applied to the Newton-Raphson algorithm 405
Table 13.4 Error masking for each ATMR configuration variating thresholds 411
Table 14.1 Neural network configurations: the convolutional layer parameters are denoted as "conv<receptive field size>-<number of channels>". The ReLU activation function is not shown for brevity. The building blocks of Resnet and the number of times they are applied are shown in a single cell 442

Table 14.2 Performance profiles of both the A^2DTWP and the 32-bit floating point approaches expressed in milliseconds on the x86 system. We consider the VGG network model with batch size 64 450
Table 14.3 Performance profiles of both the A^2DTWP and the 32-bit Floating Point approaches expressed in milliseconds on the POWER system. We consider the VGG network model with batch size 64 450
Table 14.4 Established reference network architectures 454
Table 15.1 Recent evolution of DNNs for image classification on the ImageNet dataset 475
Table 15.2 Estimated cost of training recent NLP models in terms of power, time, and CO_2 emissions. 475

Chapter 1
General Introduction

Alberto Bosio, Daniel Ménard, and Olivier Sentieys

1.1 Introduction

Energy efficiency is definitely one of the major driving forces of current computer industry, which is moreover relevant from supercomputers and clouds to small portable personal electronics and sensors. A good picture of the current situation and trend is illustrated in Fig. 1.1, which shows that by 2040 computers will need more electricity than the world energy resources can generate [1]. A similar trend exists on the communications side where energy consumption in mobile broadband networks and mobile terminals is comparable to datacentres. In addition to the traditional personal communications, the Internet-of-Things (IoT) will soon connect up to 50 billion devices [2] through wireless networks to the cloud, which will accelerate these trends.

To better understand such trends, it is interesting to deeper analyse the root causes of the energy consumption of computing systems. To do that, we can refer to the example depicted in Fig. 1.2 where the energy consumption of several components in a processor is we reported (figure adapted from [4]). Interestingly, the energy depends on the size of the manipulated data (i.e., the higher the bit-width, the higher the energy required to manipulate it), on the type of operations executed on given data (i.e., floating-point operations are more expensive than fixed-point or integer

A. Bosio (✉)
University of Lyon, ECL, INSA Lyon, CNRS, UCBL, CPE Lyon, INL, UMR5270, Écully, France
e-mail: alberto.bosio@ec-lyon.fr

D. Ménard
Univ. Rennes, INSA Rennes, IETR, Rennes, France
e-mail: daniel.menard@insa-rennes.fr

O. Sentieys
University of Rennes, INRIA/IRISA, Rennes, France
e-mail: olivier.sentieys@inria.fr

A. Bosio et al. (eds.), *Approximate Computing Techniques*,
https://doi.org/10.1007/978-3-030-94705-7_1

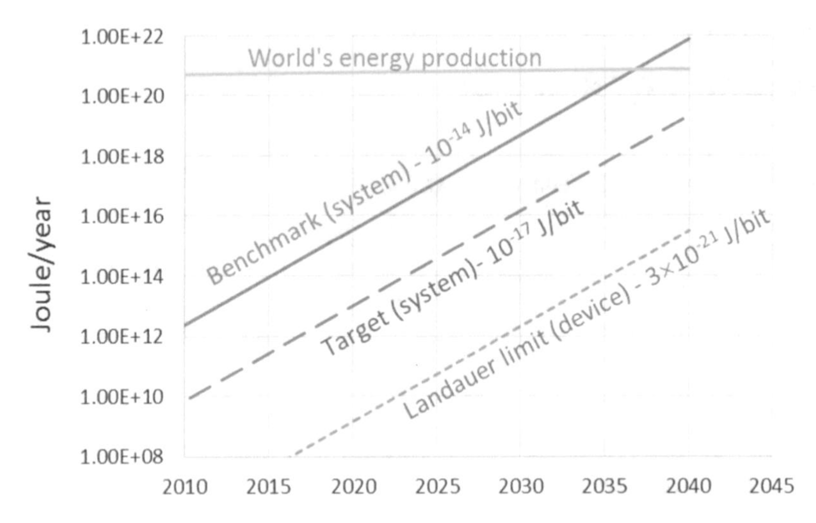

Fig. 1.1 Energy consumption trends in computing versus the world energy production. Source: SIA/SRC [3]

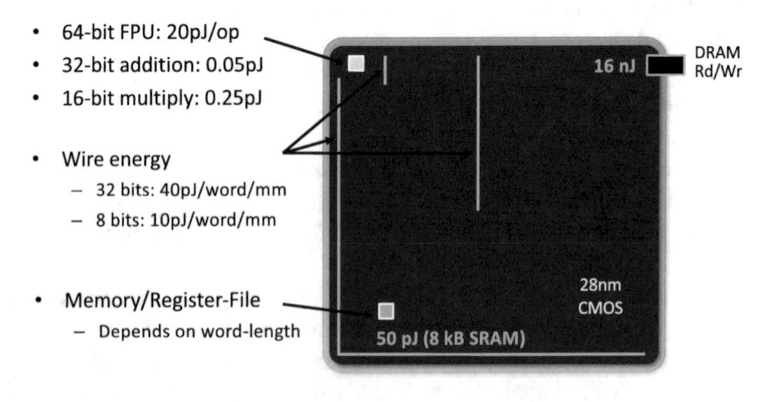

Fig. 1.2 Energy cost in a processor. Source: [4]

ones), and on the distance of the component where data are stored. It can be seen that every memory element has a certain energy cost and that the worst case is represented by an access to the external main memory (DRAM), for which the cost can be 3 orders-of-magnitude higher than arithmetic operations. It is also important to note that memory distance and data width will also impact performance, since the access and execution time increases with both distance and width.

Energy consumption is not only related to the processor architecture (e.g., ALU, Register File, Caches). Indeed, they are intrinsically linked to the technology used to implement computing devices composing systems. Today's computing devices are based on the CMOS technology that is subject to the famous Moore's law predicting that the number of transistors in an integrated circuit has been almost doubled every two years. The main positive consequences for the users have been the amazing performance increase reached by computing devices, as well as the dramatic reduction of cost per transistor or the energy per operation. Nevertheless, even with the advantages of the technology shrinking, we are starting to reach the physical limits of CMOS technology. Among the multiple challenges arising from technology nodes lower than 10 nm, we can highlight the high leakage current (i.e., high static power consumption), reduced performance gain, complex manufacturing process leading to low yield, complex testing process, and the extreme cost of mask fabrication [5]. In other words, computing devices manufactured with the latest technology nodes are less and less efficient (w.r.t. both performance and energy consumption) than forecast by Moore's law. Moreover, the manufactured devices are less and less reliable, meaning that errors can appear during the normal lifetime of a device with a higher probability than in previous technology nodes [6]. Fault-tolerant mechanisms are therefore required to ensure the correct behaviour of such device at the cost of extra area, power and timing overheads. Finally, process variations force the engineers to add extra guard bands (e.g., higher supply voltage or lower clock frequency than required under normal circumstances) to guarantee the correct functioning of manufactured devices.

Approximate Computing (AxC) aims at enabling the production of computing systems, which can satisfy the rising performance demands and, at the same time, improve the energy efficiency. Moreover, AxC will address the problem of maintaining reliability and thus coping with run-time errors, resorting to an acceptable amount of overheads in terms of area, performance, and energy consumption. AxC is based on the intuitive observation that, whereas performing exact computation requires a high amount of resources, allowing selective approximation or occasional violation of the specification can provide gains in efficiency. In brief, AxC exploits the gap between the level of accuracy required by the applications/users (i.e., quality of the output results of some computations) and the precision provided by the computing system, for achieving diverse optimisations.

The authors of [7] analysed different types of applications showing an intrinsic resilience to computational errors and/or noisy inputs. Such applications have the capability to provide "good enough" outputs even in the presence of errors. Let us resort to the example shown in Fig. 1.3 to explain the idea. The golden image is the output of the "reference implementation" of an image filter (median). The example focuses on the eye detail of the overall image to show the impact of computational errors on the output image. These errors correspond to the "removal" of a certain amount of instructions. In the figure, "60% instructions" (resp. 20%) corresponds to the removal of 40% instructions (resp. 80%). In other words, the image filter implementation has been modified to selectively skip a certain amount of the computations. The reader may denote the occurrences of some wrong pixels, but he/she will be able to recognise the image with no issue. For the sake of comparison, it can be noted that by randomly injecting 10% of wrong pixel into the golden image, the amount of wrong pixels is much higher than the "approximate outputs" ones. The benefits of "skipping" instructions are a reduced number of accesses to the main memory to retrieve instructions and data operands: less energy

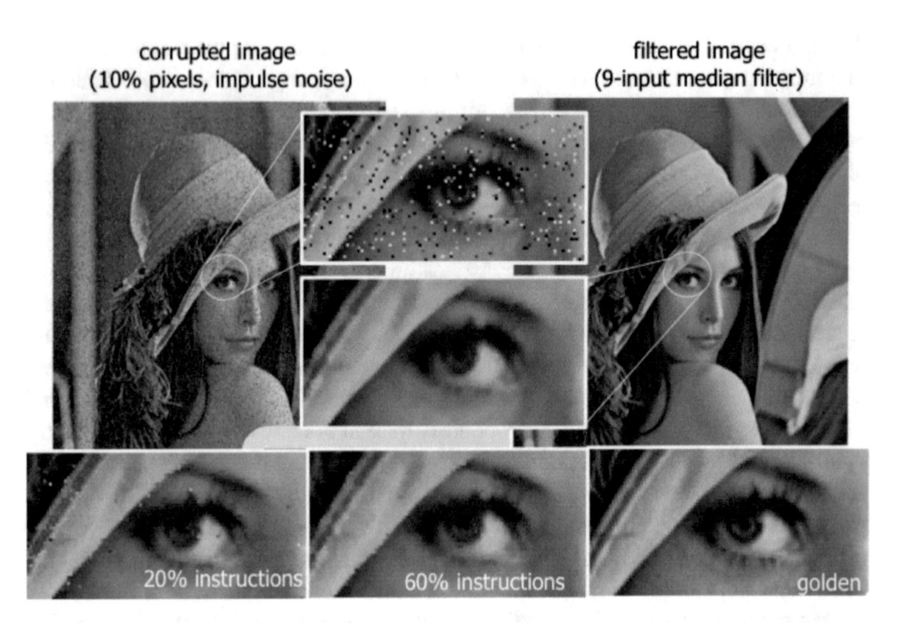

Fig. 1.3 Perceptual limitation and removal of instructions. Source: courtesy of Lukas Sekanina

is thus consumed and higher performance can be achieved. Many other examples can be found in [7].

The trade-off between accuracy degradation induced by the approximation and the cost reduction is a non-linear relation. Figure 1.4 provides a qualitative analysis of trade-off between *Accuracy* and *Cost* when the approximation is applied. The chart origin is the *Reference design* for which no approximation is applied. The reference is able to provide outputs with a given **accuracy** at a given **cost**. From the reference design, the implementation costs can be reduced, thanks to the approximation with a smaller impact on the accuracy. However, from a certain point, even a small approximation will have a huge impact on the accuracy leading to saturation. The blue curve represents this Pareto frontier where all points represent design options among the AxC design space.

Several questions arise at this point that can be summarised by the following three: (1) **Where and when the approximation has to be introduced?**, (2) **How to introduce the approximation?**, and (3) **How to quantify the impact of the introduced approximation on the system accuracy**.

Some books in the field of approximate computing [8, 9] already exist, but they target only the hardware layer, so they do not cover all three questions. Therefore, the aim of the book is to answer these three questions by presenting a global picture of the approximate computing paradigm. To accomplish the task, it covers an emerging research field shared among different communities (e.g., electrical engineer, computer engineer, computer science). In the literature, many publications also exist in this field of approximate computing. However, only a single context is

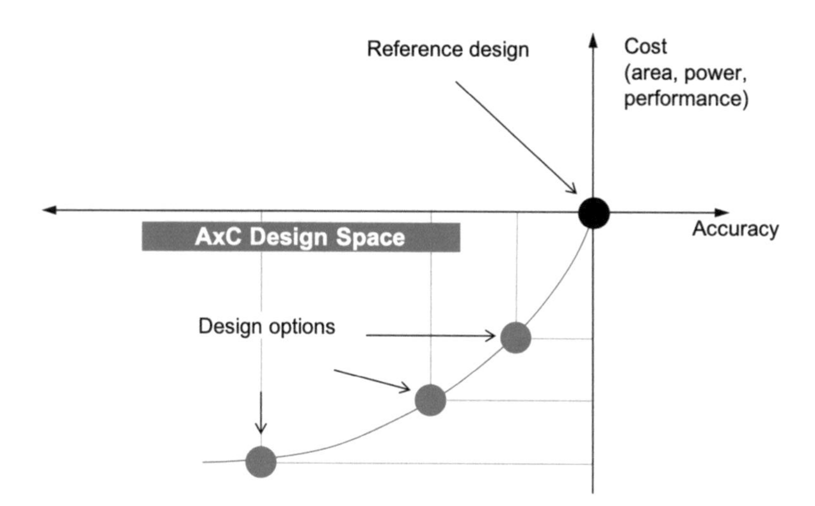

Fig. 1.4 AxC design space exploration: accuracy vs. cost

usually targeted, e.g., hardware approximation, software approximation, component level, etc. This book has the ambition to cover the existing gap by proposing a comprehensive work presenting the different AxC techniques, their integration, as well as real applications spanning from Artificial Intelligence, high performance computing, and approximation for security and safety-critical applications that can take strong advantage of AxC. It aims at serving as an important reference for researchers, students, and engineers.

1.2 Book Structure and Contributions

This section describes the book structure and the contributions within. The book structure reflects the abstraction levels composing the computing stack. A typical computing system, either a supercomputer or a small object, is composed of several layers. Figure 1.5 depicts a schematic view of these layers. The main classification is done by considering two macro-layers: the **Software** and the **Hardware**. In more detail, the hardware is further composed of:

- The **Technology** layer specifies the technology used to build the hardware components (e.g., CMOS transistors).
- The **Circuit** layer specifies the basic components of the hardware in terms of logic/memory elements (e.g., logic gates, Flip-Flops).
- The **Architecture** layer specifies the hardware components and their interconnections (e.g., CPUs, cache memory, I/O system).

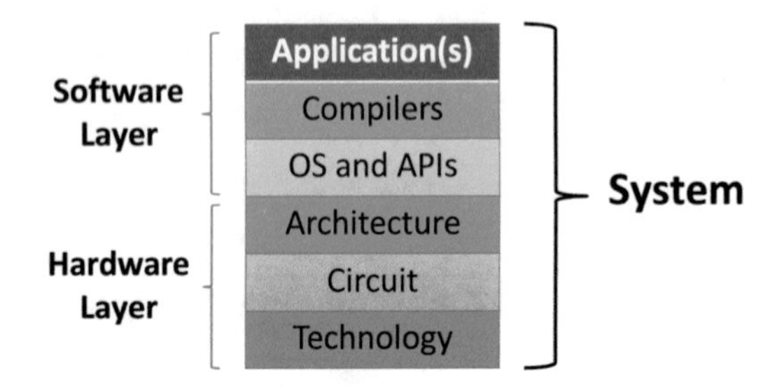

Fig. 1.5 Computing system layers

The Software layer is further divided into:

- The **OS and APIs** specify the main Operating System components and the set of the Application Programming Interfaces (e.g., process scheduler, file system access).
- The **Compiler** translates a software specification from high-level to low-level language (e.g., assembly).
- The **Application(s)** layer is the user application(s) running on the computing system.

This book aims at proposing a comprehensive analysis of approximate computing and its application to computing system layers. Moreover, it discusses the usage of AxC in emerging field of applications such as cyber-security, dependable computing for safety/mission critical systems and machine learning. It is structured into three parts: (1) techniques for approximate computing, (2) methods and tools for approximate computing, and (3) approximate computing applied to real-life applications. The first part, composed of four chapters (from Chaps. 2 to 4), presents the main techniques for approximate computing depending on the abstraction layer: number representation, hardware, software, and application. The second part discusses the main methodologies and tools available for applying the approximate computing techniques discussed in the first part. It is composed of four chapters (from Chaps. 5 to 9), and it overviews the analysis of the approximation impact on a given application, compilers, and design space exploration frameworks. The third, and last part, is devoted to newer application field of the approximate computing, with a special emphasis to real applications. It is composed of six chapters (from Chaps. 10 to 15) and it encompasses the use of approximate computing in real-time scheduling, security, digital circuit testing, and safety-critical applications. This part ends with two chapters devoted to the use of approximate computing in scientific and deep learning applications.

References

1. By 2040, computers will need more electricity than the world can generate. Retrieved May 04, 2021, from https://www.theregister.com/2016/07/25/semiconductor_industry_association_international_technology_roadmap_for_semiconductors/

2. IoT connections to grow 140% to hit 50 billion by 2022, as edge computing accelerates ROI. Retrieved May 04, 2021, from https://www.juniperresearch.com/press/iot-connections-to-grow-140pc-to-50-billion-2022
3. Rebooting the IT revolution. Retrieved May 04, 2021, from https://www.src.org/newsroom/rebooting-the-it-revolution.pdf
4. Power, programmability, and granularity: The challenges of exascale computing, IPDPS 2011. Retrieved May 04, 2021, from http://techtalks.tv/talks/54110/
5. Borkar, S. (2009). Design perspectives on 22nm CMOS and beyond. In *2009 46th ACM/IEEE Design Automation Conference* (pp. 93–94)
6. Gielen, G., De Wit, P., Maricau, E., Loeckx, J., Martin-Martinez, J., Kaczer, B., Groeseneken, G., Rodriguez, R., Nafria, M. (2008). Emerging yield and reliability challenges in nanometer CMOS technologies. In *2008 Design, Automation and Test in Europe* (pp. 1322–1327)
7. Chippa, V. K., Chakradhar, S. T., Roy, K., & Raghunathan, A. (2013). Analysis and characterization of inherent application resilience for approximate computing. In *2013 50th ACM/EDAC/IEEE Design Automation Conference (DAC)* (pp. 1–9)
8. Chandrasekharan, A., Große, D., & Drechsler, R. (2019). *Design automation techniques for approximation circuits*. Berlin: Springer.
9. Reda, S., & Shafique, M. (Eds.) (2019). *Approximate circuits*. Berlin: Springer.

Part I
Techniques for Approximate Computing

Chapter 2
Customizing Number Representation and Precision

Olivier Sentieys and Daniel Ménard

2.1 Introduction

There is a growing interest in the use of reduced-precision arithmetic, exacerbated by the recent interest in artificial intelligence, especially with deep learning. CPU, GPU, and TPU architectures already provide interesting, but limited, reduced-precision capabilities. 8-bit integer and 16-bit floating-point (e.g., `float16`, `bfloat16`) are typical examples of low-precision computations included in the architectures. Through the use of hardware acceleration on FPGA architectures, and thanks to their reconfiguration features, arithmetic customization can be further extended and almost any number format and word-length can be leveraged in the accelerator. All these examples illustrate the growing interest in the use of custom arithmetic.

In computer arithmetic, the representation of real numbers is a major issue. Indeed, most algorithms are using mathematical functions, and their accuracy and stability are directly related to the accuracy of the number representation they use. To represent real numbers, there exist two main formats: fixed-point and floating-point. Fixed-point (FxP) representation encodes real numbers as an integer value scaled by a fixed factor, thus leading to a format comprising an integer part and a fractional part, the point of the real number being at a fixed position. In the floating-point (FlP) representation, the scaling factor is encoded in the format, which comprises a mantissa (or significand) and an exponent, the point being floating along with the computations.

O. Sentieys (✉)
Univ. Rennes, Inria, IRISA, Rennes, France
e-mail: Olivier.Sentieys@inria.fr

D. Ménard
Univ. Rennes, INSA Rennes, IETR, Rennes, France
e-mail: Daniel.Menard@insa-rennes.fr

A. Bosio et al. (eds.), *Approximate Computing Techniques*,
https://doi.org/10.1007/978-3-030-94705-7_2

Fixed-point arithmetic is sometimes favoured due to its high efficiency in terms of energy consumption, cost, and performance, with a reputed clear advantage compared to floating-point. This comes at the cost of the pain of the programmer, who needs to manage all scaling operations to respect the rules imposed by FxP arithmetic. Floating-point representation can be considered as the main representation for real numbers, especially in high-performance computing. In contrast to FxP, FlP provides a high dynamic range, is able to represent with high accuracy both small and large numbers, and is very easy from a programmer point of view, since all scaling and rounding operations are totally managed by the hardware. However, this ease of use comes with relatively important area, delay, and energy penalties when compared to FxP.

This chapter presents both number representations and tries to draw a fair comparison between customized fixed-point and floating-point arithmetic. One conclusion is that the choice between FxP and FlP is not obvious and depends on the application considered. It is shown that, in some cases, low-precision floating-point arithmetic can be the most effective and provides some benefits over the classical fixed-point choice for energy-constrained applications. Indeed, combining the ease of use of floating-point representation associated with low-energy benefits of small bit-width makes reduced-precision floating-point arithmetic very promising, but not always useful.

Section 2.2 presents in detail the fixed-point representation, the rules governing the propagation of the fixed-point formats through operations, the quantization error process associated with computations relying on reduced-precision fixed-point arithmetic, and how overflow should also be considered. Overflow is critical in FxP since the dynamic range to represent real values is very limited in this format.

As already mentioned, a fixed-point number is composed of an integer and a fractional part. The aim of the fixed-point conversion process is to determine for each data the binary-point position and more specifically the number of bits for the integer part and the fractional part. This process is explained in Sect. 2.3; more details can also be found in Chap. 9.

Section 2.4 details the floating-point representation, the principle of FlP addition and multiplication, and provides some fair comparisons of their cost and performance with regard to FxP. Section 2.4 also presents some opportunities to reduce the cost of FlP operators as well as some libraries that can be used to simulate and perform hardware synthesis of customized, low-precision floating-point computations.

Finally, Sect. 2.5 gives some comparison results in terms of area, delay, and energy between the two number representations FxP and FlP, first at the operator level and then in the context of their use in applications, thus considering the errors due to low-precision computations.

2.2 Fixed-Point Arithmetic

2.2.1 *Fixed-Point Representation*

Fixed-point (FxP) representation is a way to encode real numbers with a virtual binary-point (BP) located between two bit locations as shown in Fig. 2.1. A fixed-point number is made up of an integer part (left to the BP) and a fractional part (right to the BP). The term m designates the integer part word-length (IWL) and corresponds to the number of bits for the integer part when this term is positive. This IWL includes the sign bit for signed numbers. The term n designates the fractional part word-length (FWL) and corresponds to the number of bits for the fractional part when this term is positive. The fixed-point value x_{fxpt} is computed from the following relation:

$$x_{fxpt} = -2^{m-1}.S + \sum_{i=-n}^{m-2} b_i 2^i \tag{2.1}$$

Numbers in the dynamic range $[-2^{m-1}, 2^{m-1} - 2^{-n}]$ can be represented in this fixed-point format with a precision of $q = 2^{-n}$. The term q corresponds to the quantization step and is equal to the weight of the least significant bit b_{-n}. The *Q-format* notation can be used to specify fixed-point numbers. For a fixed-point number having an IWL and FWL equal to m and n, respectively, the notation $Q_{m.n}$ is used for signed numbers and $uQ_{m.n}$ for unsigned numbers. The total number of bits w is equal to $m + n$. In fixed-point arithmetic, m and n are fixed and lead to an implicit scaling factor equal to 2^{-n} which does not change during the processing. The fixed-point value x_{fxpt} of the data x can be computed from the integer value x_{int} of the data x such as $x_{fxpt} = x_{int}.2^{-n}$.

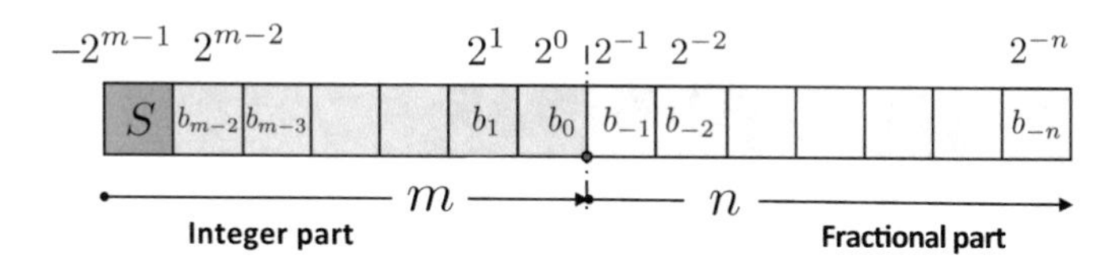

Fig. 2.1 Fixed-point specification

2.2.2 Format Propagation

In this section, the rules governing the propagation of the fixed-point formats through operations are described for the different arithmetic operations. Let us consider an operation $\diamond$ having x and y as input operands and z as output operand. Let $Q_{m_x.n_x}$, $Q_{m_y.n_y}$, and $Q_{m_z.n_z}$ be the *Q-format* of the operand x, y, and z, respectively.

Addition–Subtraction
The addition or the subtraction of two fixed-point numbers x and y can lead to an overflow if the operation result is not in the dynamic range of x and y. In this case, one more bit must be used to represent the integer part. Thus, dynamic range of the output result must be taken into account. A common IWL, m_c, must be defined to represent the input and the output

$$m_c = \max(m_x, m_y, m_z) \tag{2.2}$$

where m_z is computed from the dynamic range of the variable z. This IWL allows aligning the binary-point of the two input operands before computing the addition or the subtraction. The fixed-point format of the operation output is as follows:

$$\begin{cases} m_z = m_c \\ n_z = \max(n_x, n_y) \end{cases} \tag{2.3}$$

Multiplication
In contrast to the addition or the subtraction, there is no risk of overflow for the multiplication if the format of the output respects the following conditions. Thus, the fixed-point format of the output $z = x \times y$ is obtained from the input x and y fixed-point format with the following expression:

$$\begin{cases} m_z = m_x + m_y \\ n_z = n_x + n_y \end{cases} \tag{2.4}$$

The multiplication leads to an increase of the number of bits to represent the operation output. The total number of bits w_z is equal to $w_x + w_y = m_x + n_x + m_y + n_y$.

Division
For the division operation $z = x/y$, the value 0 must be excluded of the divisor y interval $[\underline{y}, \overline{y}]$ leading to the interval $[\underline{y}, -2^{-n_y}] \cup [2^{-n_y}, \overline{y}]$ if we consider the case that $\underline{y}$ is strictly negative and $\overline{y}$ is strictly positive. The IWL of the division output must be able to represent the largest value of the division result. This one is obtained by dividing the largest dividend by the smallest divisor. The largest possible dividend is -2^{m_x-1}, while the smallest divisor is 2^{-n_y}.

The FWL of the division output must be able to represent the smallest absolute value of the division result. This one is obtained by dividing the smallest dividend by the largest divisor. The smallest dividend is 2^{-n_x}, while the largest divisor is -2^{m_y-1}.

Thus, the fixed-point format of the output $z = x/y$ is obtained from the input x and y fixed-point format with the following expression:

$$\begin{cases} m_z = m_x + n_y \\ n_z = n_x + m_y \end{cases}. \tag{2.5}$$

Like for the multiplication, the total number of bits w_z is equal to $w_x + w_y = m_x + n_x + m_y + n_y$.

2.2.3 *Quantization Process and Rounding Modes*

In DSP applications, a sequence of arithmetic operations leads to an increase of data word-length when multiplication and division operations are involved. To maintain data word-lengths in reasonable range, the number of bits must be reduced. In fixed-point arithmetic, the least significant bits are discarded. Let x' be a fixed-point variable with a word-length of $w_{x'}$ bits. The quantization process $Q()$ leads to the variable x, depicted in Fig. 2.1, and having a word-length $w = w_{x'} - d$. Let S_x be the set containing all the values which can be represented in the format after quantization.

Truncation
In the case of truncation, the data x is always rounded towards the lower value available in the set S_x:

$$x = \lfloor x \cdot q^{-1} \rfloor \cdot q = kq \; \forall x \in [k \cdot q; (k+1)q[\tag{2.6}$$

with $\lfloor \cdot \rfloor$, the floor function defined as $\lfloor x \rfloor = \max(n \in \mathbb{Z} | \, n \leq x)$ and $q = 2^{-n}$ the quantization step. The value x after quantization is always lower than or equal to the value x before quantization. Thus, the truncation adds a bias on the quantized signal and the output quantization error will have a non-zero mean. Truncation rounding is widely used because of its cheapest implementation. The d LSBs of x' are discarded and no supplementary operation is required.

Conventional Rounding
To improve the precision after the quantization, the rounding quantization mode can be used. The latter significantly decreases the bias associated with the truncation.

This quantization mode rounds the value x to the nearest value available in the set S_x:

$$x = \left\lfloor \left(x + \frac{1}{2}q \right) \cdot q^{-1} \right\rfloor \cdot q = \begin{cases} kq & \forall x \in [k \cdot q; (k + \frac{1}{2})q[\\ (k + 1)q & \forall x \in [(k + \frac{1}{2})q; (k + 1)q] \end{cases} \tag{2.7}$$

The midpoint $q_{1/2} = (k + \frac{1}{2})q$ between kq and $(k + 1)q$ is always rounded up to the higher value $(k + 1)q$. Thus, the distribution of the quantization error is not exactly symmetrical and a small bias is still present.

The conventional rounding can be directly implemented from (2.7). The value 2^{-n-1} is added to x', and then the result is truncated on w bits. In the technique presented in [1], the conventional rounding is obtained by the addition of x' and the value $b_{-n-1}.2^{-n}$, and then the result is truncated on w bits. This implementation requires an adder of w bits.

Convergent Rounding

To reduce the small bias associated with the conventional rounding, the convergent rounding can be used. To obtain a symmetrical quantization error, the specific value $q_{1/2}$ must be rounded up to $(k + 1)q$ and rounded down to kq with the same probability. The probabilities that a particular bit is 0 or 1 are assumed to be identical and thus the rounding direction can depend on the bit b_{-n} value.

$$x = \begin{cases} kq & \forall\, x \in [k.q; (k + \frac{1}{2})q[\\ (k + 1)q & \forall\, x \in](k + \frac{1}{2})q; (k + 1)q] \\ kq & \forall\, x = q_{1/2} \quad \text{and} \quad b_{-n} = 0 \\ (k + 1)q & \forall\, x = q_{1/2} \quad \text{and} \quad b_{-n} = 1 \end{cases} \tag{2.8}$$

The specific value $q_{1/2}$ has to be detected to modify the computation in this case. For this specific value, the addition of the data x with the value 2^{-n-1} has to be done only if the bit b_{-n} is equal to one.

The alternative to this conditional addition is to add the value $b_{-n-1}.2^{-n}$ in every case. Then, for the specific value $q_{1/2}$, the least significant bit b_{-n} of the data x is forced to 0 to obtain an even value. This last operation does not modify the result when b_{-n} is equal to 1 and discard the previous addition operation if b_{-n} is equal to 0. The convergent rounding requires a supplementary addition operation and an operation (DTC) to detect the value 2^{-n-1} and then to force bit b_{-n} to zero.

2.2.4 Overflow Modes

In DSP applications, numerous processing kernels involve summations requiring to accumulate intermediate results. Consequently, the dynamic range of the accumulation variable grows and can exceed the bounds of the values that can be represented leading to overflows. When an overflow occurs, if no supplementary hardware is used, the wrap-around overflow mode is considered. For the wrap-around overflow mode, the value x_{wa} of variable x coded with m bits for the IWL is equal to

$$x_{wa} = \left((x + 2^{m-1}) \bmod 2^m\right) - 2^{m-1} \tag{2.9}$$

with mod the modulo operation. To avoid overflow, the fixed-point conversion process described in the rest of this chapter must be followed conscientiously. Especially, the dynamic range of the different variables must be carefully evaluated for sizing the IWL. For variables having a long tail for its probability density function, the IWL can be large and thus leading to an over-estimation for numerous values. In this case, saturation arithmetic can be used to reduce the IWL. Let us consider a variable x coded with m bits for the IWL. In saturation arithmetic, when the value x is lower than -2^{m-1}, the value x is set to -2^{m-1}. When the value x is higher than -2^{m-1}, the value x is set to $-2^{m-1} - 2^{-n}$.

2.3 Fixed-Point Conversion Process

As described in Sect. 2.2, a fixed-point number is made up of an integer part and a fractional part. The aim of the fixed-point conversion process is to determine for each data the binary-point position and more specifically the number of bits for the integer part and the fractional part.

The total number of bits $w_i = m_i + n_i$ to encode a data influences the implementation cost C. The implementation cost reduction implies to minimize the integer and fractional part word-lengths. The reduction of the number of bit leads to unavoidable error between the finite precision values and the infinite precision ones and thus degrades the quality of the application output. Consequently, the implementation cost minimization through word-length optimization is achieved with the constraint that the output quality degradation $\Delta\lambda$ is limited and below a maximal value $\lambda_{\max}$. This fixed-point conversion process can be modelled by the following optimization process:

$$\min_{\mathbf{w}} (C(\mathbf{w})) \quad \text{subject to} \quad \Delta\lambda(\mathbf{w}) \leq \lambda_{\max} \tag{2.10}$$

where $\mathbf{w}$ is an N-length vector containing the word-lengths of the N data inside the application, and $C(\cdot)$ is an implementation cost function that models the cost such as area or energy consumption according to the data word-lengths. $\Delta\lambda(\cdot)$ computes the quality degradation due to the tested word-length configuration $\mathbf{w}$ [2–4], and $\lambda_{\max}$ represents the maximal quality degradation tolerable by the application.

Reducing the number of bits for the integer part or for the fractional part leads to different effects. The integer part word-length m defines the range of values that can be represented. When m is too low, overflows occur, leading to non-linearity in the processing and a significant amplitude for the error compared to infinite precision.

As long as m is greater than $m_{\min}$, the minimum value that guarantees no overflow, modifying m will have no effect on the quality of the output. When m is lower than $m_{\min}$, overflow occurs and quickly the output quality is highly degraded. To determine the integer word-length, the data dynamic range is evaluated and the minimal value of m ensuring no overflow or a sufficiently low overflow probability is selected.

The fractional part word-length n defines the accuracy. The larger n is, the smaller the error between finite and infinite precisions is and the higher the accuracy is improved. Unlike for the integer part, reducing the number of bit the fractional part will progressively reduce the accuracy and increase the output quality degradation. Thus, the determination of the fractional part word-length is a trade-off between the implementation cost and the quality degradation. This trade-off is explored through the solving of the optimization process described in Eq. (2.10). The word-length of each data is optimized through the minimization of the implementation cost under quality degradation constraint. This optimization process requires three elements, an optimization algorithm, a cost function $C(\cdot)$, and a quality degradation function $\Delta\lambda(\cdot)$. The quality degradation function depends on the fractional part word-length of each data. The cost function $C(\cdot)$ requires the knowledge of the total word-length of each data. Thus, for this optimization process, the integer part word-length has to be known.

Consequently, the fixed-point conversion process is split into two parts. Firstly, the integer part word-length is determined from the results of the data dynamic range evaluation. Secondly, the fractional part word-length is optimized by solving the optimization process described in Eq. (2.10).

2.3.1 Integer-Part Word-Length Determination

The first stage of the fixed-point conversion process aims at determining the number of bits for the integer part of each data in the considered application. The goal is to minimize the number of bits while protecting against overflows that degrade significantly the application quality. Firstly, the dynamic range of each signal is evaluated. The different types of techniques available to estimate the dynamic range are presented in Sect. 2.3.1.1. Secondly, the IWL is determined from the dynamic range and the fixed-point format propagation rules. Scaling operations are inserted to adapt fixed-point formats. This process is described in Sect. 2.3.1.2.

2.3.1.1 Dynamic Range Evaluation

The determination of the number of bits for the integer part requires to evaluate the signal dynamic range. Existing techniques to evaluate the dynamic range can be classified according to the targeted applications. Critical systems do not tolerate high computational errors. Any overflow occurrence may lead to a system failure or a serious quality degradation. For example, in 1996, the first launch of the

Ariane 5 rocket ended in explosion due to software failure. This failure was caused by the overflow of the variable representing the rocket acceleration. Thus, for critical systems, the integer part word-length has to cover the entire range of possible values. In this case, the bounds should be determined by techniques that guarantee the absence of overflow occurrence and allowing to certify the data dynamic range. Techniques based on interval arithmetic or affine arithmetic satisfy these constraints, but at the expense of an overestimation of the bounds. Statistical approaches that determine bounds from a set of simulation results can reduce the overestimation but cannot ensure the absence of overflow occurrence.

Overflows occur when the number of bits of the integer part is not sufficient. Overflow occurrence degrades the result quality at the system output. However, the hardware implementation cost is unnecessarily increased if the number of bits exceeds the needs. Many systems are tolerant to overflows if the probability of overflow occurrence is low enough. In this case, determining the number of bits of the integer part is a trade-off between the implementation cost and the output system quality degradation. This is translated into an optimization problem where the integer word-length of each variable of the system is reduced while maintaining an overflow probability lower than the accepted probability [5]. The challenge is to estimate the probability density function (PDF) of the data in order to be able to compute the overflow probability. Stochastic approaches, which model the variable PDF by propagating data PDF model from the inputs to the system output, can be considered.

2.3.1.2 IWL Determination and Insertion of Scaling Operations

The IWL is determined by propagating the IWL through the operations from the inputs to the outputs with the help of the dynamic range evaluation results. This propagation process uses the format propagation rules provided in Sect. 2.2.2.

To illustrate this stage, let us consider the sequence of operations depicted in Fig. 2.2. The data d is the output of the operation O_j and the input of operation O_k. Let m_s, m_d, and m_i be, respectively, the IWL for the operation O_j output, the data d, and the operation O_k input.

The IWL of the signed data d is computed from its dynamic range $[\underline{x}, \overline{x}]$ with the following expression:

$$m_x = \max\left(\lfloor \log_2(|\ \overline{x}\ |) \rfloor + 2, \lceil \log_2(|\ \underline{x}\ |) \rceil + 1\right) \tag{2.11}$$

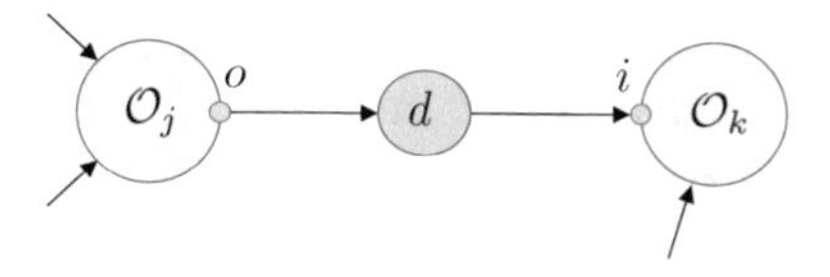

Fig. 2.2 Example of a sequence of operations: data d is the output of the operation O_j and the input of operation O_k

The IWL m_o is computed from the propagation of the IWL of the operation O_j inputs with the help of the rules provided in Sect. 2.2.2. A scaling operation is inserted at the operation O_j output if $s_o = m_d - m_{o_j}$ is strictly negative. In this case, a left shift of s_o bits is required to modify the IWL. It means that the IWL of the operation O_j was too important compared to the data dynamic range of d and the s_o most significant bits of x are a copy of the sign bit and can be discarded.

For multiplication and division, the IWL m_i is equal to m_d and no scaling operation is required at the operation O_j input. For addition and subtraction, a common IWL m_c for the input and the output must be determined and $m_i = m_c$. A scaling operation is inserted at the operation O_k input if $s_i = m_{o_i} - m_d$ is strictly positive. In this case, a right shift of s_i bits is required to modify the IWL of the data d. It means that supplementary bits are required for the addition or subtraction O_j to avoid overflow.

2.3.2 *Fractional Part Word-Length Determination*

The fractional part word-length is optimized by solving the optimization process described in Eq. (2.10). The search space for this combinatorial optimization problem is huge, and numerous algorithm have been proposed to find an optimized solution in a reasonable execution time. In Chap. 9 (Word-Length Optimization of Fixed-Point Algorithms), a survey of the different optimization algorithms is proposed. These optimization algorithms are iterative process, testing different combinations of word-length and moving in the search space. Consequently, the cost function $C(\cdot)$ and the quality degradation function $\Delta\lambda(\cdot)$ are evaluated numerous time. The challenge is to develop techniques able to evaluate efficiently and accurately the quality degradation. In Chap. 6, a survey of the different existing techniques to evaluate the quality degradation is proposed.

2.4 Floating-Point Arithmetic

Floating-point (FlP) representation is today the main representation for real numbers in computing, thanks to a potentially high dynamic range and to its ease of use since all scaling and rounding operations are totally managed by the hardware, contrary to fixed-point arithmetic. However, this ease of use comes with relatively important area, delay, and energy penalties. The floating-point representation is presented in Sect. 2.4.1. Section 2.4.2 details the principle of FlP addition and multiplication and provides some fair comparisons of their cost and performance with regard to FxP. Then, Sect. 2.4.3 presents some opportunities to reduce the cost of FlP operators, without jeopardizing the accuracy of the computations too much. Finally, Sect. 2.4.4 describes some libraries that can be used to simulate and perform hardware synthesis of customized, low-precision floating-point computations.

2.4.1 Floating-Point Representation for Real Numbers

In computer arithmetic, the representation of real numbers is a major issue. Indeed, most algorithms are using mathematical functions, and their accuracy and stability are directly related to the accuracy of the number representation they use. The floating-point (FlP) representation is a way to encode real numbers with a scaling factor encoded in the data. Given an unsigned M-bit mantissa m, a signed integer exponent of value e coded on E bits, often represented in biased representation, and a sign bit s, the radix-2 floating-point value x_{flpt} is represented as

$$x_{flpt} = (-1)^s \times 1.m \times 2^e \tag{2.12}$$

Contrary to fixed-point representation, the point in the FlP representation of the number is *"floating"* and scaled by the exponent, similarly to the scientific representation in decimal arithmetic that we use in our daily life. The mantissa m—or the significand of the representation—is used to generate a normalized number with an implicit "1" conforming the integer part belonging to [1, 2[. This "1" being implicit, it is not represented in the format, freeing space for one more digit. This number is then scaled by means of the exponent e, and the sign is controlled by the value of the sign bit s. e being a signed number, the exponent is usually represented in the number format as biased by a constant value b. With this representation, any number under this format can be represented using $M + E + 1$ bits as shown in Fig. 2.3.

Nevertheless, automatically keeping the *floating* point at the right position along computations requires an important hardware overhead, as discussed in Sect. 2.4.2. Managing subnormal numbers (numbers between 0 and the smallest positive possible representable value) and the values 0 and infinity also represents an overhead. Despite this additional cost, FlP representation is today established as the *de facto* standard for real number representation. Indeed, besides its high accuracy and high dynamic range, it has the huge advantage of leaving the whole management of the representation to the hardware instead of leaving it to the software designer, significantly diminishing developing and testing time. This domination is sustained by IEEE 754 standard, lastly revised in 2008 [6], which sets the conventions for floating-point number possible representation, subnormal numbers management, and the different cases to be handled, ensuring a high portability of programs. Table 2.1 gives the representations of the FlP numbers following the IEEE 754-2008

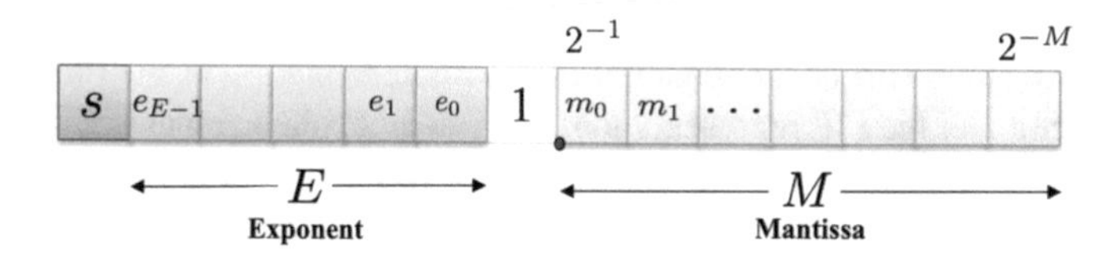

Fig. 2.3 Floating-point representation

Table 2.1 IEEE 754 normalized floating-point representation

Precision	Mantissa width (M)	Exponent width (E)	Minimum exponent value (e_{min})	Exponent bias (b) (also e_{max})
Half precision (16 bits)	10	5	−14	15
Single precision (32 bits)	23	8	−126	127
Double precision (64 bits)	52	11	−1022	1023
Quadruple precision (128 bits)	112	15	−16382	16383

standard. Mantissa width M is without the implicit 1. The bias b is equivalent to the maximum exponent value e_{max}. However, such a strict normalization implies:

- An important overhead for throwing flags for the many special cases and even more for the management of these special cases
- A low flexibility in the widths of the mantissa and exponent, which have to respect the rules of Table 2.1 for 16-, 32-, 64-, and 128-bit precisions.

2.4.2 *Floating-Point Operators*

Integer addition (or subtraction) is the simplest arithmetic operator. However, in floating-point arithmetic, addition suffers from a high control overhead, which requires several steps to be performed:

- First, the difference of the exponents is computed.
- Depending on the difference of the exponents, one among two computing paths may be selected [7]: the *close path* is for situations where a massive cancellation (more than 1 bit) may occur or effective subtractions of inputs with exponents that differ by at most 1. The *far path* is for distant exponents, where their difference is at least 2 bits. The following computations may slightly vary depending on the chosen path.
- The addition of the mantissas is performed.
- Then, rounding is performed on the mantissa, depending on the dropped bits and the rounding mode (to zero, to nearest, etc.) selected.
- Special cases are then handled (zero, infinity, subnormal results) and the output sign.
- Then, mantissa is shifted, so it represents a value in [1, 2[.
- And the exponent is modified depending on the number of shifts.

More control can be needed, depending on the implementation of the FlP adder and the specificities of the FlP representation. For instance, management of the implicit 1 implies to add 1s to the mantissas before addition, and an important overhead can be dedicated to exception handling. For a figure illustrating the FlP addition

Table 2.2 Cost, delay, and power of FlP addition vs. integer addition

	Area (μm^2)	Total power (mW)	Critical path (ns)	Power delay Product (fJ)
32-bit float	653	4.39E−4	2.42	1.06E−3
64-bit float	1453	1.12E−3	4.02	4.50E−3
32-bit int	189	3.66E−5	1.06	3.88E−5
64-bit int	373	7.14E−5	2.10	1.50E−4

principle and more details on its hardware implementation, the reader can refer to Figure 8.13 of [7] and the related chapter of the book.

For cost, delay, and power comparison, Table 2.2 shows the performance of 32-bit and 64-bit FlP addition compared with 32-bit and 64-bit integer addition, synthesized using Synopsys Design Compiler targeting 28nm FDSOI with a 200 MHz clock. Power is estimated using 10, 000 uniform input samples. FlP addition power was estimated activating in an equivalent way the *close* and *far* paths 50% of the time. These results clearly show the overhead of floating-point addition. For 32 bits, the FlP addition is 3.5× larger and 2.3× slower and consumes 27× more energy than integer addition. For 64 bits, the FlP addition is 3.9× larger and 1.9× slower and consumes 30× more energy. The overhead seems to be roughly linear with the size of the operator, and the impact of numbers representation is highly impacting performance. However, it is showed later in this chapter that this high difference reduces with the bit-width of the operands.

FlP multiplication is less complicated than addition as only a low control overhead is necessary to perform the operation. Input mantissas are multiplied using a classical integer multiplier, while exponents are simply added. At worse, a final +1 on the exponent can be needed, depending on the result of the mantissas multiplication and the related rounding and normalization required. For a figure illustrating the basic architecture of an FlP multiplier, the reader can refer to Figure 8.14 of [7] and the related chapter of the book.

Obviously, all classical hardware overheads needed by FlP representation are again necessary (rounding logic, normalization, management of particular cases), but the overhead for the multiplication is less than for addition. Table 2.3 shows the difference between 32-bit and 64-bit floating-point multiplication and 32-bit and 64-bit fixed-width integer multiplication, with the same experimental setup as discussed before for the addition.

A first observation on the area shows that the integer multiplier is 48% larger than the FlP version for 32 bits and 37% larger for 64 bits. This difference is due to the smaller size of the integer multiplier in the FlP multiplier, since it is limited to the size of the mantissa (24 bits for 32-bit version and 53 bits for 64-bit version). Despite the management of the exponent, the overhead is not large enough

Table 2.3 Cost, delay, and power of floating-point multiplication vs. integer multiplication

	Area (μm^2)	Total power (mW)	Critical path (ns)	Power delay Product (fJ)
32-bit float	1543	8.94E−4	2.09	1.87E−3
64-bit float	6464	6.56E−3	4.70	3.08E−2
32-bit int	2289	6.53E−5	2.38	1.55E−4
64-bit int	8841	1.84E−4	4.52	8.31E−4

to produce a larger operator. However, the 32-bit FlP multiplication energy is 11× higher than for the integer version, while 64-bit version consumes even 37× more energy. This can be justified by the higher activity of the logic in the FlP operator due to the management of the exponent and special cases. It is interesting to note that the difference of energy consumption between addition and multiplication is much more important for integer operators than for FlP. As an example, for 32-bit, integer multiplication consumes 4.7× more energy than integer addition, while this factor is only 1.4× for 32-bit FlP multiplier compared to 32-bit FlP adder. Therefore, using multiplication in FlP computing is relatively less penalizing than for integer multiplication, typically used in fixed-point arithmetic.

2.4.3 Low-Precision Floating-Point Arithmetic

There is a growing interest in the use of reduced-precision arithmetic, exacerbated by the recent interest in artificial intelligence, especially with deep learning. CPU, GPU, and TPU architectures already provide interesting, but limited, reduced-precision capabilities. 8-bit integer and 16-bit floating-point (e.g., `float16`, `bfloat16`) are typical examples of low-precision computations included in the architectures. Through the use of hardware acceleration on FPGA architectures, and thanks to their reconfiguration features, arithmetic customization can be further extended and almost any number format and word-length can be leveraged in the accelerator. All these examples illustrate the interest of customizable floating-point architectures. Indeed, combining the ease of use of floating-point representation associated with low-energy benefits of small bit-width makes reduced-precision floating-point arithmetic very promising.

There are several possible opportunities to relax accuracy in floating-point arithmetic to increase performance and save power and hardware cost. Of course, the main technique is to reduce the size of the mantissa and exponent (i.e., smaller operand bit-width or word-length). With a mantissa normalized in $[1, 2[$, reducing the word-length corresponds to pruning the LSBs, which comes with no overhead, except if faithful rounding is performed. For the exponent, the transformation can be more complicated if it is represented with a bias. Indeed, if E is the exponent width, an implicit bias of $2^E - 1$ applies to the exponent in classical exponent representation. Therefore, reducing the exponent to a width E' means that a new bias must be applied. The original exponent must be added $2^{E'} - 2^E$ (< 0) before pruning the MSBs, implying hardware overhead at conversion. The original exponent must represent a value in $\left[-2^{E'-1} + 1, 2^{E'-1}\right]$ to avoid overflow. In practice, it is better to keep a constant exponent width to avoid useless overhead and conversion overflows, which would have a huge impact on the quality of the computations.

A second way to improve computation at reduced precision is to play with the implicit bias of the exponent. Indeed, increasing the exponent width increases the dynamic towards infinity, but the accuracy towards zero. Thus, if the absolute maximum values to be represented are known, the bias can be chosen, so it is just large enough to represent these values. This way, the exponent gives more accuracy to very small values, increasing accuracy. However, using a custom bias means that the arithmetical operators (addition and multiplication) must consider this bias in the computation of resulting exponent, and the optimal bias along computation may diverge to $-\infty$. To avoid this, if the original $2^E - 1$ exponent bias is kept, exponent bias can be simulated by biasing the exponents of the inputs of each or some computations using shifting. For the addition, biasing both inputs adding $2^{E_{\text{in}}}$ to the exponent implies that the output will also be biased by $2^{E_{\text{in}}}$. For the multiplication, the output will be biased by $2^{E_{\text{in}}+1}$. Keeping an implicit track of the bias along computations allows to know any algorithm output bias and to perform a final rescaling of the outputs.

Finally, accuracy can be relaxed in the integer operators composing the considered FlP operators, e.g., the integer adder adding the mantissas in FlP addition or the integer multiplier in the FlP multiplication. Indeed, they can be replaced by approximate adders and multipliers as described in other chapters of this book, to improve performance while relaxing accuracy. However, as most of the cost relies in control hardware, the impact on accuracy would be strong for a very small cost or performance benefit. The same approximation can be applied on the exponent management, but the impact of approximate arithmetic would be too high on the accuracy and this is therefore strongly unadvised.

2.4.4 *Reduced-Precision Floating-Point Libraries*

The past years have hosted the creation of several customizable floating-point libraries. As part of the synthesizable C++ libraries AC Datatypes [8], Mentor Graphics proposes the custom floating-point class AC_FLOAT. Based on the fixed-point library AC_FIXED, AC_FLOAT allows for light floating-point computation, thanks to simple operators. The mantissa in the representation is not normalized and has no implicit 1. This allows for easy management of subnormals but induces a potential loss of accuracy in computations. The mantissa is represented in signed two's complement, so the sign information is contained in the mantissa instead of using an extra sign bit. However, there is no benefit to this choice since two's complement represents a loss of 1 bit of precision compared to unsigned representation. The choice of two's complement representation on the mantissa also turns comparison operator more complex. Moreover, many cases are not handled such as zero or infinity. AC_FLOAT also supports custom exponent bias, but managing the exponent bias comes with an overhead.

FLOPOCO (for Floating-Point Cores, but not only) is a generator of arithmetic cores [9]. Also based on C++, it has its own synthesis engine and directly returns VHDL. More than simple arithmetic operators, it is able to generate optimized floating-point computing cores performing complex arithmetic expressions. In this section, we will only get interested in FLOPOCO's custom floating-point addition and multiplication. The main difference of FLOPOCO's floating-point representation is the extra 2-bit exception field transported in data. Like for CT_FLOAT, subnormals are not handled by FLOPOCO. Unlike AC_FLOAT, both CT_FLOAT and FLOPOCO do not support custom exponent bias.

Other alternatives such as VFLOAT [10, 11] or OptiFEX [12] do exist but are not taken into account in the study led in this chapter. VFLOAT proposes IEEE 754-2008 compliant customizable computing cores for existing FPGA. OptiFEX generates floating-point computing cores targeting FPGA like FLOPOCO.

CT_FLOAT [13][1] offers a balance between computational safety and simplicity. Inspired by AC_FLOAT, it is provided as C++ template for High-Level Synthesis (HLS), compatible with Mentor Graphics CatapultHLS and Xilinx Vivado HLS. As CT_FLOAT will be used for comparison with fixed-point representation in the rest of this chapter, we provide below more details on the library.

The declaration of an instance of CT_FLOAT requires three template parameters: the exponent width E, the mantissa width M, and the rounding mode. The mantissa also includes a sign bit and is represented as sign plus absolute value, as in standard FlP. The total number of bits in memory is therefore equal to $E + M$. Currently, two rounding modes are supported: `CT_RN` rounding to nearest with halfway-to-even tie-breaking rule and `CT_RZ` rounding towards 0 or truncation. CT_FLOAT representation and arithmetic operators were created to remain simple and energy efficient, thanks to the combination of several implementation choices. CT_FLOAT mantissa is represented in $[1, 2[$ with an implicit 1. However, subnormal numbers are not handled, which implies that a certain range of numbers are not representable around 0. The exponent is represented in a biased representation. The bias is set at the centre of the exponent range, similar to the IEEE 754 representation. Using biased representation instead of two's-complement results in simpler exponent value comparisons, which are omnipresent in arithmetic operators. In variants of the CT_FLOAT library, the bias can also be customized.

[1] https://gitlab.inria.fr/sentieys/ctfloat.

The library provides a rich set of synthesizable operator overloading: unary operators (unary $-$, !, $++$, $--$), relational operators ($<$, $>$, $<=$, $>=$, $==$, $!=$), binary operators ($+$, $+=$, $--=$, $*$, $*=$, $<<$, $<<=$, $>>$, $>>=$), and assignment operator from/to another instance of CT_FLOAT. It also provides non-synthesizable operator overloading features, such as conversion from/to C++ native datatypes (float, double), and output operator $<<$ for easy display and writing in files. Other built-in functions allow easy manipulation of floating-point values, such as functions to get information about the extreme representable values for a given floating-point representation, to test if a given value is representable, etc.

An example (not including all statements and declarations) of the use of CT_FLOAT is given below:

```
ct_float<7, 9, CT_RN> h = 1.046978e-3;
ct_float<7, 9, CT_RN> x, y;
x = -0.02266398;
y = x * y + 0.55;
cout << y << endl;
```

This example can be simulated and synthesized to hardware using HLS. In the example, all variables have the same representation (i.e., $E = 7$ and $M = 9$, rounding mode is CT_RN). It is also possible to deal with various representations. If the inputs are on (E_1, M_1) and (E_2, M_2) representation, the output representation (E_o, M_o) is given by

$$
\begin{aligned}
E_o &= \max(E_1, E_2) \\
M_o &= \max(M_1, M_2)
\end{aligned}
\tag{2.13}
$$

Moreover, as subnormals are not representable by CT_FLOAT, the output is always saturated to the smallest absolute possible representable value with the same sign. Towards infinity, the operators do not under/overflow. Saturation to the highest absolute representable value of the same sign is returned.

Table 2.4 recapitulates the different known properties of AC_FLOAT, CT_FLOAT, and FLOPOCO floating-point representation. In this table, the number of additional bits in the representation is taking for reference a representation with implicit 1 in the mantissa and with one bit of sign in the representation. For an equal general accuracy, AC_FLOAT needs one more bit on the mantissa than CT_FLOAT and FLOPOCO. However, with its 2-bit exception field, FLOPOCO has the representation requiring not only the largest width but also the highest computing reliability.

Then, the hardware performance comparison process for AC_FLOAT, CT_FLOAT, and FLOPOCO is as follows. All operators are characterized for a 28nm FDSOI @ 1.0V, 25C ASIC library. All designs are synthesized and estimated with a clock of 200 MHz. For power analysis, the random inputs generated for adder/subtracter characterization are ensuring an activation of the close path for at least 50% of the computations. However, the benchmark generated by FLOPOCO does not insure any proportion of activation of the close path, so the dynamic power could be

Table 2.4 Main properties of the custom floating-point libraries AC_FLOAT, CT_FLOAT, and FLOPOCO

	AC_FLOAT	CT_FLOAT	FLOPOCO
Custom exp. bias	✓	✗ (✓)	✗
Mantissa Implicit 1	✗	✓	✓
Zero and inf. exception flags	✗	✗	✓
Zero and inf. internal handling	✗	✓	✗
Subnormal exception flag	✗	✗	✓
Subnormal internal handling	✓	✗	✗
Additional bits in representation	+1	+0	+2

Table 2.5 Comparative results for 16-bit FlP addition/subtraction with $F_{\texttt{clk}} = 200\text{MHz}$

	Area (μm^2)	Critical path (ns)	Total power (mW)	Energy per operation (pJ)
AC_FLOAT	312	1.44	1.84 E-1	9.07 E-1
CT_FLOAT	318	1.72	2.13 E-1	1.05
FLOPOCO	361	2.36	1.84 E-1	9.06 E-1
CT_FLOAT /AC_FLOAT	+2.15%	+19.**4%**	+15.4%	+15.7%
CT_FLOAT /FLOPOCO	-11.8%	-27.0%	+15.7%	+15.8%

underestimated. Moreover, FLOPOCO's benchmark does not consider any input and output data registers, whereas AC_FLOAT and CT_FLOAT, synthesized with HLS, do. This may represent about 5–10% underestimation in the total power for FLOPOCO operators, which has to be kept in mind for the analysis of results. All operators are generated, so they execute in 1 cycle. It may not be the most efficient implementation because of possible glitches, but it is a good starting point for a fair comparison.

For this comparative study, half-precision ($E = 5$, $M = 11$) and single-precision ($E = 8$, $M = 24$) floating-point representations are considered. Results for 16-bit addition/subtraction, 16-bit multiplication, 32-bit addition/subtraction, and 32-bit multiplication are given in Tables 2.5, 2.6, 2.7, and 2.8, respectively. The two last lines of the tables refer to the relative performance of CT_FLOAT vs. AC_FLOAT (respectively, FLOPOCO) (e.g., CT_FLOAT area is 2.15% higher than AC_FLOAT).

The main conclusion is that the three custom floating-point libraries provide results in the same order of magnitude. For 16-bit addition/subtraction, CT_FLOAT consumes 15% more energy than both AC_FLOAT and FLOPOCO, despite an area being equivalent to AC_FLOAT and 12% smaller than FLOPOCO. The fastest 16-bit adder/subtracter is AC_FLOAT, followed by CT_FLOAT, which is 19% slower

Table 2.6 Comparative results for 16-bit FlP multiplication with $F_{\texttt{clk}} = 200\text{MHz}$

	Area (μm^2)	Critical path (ns)	Total power (mW)	Energy per operation (pJ)
AC_FLOAT	488	1.18	2.15 E-1	1.05
CT_FLOAT	389	1.13	1.76 E-1	8.59 E-1
FLOPOCO	361	1.52	1.34 E-1	6.50 E-1
CT_FLOAT /AC_FLOAT	-20.4%	-4.24%	-18.2%	-18.2%
CT_FLOAT /FLOPOCO	+7.68%	-25.6%	+31.7%	+32.1%

Table 2.7 Comparative results for 32-bit FlP addition/subtraction with $F_{\texttt{clk}} = 200\text{MHz}$

	Area (μm^2)	Critical path (ns)	Total power (mW)	Energy per operation (pJ)
AC_FLOAT	678	2.49	4.46 E-1	2.21
CT_FLOAT	720	2.84	4.86 E-1	2.41
FLOPOCO	772	4.10	5.05 E-1	2.51
CT_FLOAT /AC_FLOAT	+6.06%	+14.1%	+8.92%	+9.12%
CT_FLOAT /FLOPOCO	-6.85%	-30.8%	-3.69%	-4.15%

Table 2.8 Comparative results for 32-bit FlP multiplication with $F_{\texttt{clk}} = 200\text{MHz}$

	Area (μm^2)	Critical path (ns)	Total power (mW)	Energy per operation (pJ)
AC_FLOAT	1689	2.19	1.02	5.03
CT_FLOAT	1469	2.30	5.84 E-1	2.70
FLOPOCO	2890	3.20	1.03	5.07
CT_FLOAT /AC_FLOAT	-13.0%	+5.02%	-42.8%	-46.3%
CT_FLOAT /FLOPOCO	-49.2%	-28.2%	-43.3%	-46.8%

but 27% faster than FLOPOCO. All metrics are slightly in favour of AC_FLOAT for 16-bit addition/subtraction. For 16-bit multiplication, FLOPOCO's multiplier is the smallest and with the lowest energy consumption. However, CT_FLOAT is 25% faster but consumes 32% more energy.

32-bit addition/subtraction gives similar energy for AC_FLOAT, CT_FLOAT, and FLOPOCO. FLOPOCO is the slowest operator, CT_FLOAT being 27% faster. The energy of 32-bit multiplication is strongly in favour of CT_FLOAT, saving more than 45% more energy than both AC_FLOAT and FLOPOCO. CT_FLOAT is 13% smaller than AC_FLOAT and 49% smaller than FLOPOCO. However, AC_FLOAT is 5% faster.

As a conclusion, AC_FLOAT, CT_FLOAT, and FLOPOCO addition/subtraction and multiplication provide similar results. Though they all have different features (implicit 1 or not, particular cases management, etc.), they all are quite close in terms of performance. In the following section, CT_FLOAT alone is then used as a reference for the comparison with fixed-point arithmetic.

2.5 Comparison Between Fixed-Point and Custom Floating-Point

This section draws a comparison between customized fixed-point and floating-point arithmetic. Section 2.5.1 compares FxP and FlP in terms of area, delay, and energy at the operator level. Then, Sect. 2.5.2 compares the two number representations in the context of their use in applications, thus considering the errors due to low-precision computations. One conclusion is that the choice between FxP and FlP is not obvious and depends on the application considered. It is shown that, in some cases, low-precision floating-point arithmetic can be the most effective, providing some benefits over the classical fixed-point choice for energy-constrained applications.

2.5.1 Operator-Level Comparison

This section compares FxP and FlP operators in terms of area, delay, and energy and does not consider computing errors at the operator level. Indeed, floating-point error magnitude is related to data values. Low-amplitude data have low error magnitude, whereas high-amplitude data have much higher error magnitude. Oppositely, fixed-point has a very homogeneous error magnitude, uniformly distributed between fixed bounds. Therefore, its relative error depends on the amplitude of the represented data. It is low for high-amplitude data and high for low-amplitude data. This duality makes these two paradigms impossible to be atomically compared using the same error metric. The only interesting error comparison that can be performed is to compare the error behaviour inside the same application, which is reported in Sect. 2.5.2 on FFT and K-means clustering.

The study in Sect. 2.5 uses the CT_FLOAT library for custom floating-point and AC_FIXED datatypes. A 100 MHz clock is set for synthesis and power estimation. All the other parameters are the same as for the previous section.

In this section, 8-, 10-, 12-, 14-, and 16-bit operators are compared. For each of these bit-widths, several versions of the floating-point operators are estimated with different exponent widths and compared with fixed-point. 25.10^3 uniformly distributed inputs are used for each operator characterization. For the floating-point adder, inputs are distributed such that the close path, which has the highest energy by nature, is activated 25% of the time. Adders and multipliers are all tested in their fixed-width version, meaning their number of input and output bits is the same. The output is truncated.

Figure 2.4 (respectively, Fig. 2.5) shows the area, delay, and energy of adders (respectively, multipliers) for different bit-widths, relative to the corresponding fixed-point operator. $\mathrm{FlP}_N(E)$ represents N-bit floating-point with E bits for exponent. As discussed before in this chapter, the floating-point adder shows an important overhead compared to fixed-point. For any configuration, area and delay

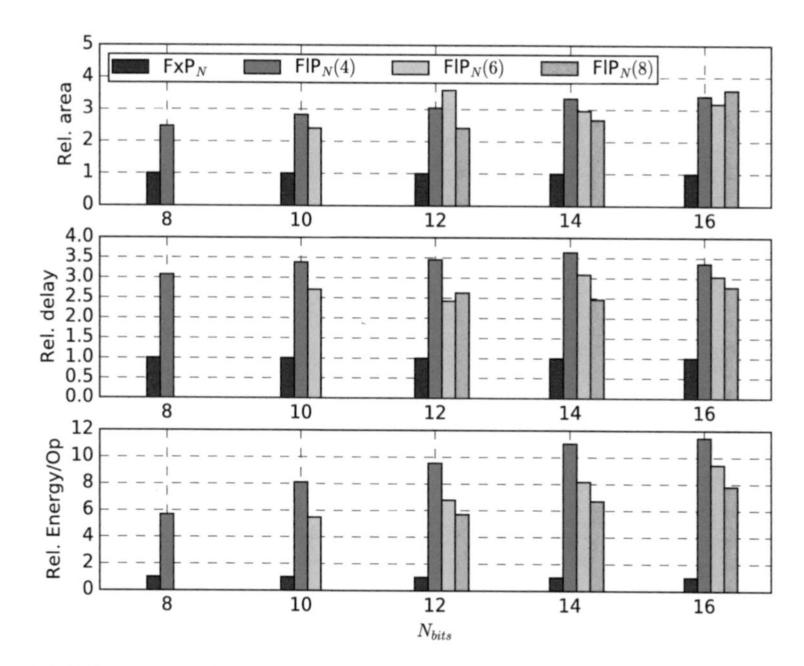

Fig. 2.4 Relative area, delay, and energy per operation between fixed-point and floating-point of adders for different bit-widths

are around 3× higher for floating-point. As a consequence, the higher complexity of the floating-point adder leads to 5× to 12× more energy per operation.

Results for the multipliers are very different. Floating-point multipliers are 2-3× smaller than fixed-point. The control part of floating-point multiplier being less complex than for adder and as multiplication is performed only on the mantissa, the area gets smaller. Timing is also slightly better for floating-point but still constrained by operand shifts during computations. These shifts also significantly impact energy per operation, especially for large mantissas, which results in an overhead of 2× to 10× on the energy per operation for floating-point multiplication.

However, it must be kept in mind that, when using fixed-point numbers in an application, shifting is often needed at many steps during execution to align the number formats. The cost of shifting in the case of FxP is not considered in the results presented here, whereas it is already present in the case of floating-point. Thus, the advantage of fixed-point highlighted by Figs. 2.4 and 2.5 is expected to be tempered when full applications are considered. This is the main objective of the next section.

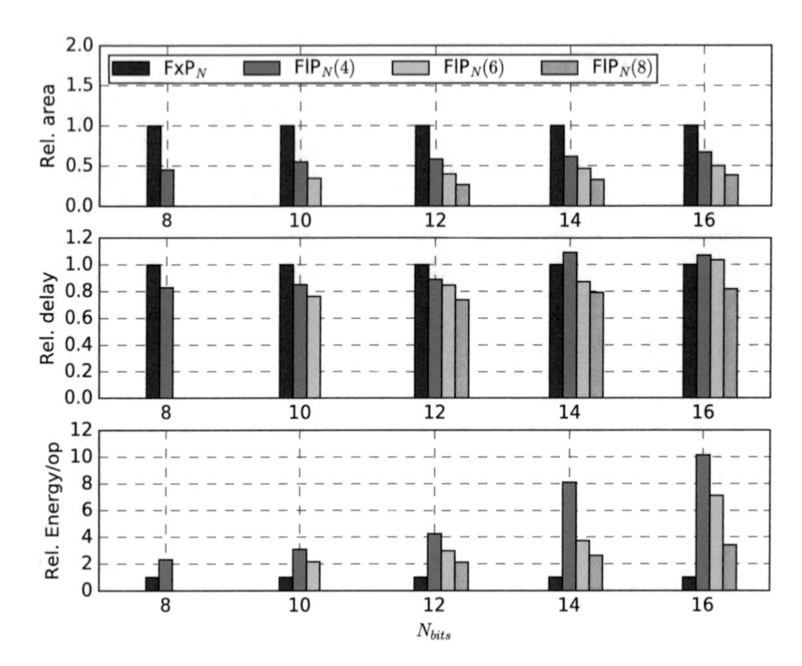

Fig. 2.5 Relative area, delay, and energy per operation between fixed-point and floating-point of multipliers for different bit-widths

2.5.2 Application-Level Comparison

In this section, floating-point and fixed-point operators are compared in the context of their use in applications. Indeed, as stated below, they have very different error nature, and thus their error cannot be fairly compared when considering only a single operation. Both number representations are compared first using the K-means clustering algorithm (also in [13]) and then on the Fast Fourier Transform (FFT).

2.5.2.1 Results on K-Means Clustering

This section first describes the K-means clustering algorithm before to provide comparative results between FxP and FlP.

K-Means Clustering Principle, Algorithm, and Experimental Setup

K-means clustering is a well-known method for vector quantization, which is mainly used in data mining, image classification, or voice identification. It consists in organizing a multidimensional space into a given number of clusters, each being totally defined by its centroid. A given vector in the space belongs to the cluster in which it is nearest from the centroid. The clustering is optimal when the sum of the distances of all points to the centroids of the cluster they belong to is minimal, which corresponds to finding the set of clusters $S = \{S_i\}_{i \in [0,k-1]}$ satisfying

$$\arg\min_S \sum_{i=1}^{k} \sum_{x \in S_i} \|x - \mu_i\|^2 \tag{2.14}$$

where μ_i is the centroid of cluster S_i. Finding the optimal centroid position of a vector set is NP-hard. However, some iterative algorithms find good approximations of the optimal centroids by an *estimation-maximization* process, with a linear complexity (linear with the number of clusters, the number of data to process, the number of dimensions, and the number of iterations). Lloyd's iterative algorithm [14] is used in our case study. It is applied to bidimensional sets of vectors to ease display and interpretation of the results. From now, we only refer to Lloyd's algorithm in two dimensions. Figure 2.6 shows results of K-means on a random set of input vectors, obtained using double-precision floating-point computations with a very restrictive stopping condition; these values are then considered as the reference golden outputs.

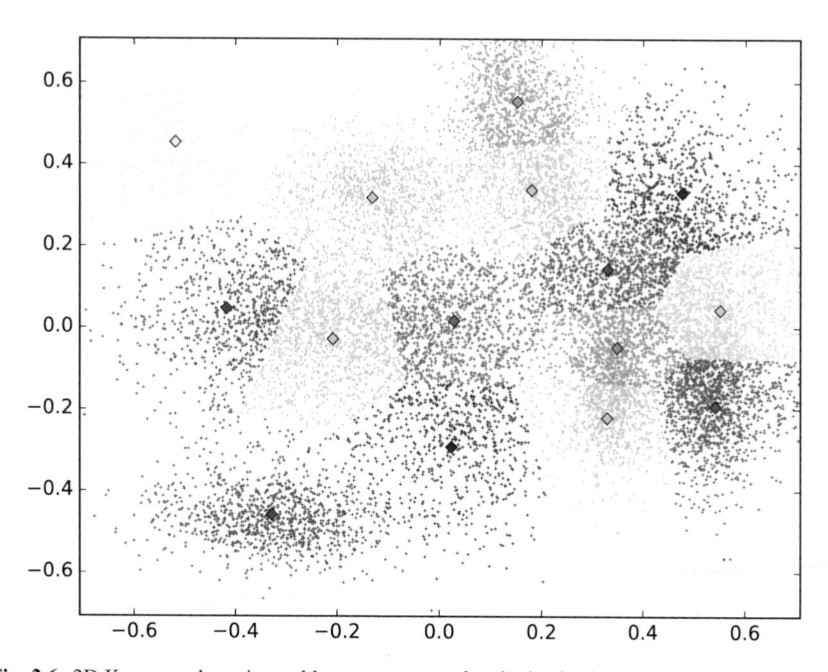

Fig. 2.6 2D K-means clustering golden output example, obtained using double-precision (64-bit) floating-point

The algorithm consists of three main steps:

1. Initialization of the centroids.
2. Data labelling.
3. Centroid position update.

Steps 2 and 3 are iterated until a stopping condition is met. In our case, the main stopping condition is when the difference of the sums of all distances from data points to their cluster's centroid between two iterations is less than a given threshold. A second stopping condition is the maximum number of iterations, required to avoid the algorithm getting stuck when the computations are too approximated to converge. The detailed algorithm for one dimension is given by Algorithm 1. Inputs are represented by the vector $data$ of size N_{data} and output centroids by the vector c of size k. The accuracy target for stopping condition is defined by acc_target and the maximum allowed number of iterations by max_iter. In our study, we use

Algorithm 1 K-Means Clustering algorithm in one dimension

```
Require: k ≤ N_data
  err ← +∞
  cpt ← 0
  c ← init_centroids(data)
  do                                                        ▷ Main loop
     old_err ← err
     err ← 0
     c_tmp[0 : k − 1] ← 0
     min_distance ← +∞
     for d ∈ {0 : N_data − 1} do
        min_distance ← +∞
        for i ∈ {0 : k − 1} do                              ▷ Data labelling
           distance ← distance_comp(data[d], c[i])
           if distance < min_distance then
              min_distance ← distance
              labels[d] ← i
           end if
        end for
        c_tmp[labels[d]] ← c_tmp[labels[d]] + data[d]
        counts[labels[d]] ← counts[labels[d]] + 1
        err ← err + min_distance
     end for
     for i ∈ {0 : k − 1} do                                 ▷ Centroid position update
        if counts[i] ≠ 0 then
           c[i] ← c_tmp[i]/counts[i]
      else
      end
      c[i] ← c_tmp[i]
        end if
     end for
     cpt ← cpt + 1
  while (|err − old_err| > acc_target) ∨ (cpt < max_iter)
```

several values for acc_target, and max_iter is set to 150, which is never reached in practice.

The impact of fixed-point and floating-point arithmetic on performance and accuracy is evaluated considering the distance computation function distance_comp, defined by

$$d \leftarrow (x - y) \times (x - y) \tag{2.15}$$

In the 2D case, the distance computation becomes

$$d \leftarrow (x_0 - y_0) \times (x_0 - y_0) + (x_1 - y_1) \times (x_1 - y_1) \tag{2.16}$$

which is equivalent to 1 addition, 2 subtractions, and 2 multiplications. However, as distance computation is cumulative on each dimension, the hardware implementation relies only on 1 adder (accumulation), 1 subtracter, and 1 multiplier.

The experimental setup is divided into two parts: accuracy evaluation and cost/performance/energy estimation. Accuracy estimation is performed on 20 data sets composed of 15.10^3 bidimensional data samples, all generated in a square delimited by $\{\pm\sqrt{2}, \pm\sqrt{2}\}$, using Gaussian distributions with random covariance matrices around 15 random mean (centroid) points. Several accuracy targets are used to set the stopping condition: 10^{-2}, 10^{-3}, 10^{-4}. As stated, the reference for accuracy estimation is IEEE-754 double-precision floating-point (Fig. 2.6). The error metrics for the accuracy estimation are (i) the Mean Square Error of the resulting cluster Centroids (CMSE) and (ii) the classification Error Rate (ER), which is defined as the proportion of points not being tagged by the right cluster identifier. The lower the CMSE, the better the estimated position of centroids compared to golden output. Energy estimation is performed using the first of these 20 data sets, limited to 20.10^3 iterations of distance computation for time and memory purposes. As data sets were generated around 15 points, the number of clusters researched is also set to 15. Area, latency of execution and energy are estimated using the same library and tools as in the previous section. Iterative distance computation is specified in C++ and HLS is used to generate the hardware under evaluation.

Experimental Results on K-Means clustering

Section 2.5.1 showed that FxP additions and multiplications consume less energy than their FlP counterparts for the same bit-width. However, these results do not yet consider the impact of the number formats on accuracy. This section details the impact of accuracy on the 2D K-means clustering algorithm.

A first qualitative study on the K-means clustering showed that, to get correct results (no artefacts), FlP data must have a minimal exponent width of 5 bits in distance computation (smaller exponents are too inaccurate in low distance computations) and fixed-point data a minimal number of 3 bits for its integer part.

Table 2.9 8- and 16-bit area, energy, and accuracy for K-means clustering experiment

	ct_float$_8$(5)	ct_float$_{16}$(5)	ac_fixed$_8$(3)	ac_fixed$_{16}$(3)
Area (μm^2)	392.3	1148	180.7	575.1
N_{cycles}	3	3	2	2
E_{dc} (nJ)	1.23E-4	5.99E-4	5.03E-5	3.25E-4
N_{it}	8.35	59.3	14.9	65.1
$E_{\text{K-means}}$ (nJ)	38.24	1100	23.90	644.34
CMSE	1.75E-3	3.03E-7	1.85E-2	3.28E-7
Error Rate	35.1 %	2.94 %	62.3 %	0.643 %

Thus, all the following results use these two configurations and vary the mantissa and fractional part for FlP and FxP, respectively. The total energy is defined as

$$E_{\text{K-means}} = E_{\text{dc}} \times \left(N_{\text{it}} + N_{\text{cycles}} - 1\right) \times N_{\text{data}} \tag{2.17}$$

where E_{dc} is the energy per distance computation estimated as in the previous section, N_{it} the average number of iterations necessary to reach K-means stopping condition, N_{cycles} the number of pipeline stages in the distance computation core, as determined by HLS, and N_{data} the number of processed data per iteration.

Results for 8-bit and 16-bit FlP and FxP arithmetic operators are detailed in Table 2.9, with a stopping condition set to 10^{-4}. For the 8-bit version of the algorithm, several interesting results can be highlighted. First, the custom FlP version is 2× larger than FxP version, and FlP distance computation consumes 2.44× more energy than FxP. However, the FlP version of K-means converges in 8.35 cycles on average, against 14.9 cycles for FxP. This results in making the floating-point version for the whole K-means algorithm consuming only 1.6× more energy than fixed-point. Moreover, the FlP version provides a huge advantage in terms of accuracy of results. Indeed, CMSE is 10× better for FlP and ER is 1.8× better. Figure 2.7a, b shows the output for floating-point and fixed-point 8-bit computations, applied on the same inputs as the golden output of Fig. 2.6. A very neat stair-effect on data labelling is clearly visible, which is due to the high quantization levels of the 8-bit representation. However, in the floating-point version, the positions of clusters' centroid are very similar to the reference, which is not the case for fixed-point.

For the 16-bit version, all results are in favour of fixed-point, floating-point being twice bigger and consuming 1.7× more energy. FxP also provides slightly better error results (2.9% for ER vs. 0.6%). Figure 2.7c, d shows output results for 16-bit floating-point and fixed-point. Both are very similar and nearly equivalent to the reference, which reflects the high success rate of clustering.

The competitiveness of FlP over FxP on small bit-widths and the higher efficiency of FxP on larger bit-widths are confirmed by Fig. 2.8 depicting energy vs. classification error rate. Indeed, for different accuracy targets ($10^{-\{2,3,4\}}$), only 8-bit FlP provides higher accuracy for a comparable energy cost, whereas 10- to 16-bit FxP versions reach an accuracy equivalent to FlP with much less energy.

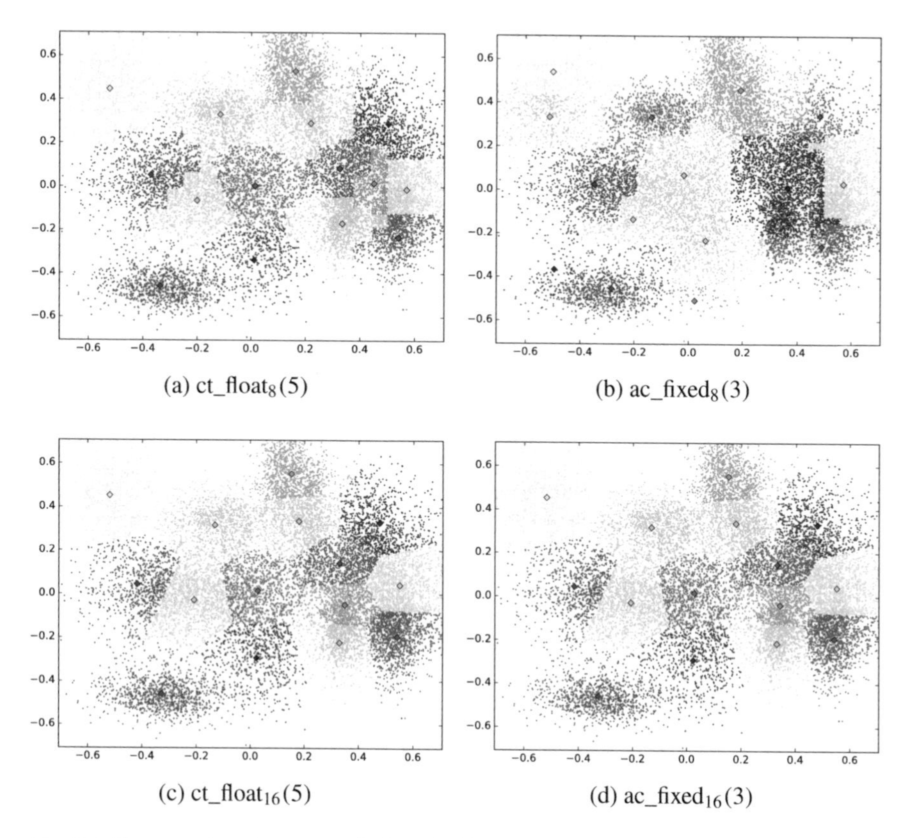

(a) ct_float$_8$(5)

(b) ac_fixed$_8$(3)

(c) ct_float$_{16}$(5)

(d) ac_fixed$_{16}$(3)

Fig. 2.7 K-means clustering outputs for 8- and 16-bit floating-point and fixed-point, with accuracy target of 10^{-4}. (**a**) ct_float$_8$(5). (**b**) ac_fixed$_8$(3). (**c**) ct_float$_{16}$(5). (**d**) ac_fixed$_{16}$(3)

The stopping condition does not seem to have a major impact on the relative performance.

2.5.2.2 Results on Fast Fourier Transform

In the previous section, a comparative study between custom FxP and FlP was performed on K-means, showing that, contrary to what could be expected, floating-point can be very competitive for small bit-widths. In this section, a similar study is performed on the Fast Fourier Transform (FFT).

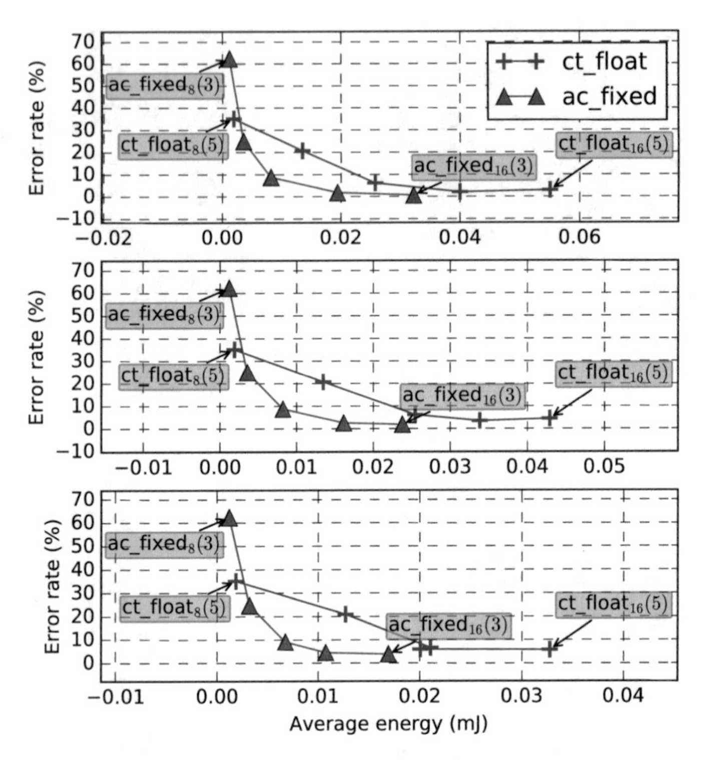

Fig. 2.8 Energy vs. classification error rate for K-Means clustering with stopping conditions of 10^{-4} (top), 10^{-3} (centre) and 10^{-2} (bottom)

The implementation of the FFT is Radix-2 Decimation-in-Time (DIT), which is the most common form of the Cooley–Tukey algorithm [15]. For the hardware estimation, only the kernel computations of the FFT are considered, i.e.,

$$X_k = E_k + e^{-\frac{2\pi i}{N}k} O_k,$$
$$X_{k+\frac{N}{2}} = E_k - e^{-\frac{2\pi i}{N}k} O_k \tag{2.18}$$

equivalent to 6 additions/subtractions and 4 multiplications. For each version of the FFT, all constants and variables are represented with the same parameters (same bit-width, same integer part width for FxP, and same exponent width for FlP). The absence of over/underflow for the FxP version is ensured. For the FlP version, the repartition of the exponent and mantissa widths is chosen for giving the smallest error after an exhaustive search. For hardware performance estimation, only FFT-16 (FFT on $N = 16$ samples) was characterized. The error metric is the Mean Square Error (MSE) at the output compared to double-precision floating-point.

Energy per operation (pJ) related to error (MSE in dB) for FFT-16 is depicted in Fig. 2.9. The error-energy trade-off is better when reaching the bottom-left corner.

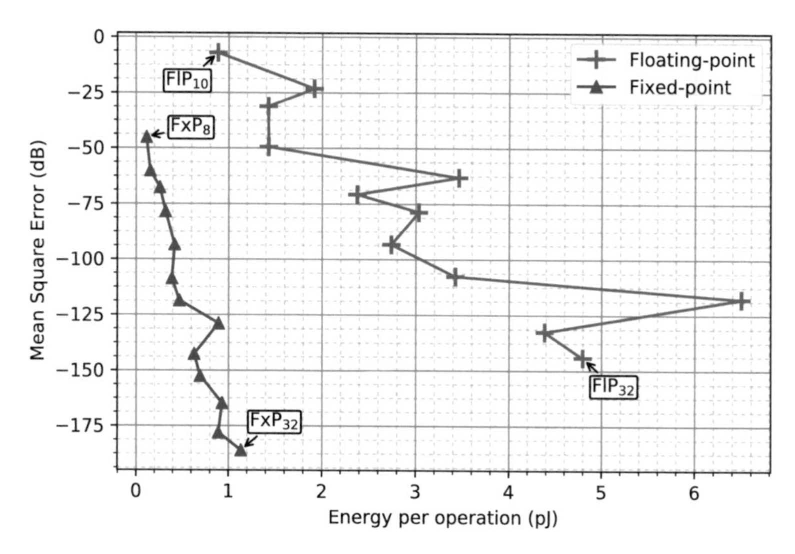

Fig. 2.9 Fixed-point and floating-point energy per operation (pJ) vs MSE for FFT-16 for different bit-widths

For each curve, each point going from the top left to the bottom right represents an increase of two digits in the bit-width.

For this application, the advantage is clearly in favour of fixed-point. Indeed, for any identical bit-width, FxP outperforms FlP in both energy and accuracy. As already showed in Sect. 2.5.1, FlP operations, and additions in particular, are much more expensive than FxP. However, FFT output quality is not as dependent on accuracy on a dynamic as large as for K-means clustering. This makes FlP even less accurate than FxP at equal bit-width, because of a smaller significant part, mantissa for floating-point, all bits for fixed-point. Indeed, in the experiment, the exponent takes 7 bits of the total width, which are not assigned to more accuracy on the significant part. Another interesting point is the data points presenting an *energy peak*, which are occurring for 12-, 18-, and 28-bit floating-point and 22-bit fixed-point. These peaks are most probably due to differences of implementation in the HLS process. For example, larger adder or multiplier structures may have been selected by the tool to meet constraints of delays, leading to energy overhead.

2.6 Conclusion

Computing with low-precision arithmetic is an efficient way to maximize performance per Watt [16]. Customization is then ruled by finding a trade-off between reducing precision to improve energy efficiency and respecting constraints on application output quality. This chapter mainly focused on floating-point and fixed-point number representations, presented their principle and some opportunities for arithmetic customization on both formats, and provided a comparison between their cost, performance, and energy, as well as their impact on accuracy during computations.

Comparing floating-point and fixed-point arithmetic at the operator level gives a clear advantage in area, delay, and energy efficiency for fixed-point. However, when considering real applications, e.g., the study on K-means clustering algorithm in this chapter, custom floating-point arithmetic can provide interesting features and tends to show a better energy/accuracy trade-off for very small bit-widths (8 bits in this study). However, the advantage comes back to fixed-point when the considered application is an FFT (the same is true for digital filters or most classical signal and image processing algorithms). One explanation is that applications requiring both large dynamic range and high accuracy are more tolerant to low precision when the floating-point representation is used. An interesting follow-up of this study would be to consider larger FFT, which would lead to larger dynamic range, and to see how the MSE vs. energy per operation would scale for both representations.

Another important aspect of the study is from a hardware design point of view. Floating-point is very complex compared to fixed-point arithmetic for large bit-widths, but the overhead is shrinking when lowering precision. Moreover, the cost of multiplication can be considered at the advantage of floats. Also, the overhead of scaling instructions required to be added to deal with fixed-point data types is often not studied. It is an interesting perspective to include this overhead in the choice of the right representation.

Hence, in the aim of designing general-purpose low-energy processors, low-precision floating-point arithmetic can provide major advantages compared to classical integer operators embedded in microcontrollers, with a better compromise between ease of programming, energy efficiency, and computing accuracy. In the context of inference and training of deep neural networks, custom float is also a serious competitor. This is one of the objectives of Chap. 15 presenting opportunities for approximations in deep learning.

References

1. Lapsley, P., Bier, J., Shoham, A., & Lee, E. A. (1996). *DSP Processor Fundamentals: Architectures and Features*. Fremont, CA: Berkeley Design Technology, Inc.
2. Menard, D., Rocher, R., Scalart, P., & Sentieys, O. (2004). SQNR determination in non-linear and non-recursive fixed-point systems. In *XII European Signal Processing Conference (EUSIPCO 2004)*, Vienna, September 2004 (pp. 1349–1352).
3. Rocher, R., Menard, D., Scalart, P., & Sentieys, O. (2004). Accuracy evaluation of fixed-point LMS algorithm. In *IEEE International Conference on Acoustics, Speech and Signal Processing (ICASSP 2004)*, Montreal, May 2004 (pp. 237–240).
4. Menard, D., Rocher, R., & Sentieys, O. (2008). Analytical fixed-point accuracy evaluation in linear time-invariant systems. *IEEE Transactions on Circuits and Systems I: Regular Papers, 55*, 3197–3208.

5. Nehmeh, R., Menard, D., Nogues, E., Banciu, A., Michel, T., & Rocher, R. (2015). Fast integer word-length optimization for fixed-point systems. *Journal of Signal Processing Systems, 85*, 113–128.
6. IEEE standard for floating-point arithmetic, *IEEE Std 754-2008* (pp. 1–70), Aug 2008.
7. Muller, J.-M., Brisebarre, N., De Dinechin, F., et al. (2009). *Handbook of floating-point arithmetic*. New York: Springer Science & Business Media.
8. Graphics, M. (2021). *Algorithmic c (ac) datatypes v4.2*. https://github.com/hlslibs/ac_types, Apr. 2021.
9. de Dinechin, F., & Pasca, B. (2011). Designing custom arithmetic data paths with FloPoCo. *IEEE Design Test of Computers, 28*, 18–27.
10. Fang, X., & Leeser, M. (2016). Open-source variable-precision floating-point library for major commercial FPGAs. *ACM Transactions on Reconfigurable Technology and Systems, 9*, 20:1–20:17.
11. Fang, X., & Leeser, M. (2016). Open-source variable-precision floating-point library for major commercial FPGAs. *ACM Transactions on Reconfigurable Technology and Systems, 9*, 20:1–20:17.
12. Mahzoon, A., & Alizadeh, B. (2017). Optifex: A framework for exploring area-efficient floating point expressions on FPGAs with optimized exponent/mantissa widths. *IEEE Transactions on Very Large Scale Integration (VLSI) Systems, 25*, 198–209.
13. Barrois, B., & Sentieys, O. (2017). Customizing fixed-point and floating-point arithmetic - A case study in K-means clustering. In *IEEE International Workshop on Signal Processing Systems (SiPS)*, October 2017.
14. Lloyd, S. (1982). Least squares quantization in PCM. *IEEE Transactions on Information Theory, 28*, 129–137.
15. Cooley, J. W., & Tukey, J. W. (1965). An algorithm for the machine calculation of complex Fourier series. *Mathematics of Computation, 19*(90), 297–301.
16. Barrois, B., Sentieys, O., & Menard, D. (2017). The hidden cost of functional approximation against careful data sizing a case study. In *IEEE/ACM Design Automation and Test in Europe (DATE)*, Lausanne (p. 6).

Chapter 3
Hardware Level Approximations

Ioannis Tsiokanos, George Papadimitriou, Dimitris Gizopoulos, and Georgios Karakonstantis

3.1 Introduction

Approximate computing has recently emerged as a concept that takes advantage of the error-resilient properties of applications to tolerate a number of errors or approximate a number of less critical operations for reducing the power consumption. Existing work has showcased the inherent resilience of various signal/image processing [1–4], machine learning [1, 5], and scientific computation algorithms [2] to faults or inaccurate operations. To better understand the involved trade-offs, let us step back and look at the main components constituting the power dissipation of digital circuits. On-chip power consumption is characterized by static and dynamic components. Static, sometimes called leakage power, can be attributed to junction leakage in the transistor when it is inactive (not in the process of switching). Static/leakage power (P_{static}) is a function of the supply voltage V_{dd} and static current of each transistor (I_S) and is determined by the formula:

$$P_{static} = V_{dd} \times I_s \times N_{tr} \times k_d \tag{3.1}$$

where N_{tr} denotes the number of transistors and k_d is a device-specific constant. While static power consumption due to leakage current is almost continuous, dynamic power, sometimes called switching power, is only dissipated when there is switching activity at some nodes/transistors in a circuitry. Dynamic power consumption is associated with circuit activity (i.e., transistor switches, changes of

I. Tsiokanos · G. Karakonstantis (✉)
Queen's University Belfast, Belfast, UK
e-mail: G.Karakonstantis@qub.ac.uk

G. Papadimitriou · D. Gizopoulos
University of Athens, Athens, Greece
e-mail: dgizop@di.uoa.gr

A. Bosio et al. (eds.), *Approximate Computing Techniques*,
https://doi.org/10.1007/978-3-030-94705-7_3

values in registers, etc.) and strongly depends on the executed workload; therefore, it is formally defined as:

$$P_{dynamic} = a \times C \times V_{dd}^2 \times F \tag{3.2}$$

where a is the switching activity, C is the physical capacitance, and F denotes the clock frequency. Total power consumption (P_{total}) of a system is the sum of the dynamic and static power:

$$P_{total} = P_{static} + P_{dynamic} \tag{3.3}$$

In the following, we discuss about possible approaches to provide low-power operations through exploring hardware-level applications of approximate computing.

As can been seen from Eqs. (3.1) and (3.2), the power dissipated can be reduced by reducing either the clock frequency F, or the supply voltage V_{dd}, or the switching activity parameter a. The above are common techniques that the approximate computing paradigm exploits to significantly reduce the power consumption of a system. Existing approximate computing techniques for achieving power/energy efficiency can be categorized into three groups.

(1) Approximate Circuits Using Less Hardware. First, designers use inexact hardware using less components/bits in different processing units (e.g., CPU, GPU, FPGA) and memory technologies for saving power. This decreases the number of transistors and thus the switching activity, leading to reduction of both static and dynamic power dissipation.

(2) Heterogeneous Architectures Combining Reliable and Approximate (Inaccurate) Cores/Memories. Although effective, approximate computing is not a panacea. Several current studies have indicated that any approximation should be applied only to error-resilient code or less-significant data regions in applications, since uniform approximation of all data may result in catastrophic quality degradation [3, 4, 6, 7]. For instance, approximating few Less Significant Bits (LSBs) in the mantissa part of floating-point operands leads to insignificant quality loss as opposed to approximation of any bit of the exponent part [8, 9]. Taking this into account, the second group of low-power, approximation-based techniques focuses on the principles of trying to distinguish the significant from the so-called less significant operations [3, 4]. Particularly, memory [4, 10–12] and multicore [13, 14] heterogeneous frameworks could be explored by allocating data in different cores and/or memory domains based on their vulnerability to approximations. For example, the critical (not amenable to inaccurate results) data/code could be assigned to the cores/memories that have been implemented to be highly reliable, while the off-critical data could be assigned to approximate cores/memories that operate at relaxed parameters/settings [4, 10].

(3) Approximate Techniques Facilitating Voltage Scaling. One of the most effective methods to reduce the total power consumption is to scale down the supply voltage [15] (see Eqs. (3.2) and (3.1)). As we will explain later in this chapter, voltage scaling reduces energy consumption of circuits at the cost of possible delay increase [15–18]. Such a delay increase induced by voltage under-scaling may result in errors. For instance, reducing SRAM supply voltage saves leakage energy but also increases probability of read upset (flipping of a bit during read operation) and write failure (writing a wrong bit) [19]. Another study on an 80-core Intel processor indicates that operations at 0.8 V (i.e., 30% lower than the nominal supply voltage) can lead to up to 50% delay fluctuations [20], resulting in delay-induced timing errors [21]. The third group of approximate computing techniques facilitates voltage scaling while accounting for this delay uncertainty and the resultant errors. In this scope, one of the most popular and simple, yet effective, approximation approach truncates the bitwidth of all the input operands in simple data-paths, reducing the delay of those designs and enabling voltage reduction without delay-induced errors [9, 22]. Operand truncation is realized by setting a number of LSBs to a constant value of zero ('0'). To elucidate the impact of operand truncation on computational delay, let us consider a simple 4-bit ripple carry adder (RCA), as shown in Fig. 3.1. The depicted adder consists of four full adders (FAs). FA is a logic circuit that adds two input operand bits (Ai, Bi) plus a Carry in bit (Ci,i) and generates a Carry out bit (Co, i) and a Sum bit (S, i). In such a design, the most timing critical path $LLP1$ (emphasized in red dotted line) will be activated when the carry propagates all the way from Ci,0 to $Co, 3$. If we define the gate delay as T, then $LLP1$ requires a delay equal to $8T$ to be completed, since $LLP1$ will have to travel from the AND gate (emphasized in red) in FA0 down to XOR gate (emphasized in red) in FA3. Note that such a delay will be activated only in case of a suitable combination of operands. For instance, when $A = 1111$ and $B = 0001$, the carry generated in the first bit position from the right (i.e., LSB) is propagated all the way to the final bit position, exciting the error-prone $LLP1$. Under an assumed

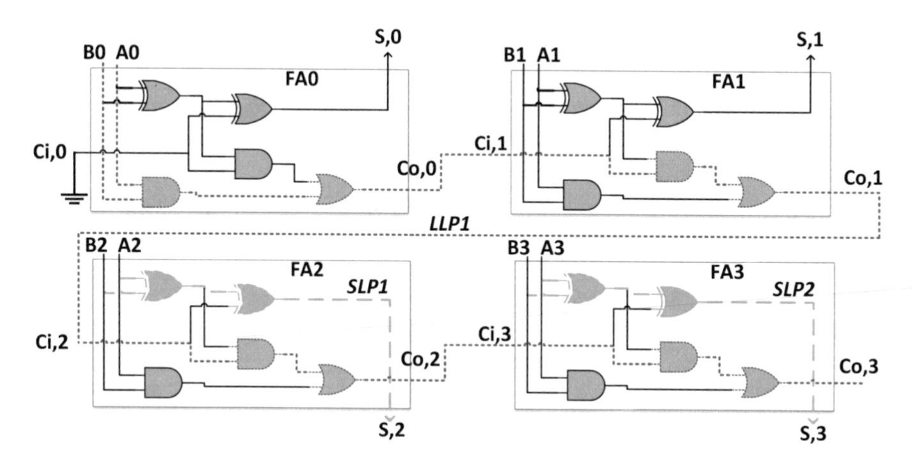

Fig. 3.1 Data-paths in a 4-bit ripple carry adder, highlighting the longest latency path (*LLP1*) and two short latency paths (*SLP1* and *SLP2*)

delay increase (induced by voltage down-scaling), this path may fail, leading to a timing error. However, by modifying the inputs and inserting 0s in the last 2 bits, such as: $A = 1100$ and $B = 0000$, there is no carry propagation and thus only less critical short latency paths $SLPs$ (e.g., $SLP1$ and $SLP2$ highlighted in green dashed lines) will be excited. By truncating the last 2 bits of the input operands, the delay of the critical paths that are excited is reduced to $2T$, thus providing enough timing slack to address any potential delay increase due to supply voltage reduction. This allows the manufacturer to provide operations beyond the always correct voltage margins.

This chapter mainly focuses on the last two groups of approximate techniques and reviews the state-of-the-art for approximating variables/operations at circuit and microarchitecture layer or at real processors (after manufacturing), facilitating low-power, error-resilient operations.

3.2 Circuit Level Functional Approximation

In this section, we first discuss the design of inexact/approximate circuits. Then, approximate-based approaches that allow efficient synthesis are also introduced.

3.2.1 Inexact Units

Several approximation methodologies have been proposed for the implementation of inaccurate arithmetic units. Arithmetic circuits such as adders, multipliers, and divisions are key components for several error-tolerant applications (e.g., image and video processing and artificial neural networks). Therefore, there exist many inaccurate circuits that replace the exact calculations to meet the energy requirements.

D. Celia et al. [23] present the design of an inexact adder that approximates the lower k bits with a fixed value (i.e., $2^k - 1$). C. K. Jha et al. [24] develop an approximate adder that performs either m single n-bit exact addition or two n-bit approximate additions on the same hardware. The work provides runtime configuration for dynamically altering approximate or accurate addition operations. Another set of studies [25–28] has also proposed the implementation of inexact adders to enhance performance and/or energy efficiency.

Multipliers and divisions are fundamental components of various signal processing applications and are among the most power- and energy-hungry blocks of a core [29, 30]. Therefore, inexact (approximate) multiplication and division operations have gained a lot of attention in the literature [31–35]. R. Zendegani et al. [35] introduce RoBA multiplier, an approximate multiplier suitable for DSP applications. Such an inexact block relies on rounding the input operands to the nearest exponent of two. By doing so, the multiplication operation is simplified (lesser hardware blocks are used) and the computation intensive part is omitted. This results in delay reduction and power/energy gains at a small output quality loss. D. Esposito et al. [31] implement a new approximate compressor and an algorithm for utilizing in the design of inexact multipliers. S. Vahdat et al. [34] present a fast, approximate divider where the division operation is performed by multiplying the dividend by the inverse of the divisor. J. Melchert et al. [32] devise an iterative approximate reciprocal process with scalable accuracy to devise an inexact, energy-efficient divider. W. Liu et al. [33] combine restoring array and logarithmic dividers into an approximate, unsigned, hybrid divider.

Floating-point operations emerge as a major contributor to the energy consumption and typically determine the clock frequency. In particular, existing experimental results [36, 37] indicate that 30% of the energy consumption of a target core is due to floating-point operations. Moreover, floating-point instructions have higher energy-per-instruction costs than their integer counterparts [37, 38]. Taking into account these trends, there are several frameworks that provide efficiency–accuracy trade-offs in floating-point operations. Design and analysis of inexact floating-point adders [8, 39], multipliers [40, 41], and dividers [42, 43] have indicated significant energy and performance gains in the overall system. I. Tsiokanos et al. present a fully automated, evolutionary computing-based framework [44] that inspects any application's binary and identifies approximable floating-point instruction sequences. It can be very effective in facilitating approximation schemes (e.g., dynamic precision scaling, voltage down-scaling) at runtime, without compromising reliability.

3.2.2 *Automated Synthesis of Approximate Circuits*

However, the optimizations performed by commercial Electronic Design Automation (EDA) tools may negatively affect the gains from the dynamic voltage and accuracy scaling techniques. To help the design of efficient approximate-based approaches, tools for the automatic synthesis of approximate logic circuits have been lately investigated [45–50]. Those tools focus on improving the analysis of approximate circuits, such as evaluating the output error that can be caused due to circuit simplifications rather than proposing new techniques that enhance the design of approximate units and limit the impact of approximations on output quality loss. Particularly, they propose systematic methodologies that can be leveraged during the synthesis step of the conventional EDA flow [51] in order to synthesize an approximate version of the circuit to meet the predefined quality bound. The above tools pave the way to incorporate the approximate circuit design (which typically requires manual effort) into the design automation by suggesting general and scalable means towards the automated synthesis of approximate circuits.

3.3 Voltage Down-Scaling and Hardware Approximations

As discussed in Sect. 3.4, modern platforms allow operations beyond the nominal voltage values for boosting energy efficiency [52–54]. Reducing the supply voltage is a common technique that the approximate computing paradigm [1, 55] exploits to trade off quality of desired output for energy improvements. B. Moons et al. [56] accommodates voltage and accuracy scaling in multipliers for saving energy by disabling some of the input bits. H. Afzali-Kusha et al. [57] explore voltage down-scaling and different approximation settings on a Dadda multiplier to improve energy consumption as well as the reliability and lifetime of the block under test. To make the multiplier more area and energy efficient, the truncation of 4 LSBs of multiplication output is also studied.

3.3.1 Approximate Application-Specific Processing Cores

Uniform approximations may be effective but incur a quality degradation that may be significant for specific applications. Thus, recent schemes determine the applied approximation strategies on a per-application basis. In this content, dynamic voltage and precision scaling are exploited to provide energy-efficient processing cores suitable for convolutional neural network [58–60]. The increased computation demands of neural networks drive several industry efforts towards custom-developed, approximate ASICs that are designed to accelerate machine learning workloads. In this scope, Google designed a Tensor Processing Unit (TPU) to perform fast, bulky matrix multiplication, which is considered the most computationally intensive part of running a trained ML model. To achieve this, Google TPU leverages a novel number encoding format, namely Brain floating-point format (bfloat16 or BF16) occupying 16 bits representing a floating-point number, to improve the training and inference throughput of a wide range of advanced NNs. Goya, a microarchitecture for inference neural processors designed by Habana Labs, supports mixed-precision operations including 8-bit, 16-bit, and 32-bit vector operations for both integer and floating-point. Additionally, NVIDIA Tensor Cores offer a full range of precisions "TF32, bfloat16, FP16, INT8, and INT4" to accelerate AI training and inference. Apart from approximate ASICs, application-specific FPGAs that operate at reduced voltages and/or operations have been also investigated. B. Salami et al. [59] evaluated an FPGA-based NN accelerator under low-voltage operations. The variable-precision digital signal processing (DSP) blocks within Intel Stratix 10 FPGA devices support fixed-point arithmetic and single-precision floating-point arithmetic in order to optimize DSP applications. Finally, prior studies [61, 62] propose energy-efficient, large-scale neuromorphic system architectures through approximate computing. This is achieved by selectively approximating artificial neurons that have been characterized as less critical in estimating the network output quality.

3.3.1.1 Significance Driven Design Methodology

There is a design methodology, the so-called significance-driven, that was introduced by Karakonstantis et al. in 2007 [3, 7] and was exploited in the past decade for the design of various application-specific cores for popular signal processing kernels used in multimedia, biomedical, and wireless applications [3, 7, 63, 64]. The main principle of such a methodology is the initial classification of operations into significant and less-significant based on their contribution to the output quality. Afterward, via algorithm and architecture co-design, the significant operations are given higher priority and provided adequate timing slack as opposed to the less-significant operations. By doing so, the significant operations are ensured to not fail under delay variations and aggressive voltage scaling, thus allowing the maintenance of high output quality, since only less-significant computations can dynamically be pruned or skipped if necessary.

3.3.2 Approximations in General-Purpose Processing Cores

Any application involves a large amount of arithmetic calculations, which are typically performed in full precision on general-purpose processing cores. Since floating-point operations are one of the most flexible and dynamic numerical computations, floating-point units that support several formats have been proposed [36, 65–67]. Work in this domain departs from the standard IEEE 754 single (32-bit format) and double (64-bit format) precision formats and extends the target instruction set architecture (ISA) with new, custom formats (e.g., 16- and 8-bit formats). Lower than the conventional double and single precision allows the designer to reduce the supply voltage to a wide range, leading to energy/power gains at insignificant output quality loss.

Device's variation during fabrication, known as static variation, remains constant during the chip lifetime. On top of that, transistor ageing and dynamic variation in supply voltage and temperature, caused by different workload interactions, is also of primary importance. Both static and dynamic variations lead microprocessor architects to apply conservative guardbands (operating voltage and frequency settings) to avoid timing failures and guarantee correct operation, even in the worst-case conditions excited by unknown workloads or the operating environment [68, 69]. However, these guardbands increase the power consumption. To bridge the gap between energy efficiency and performance improvements, several hardware and software techniques have been proposed, such as Dynamic Voltage and Frequency Scaling (DVFS) [70]. The premise of DVFS is that the microprocessor's workloads as well as the cores' activity vary. Voltage and frequency scaling during epochs where peak performance is not required enables a DVFS-capable system to achieve average energy-efficient gains without affecting peak performance adversely. However, energy-efficient gains are limited by the pessimistic guardbands. Revealing and harnessing the pessimistic design-time voltage margins offers a significant opportunity for energy-efficient computing in multicore CPUs. The full energy saving potential can be exposed only when accurate core-to-core, chip-to-chip, and workload-to-workload voltage scaling variation is measured. When all these levels of variation are identified, system software can effectively allocate hardware resources to software tasks matching the capabilities of the former (undervolting potential of the CPU cores) and the requirements of the latter (for energy or performance).

To this end, Papadimitriou et al. [54] propose a fully automated system-level framework built around Applied Micro's (APM) X-Gene 2 micro-server. The automated infrastructure aims to increase the throughput of massive undervolting campaigns that require multiple benchmarks execution at several voltage supply levels of all individual cores. The primary goals of the proposed framework are (1) to identify the target system's limits when it operates at scaled voltage and frequency conditions, and (2) to record/log the effects of a program's execution under these conditions. The framework provides the following features:

- It compares the outcome of the program with the correct output of the program when the system operates in nominal conditions to record Silent Data Corruptions (SDCs).
- It monitors the exposed corrected and uncorrected errors from the hardware platform's error reporting mechanisms.
- It recognizes when the system is unresponsive to restore it automatically.
- It monitors system failures (crash reports, kernel hangs, etc.).
- It determines the safe, unsafe, and non-operating voltage regions for each application for all frequencies.
- It performs massive repeated executions of the same configuration.

This automated characterization process requires minimal human intervention and records all possible abnormalities due to undervolting: silent data corruptions (SDCs, e.g., program output mismatches without any hardware error notification), corrected errors, uncorrected (but detected) errors (provided by Linux EDAC driver [71]), as well as application and system crashes [72].

Towards the formalization of the behaviour in undervolting conditions, the authors also present a simple consolidated function, the Severity function. Severity function aggregates the effects of reduced voltage operation in the cores of a multicore CPU by assigning values to the different abnormal observations. The lower the voltage level, the higher the value of the severity function. The severity function assists an undervolting classification of the cores of a CPU chip for a given benchmark: different core, benchmark, and voltage values lead to different severity patterns, some with an abrupt increase to the severity (e.g., the benchmark keeps executing correctly until a voltage level at which the system crashes), while others have a "smooth" severity increase while voltage is reduced (the system remains responsive throughout a range of voltage values but it generates ECC errors or produces SDCs). The fine-grained analysis of the behaviour of a microprocessor using the severity function can assist energy efficiency decisions for task-to-core allocation by the system software.

Such a comprehensive characterization for ARMv8-based multicore CPUs [54] confirms that a different microarchitecture, circuit design, or manufacturing technology exhibits different abnormal behaviour when operating beyond nominal voltage conditions. Understanding the behaviour in non-nominal conditions is very important for making software and hardware design decisions, even using the approximate paradigm, for improved energy efficiency that preserves the correctness of operation, or the acceptable correct rates of the approximate application's execution.

3.3.2.1 Abnormal Behaviours Below Safe Voltage (Vmin)

Variation can also cause circuits to malfunction, especially at low voltage. Previous studies on Intel Itanium CPUs [53, 73] have shown a large region of voltage values that contains only ECC corrected errors during undervolting. By reducing the voltage on those chips, the number of corrected errors increases gradually for quite many voltage steps until it exposes other types of abnormal behaviour (SDCs, uncorrected errors, crashes). In such systems, ECC corrected errors can serve as proxies for the effects of undervolting. In contrast to these studies, silent data corruptions (SDCs) appear at higher voltage levels than corrected errors alone, in ARMv8-compliant multicore CPUs [54, 72, 74–78]. More specifically, (1) SDCs occur when the pipeline gets stressed, and (2) the cache bit-cells safely operate at higher voltages. This observation indicates that the ARMv8-compliant CPUs are more susceptible to timing-path failures than to SRAM array failures, and thus, approximation paradigms, which can tolerate potential SDCs during subnominal voltage conditions, become necessary in these microprocessor models. In [73] and [53] the reported range of voltage levels with corrected errors alone offers a significant opportunity for energy savings without jeopardizing the correctness of operation. Further, ECC corrections appear at a higher voltage on the Itanium compared to SDCs and system crashes. In [54], the authors attribute the increased robustness to timing errors/failures on the Itanium to circuit-level dynamic-margin mitigation techniques such as the capability to perform continuous clock-path de-skewing during dynamic operation [79]. The X-Gene 2, which is an ARMv8-compliant microprocessor, does not deploy such circuit-level techniques and, thereby, generates SDCs due to timing-path failures. High correctable error rate is helpful to an ECC guided voltage speculation but this is not the case in the X-Gene 2.

For example, Fig. 3.2 shows the Vmin characterization results for one benchmark and three different X-Gene 2 chips (TTT, TFF, and TSS). There are significant divergences among cores for the same benchmark due to process variations. Process variations can affect transistor dimensions (length, width, oxide thickness, etc.), which have direct impact on the threshold voltage of a MOS device, and thus, on the guardband of each core. This variation among cores of the same chip can result in high energy savings. Moreover, there is a large unsafe region for any core and chip (the grey-coloured bars in the graph), which indicates that by reducing the voltage on these chips, incorrect program output due to any abnormal behaviour (SDCs,

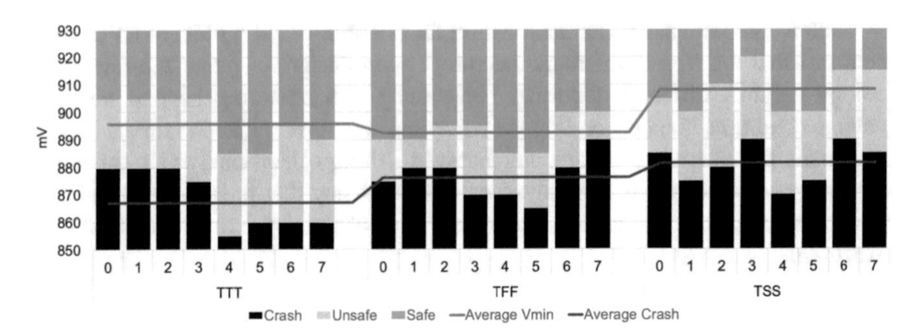

Fig. 3.2 X-Gene 2 characterization results for one benchmark on three different chips (TTT, TFF, and TSS). Blue colour in bars represents the Safe region, grey represents the Unsafe region, and black represents the Crash region

uncorrected errors) can be introduced in the microprocessor's execution during low-voltage operation.

3.3.2.2 Suggestions for Undervolting Effects' Mitigation

Depending on the actual characterization findings for a CPU core during undervolting, certain hardware-based or software-based mitigation approaches can be employed to maximize the energy savings while preserving the correctness of program execution. The primary aspect that determines the most suitable approach is the first observed effect as undervolting goes down the voltage levels. A fine-grained categorization is presented below, which describes the behaviour and discusses the potential corresponding mitigation approaches:

- **Nothing abnormal**. The voltage range is absolutely safe (above the Vmin of a core); no mitigation action is required. System operation in this range is the most conservative option and no mitigation provision is needed. Energy savings are the minimum.
- **Corrected errors first**. This is a voltage range with the behaviour as the one observed in [53, 73] for Intel's Itanium. In such a case, ECC hardware serves as a proxy for abnormal behaviour due to undervolting but program operation is still correct. Significant energy savings can be obtained without any mitigation other than the ECC correction but going further down the voltage is risky.
- **SDCs alone or with corrected and uncorrected errors**. Voltage ranges with these behaviours can potentially generate incorrect program outputs and require extra mitigation approaches. In [54], the authors show that the first abnormal behaviour generated by undervolting belongs in this category for the majority of benchmarks and corrected errors as observed in [53, 73] do not appear first alone in the system under characterization. In particular, the cases where SDCs appear alone are the worst ones since there is no indication about the malfunction of the

system; these areas should be avoided. When an eventual SDC (output mismatch) is accompanied by corrected or uncorrected error notifications, recovery actions can be employed such as rollback to a previously stored check-point or program re-execution in safe voltage and frequency combinations. There are also many applications that can tolerate SDCs and benefit from the severity function. These applications are (1) approximate computing algorithms, (2) video streaming and other image and video processing, and (3) security-oriented applications such as jammer attacks detectors, etc. These applications are tolerant to faults, as they have minor impact on the returned output.

- **Application and system crashes (or application timeouts) with or without corrected and uncorrected errors**. Voltage levels with this behaviour (the result of massive hardware malfunction) are well beyond the limits of cores operation in undervolted conditions. Application or system unresponsiveness is systematic in these ranges and unless serious hardware re-design is employed these ranges are unusable.

3.3.2.3 Design Enhancements

Undervolting characterization studies can be used to provide hardware design recommendations for enhancements if the system (or its future revisions) is to be used in scaled voltage conditions for energy efficiency. There are some key hardware design guidelines, which are presented below:

- **Stronger error protection**. SECDEC ECC protection at the lower levels of the memory hierarchy does not provide enough protection at lower voltages. If (a) stronger ECC codes are employed [80, 81] and (b) more blocks are protected, SDC behaviour with or without errors will have significant probability to be transformed to corrected errors' behaviour similarly to [73] and [53]. Employing stronger ECC protection has been also reported in [82] for scaled voltage operation.
- **Hardware detectors**. If stronger ECC protection is too costly, other types of hardware support can be employed for voltage emergencies' detection such as the skitter circuit [83–85] (also cited in [86]) or the monitoring circuits used in Power7+ designs [87].
- **Finer-grained voltage domains**. Following the previous discussion, coarse-grained voltage domain design of X-Gene 2 (a single voltage domain for all 8 cores) reduces the potential of energy savings since the voltage value of the domain is determined by its weakest core (the one with the higher safe Vmin). If each core or couple of cores was designed to operate on a separate voltage domain, more aggressive voltage scaling (and energy savings) would have been possible.

Of course, all the above hardware design modifications have their own design complexity, area, and performance implications, which must be jointly considered with the potential of energy savings through undervolting.

3.4 Approximation Strategies for Voltage Down-Scaled and Timing Error-Resilient Designs

Although effective in saving power and energy, voltage down-scaling worsens performance variations of nanometre circuits and threatens their reliability [20, 89, 90]. Figure 3.3 illustrates that near-threshold computing (NTC) at 400 mV increases the delay variability by 5× compared to 30% at the nominal voltage [88]. Such a delay variability is typically manifested in terms of timing errors the rate of which is expected to further increase as we move towards the atomic-scale fabrication [4, 17, 91]. Timing errors refer to the discrepancy between a computed, observed, or measured value and the true, specified, or theoretically correct value and threaten system functionality and output correctness. Such errors mainly occur within a processor and can cause insidious application failures known as silent data corruption (SDC). For example, when you perform the following instruction 2 × 3, the CPU may give a result of 4 instead of 6 silently under certain microarchitectural conditions, without an indication of the miscomputation in system event or error logs. As a result, a service utilizing the CPU is potentially unaware of the computational accuracy and keeps consuming the incorrect values in the application.

Several approximate computing techniques use voltage scaling while accounting for the timing errors induced by overscaled voltage and/or process variations. Below, we discuss approximation strategies employed for different memory technologies,

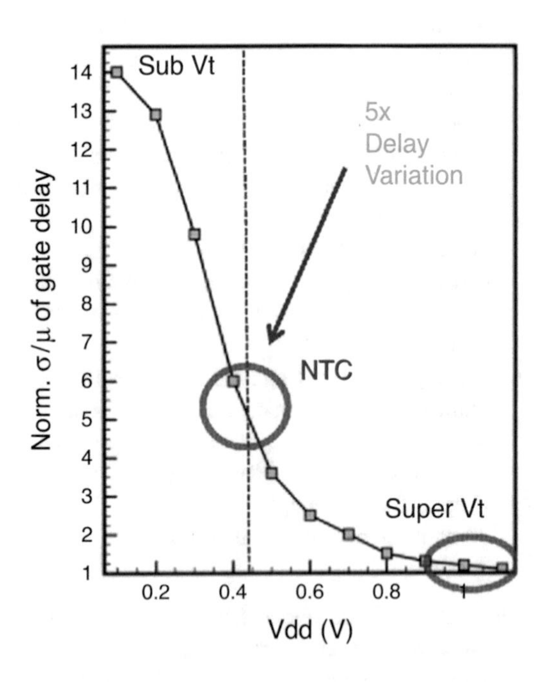

Fig. 3.3 Impact of voltage down-scaling on gate delay variation [88]

system components, and processing units to facilitate low-power and error-resilient operations.

3.4.1 Special Storage Modules

A. Rahimi et al. [92] present an approximation technique for saving energy by reducing supply voltage and tolerating few timing errors suitable for GPU applications. In particular, each FPU in the GPU uses a special storage module, which stores frequent redundant GPU computations. Reusing these values removes the need of FPU re-execution of these computations. When supply voltage is reduced into a coarse granularity (e.g., from 1.0 to 0.725 V), the error behaviour of the module remains controllable; an input search pattern is matched with any of the stored computations within a low quality loss in terms of hamming distance (Hamming distance between an input item and a pre-stored value in the module is 0, 1 or 2.). Such a technique is very effective in error-resilient GPU applications, such as image processing and artificial neural networks [1, 2, 5]. Papagiannopoulou et al. [93] opportunistically ignore timing errors, enabling aggressive voltage scaling at an error-induced quality loss. To recover from critical errors, i.e., timing errors that affect significant parts of arithmetic computations, an error management scheme based on hardware transactional memory is proposed.

3.4.2 Precision Scaling and Operand Truncation

Recently, approximate computing has emerged as an alternative approach for addressing potential timing errors with less overheads than the ones incurred by the conventional guardband-based techniques [4, 22]. Several approximation-based schemes try to reduce timing errors by reducing the precision and thus the time required to complete computation [9, 94, 95]. A representative approach [94] uses precision scaling to limit timing errors in the context of transistor ageing and thus mitigating the estimated delay increase over few years. A post-silicon technique in [9] truncates the bitwidth of all the input operands to prevent timing errors in typical DSP hardware modules. I. Tsiokanos et al. [22, 96] jointly consider a design-centric error mitigation scheme with significance-driven operand truncation to minimize timing errors and output quality degradation while saving power. The basis of these schemes is the delay reduction resulted by the applied precision scaling or operand truncation. While considering arithmetic units such as the adder in Fig. 3.1, carry propagation is the main bottleneck in the performance/delay of these systems. As explained in Sect. 1.1, it is possible to break the carry propagation chain by setting a number of LSBs of the input operands to a constant value of zero. The truncation of a number of LSBs from input operands provides a slack (timing margin) and reduces the path excitation as discussed above, but this comes at a quality loss. However, such a loss can be controlled by appropriately selecting the number of truncated LSBs, ensuring that it is not as catastrophic as the loss incurred by random timing errors when these affect the most significant bits (MSBs). For instance, let us consider the addition of two floating-point operands A and B, which results to an output C. These operands follow the IEEE-754 [97] double precision format in which the first bit from the left represents the sign, the next 11 bits represent the exponent, while the rest 52 bits represent the mantissa. As illustrated in Table 3.1, a random bit flip in the exponent part, e.g., in the 10th bit of the output C (highlighted in red), induced by a timing error will lead to a completely different number than the one expected, resulting in high Relative Error of $\sim$0.9375 (Relative Error is defined in Eq. (3.6)). Conversely, in case of 32 LSBs truncation in the mantissa part of each operand, the resulting output value is very close to the reference value with very low RE equal to $\sim 3.6556 \cdot 10^{-7}$. Such a low RE is attributed to the fact that we truncate LSBs in the mantissa part that are not critical for determining the output value in floating-point operations. On the other hand, the exponent plays a significant role in determining the range of the output and any error either due to random bit flip or truncation in that part is likely to result in catastrophic outcomes [8].

3.4.3 Dynamic Prediction of Error-Prone Instructions

The former approximation strategy that relies on static operand truncation/precision scaling induces an output quality loss, a degree of which may be tolerated by some applications [1, 95, 98], but there is room to limit it. Any reduction of the truncation/precision scaling-induced quality loss may be beneficial for many applications, especially for those, where few operations play a significant role in determining the output quality [5, 99]. The unnecessary quality loss is attributed to the fact that the majority of the existing approximation-based schemes reduce the precision of all operands statically, neglecting the fact that only few instructions/operands activate timing-critical paths, which are error-prone [100–102]. Conversely, the majority of the proceed operands are off-critical (i.e., error free), exciting paths that have sufficient timing slack to handle any delay increase without obtaining errors. Hence, there is no need to apply precision scaling or bitwidth truncation to those operands and inflate the quality loss. The challenge in limiting any unnecessary quality loss lies in the need to dynamically identify operands that activate error-prone paths and apply bitwidth truncation or any other approximation only to them.

To achieve this, recent techniques identify critical instructions and operands (i.e., instruction that activates error-prone timing paths) at runtime by exploiting the dynamic data-dependent sensitization of combinational paths. These design-centric schemes integrate special units/registers to monitor the critical long latency paths (LLPs) and either change the cycle time [100, 101] or utilize different approximation mechanisms (e.g., operand truncation) [67] to prevent timing errors under supply voltage reduction. Existing approaches show that monitoring the carry propagate

Table 3.1 Quality degradation in terms of Relative Error (RE) under "random bit errors" induced by timing errors and "deterministic errors" induced by deliberately setting the last 32 bits of each operand to 0 in a floating-point addition

	Operand A (64 bits)	Operand B (64 bits)	Output C (64 bits)	Relative error (RE)
Reference	0100000010001111 0011111000000101 0001111010111000 0101000111101100 (decimal: 999.7525)	0100000011010000 0001111011011110 0110011001100110 0110011001100110 (decimal: 16507.475)	0100000011010001 0001100011001110 1000111101011100 0010100011110101 (decimal: 17507.2275)	0
"Random bit error" (Bit flip in the 10th MSB)	0100000010001111 0011111000000101 0001111010111000 0101000111101100 (decimal: 999.7525)	0100000011010000 0001111011011110 0110011001100110 0110011001100110 (decimal: 16507.475)	010000001*0* 010001 0001100011001110 1000111101011100 0010100011110101 (decimal: ~1094.2017)	~0.9375
"Deterministic error" (truncating 32 LSBs)	0100000010001111 0011111000000101 **0000000000000000** **0000000000000000** (decimal: 999.7524)	0100000011010000 0001111011011110 **0000000000000000** **0000000000000000** (decimal: 16507.468)	0100000011010001 0001100011001110 0010100000000000 0000000000000000 (decimal: ~17507.2211)	$\sim 3.6556 \cdot 10^{-7}$

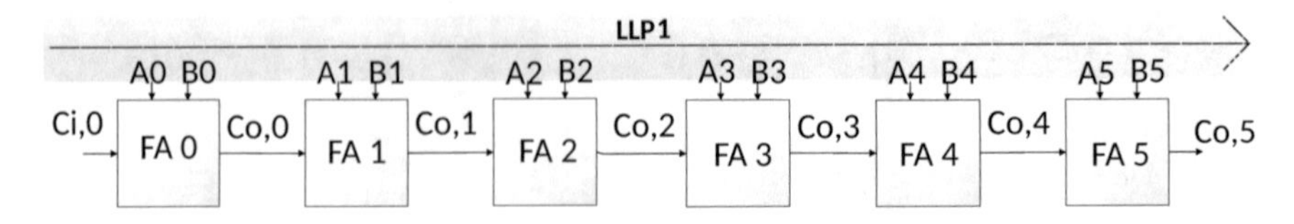

Fig. 3.4 Carry propagation in a 6-bit ripple carry adder (RCA)

signal at the middle of arithmetic units (such as the adder depicted in Fig. 3.1), we can identify if any LLP is going to be activated. To elucidate the basic concept, let us provide a simple example of a 6-bit RCA [101, 103], as shown in Fig. 3.4. In such a design, the most timing critical path ($LLP1$) will be activated when the carry propagates all the way from $Ci, 0$ to $Co, 5$. By monitoring the carry propagate signal at the middle of such a data-path, i.e., at FA2, which is given by $(A0 \oplus B0) \cdot (A1 \oplus B1) \cdot (A2 \oplus B2)$, we can identify if any LLP is going to be activated. Upon detection of operands that activate LLPs, a number of LSBs is deliberately set to 0 [67] or an extra clock cycle is provided [101, 104], allowing the path to be completed correctly even under potential delay increase due to voltage scaling. In any other case, only off-critical paths will be excited, which have sufficient timing slack to address potential delay variations, and thus no approximations needed. The LLP prediction unit ($LLPPU$) is depicted in Fig. 3.5. It monitors $(m - n)$ bits (with $m > n$) of the two addition operands ($M1$, $M2$) and detects if carry propagates across the mth bit, implementing the following logic (as explained in Sect. 3.2):

$$F(m, n) = (M1_m \oplus M2_m) \cdot (M1_{m-1} \oplus M2_{m-1}) \cdot ..(M1_n \oplus M2_n). \tag{3.4}$$

Only when F evaluates a value of 1, the carry bit propagates from the nth to the mth bit into the addition result. The probability of a carry propagation across the mth bit is essentially equivalent to the possibility that operands of an executed instruction will activate the error-prone $LLPs$. In case of F equals to 0, short latency paths ($SLPs$) are activated as there is no carry propagation across the mth bit and the effective computation time is maximum of the two delays: one from the 0 to mth bit and the other from mth to the MSB.

Note that similar prediction units have been applied to different voltage-scaled integer adder and multiplier designs [104–106] as well as floating-point instructions [67, 101, 107].

3.4.4 Path Redistribution for Voltage Down-Scaling

Several hardware-oriented approximate computing strategies may have large implementation overheads; for instance, reducing voltage while dealing with timing errors

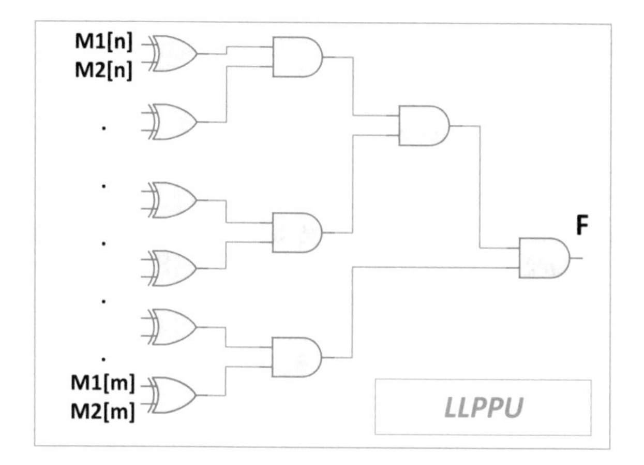

Fig. 3.5 Block diagram of the long latency paths prediction unit (LLPPU)

may require several error detection and correction mechanisms. Similarly, uniform precision scaling or operand truncation is likely to lead to significant output quality degradation. The approximate strategies explained above can be very effective in simple arithmetic units, but their application to pipelined designs is not so straightforward. Here, we provide the main source of overheads in existing hardware approximation techniques. This is the performance-centric design implementation of pipelined processing units.

3.4.4.1 Timing Properties of Pipelined Designs

In particular, instruction pipelining is a common technique to improve the execution throughput of a CPU by allowing the simultaneous execution of several instructions [108]. Typically, a pipelined, computational core consists of a set of N unique timing paths $P = \{P_1, P_2, \ldots, P_N\}$, which are characterized by their delays $D(P_i)$ for $i = 1, 2, \ldots, N$ ($D(P_i)$ also considers the clock-to-output delay and the setup time of a register [21]). In such a core, each of these paths can be found within exactly one pipeline stage s, with $s = 1, 2, \ldots, S$, and only few of them will be excited at every instance depending on the executed instruction. Note that each pipeline stage processes a specific part of one instruction at a time, allowing the parallel execution of multiple instructions. By the terms parallel or concurrent execution of instructions, we mean that up to S instructions share the same hardware circuitry (i.e., pipeline) in a time-sharing fashion.

At design time, the conventional static timing analysis (STA) evaluates the longest timing path across all S pipeline stages and determines the timing bound of the operation (i.e., the clock period), such as:

$$CP_{STA} = \max_{s=1\ldots S}\left\{\max_{p\in P^s}\{D(p)\}\right\} = \max_{p\in P}\{D(p)\}, \tag{3.5}$$

where P^s is the set of unique path-groups in any of the S pipeline stages for $s = 1, 2\ldots, S$ such that $\cup_{s=1}^{S} P^s = P$ and $P^s \cap P^{s'} = \emptyset$ for s $\neq$ s'. During the circuit operation, only few of these paths get activated depending on the instruction type and its operands. Any excited path P_i has a positive timing slack, $slack_i = CP_{STA} - D(P_i)$, until the so-called point of failure (PoF). In case of any delay increase or clock reduction, which exceeds the available slack $slack_i$, the activated path P_i will fail since $D(P_i) > CP_{STA}$, leading to a setup timing error [21, 109]. Figure 3.6 provides an example where S instructions ($I\_1, I\_2, \ldots, I\_S$) are executed on a pipelined core with S stages.

3.4.4.2 Timing Wall Phenomenon

Figure 3.7 depicts a typical distribution of all the path delays $D(P)$, which is obtained by applying conventional design flows and STA [110]. As it can be seen, such a distribution is characterized by a so-called timing wall, with many $LLPs$ across all pipeline stages close to CP_{STA}. Such a wall of paths is a consequence of how modern designs are optimized for power and area, subject to a global frequency constraint. In particular, current design flows minimize the delay of $LLPs$ by (area/power hungry) gate up-sizing [111], while the inherently short latency paths ($SLPs$) are allowed to become near critical for recovering any area or power costs [112]. This "timing wall" does not have any negative impact on the adopted CP_{STA}

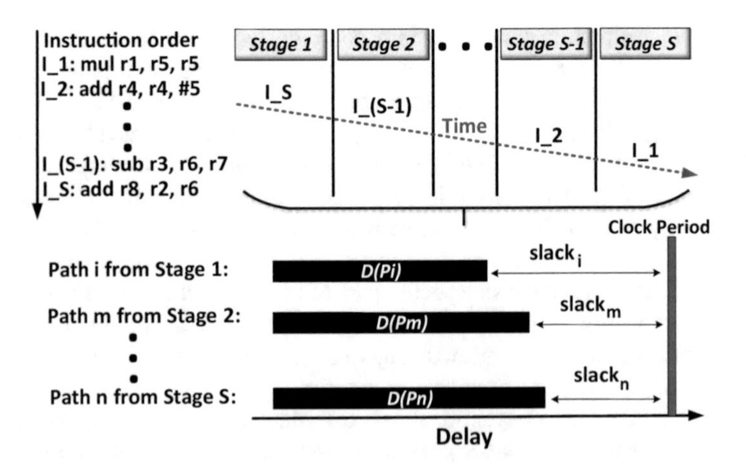

Fig. 3.6 Instruction flow and delay requirements across different paths and stages in a pipelined core

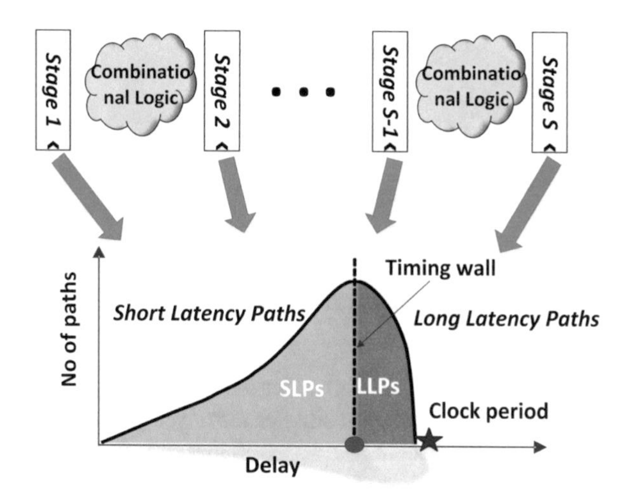

Fig. 3.7 Conventional path distribution of a pipelined design

of the design; however, it critically affects the probability of timing errors since under any (even small) delay increase many paths are likely to fail [113].

3.4.4.3 Path Shaping

A recent work [114] on a six-stage pipelined ARMv7 microprocessor indicated that every stage contributes to the critical paths with 74% of the total delay paths to be categorized as $LLPs$ (i.e., paths with delay more than 75% of the clock period). To minimize the registers or pipeline stages where any approximate scheme would need to be included, current research efforts propose a path redistribution technique, called "Path Shaping" [22, 94, 96, 112] that isolates the critical $LLPs$ to as few pipeline stages as possible. This facilitates the adoption of any approximation strategy within pipelined cores, by applying it only to the specific stage(s) where the $LLPs$ are isolated.

Path shaping targets to move away from a path distribution with many $LLPs$ close to the CP_{STA}, which is typical in performance-centric designs (see Fig. 3.8a). The primary goal of path shaping is to minimize $LLPs$ subject to the target clock period CP_{STA} and power/area constraints. To achieve this, appropriate timing constraints for different path-groups are imposed on the design under test. By introducing such path constraints, it is ensured that the inherently fast paths (i.e., $SLPs$) are not made slower, as opposed to the conventional approach. The end goal is to obtain a path distribution similar to the one depicted in Fig. 3.8b, where $|P_{SLP}| >> |P_{LLP}|$. Note that path shaping does not only reduce the number of the error-prone paths but also enables the isolation of P_{LLP} to specific stage(s) accessed by few specific instructions. This allows the developer to apply any approximation

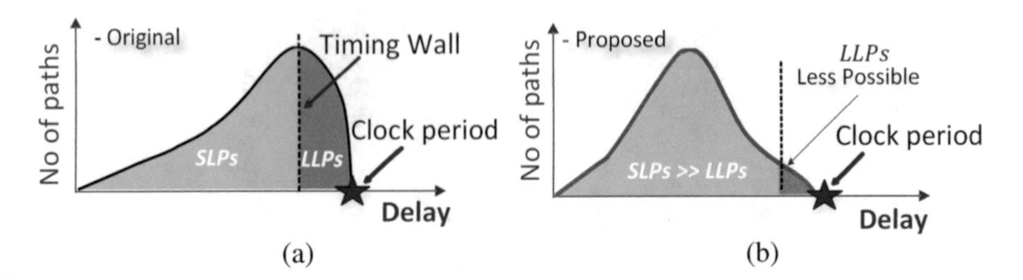

Fig. 3.8 (**a**) Conventional path distribution and (**b**) path distribution after applying path shaping

technique to few stages rather than using it for the whole design, which is far more complicated and costly. In addition, it facilitates the development of error mitigation mechanisms tailored for the specific instruction(s) that activate(s) the few remaining $LLPs$. It is also important to note that the path redistribution technique does not change the clock period (iso-performance), since any path shaping is made subject to maintaining the conventional speed achieved by STA. Path shaping complements existing approximation techniques and can be jointly considered with any of the approximation strategies explained above.

3.4.4.4 Comparison Between Strategies

The efficacy of the presented strategies in achieving energy/power efficiency while limiting timing error-induced quality degradation under different levels of potential delay increase is presented next. The levels of the assumed delay increase range from 0% (nominal conditions) to 9%, representing potential degrees of delay increase caused by supply voltage reduction [17, 20, 115–117]. For the case study of using the approximate strategies mentioned in this section, we focus on an out-of-order (OoO), six-stage, IEEE-754 compatible, double precision FPU, following the representation: $-1^S \times M \times 2^E$, where S is the sign, E is the exponent, and M is the mantissa. This section compares the redesigned FPUs (i.e., the FPU after applying the approximation strategies described below) with the original (reference) FPU. For a fair comparison, we applied the truncation of 44 LSBs of the error-prone floating-point operands to the original (Orig), unmodified FPU, and also to the one when only path shaping is enabled (i.e., $Strat1 + Strat3$). Then, the proposed framework when path shaping, LLPPU and dynamic operand truncation are jointly considered (i.e., $Strat1 + Strat2 + Strat3$), is evaluated. The latter design eliminates timing errors and further reduces truncation-induced quality loss by dynamically truncating 44 LSBs of operands that actually trigger the isolated LLPs.

To acquire floating-point instructions from real-world applications running on an `ARM A7`-based system, an open-source profiling tool [118] is utilized. In this experimental analysis, the `K-means`, `CFD`, `Heartwall` benchmarks from the Rodinia suite [119] and the `Raytrace` benchmark from the Parsec suite [120] are used. This set of benchmarks represents a variety of algorithms that have many floating-point operations and covers a wide range of domains, i.e., Data Mining, Fluid Dynamics, Medical Imaging, and Computer Graphics. Specifically, $10K$ operands from the most frequently executed floating-point instructions for each application are extracted and used to evaluate the efficacy of the comparable FPU designs.

Quality Degradation. As already explained, random timing errors as well as static and dynamic operand truncation impose an output quality degradation. To evaluate this, we estimate the average relative error (RE) achieved by the approximate designs and compare it with the RE of the original design for all considered applications. The average relative error, a common metric for estimating the output quality [8, 48], is defined as:

$$RE = \frac{\sum_{i=1}^{K} \left| \frac{O_{gold}(i) - O_{sim}(i)}{O_{gold}(i)} \right|}{K}, \tag{3.6}$$

where O_{gold}(i) denotes the exact error-free output value obtained from the reference FPU design and O_{sim}(i) represents the output obtained by the gate-level simulation for a specific (i) floating-point instruction and the associated operands. The O_{sim}(i) value is extracted by the output register of the considered FPU after simulating both designs (original and proposed) under a specific delay increase and bitwidth truncation range. For these experiments, $10K$ floating-point instructions are extracted for each benchmark and thus i varies from 1 up to $K = 10{,}000$.

To visually illustrate the benefits of the target approximation strategies in minimizing quality loss, Fig. 3.9 plots the RE of the original unmodified FPU (referred to as $Orig$), $Orig$ with static truncation of 44 LSBs (referred to as $Strat1$), $Strat1 + Strat3$ with 44 LSBs truncation, and the FPU design that combines all the explained approximation strategies ($Strat1 + Strat2 + Strat3$). As showcased, the combination of path shaping and operand truncation ($Strat1 + Strat3$) minimizes errors, but, depending on the number of truncated bits, it may result in unnecessary quality loss. This is because the bitwidth truncation is operand agnostic and is applied to all the operands of the error-prone instruction types. To further reduce the quality degradation, $Strat1 + Strat2 + Strat3$ initially detects the infrequent operands that actually trigger the isolated LLPs (using the $LLPPU$ explained in Sect. 3.4.3) and then truncates the bitwidth of these operands.

The original FPU design without any truncation under the nominal conditions (i.e., 0%) introduces no quality degradation since no errors have been manifested. In the case of CFD program, Orig leads to unacceptable (>2) RE levels even under a small worst-case delay increase (i.e., 3%). Static truncation ($Strat1$) helps to reduce the RE to approximately 0.001 under 3 and 6%, while for $Strat1 + Strat3$ the RE remains constant (RE $= 5 \cdot 10^{-3}$) for the considered delay increase levels. On the other hand, $Stat1 + Stat2 + Stat3$ limits this quality loss (RE $= 4 \cdot 10^{-6}$) under the voltage scaling induced delay increase. Similar results are obtained for the Raytrace program, where Prop2 introduces a negligible RE (RE $= 5 \cdot 10^{-7}$) for all the delay

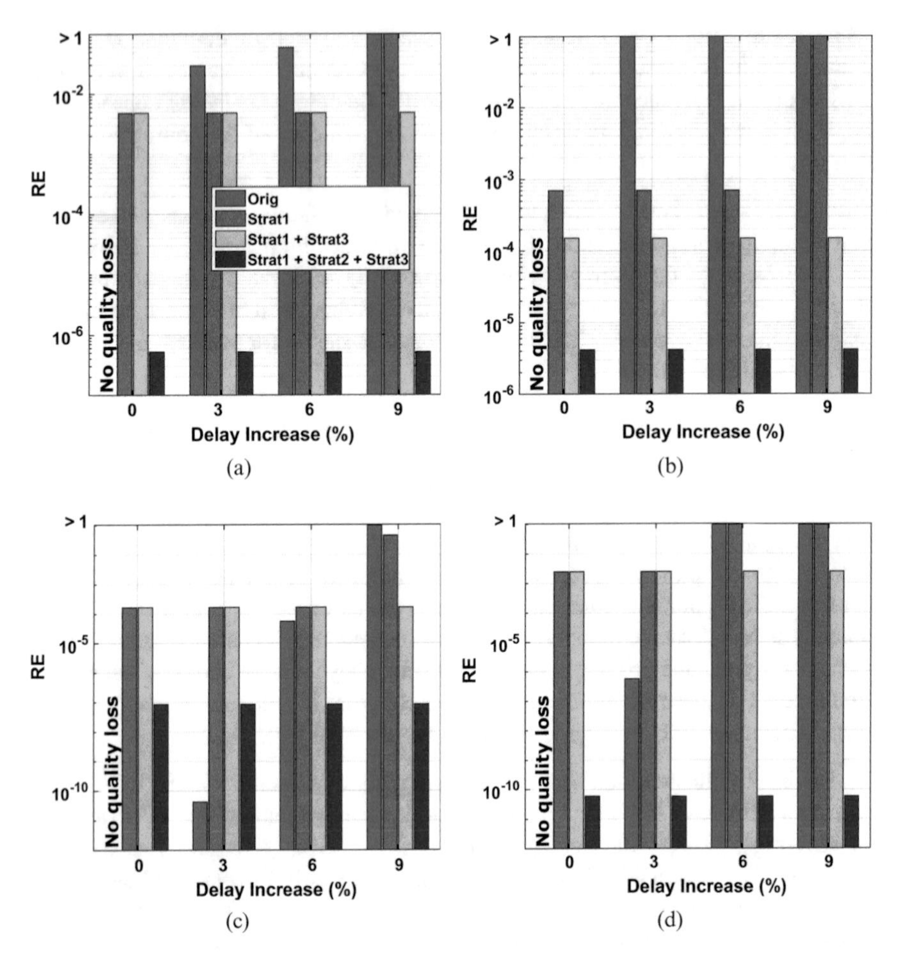

Fig. 3.9 Average relative error (RE) under different delay increase levels for the original FPU with ($Strat1$) and without statically truncating the 44 LSBs (Orig), the FPU when path shaping and static truncation of 44 LSB are enabled ($Strat1 + Strat3$), and the FPU when path shaping, timing error prediction, and dynamic truncation of 44 LSBs are jointly considered ($Strat1 + Strat2 + Strat3$). (**a**) Raytrace. (**b**) CFD. (**c**) K-means. (**d**) Heartwall

increase levels. In the case of k-means program, we observe the lowest RE among all benchmarks (RE $= 6 \cdot 10^{-11}$) for $Strat1 + Strat2 + Strat3$, whereas for Orig and $Strat1$, the quality degrades significantly after 3% delay increase. Finally, in the case of Heartwall program, it is shown that Orig incurs a lower RE than the static or dynamic truncation under 3% delay increase. Nonetheless, under 6 and 9% delay increase the combination of approximation strategies ($Strat1 + Strat3$, and $Strat1 + Strat2 + Strat3$) obtains at least 99.9% lower RE when compared to Orig and $Strat1$. Overall, $Strat1 + Strat2 + Strat3$ exhibits a significantly lower quality loss than Orig under 3–9% delay increase. Compared to $Strat1$, $Strat1 + Strat2 +$

$Strat3$ achieves up to $4.2 \cdot 10^7\times$ lower RE across the considered benchmarks. Additionally, when compared to $Strat1+Strat3$, $Strat1+Strat2+Strat3$ reduces quality loss by an averaged 10^7 in terms of RE.

3.4.4.5 Approximation Strategies and Power Savings

Those approximate-based strategies not only deal with timing errors but are very effective in voltage-scaled designs (i.e., operations at a lower than the nominal supply voltage). For instance, $Strat1 + Strat3$ facilitates operation at 0.95 V, while the nominal supply voltage is set to 1.1 V. This means that $Strat1 + Strat3$ operating at 0.95 V enables the designer to save 59.6% of power on average under the considered benchmarks compared to the original FPU operating at 1.1 V. $Strat1 + Strat3$ results in significant power gains at reduced voltages, but it comes with a notable quality loss that may be important for some applications.

$Strat1 + Strat2 + Strat3$ reduces such loss by reducing the number of times that the data truncation is applied. When compared to the original FPU (@ 1.1 V), $Strat1+Strat2+Strat3$ at 0.95 V leads to 26.8% power savings. Therefore, those approximation strategies not only prevent timing errors, but they also enable low-voltage operations. $Strat1 + Strat3$ is suitable for ultra low-power applications that allow a deterministic output quality loss, while $Strat1 + Strat2 + Strat3$ can be extremely useful for applications requiring ultra low-quality loss and low-power consumption.

3.5 Approximate Memories

Apart from inexact functional units and processor components, approximate computing techniques are applied to different memory technologies, such as SRAM and DRAM. Memories play a crucial role in the overall power/energy consumption of a server [121–123]. One of the reasons for a high energy consumed by the memory devices is the usage of pessimistic operating parameters, such as voltage, refresh rate (T_{REFP}), and temperature, set by the vendors. To improve the energy efficiency of these memories, approximate techniques by reducing the supply voltage, operating temperature, and refresh-rate in SRAM and DRAM have been proposed.

3.5.1 Approximate SRAMs

M. Shoushtari et al. [124] explore how partially forgetful memories can be used in the context of approximate computing. Particularly, they propose an approximate computing technique that saves cache energy by reducing the supply voltage of

SRAM caches while still generating acceptable quality results. Another example of approximations introduced at an SRAM is presented by S. Ganapathy et al. [125] who minimize the magnitude of an SRAM error (caused due to a faulty cell) instead of correcting the faults (like ECC). Authors in this study approximate lower-order bits that have smaller significance than the high-order bits. This is achieved by placing bits of lower significance into the error-prone cells (i.e., SRAM cells that operate at reduced supply voltage), leading to a tolerable loss in output quality. When compared to using ECC, their technique achieves considerable improvement in latency, power, area, and yield at a cost of insignificant output quality degradation. In this direction, H. Esmaeilzadeh et al. [126] propose an approximate SRAM, which relies on a dual voltage supply: a high/nominal V_{dd} (accurate yet power-hungry operations) and a low V_{dd} (inaccurate but lower-power operations). Authors showcase that many programs have non-critical portions and thus small errors have a minor impact on estimating the final program output. For instance, it is shown that in a 3D raytracer application, 91% of data accesses are approximable (can be placed in the SRAM operating at reduced V_{dd}) [126].

During the last years, the goal for improving microprocessors' energy efficiency while reducing their power supply voltage is a major concern of many scientific studies that investigate the chips' operation limits in nominal and off-nominal conditions. Whilkerson et al. [80] go through the physical effects of low-voltage supply on SRAM cells and the types of failures that may occur. After describing how each cell has a minimum operating voltage, they demonstrate how typical error protection solutions start failing far earlier than a low-voltage target (set to 500 mV) and propose two architectural schemes for cache memories that allow operation below 500 mV. The word-disable and bit-fix schemes sacrifice cache capacity to tolerate the high failure rates of low-voltage operation. While both schemes use the entire cache on high voltage, they sacrifice 50 and 25% accordingly in 500 mV. Compared to existing techniques, the two schemes allow a 40% voltage reduction with power savings of 85%. Chishti et al. [81] propose an adaptive technique to increase reliability of cache memories, allowing high tolerance on multi-bit failures that appear on low-voltage operation. The technique sacrifices memory capacity to increase the error-correction capabilities, but unlike previously proposed techniques, it also offers soft and non-persistent error tolerance. Additionally, it does not require self-testing to identify erratic cells in order to isolate them. The MS-ECC design can achieve a 30% supply voltage reduction with 71% power savings and allows configurable ECC capacity by the operating system based on the desired reliability level. Duwe et al. [82] propose an error-pattern transformation scheme that re-arranges erratic bit-cells that correspond to uncorrectable error patterns (e.g., beyond the correctable capacity) to correctable error patterns. The proposed method is low latency and allows the supply voltage to be scaled further that it was previously possible. The adaptive rearranging is guided using the fault patterns detected by self-test. The proposed methodology can reduce the power consumption up to 25.7%, based on simulated modelling that relies on literature SRAM failure probabilities.

There are also several studies that explore methods to eliminate the effects of voltage noise. Gupta et al. [83] and Reddi et al. [86] focus on the prediction of critical parts of benchmarks, in which large voltage noise glitches are likely to occur, leading to malfunctions. In the same context, several studies either in the hardware or in the software level were presented to mitigate the effects of voltage noise [68, 83, 127–129] or to recover from them after their occurrence [130]. Ketkar et al. [131] and Kim et al. [132, 133] propose methods to maximize voltage droops in single core and multicore chips in order to investigate their worst-case behaviour due to the generated voltage noise effects. There are also several characterization studies of commercial chips in off-nominal voltage conditions [53, 73, 87, 134–136].

3.5.2 Approximate DRAMs

Recent projections forecast that the DRAM subsystem will soon be responsible for more than 40% of the overall power consumption within most servers [121]. This reality has led researchers to question if the pessimistic DRAM parameters can be relaxed exploiting the error resilience of some applications. In this scope, existing studies [137–140] store only few critical data structures in well-protected arrays operating under the nominal parameters that are consuming increased power while allowing resilient/non-critical data to be stored on low energy, less reliable arrays [141, 142]. A scheme [138] that gained a lot of attention in this context splits the memory space into a region operated under nominal circuit parameters and a region operated under relaxed parameters, which may be more prone to errors but consumes less energy. Therefore, the use of heterogeneous reliability memory and placement of data in each type of memory based on the required reliability has gained a lot of attention in recent years and can be also used to enable energy-efficient approximate computing [98, 141–143].

The benefits of HRM have been mainly showcased on simulators [123, 138] but state-of-the-art studies implemented and evaluated HRM on real systems with a complete virtualization stack [10, 144]. Those characterization studies on workload-dependent DRAM error behaviour reveal that efficient DRAM refresh rates vary per application basis. In this work, authors use a 64-bit ARMv8-based server (see Fig. 3.10), APM X-Gene2, which is a typical Edge server. The X-Gene2 SoC consists of eight 64-bit ARMv8 cores running at 2.4GHz. The X-Gene2 has four DDR3 Memory Controller Units (MCUs). In the experimental campaign, authors are experimenting with 4 Micron DDR3 8GB DIMMs at 1866 MHz, with one DIMM per MCU. In total, 72 chips of 4Gb x8 DDR3 are characterized since each DIMM includes 16 and 2 DRAM chips for data storage and ECC, respectively. Notably, the X-Gene2 provides access to a separate light-weight intelligent processor (SLIMpro), which is a special management core that is used to boot the system and provide access to the on-board sensors to measure the temperature and the power of the SoC and DRAM. The SLIMpro also reports all memory errors corrected or detected by SECDED ECC to the Linux kernel, providing information about the DIMM,

Fig. 3.10 X-Gene 2 with the custom thermal adapters

bank, rank, row, and column in which the error occurred. Finally, SLIMpro allows the configuration of the parameters of the MCUs, such as refresh-rate and V_{DD}. Specifically, refresh-rate may be changed from the nominal 64 ms to 2.283 s, which is the maximum on the X-Gene2 server. The server runs a fully fledged OS based on CentOS 7 with the default Linux kernel 4.3.0 for ARMv8 and support for 64KB pages. Evaluation results with DRAM operating under variable refresh-rate (from 64 ms to 2.283 s) and lowered V_{DD} (1.428 V) at 50 °C demonstrate that memory errors differ significantly across workloads and DRAM refresh-rate. To measure the rate of memory errors, authors introduce a specific metric, WER, which is defined as: $WER = \frac{N_{CE}}{MEM_{SIZE}}$, where N_{CE} is the number of unique 64-bit word locations where CEs have manifested and MEM_{SIZE} is the size (in 64-bit words) of memory allocated by the application. WER shows the probability of a word being erroneous regardless of the size of memory allocated by the application. In their study, authors investigate how WER varies across benchmarks when DRAM operates under different T_{REFP} at 50 °C and lowered V_{dd}. Authors run benchmarks for DRAM operating under 0.618 s, 1.173 s, 1.727 s, 2.283 s T_{REFP} and lowered V_{DD}. Figures 3.11 and 3.12 illustrate how WER changes with scaling T_{REFP} at 50 °C. These results show that WER varies across benchmarks significantly. Such a variation of DRAM error behaviour across workloads, including single-threaded and multi-threaded programs, is attributed to different types of access and data patterns in the workloads.

Apart from the above circuit parameters, one of the main environmental conditions affecting the reliability of DRAM and used in approximate memories is temperature [145, 146]. To test DRAM reliability under different temperatures, previous studies develop specific thermal testbeds [146, 147]. There are two alternative approaches for designing such testbeds: (1) use a heating chamber, in which an entire experimental server is placed and (2) use heating elements that are explicitly connected to DRAM devices. For example, the second approach

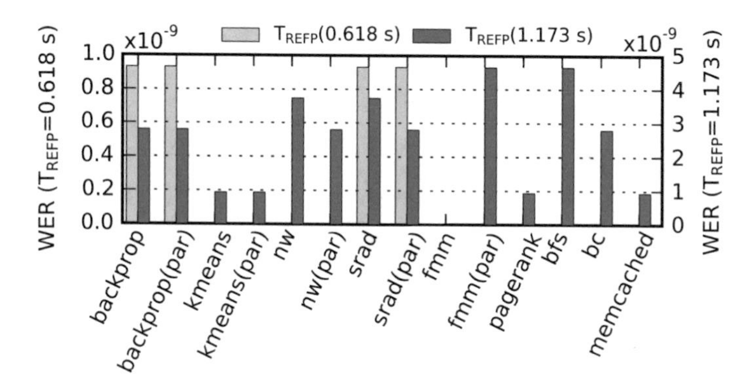

Fig. 3.11 WER for DRAM operating under 0.618 s and 1.173 s at 50 °C

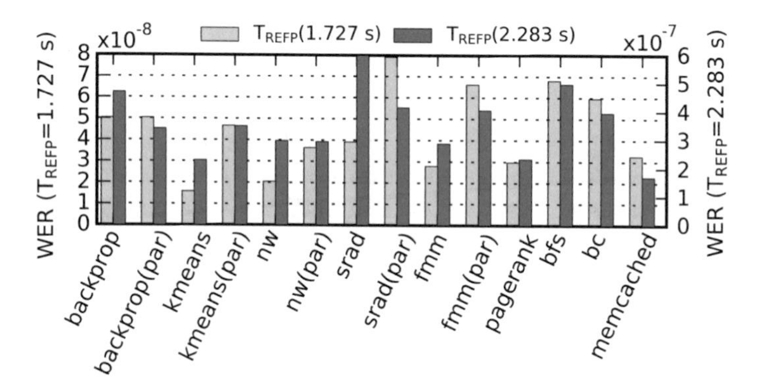

Fig. 3.12 WER for DRAM operating under 1.727 s and 2.283 s at 50 °C

is discussed in a recent study [147]. In this study, to perform the experiments under controlled temperatures, authors implement a unique temperature-controlled server testbed using heating elements [147]. Figure 3.10 shows the APM X-Gene 2 server with four DIMMs fitted with custom adapters. The temperature of each element is regulated by a controller board, as shown in Fig. 3.13, which contains a Raspberry Pi 3, four closed-loop PID controllers, and eight solid-state relays controlling the resistive elements of each DIMM and rank independently. This work can be used in determining efficient temperature guardbands in approximate memories.

Noteworthy, approximate memories actually are realized on a real server with a full virtualization software stack as shown in [10]. The so-called HaRMony scheme realized a heterogeneous-reliability memory framework, in conjunction with QoS-aware energy management policies. HaRMony exposed to the QEMUKVM hypervisor two unique policies. The first policy enables the hypervisor to seek the most power-efficient DRAM circuit parameters based on the server availability requested by the user. The second policy enables users to exploit the inherent application error

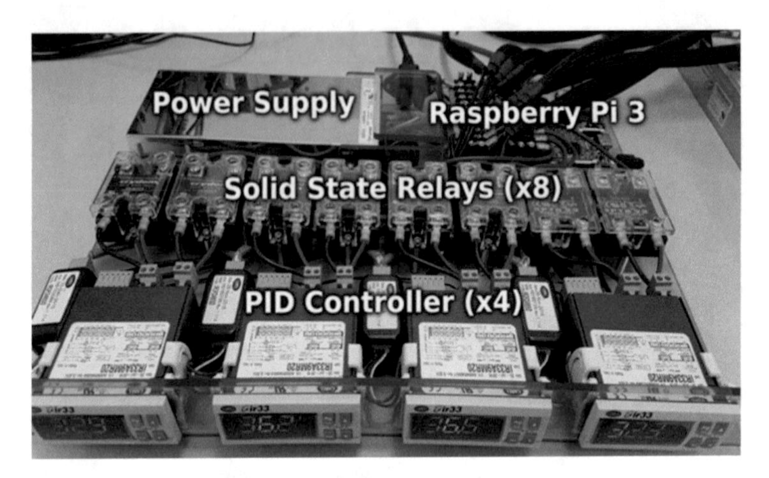

Fig. 3.13 Temperature controller board [147]

resiliency by allowing them to limit the error protection mechanisms and allocate data structures on variably-reliable memory domains, thus realizing approximate storage on real devices. Results show that HaRMony reduces the performance overhead incurred due to disabling hardware interleaving from 29.3% down to 1.1% and leads to 17.7% DRAM energy savings and 8.6% total system energy savings on average in case of native execution of 28 benchmarks on an ARMv8-based server [10]. It was also shown that the developed QoS-aware scaling governor integrated with QEMU-KVM can dynamically scale the DRAM parameters while reducing the system energy by 8.4% and meeting the targeted QoS even under extreme DRAM temperatures.

References

1. Zhang, H., Putic, M., & Lach, J. (2014). Low power GPGPU computation with imprecise hardware. In *The 51st Annual Design Automation Conference 2014, DAC '14, San Francisco, CA, June 1–5, 2014* (pp. 99:1–99:6). ACM.
2. Grigorian, B., Farahpour, N., & Reinman, G. (2015). BRAINIAC: Bringing reliable accuracy into neurally-implemented approximate computing. In *21st IEEE International Symposium on High Performance Computer Architecture, HPCA 2015, Burlingame, CA, February 7–11, 2015* (pp. 615–626). IEEE Computer Society.
3. Karakonstantis, G., Banerjee, N., & Roy, K. (2010). Process-variation resilient and voltage-scalable DCT architecture for robust low-power computing. *IEEE Transactions on Very Large Scale Integration (VLSI) Systems, 18*(10), 1461–1470.
4. Karakonstantis, G., Chatterjee, A., & Roy, K. (2011). Containing the nanometer "pandora-box": Cross-layer design techniques for variation aware low power systems. *IEEE Journal on Emerging and Selected Topics in Circuits and Systems, 1*(1), 19–29.
5. Esmaeilzadeh, H., Sampson, A., Ceze, L., & Burger, D. (2012). Architecture support for disciplined approximate programming. In T. Harris & M. L. Scott (Eds.), *Proceedings of the*

17th International Conference on Architectural Support for Programming Languages and Operating Systems, ASPLOS 2012, London, March 3–7, 2012 (pp. 301–312). ACM.
6. Venkataramani, S., Ranjan, A., Roy, K., & Raghunathan, A. (2014). Axnn: energy-efficient neuromorphic systems using approximate computing. In Y. Xie, T. Karnik, M. M. Khellah, & R. Mehra (Eds.), *International Symposium on Low Power Electronics and Design, ISLPED'14, La Jolla, CA—August 11–13, 2014* (pp. 27–32). ACM.
7. Karakonstantis, G., Banerjee, N., Roy, K., & Chakrabarti, C. (2007). Design methodology to trade-off power, output quality and error resiliency: Application to color interpolation filtering. *IEEE ICCAD*.
8. Liu, W., Chen, L., Wang, C., O'Neill, M., & Lombardi, F. (2016). Design and analysis of inexact floating-point adders. *IEEE Transactions on Computers, 65*(1), 308–314.
9. Narasimhan, S., Kunaparaju, K., & Bhunia, S. (2012). Healing of DSP circuits under power bound using post-silicon operand bitwidth truncation. *IEEE Transactions on Circuits and Systems, 59-I*(9), 1932–1941.
10. Tovletoglou, K., Mukhanov, L., Nikolopoulos, D. S., & Karakonstantis, G. (2020). Harmony: Heterogeneous-reliability memory and QoS-aware energy management on virtualized servers. In *Proceedings of the Twenty-Fifth International Conference on Architectural Support for Programming Languages and Operating Systems*, ASPLOS '20 (pp. 575–590). New York, NY: Association for Computing Machinery.
11. Tovletoglou, K., Mukhanov, L., Nikolopoulos, D. S., & Karakonstantis, G. (2019). Shimmer: Implementing a heterogeneous-reliability DRAM framework on a commodity server. *IEEE Computer Architecture Letters, 18*(1), 26–29.
12. Gupta, M., Roberts, D., Meswani, M. R., Sridharan, V., Tullsen, D. M., & Gupta, R. K. (2016). Reliability and performance trade-off study of heterogeneous memories. In B. Jacob (Ed.), *Proceedings of the Second International Symposium on Memory Systems, MEMSYS 2016, Alexandria, VA, October 3–6, 2016* (pp. 395–401). ACM.
13. Kumar, R., Farkas, K. I., Jouppi, N. P., Ranganathan, P., & Tullsen, D. M. (2003). Single-ISA heterogeneous multi-core architectures: the potential for processor power reduction. In *Proceedings. 36th Annual IEEE/ACM International Symposium on Microarchitecture, 2003. MICRO-36.* (pp. 81–92).
14. Augonnet, C., Thibault, S., Namyst, R., & Wacrenier, P.-A. (2011). Starpu: A unified platform for task scheduling on heterogeneous multicore architectures. *Concurrency and Computation: Practice and Experience, 23*(2), 187–198.
15. Karakonstantis, G., & Roy, K. (Aug. 2011). Voltage over-scaling: A cross layer design perspective for energy efficient systems. *IEEE European Conference on Circuit Theory and Design (ECCTD)*.
16. Bowman, K. A., Tschanz, J. W., Lu, S. L., Aseron, P. A., Khellah, M. M., Raychowdhury, A., Geuskens, B. M., Tokunaga, C., Wilkerson, C. B., Karnik, T., & De, V. K. (2011). A 45 nm resilient microprocessor core for dynamic variation tolerance. *IEEE Journal of Solid-State Circuits, 46*(1), 194–208.
17. Whatmough, P. N., Das, S., Hadjilambrou, Z., & Bull, D. M. (2017). Power integrity analysis of a 28 nm dual-core arm cortex-a57 cluster using an all-digital power delivery monitor. *IEEE Journal of Solid-State Circuits, 52*(6), 1643–1654.
18. Mittal, S. (2016). A survey of architectural techniques for near-threshold computing. *ACM Journal on Emerging Technologies in Computing Systems, 12*(4), 46:1–46:26.
19. Sampson, A., Dietl, W., Fortuna, E., Gnanapragasam, D., Ceze, L., & Grossman, D. (2011). EnerJ: Approximate data types for safe and general low-power computation. In *PLDI '11: Proceedings of the 32nd ACM SIGPLAN Conference on Programming Language Design and Implementation* (pp. 164–174). ACM.
20. Dighe, S., Vangal, S. R., Aseron, P., Kumar, S., Jacob, T., Bowman, K. A., Howard, J., Tschanz, J., Erraguntla, V., Borkar, N., De, V. K., & Borkar, S. (2011). Within-die variation-aware dynamic-voltage-frequency-scaling with optimal core allocation and thread hopping for the 80-core teraflops processor. *IEEE Journal of Solid-State Circuits, 46*(1), 184–193.

21. Bhasker, J., & Chadha, R. (2009). *Static timing analysis for nanometer designs: A practical approach*. New York, NY, USA: Springer.
22. Tsiokanos, I., Mukhanov, L., Nikolopoulos, D. S., & Karakonstantis, G. (2019). Significance-driven data truncation for preventing timing failures. *IEEE Transactions on Device and Materials Reliability, 19*(1), 25–36.
23. Celia, D., Vasudevan, V., & Chandrachoodan, N. (2018). Optimizing power-accuracy trade-off in approximate adders. In *2018 Design, Automation Test in Europe Conference Exhibition (DATE)* (pp. 1488–1491).
24. Jha, C. K., & Mekie, J. (2019). SEDA - single exact dual approximate adders for approximate processors. In *2019 56th ACM/IEEE Design Automation Conference (DAC)* (pp. 1–2).
25. Kahng, A. B., & Kang, S. (2012). Accuracy-configurable adder for approximate arithmetic designs. In *DAC Design Automation Conference 2012* (pp. 820–825).
26. Hanif, M. A., Hafiz, R., Hasan, O., & Shafique, M. (2017). Quad: Design and analysis of quality-area optimal low-latency approximate adders. In *2017 54th ACM/EDAC/IEEE Design Automation Conference (DAC)* (pp. 1–6).
27. Xu, W., Sapatnekar, S. S., & Hu, J. (2018). A simple yet efficient accuracy-configurable adder design. *IEEE Transactions on Very Large Scale Integration (VLSI) Systems, 26*(6), 1112–1125.
28. Rezaalipour, M., Rezaalipour, M., Dehyadegari, M., & Bojnordi, M. N. (2020). Axmap: Making approximate adders aware of input patterns. *IEEE Transactions on Computers, 69*(6), 868–882.
29. Horowitz, M. (2014). 1.1 computing's energy problem (and what we can do about it). In *2014 IEEE International Solid-State Circuits Conference Digest of Technical Papers (ISSCC)* (pp. 10–14).
30. Amanollahi, S., & Jaberipur, G. (2017). Energy-efficient VLSI realization of binary64 division with redundant number systems. *IEEE Transactions on Very Large Scale Integration (VLSI) Systems, 25*(3), 954–961.
31. Esposito, D., Strollo, A. G. M., Napoli, E., De Caro, D., & Petra, N. (2018). Approximate multipliers based on new approximate compressors. *IEEE Transactions on Circuits and Systems I: Regular Papers, 65*(12), 4169–4182.
32. Melchert, J., Behroozi, S., Li, J., & Kim, Y. (2019). SAADI-EC: A quality-configurable approximate divider for energy efficiency. *IEEE Transactions on Very Large Scale Integration (VLSI) Systems, 27*(11), 2680–2692.
33. Liu, W., Xu, T., Li, J., Wang, C., Montuschi, P., & Lombardi, F. (2020). Design of unsigned approximate hybrid dividers based on restoring array and logarithmic dividers. *IEEE Transactions on Emerging Topics in Computing* 1–1
34. Vahdat, S., Kamal, M., Afzali-Kusha, A., Pedram, M., & Navabi, Z. (2017). Truncapp: A truncation-based approximate divider for energy efficient dsp applications. In *Design, Automation Test in Europe Conference Exhibition (DATE), 2017* (pp. 1635–1638).
35. Zendegani, R., Kamal, M., Bahadori, M., Afzali-Kusha, A., & Pedram, M. (2017). Roba multiplier: A rounding-based approximate multiplier for high-speed yet energy-efficient digital signal processing. *IEEE Transactions on Very Large Scale Integration (VLSI) Systems, 25*(2), 393–401.
36. Tagliavini, G., Mach, S., Rossi, D., Marongiu, A., & Benin, L. (2018). A transprecision floating-point platform for ultra-low power computing. In *2018 Design, Automation Test in Europe Conference Exhibition (DATE)* (pp. 1051–1056).
37. Rahimi, A., Marongiu, A., Gupta, R. K., & Benini, L. (2013). A variability-aware openmp environment for efficient execution of accuracy-configurable computation on shared-fpu processor clusters. In *2013 International Conference on Hardware/Software Codesign and System Synthesis (CODES+ISSS)* (pp. 1–10).
38. Salehi, S., & DeMara, R. F. (2015). Energy and area analysis of a floating-point unit in 15nm cmos process technology. In *SoutheastCon* (pp. 1–5).
39. Garg, P., & Suneja, K. (2020). Hardware design of high speed 1-D DCT module using approximate floating point adder. In *2020 7th International Conference on Signal Processing and Integrated Networks (SPIN)* (pp. 623–625).

40. Saadat, H., Bokhari, H., & Parameswaran, S. (2018). Minimally biased multipliers for approximate integer and floating-point multiplication. *IEEE Transactions on Computer-Aided Design of Integrated Circuits and Systems, 37*(11), 2623–2635.
41. Peroni, D., Imani, M., & Rosing, T. S. (2020). Runtime efficiency-accuracy tradeoff using configurable floating point multiplier. *IEEE Transactions on Computer-Aided Design of Integrated Circuits and Systems, 39*(2), 346–358.
42. Saadat, H., Javaid, H., & Parameswaran, S. (2019). Approximate integer and floating-point dividers with near-zero error bias. In *2019 56th ACM/IEEE Design Automation Conference (DAC)* (pp. 1–6).
43. Jha, C. K., Prasad, K., Srivastava, V. K., & Mekie, J. (2020). Fpad: A multistage approximation methodology for designing floating point approximate dividers. In *2020 IEEE International Symposium on Circuits and Systems (ISCAS)* (pp. 1–5).
44. Tsiokanos, I., Mukhanov, L., Georgakoudis, G., Nikolopoulos, D. S., & Karakonstantis, G. (2020). DEFCON: generating and detecting failure-prone instruction sequences via stochastic search. In *DATE* (pp. 1121–1126). IEEE.
45. Venkataramani, S., Kozhikkottu, V. J., Sabne, A., Roy, K., & Raghunathan, A. (2020). Logic synthesis of approximate circuits. *IEEE Transactions on Computer-Aided Design of Integrated Circuits and Systems, 39*(10), 2503–2515.
46. Scarabottolo, I., Ansaloni, G., Constantinides, G. A., Pozzi, L., & Reda, S. (2020). Approximate logic synthesis: A survey. In *Proceedings of the IEEE, 108*, 1–19.
47. Pagliari, D. J., Macii, E., & Poncino, M. (2019). Automated synthesis of energy-efficient reconfigurable-precision circuits. *IEEE Access, 7*, 172030–172044.
48. Venkataramani, S., Sabne, A., Kozhikkottu, V. J., Roy, K., & Raghunathan, A. (2012). SALSA: systematic logic synthesis of approximate circuits. In P. Groeneveld, D. Sciuto, & S. Hassoun (Eds.), *The 49th Annual Design Automation Conference 2012, DAC '12, San Francisco, CA, June 3–7, 2012* (pp. 796–801). ACM.
49. Venkataramani, S., Kozhikkottu, V. J., Sabne, A., Roy, K., & Raghunathan, A. (2020). Logic synthesis of approximate circuits. *IEEE Transactions on Computer-Aided Design of Integrated Circuits and Systems, 39*(10), 2503–2515.
50. Scarabottolo, I., Ansaloni, G., Constantinides, G. A., Pozzi, L., & Reda, S. (2020). Approximate logic synthesis: A survey. *Proceedings of the IEEE, 108*(12), 2195–2213.
51. Rodrigues, J. N., Kamuf, M., Hedberg, H., & Owall, V. (2005). A manual on asic front to back end design flow. In *2005 IEEE International Conference on Microelectronic Systems Education (MSE'05)* (pp. 75–76).
52. Parasyris, K., Koutsovasilis, P., Vassiliadis, V., Antonopoulos, C. D., Bellas, N., & Lalis, S. (2018). A framework for evaluating software on reduced margins hardware. In *2018 48th Annual IEEE/IFIP International Conference on Dependable Systems and Networks (DSN)* (pp. 330–337).
53. Bacha, A., & Teodorescu, R. (2014). Using ecc feedback to guide voltage speculation in low-voltage processors. In *Proceedings of the 47th Annual IEEE/ACM International Symposium on Microarchitecture*, MICRO-47 (pp. 306–318), Washington, DC, USA: IEEE Computer Society.
54. Papadimitriou, G., Kaliorakis, M., Chatzidimitriou, A., Gizopoulos, D., Lawthers, P., & Das, S. (2017). Harnessing voltage margins for energy efficiency in multicore cpus. In *2017 50th Annual IEEE/ACM International Symposium on Microarchitecture (MICRO)* (pp. 503–516).
55. Mittal, S. (2016). A survey of techniques for approximate computing. *ACM Computing Surveys, 48*(4), 62:1–62:33.
56. Moons, B., & Verhelst, M. (2015). DVAS: Dynamic voltage accuracy scaling for increased energy-efficiency in approximate computing. In *ISLPED*.
57. Afzali-Kusha, H., Vaeztourshizi, M., Kamal, M., & Pedram, M. (2020). Design exploration of energy-efficient accuracy-configurable dadda multipliers with improved lifetime based on voltage overscaling. *IEEE Transactions on Very Large Scale Integration (VLSI) Systems, 28*(5), 1207–1220.

58. Moons, B., & Verhelst, M. (2017). An energy-efficient precision-scalable convnet processor in 40-nm cmos. *IEEE Journal of Solid-State Circuits, 52*(4), 903–914.
59. Salami, B., Unsal, O. S., & Kestelman, A. C. (2018). Comprehensive evaluation of supply voltage underscaling in fpga on-chip memories. In *2018 51st Annual IEEE/ACM International Symposium on Microarchitecture (MICRO)* (pp. 724–736).
60. Koppula, S., Orosa, L., Yaglikçi, A. G., Azizi, R., Shahroodi, T., Kanellopoulos, K., & Mutlu, O. (2019). EDEN: Enabling energy-efficient, high-performance deep neural network inference using approximate DRAM. In *MICRO* (pp. 166–181). ACM.
61. Venkataramani, S., Ranjan, A., Roy, K., & Raghunathan, A. (2014). AxNN: energy-efficient neuromorphic systems using approximate computing. In *ISLPED* (pp. 27–32). ACM.
62. Zhang, Q., Wang, T., Tian, Y., Yuan, F., & Xu, Q. (2015). Approxann: An approximate computing framework for artificial neural network. In *2015 Design, Automation Test in Europe Conference Exhibition (DATE)* (pp. 701–706).
63. Mohapatra, D., Karakonstantis, G., & Roy, K. (Aug. 2009). Significance driven computation: A voltage-scalable, variation-aware, quality-tuning motion estimator. *IEEE ISLPED.*
64. Karakonstantis, G., Sankaranarayanan, A., Aly, M. M. S., Atienza, D., & Burg, A. (2014). A quality-scalable spectral analysis system for energy-efficient health monitoring. *IEEE DATE.*
65. Zaruba, F., Schuiki, F., & Benini, L. (2020). Manticore: A 4096-core RISC-V chiplet architecture for ultra-efficient floating-point computing. *IEEE Micro, 41*, 1–1.
66. Mach, S., Schuiki, F., Zaruba, F., & Benini, L. (2020). Fpnew: An open-source multiformat floating-point unit architecture for energy-proportional transprecision computing. *IEEE Transactions on Very Large Scale Integration (VLSI) Systems, 29*, 1–14.
67. Tsiokanos, I., Mukhanov, L., & Karakonstantis, G. (2019). Low-power variation-aware cores based on dynamic data-dependent bitwidth truncation. In *DATE* (pp. 698–703). IEEE.
68. Reddi, V. J., Kanev, S., Kim, W., Campanoni, S., Smith, M. D., Wei, G.-Y., & Brooks, D. (2010). Voltage smoothing: Characterizing and mitigating voltage noise in production processors via software-guided thread scheduling. In *2010 43rd Annual IEEE/ACM International Symposium on Microarchitecture* (pp. 77–88).
69. James, N., Restle, P., Friedrich, J., Huott, B., & McCredie, B. (2007). Comparison of split-versus connected-core supplies in the power6 microprocessor. In *2007 IEEE International Solid-State Circuits Conference. Digest of Technical Papers* (pp. 298–604).
70. Le Sueur, E., & Heiser, G. (2010). Dynamic voltage and frequency scaling: The laws of diminishing returns. In *Proceedings of the 2010 International Conference on Power Aware Computing and Systems*, HotPower'10 (page 1–8). Berkeley, CA, USA: USENIX Association.
71. The Linux Kernel Documentation (Parent Directory), Retrieved 2017 from https://www.kernel.org/doc/Documentation.
72. Papadimitriou, G., Kaliorakis, M., Chatzidimitriou, A., Gizopoulos, D., Favor, G., Sankaran, K., & Das, S. (2017). A system-level voltage/frequency scaling characterization framework for multicore CPUs. In *2017 IEEE Silicon Errors in Logic System Effects (SELSE-13)* (pp. 1–6).
73. Bacha, A., & Teodorescu, R. (2013). Dynamic reduction of voltage margins by leveraging on-chip ecc in itanium ii processors. *SIGARCH Computer Architecture News, 41*(3), 297–307.
74. Papadimitriou, G., Chatzidimitriou, A., Kaliorakis, M., Vastakis, Y., & Gizopoulos, D. (2018). Micro-viruses for fast system-level voltage margins characterization in multicore cpus. In *2018 IEEE International Symposium on Performance Analysis of Systems and Software (ISPASS)* (pp. 54–63).
75. Papadimitriou, G., Chatzidimitriou, A., & Gizopoulos, D. (2019). Adaptive voltage/frequency scaling and core allocation for balanced energy and performance on multicore cpus. In *2019 IEEE International Symposium on High Performance Computer Architecture (HPCA)* (pp. 133–146).
76. Gizopoulos, D., Papadimitriou, G., Chatzidimitriou, A., Reddi, V. J., Salami, B., Unsal, O. S., Kestelman, A. C., & Leng, J. (2019). Modern hardware margins: Cpus, gpus, fpgas recent

system-level studies. In *2019 IEEE 25th International Symposium on On-Line Testing and Robust System Design (IOLTS)* (pp. 129–134).
77. Koutsovasilis, P., Antonopoulos, C., Bellas, N., Lalis, S., Papadimitriou, G., Chatzidimitriou, A., & Gizopoulos, D. (2020). The impact of CPU voltage margins on power-constrained execution. *IEEE Transactions on Sustainable Computing*, 1–1
78. Papadimitriou, G., Chatzidimitriou, A., Gizopoulos, D., Reddi, V. J., Leng, J., Salami, B., Unsal, O. S., & Kestelman, A. C. (2020). Exceeding conservative limits: A consolidated analysis on modern hardware margins. *IEEE Transactions on Device and Materials Reliability, 20*(2), 341–350.
79. Riedlinger, R. J., Bhatia, R., Biro, L., Bowhill, B., Fetzer, E., Gronowski, P., & Grutkowski, T. (2011). A 32nm 3.1 billion transistor 12-wide-issue itanium®processor for mission-critical servers. In *2011 IEEE International Solid-State Circuits Conference* (pp. 84–86).
80. Wilkerson, C., Gao, H., Alameldeen, A. R., Chishti, Z., Khellah, M., & Lu, S.-L. (2008). Trading off cache capacity for reliability to enable low voltage operation. In *2008 International Symposium on Computer Architecture* (pp. 203–214).
81. Chishti, Z., Alameldeen, A. R., Wilkerson, C., Wu, W., & Lu, S.-L. (2009). Improving cache lifetime reliability at ultra-low voltages. In *Proceedings of the 42nd Annual IEEE/ACM International Symposium on Microarchitecture*, MICRO 42 (pp. 89–99). New York, NY, USA: Association for Computing Machinery.
82. Duwe, H., Jian, X., Petrisko, D., & Kumar, R. (2016). Rescuing uncorrectable fault patterns in on-chip memories through error pattern transformation. In *2016 ACM/IEEE 43rd Annual International Symposium on Computer Architecture (ISCA)* (pp. 634–644).
83. Gupta, M. S., Rangan, K. K., Smith, M. D., Wei, G.-Y., & Brooks, D. (2007). Towards a software approach to mitigate voltage emergencies. In *Proceedings of the 2007 international symposium on Low power electronics and design (ISLPED '07)* (pp. 123–128).
84. Franch, R., Restle, P., James, N., Huott, W., Friedrich, J., Dixon, R., Weitzel, S., Van Goor, K., & Salem, G. (2008). On-chip timing uncertainty measurements on IBM microprocessors. In *2008 IEEE International Test Conference* (pp. 1–7).
85. Restle, P., Franch, R., James, N., Huott, W., Skergan, T., Wilson, S., Schwartz, N., & Clabes, J. (2004). Timing uncertainty measurements on the power5 microprocessor. In *2004 IEEE International Solid-State Circuits Conference (IEEE Cat. No.04CH37519)* (Vol. 1, pp. 354–355).
86. Reddi, V. J., Gupta, M. S., Holloway, G., Wei, G.-Y., Smith, M. D., & Brooks, D. (2009). Voltage emergency prediction: Using signatures to reduce operating margins. In *2009 IEEE 15th International Symposium on High Performance Computer Architecture* (pp. 18–29).
87. Zu, Y., Lefurgy, C. R., Leng, J., Halpern, M., Floyd, M. S., & Reddi, V. J. (2015). Adaptive guardband scheduling to improve system-level efficiency of the power7+. In *2015 48th Annual IEEE/ACM International Symposium on Microarchitecture (MICRO)* (pp. 308–321).
88. Dreslinski, R. G., Wieckowski, M., Blaauw, D., Sylvester, D., & Mudge, T. (2010). Near-threshold computing: Reclaiming moore's law through energy efficient integrated circuits. *Proceedings of the IEEE, 98*(2), 253–266.
89. Kakoee, M. R., Loi, I., & Benini, L. (2012). Variation-tolerant architecture for ultra low power shared-l1 processor clusters. *IEEE Transactions on Circuits and Systems II: Express Briefs, 59*(12), 927–931.
90. Bull, D. M., Das, S., Shivashankar, K., Dasika, G. S., Flautner, K., & Blaauw, D. T. (2010). A power-efficient 32b ARM ISA processor using timing-error detection and correction for transient-error tolerance and adaptation to PVT variation. In *IEEE International Solid-State Circuits Conference, ISSCC 2010, Digest of Technical Papers, San Francisco, CA, 7–11 February, 2010* (pp. 284–285). IEEE.
91. The itrs website: http://www.itrs.net/links/2011itrs/home2011.htm.
92. Rahimi, A., Ghofrani, A., Cheng, K., Benini, L., & Gupta, R. K. (2015). Approximate associative memristive memory for energy-efficient gpus. In *2015 Design, Automation Test in Europe Conference Exhibition (DATE)* (pp. 1497–1502).

93. Papagiannopoulou, D., Whang, S., Moreshet, T., & Bahar, R. I. (2019). Ignoretm: Opportunistically ignoring timing violations for energy savings using htm. In *2019 Design, Automation Test in Europe Conference Exhibition (DATE)* (pp. 1571–1574).
94. Abbas, H. M., Halak, B., & Zwolinski, M. (2017). BTI mitigation by anti-ageing software patterns. *Microelectronics Reliability, 79*, 79–90.
95. Schlachter, J., Camus, V., Palem, K. V., & Enz, C. (2017). Design and applications of approximate circuits by gate-level pruning. *IEEE Transactions on Very Large Scale Integration (VLSI) Systems, 25*(5), 1694–1702.
96. Tsiokanos, I., Mukhanov, L., Nikolopoulos, D. S., & Karakonstantis, G. (2018a). Minimization of timing failures in pipelined designs via path shaping and operand truncation. In D. Gizopoulos, D. Alexandrescu, M. Maniatakos, & P. Papavramidou (Eds.), *24th IEEE International Symposium on On-Line Testing And Robust System Design, IOLTS 2018, Platja D'Aro, July 2–4, 2018* (pp. 171–176). IEEE.
97. (IEEE 754-2008. IEEE 754-2008 Standard for Floating-Point Arithmetic.)
98. Chippa, V. K., Chakradhar, S. T., Roy, K., & Raghunathan, A. (2013). Analysis and characterization of inherent application resilience for approximate computing. In *The 50th Annual Design Automation Conference 2013, DAC '13, Austin, TX, May 29 –June 07, 2013* (pp. 113.1–113:9). ACM.
99. Chippa, V. K., Chakradhar, S. T., Roy, K., & Raghunathan, A. (2013). Analysis and characterization of inherent application resilience for approximate computing. In *2013 50th ACM/EDAC/IEEE Design Automation Conference (DAC)* (pp. 1–9).
100. Constantin, J., Wang, L., Karakonstantis, G., Chattopadhyay, A., & Burg, A. (2015). Exploiting dynamic timing margins in microprocessors for frequency-over-scaling with instruction-based clock adjustment. In W. Nebel, & D. Atienza (Eds.), *Proceedings of the 2015 Design, Automation & Test in Europe Conference & Exhibition, DATE 2015, Grenoble, March 9–13, 2015* (pp. 381–386). ACM.
101. Tsiokanos, I., Mukhanov, L., Nikolopoulos, D. S., & Karakonstantis, G. (2018). Variation-aware pipelined cores through path shaping and dynamic cycle adjustment: Case study on a floating-point unit. In *Proceedings of the International Symposium on Low Power Electronics and Design, ISLPED 2018, Seattle, WA, July 23–25, 2018* (pp. 52:1–52:6). ACM.
102. Rahimi, A., Benini, L., & Gupta, R. K. (2014). Application-adaptive guardbanding to mitigate static and dynamic variability. *IEEE Transactions on Computers, 63*(9), 2160–2173.
103. Ercegovac, M. D., & Lang, T. (2008). *Digital arithmetic*. Morgan Kaufmann.
104. Mohapatra, D., Karakonstantis, G., & Roy, K. (2007). Low-power process-variation tolerant arithmetic units using input-based elastic clocking. In *Proceedings of the 2007 international symposium on Low power electronics and design (ISLPED '07)* (pp. 74–79).
105. Ghosh, S., Mohapatra, D., Karakonstantis, G., & Roy, K. (2010). Voltage scalable high-speed robust hybrid arithmetic units using adaptive clocking. *IEEE Transactions on Very Large Scale Integration Systems, 18*(9), 1301–1309.
106. Ghosh, S., Bhunia, S., & Roy, K. (2007). Crista: A new paradigm for low-power, variation-tolerant, and adaptive circuit synthesis using critical path isolation. *IEEE Transactions on Computer-Aided Design of Integrated Circuits and Systems, 26*(11), 1947–1956.
107. Ndai, P., Rafique, N., Thottethodi, M., Ghosh, S., Bhunia, S., & Roy, K. (2010). Trifecta: A nonspeculative scheme to exploit common, data-dependent subcritical paths. *IEEE Transactions on Very Large Scale Integration (VLSI) Systems, 18*(1), 53–65.
108. Gaudiot, J., Kang, J., & Ro, W. (2005). *Techniques to Improve Performance Beyond Pipelining: Superpipelining, Superscalar, and VLIW* (pp. 1–34). Advances in Computers.
109. Garyfallou, D., Tsiokanos, I., Evmorfopoulos, N., Stamoulis, G., & Karakonstantis, G. (2020). Accurate estimation of dynamic timing slacks using event-driven simulation. In *2020 21st International Symposium on Quality Electronic Design (ISQED)* (pp. 225–230).
110. Orshansky, M., Nassif, S., & Boning, D. S. (2011). *Design for manufacturability and statistical design: a comprehensive approach*. Springer.
111. Beiu, V., Tache, M., Ibrahim, W., Kharbash, F., & Alioto, M. (2013). On upsizing length and noise margins. In *CAS 2013 (International Semiconductor Conference)* (Vol. 2, pp. 219–222).

112. Kahng, A. B., Kang, S., Kumar, R., & Sartori, J. (2010). Slack redistribution for graceful degradation under voltage overscaling. In *Proceedings of the 15th Asia South Pacific Design Automation Conference, ASP-DAC 2010, Taipei, Taiwan, January 18–21, 2010* (pp. 825–831). IEEE.
113. Patel, J. (2008 [Online]). *CMOS process variations: A critical operation point hypothesis.*
114. Jia, T., Joseph, R., & Gu, J. (2019). An instruction-driven adaptive clock management through dynamic phase scaling and compiler assistance for a low power microprocessor. *IEEE Journal of Solid-State Circuits, 54*(8), 2327–2338.
115. Tziantzioulis, G., Gok, A. M., Faisal, S. M., Hardavellas, N., Ogrenci-Memik, S., & Parthasarathy, S. (2015). b-hive: A bit-level history-based error model with value correlation for voltage-scaled integer and floating point units. In *2015 52nd ACM/EDAC/IEEE Design Automation Conference (DAC)* (pp. 1–6).
116. Jiao, X., Rahimi, A., Jiang, Y., Wang, J., Fatemi, H., de Gyvez, J. P., & Gupta, R. K. (2018). Clim: A cross-level workload-aware timing error prediction model for functional units. *IEEE Transactions on Computers, 67*(6), 771–783.
117. Gupta, P., Agarwal, Y., Dolecek, L., Dutt, N. D., Gupta, R. K., Kumar, R., Mitra, S., Nicolau, A., Rosing, T. S., Srivastava, M. B., Swanson, S., & Sylvester, D. (2013). Underdesigned and opportunistic computing in presence of hardware variability. *IEEE Transactions on CAD of Integrated Circuits and Systems, 32*(1), 8–23.
118. Mukhanov, L., Nikolopoulos, D. S., & de Supinski, B. R. (2015). ALEA: fine-grain energy profiling with basic block sampling. In *2015 International Conference on Parallel Architecture and Compilation, PACT 2015, San Francisco, CA, October 18–21, 2015* (pp. 87–98). IEEE Computer Society.
119. Che, S., Boyer, M., Meng, J., Tarjan, D., Sheaffer, J. W., Lee, S., and Skadron, K. (2009). Rodinia: A benchmark suite for heterogeneous computing. In *2009 IEEE International Symposium on Workload Characterization (IISWC)* (pp. 44–54).
120. Bienia, C. (2011). *Benchmarking Modern Multiprocessors*. PhD thesis, Princeton University.
121. Giridhar, B., Cieslak, M., Duggal, D., Dreslinski, R., Chen, H. M., Patti, R., Hold, B., Chakrabarti, C., Mudge, T., & Blaauw, D. (2013). Exploring dram organizations for energy-efficient and resilient exascale memories. In *SC '13: Proceedings of the International Conference on High Performance Computing, Networking, Storage and Analysis* (pp. 1–12).
122. Buyya, R., Yeo, C. S., Venugopal, S., Broberg, J., & Brandic, I. (2009). Cloud computing and emerging IT platforms: Vision, hype, and reality for delivering computing as the 5th utility. *Future Generation Computer Systems, 25*(6), 599–616.
123. Jung, M., Zulian, É., Mathew, D. M., Herrmann, M., Brugger, C., Weis, C., & Wehn, N. (2015). Omitting refresh: A case study for commodity and wide I/O drams. In B. L. Jacob (Ed.), *Proceedings of the 2015 International Symposium on Memory Systems, MEMSYS 2015, Washington DC, DC, October 5–8, 2015* (pp. 85–91). ACM.
124. Shoushtari, M., BanaiyanMofrad, A., & Dutt, N. D. (2015). Exploiting partially-forgetful memories for approximate computing. *IEEE Embedded Systems Letters, 7*(1), 19–22.
125. Ganapathy, S., Karakonstantis, G., Teman, A., & Burg, A. (2015a). Mitigating the impact of faults in unreliable memories for error-resilient applications. In *Proceedings of the 52nd Annual Design Automation Conference*, DAC '15. New York, NY, USA. Association for Computing Machinery.
126. Esmaeilzadeh, H., Sampson, A., Ceze, L., & Burger, D. (2012b). Architecture support for disciplined approximate programming. In T. Harris, & M. L. Scott (Eds.), *Proceedings of the 17th International Conference on Architectural Support for Programming Languages and Operating Systems, ASPLOS 2012, London, March 3–7, 2012* (pp. 301–312). ACM.
127. Miller, T. N., Thomas, R., Pan, X., & Teodorescu, R. (2012). Vrsync: Characterizing and eliminating synchronization-induced voltage emergencies in many-core processors. In *2012 39th Annual International Symposium on Computer Architecture (ISCA)* (pp. 249–260).
128. Powell, M., & Vijaykumar, T. (2003). Pipeline muffling and a priori current ramping: architectural techniques to reduce high-frequency inductive noise. In *Proceedings of the 2003 International Symposium on Low Power Electronics and Design, 2003. ISLPED '03.* (pp. 223–228).

129. Leng, J., Zu, Y., & Reddi, V. J. (2015). Gpu voltage noise: Characterization and hierarchical smoothing of spatial and temporal voltage noise interference in gpu architectures. In *2015 IEEE 21st International Symposium on High Performance Computer Architecture (HPCA)* (pp. 161–173).
130. Whatmough, P. N., Das, S., Hadjilambrou, Z., & Bull, D. M. (2015). 14.6 an all-digital power-delivery monitor for analysis of a 28nm dual-core arm cortex-a57 cluster. In *2015 IEEE International Solid-State Circuits Conference - (ISSCC) Digest of Technical Papers* (pp. 1–3).
131. Ketkar, M., & Chiprout, E. (2009). A microarchitecture-based framework for pre- and post-silicon power delivery analysis. In *Proceedings of the 42nd Annual IEEE/ACM International Symposium on Microarchitecture*, MICRO 42 (pp. 179–188). New York, NY, USA. Association for Computing Machinery.
132. Kim, Y., & John, L. K. (2011). Automated di/dt stressmark generation for microprocessor power delivery networks. In *IEEE/ACM International Symposium on Low Power Electronics and Design* (pp. 253–258).
133. Kim, Y., John, L. K., Pant, S., Manne, S., Schulte, M., Bircher, W. L., & Govindan, M. S. S. (2012). Audit: Stress testing the automatic way. In *2012 45th Annual IEEE/ACM International Symposium on Microarchitecture* (pp. 212–223).
134. Lefurgy, C. R., Drake, A. J., Floyd, M. S., Allen-Ware, M. S., Brock, B., Tierno, J. A., & Carter, J. B. (2011). Active management of timing guardband to save energy in power7. In *Proceedings of the 44th Annual IEEE/ACM International Symposium on Microarchitecture*, MICRO-44 (pp. 1–11). New York, NY, USA. Association for Computing Machinery.
135. Leng, J., Buyuktosunoglu, A., Bertran, R., Bose, P., & Reddi, V. J. (2015). Safe limits on voltage reduction efficiency in GPUs: A direct measurement approach. In *Proceedings of the 48th International Symposium on Microarchitecture*, MICRO-48 (pp. 294–307). New York, NY, USA. Association for Computing Machinery.
136. Bacha, A., & Teodorescu, R. (2015). Authenticache: Harnessing cache ecc for system authentication. In *Proceedings of the 48th International Symposium on Microarchitecture*, MICRO-48 (pp. 128–140). New York, NY, USA. Association for Computing Machinery.
137. Ganapathy, S., Teman, A., Giterman, R., Burg, A., & Karakonstantis, G. (2015). Approximate computing with unreliable dynamic memories. In *2015 IEEE 13th International New Circuits and Systems Conference (NEWCAS)* (pp. 1–4).
138. Liu, S., Pattabiraman, K., Moscibroda, T., & Zorn, B. G. (2011). Flikker: Saving DRAM refresh-power through critical data partitioning. In R. Gupta, & T. C. Mowry (Eds.), *Proceedings of the 16th International Conference on Architectural Support for Programming Languages and Operating Systems, ASPLOS 2011, Newport Beach, CA, March 5–11, 2011* (pp. 213–224). ACM.
139. Raha, A., Jayakumar, H., Sutar, S., & Raghunathan, V. (2015). Quality-aware data allocation in approximate dram? In R. Iyer, & S. Garg (Eds.), *2015 International Conference on Compilers, Architecture and Synthesis for Embedded Systems, CASES 2015, Amsterdam, October 4–9, 2015* (pp. 89–98). IEEE.
140. Teman, A., Karakonstantis, G., Giterman, R., Meinerzhagen, P., & Burg, A. (2015). Energy versus data integrity trade-offs in embedded high-density logic compatible dynamic memories. In *2015 Design, Automation Test in Europe Conference Exhibition (DATE)* (pp. 489–494).
141. Luo, Y., Govindan, S., Sharma, B., Santaniello, M., Meza, J., Kansal, A., Liu, J., Khessib, B., Vaid, K., & Mutlu, O. (2014). Characterizing application memory error vulnerability to optimize datacenter cost via heterogeneous-reliability memory. In *2014 44th Annual IEEE/IFIP International Conference on Dependable Systems and Networks* (pp. 467–478).
142. Vassiliadis, V., Riehme, J., Deussen, J., Parasyris, K., Antonopoulos, C. D., Bellas, N., Lalis, S., & Naumann, U. (2016). Towards automatic significance analysis for approximate computing. In *2016 IEEE/ACM International Symposium on Code Generation and Optimization (CGO)* (pp. 182–193).

143. Venkataramani, S., Chippa, V. K., Chakradhar, S. T., Roy, K., & Raghunathan, A. (2013). Quality programmable vector processors for approximate computing. In *2013 46th Annual IEEE/ACM International Symposium on Microarchitecture (MICRO)* (pp. 1–12).
144. Mukhanov, L., Tovletoglou, K., Vandierendonck, H., Nikolopoulos, D. S., & Karakonstantis, G. (2019). Workload-aware dram error prediction using machine learning. In *2019 IEEE International Symposium on Workload Characterization (IISWC)* (pp. 106–118).
145. Hamamoto, T., Sugiura, S., & Sawada, S. (1998). On the retention time distribution of dynamic random access memory (dram). *IEEE Transactions on Electron Devices, 45*(6), 1300–1309.
146. Liu, J., Jaiyen, B., Kim, Y., Wilkerson, C., & Mutlu, O. (2013). An experimental study of data retention behavior in modern DRAM devices: implications for retention time profiling mechanisms. In A. Mendelson (Ed.), *The 40th Annual International Symposium on Computer Architecture, ISCA'13, Tel-Aviv, June 23–27, 2013* (pp. 60–71). ACM.
147. Mukhanov, L., Nikolopoulos, D. S., & Karakonstantis, G. (2020). Dstress: Automatic synthesis of dram reliability stress viruses using genetic algorithms. In *2020 53rd Annual IEEE/ACM International Symposium on Microarchitecture (MICRO)* (pp. 298–312).

Chapter 4
Inexact Arithmetic Operators

Lukas Sekanina, Zdenek Vasicek, and Vojtech Mrazek

4.1 Introduction

Developing and applying approximate implementations of arithmetic operators is currently one of the most popular approaches to reduce power consumption in compute-intensive signal, image, and video processing applications. The aim of this chapter is to summarize the operation principles of elementary approximate arithmetic circuits and the methods developed for their design. Our focus will be on selected methodological issues and practices related to the approximate circuit design, namely the understanding of the circuit approximation problem as a multi-objective optimization problem, an error analysis methodology, a correct evaluation and comparison of approximate implementations, and a fair benchmarking methodology.

We will primarily deal with *approximate adders* and *approximate multipliers* because they are the key circuits of many applications relevant for approximate computing, for example, image, video, and speech processing, deep learning, data mining, and natural language processing. No attention will be paid to approximate implementations of subtractors, dividers, and other arithmetic operations because their need in approximate computing systems is rather limited in comparison with adders and multipliers. These circuits are discussed, for example, in [1–3]. Moreover, we will not deal with common compositions of adders and multipliers in circuits such as "multiply and accumulate" (MAC) and scalar product. Their approximate implementations can be obtained either by (i) utilizing (independent) approximate adders and approximate multipliers or by (ii) designing a single block without any decomposition, where (ii) can lead to much better trade-offs than (i).

L. Sekanina (✉) · Z. Vasicek · V. Mrazek
Faculty of Information Technology, IT4Innovations Centre of Excellence, Brno University of Technology, Brno, Czech Republic
e-mail: sekanina@fit.vutbr.cz; vasicek@fit.vutbr.cz; mrazek@fit.vutbr.cz

A. Bosio et al. (eds.), *Approximate Computing Techniques*,
https://doi.org/10.1007/978-3-030-94705-7_4

Approximate implementations of digital circuits are often obtained by the so-called *functional approximation*. This method starts with an original (exact) circuit and tries to modify its logic behavior (and the subsequent implementation) in such a way that the best possible trade-off between the quality of output (the error) and electrical characteristics of the circuit is sought. By electrical characteristics, we mean one or several circuit parameters commonly used to characterize electrical circuits, for example, power consumption, area, and delay. We will not deal with voltage over-scaling and other technology exploiting approximation methods, although they are sometimes combined with functional approximation. The reason is that voltage over-scaling is very technology-dependent and hard to control, making the evaluation and fair comparison of various approximate implementations difficult.

The rest of the chapter is organized as follows. The methodological aspects that are relevant for the design of approximate arithmetic circuits are surveyed in Sect. 4.2. Some of them are then further elaborated in special chapters. In particular, Sect. 4.3 deals with error analysis methods for approximate circuits, with a special focus on formal relaxed equivalence checking. Section 4.4 presents the circuit approximation as a multi-objective optimization problem and emphasizes a correct approach enabling to compare approximate circuits under several design metrics and constraints. Problem-specific approximation methods for adders and multipliers are discussed in Sect. 4.5, while general-purpose automated approximation methods are briefly introduced in Sect. 4.6. A comprehensive open-source library of approximate circuits that was automatically generated by one of the automated methods is presented in Sect. 4.6.2. Section 4.7 includes several case studies that demonstrate some interesting aspects of the circuit approximation methods; for example, it compares the circuit simulation utilizing a subset of input vectors with an exact error analysis. Concluding remarks are given in Sect. 4.8.

4.2 Methodological Aspects

In this section, we discuss various methodological aspects that a designer has to consider before any approximate (arithmetic) circuit is created. Some of these aspects are solely connected with arithmetic circuits, while other aspects are relevant for all approximate circuits. Understanding these aspects is crucial not only for developing efficient circuit approximate methods and high-quality approximate circuits but also for performing a fair comparison of these design methods and the approximate circuits created by these methods.

4.2.1 Design Abstraction

The original (exact) circuit and its approximate implementation, which we have to devise, are usually described at the same level of abstraction. The approximation of arithmetic circuits is typically conducted at the transistor level (e.g., [4]), gate level (e.g., [3, 5]), register transfer (RT) level (e.g., [6]), behavioral level (e.g., [7]), and look-up table (LUT) level (e.g., [8]) if the target platform is a field programmable gate array (FPGA). Most approximate arithmetic circuits are combinational circuits. We will not deal with iterative or sequential implementations, but this topic is also covered in the literature, e.g., [9].

4.2.2 Target Technology

While most approximation approaches have been developed for application-specific integrated circuits (ASICs), there are some papers dealing with approximate arithmetic circuits for graphic processing units (GPUs) [10, 11] and FPGAs [8]. The target technology has to be taken into account by the approximation methodology as an approximate circuit optimized for one technology can show different electrical properties when implemented using a different technology.

4.2.3 Number Representation

Approximation strategies for arithmetic circuits are tightly coupled with the number representation utilized in a given system (Chap. 2). For the fixed-point (FX) number representation, the designers primarily decide about the number of bits used for the integer and fractional part and whether the circuit will intrinsically process signed numbers, the sign will separately be handled, or only unsigned numbers will be considered. A domain-specific quantization scheme is then used to map the input data range to the code values and vice versa. The quantization scheme should be linear and allow us to represent some important numbers (such as 0.0) exactly.

With the development of GPU-based deep learning, approximate implementations of adders and multipliers operating with the floating-point (FP) number representation have been proposed. The usual scheme for representing FP numbers (known from, e.g., the IEEE 754 standard) is simplified to reduce power consumption. For example, in the *minifloat* representation, any exponent and mantissa bit-width combination is allowed on a given total number of bits; however, to reduce the implementation overhead, the common exception handling, infinity, denormalized values, and alternative rounding modes are not supported [12].

4.2.4 Circuit Approximation Methods

Approximate implementations are usually created by (i) "manual" modifications of exact circuits, (ii) developing new application-specific approximation schemes, or (iii) automated design space exploration algorithms.

The first approach requires a skilled designer who introduces appropriate changes to the original circuit. Very specific approximation techniques were developed for particular types of arithmetic circuits such as multipliers. For example, Fig. 4.1 shows one of the first approximate multipliers created by a human expert [13]. Its implementation is based on modifying the truth table of the 2-bit multiplier in such a way that the correct results are provided for 15 out of 16 input combinations, the area is reduced to almost 50%, and the delay is also reduced by one logic level. This approximate multiplier was used as a building block of more complex multipliers. Unfortunately, the manual approximation represents a time-consuming process that is feasible for small circuits only.

The second approach does not start with a common (exact) circuit implementation. It is rather based on a new construction scheme for a given class of problems. For example, a new approximation technique for FP multipliers lies in fitting linear functions with two inputs, referred to as linear planes. The linearization of multiplication allows multiplication operations to be completely replaced with weighted addition [10].

The last method employs fully automated circuit optimization or resynthesis algorithms. We will provide a brief overview of these methods in Chap. 4.6. In addition to the evaluation of resulting approximate implementations, we are faced with a new problem—the evaluation of circuit approximation algorithms (in terms of resources and time needed to obtain a solution with desired properties).

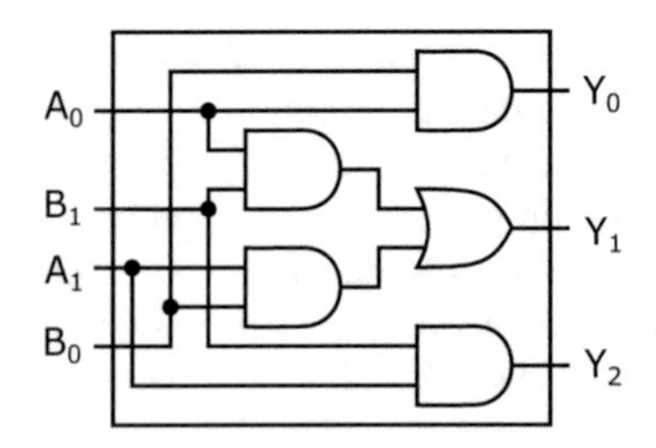

B \ A	0	1	2	3
0	0	0	0	0
1	0	1	2	3
2	0	2	4	6
3	0	3	6	7

Fig. 4.1 The 2-bit approximate multiplier proposed by Kulkarni et al. in [13] and its specification

4.2.5 *Error Metrics and Error Analysis*

Approximate arithmetic circuits are developed with the aim of minimizing one or several error metrics. We will survey commonly used error metrics and error analysis methods in Chap. 4.3. Here, we will emphasize some important aspects of the error analysis methodology. If an approximate circuit is evaluated under one error metric, then the decision of whether one circuit is better than another circuit is straightforward. If two or more error metrics are evaluated together, this relation becomes more complex and a different concept has to be employed. Chap. 4.4 will discuss the so-called Pareto dominance relation to handle this situation. Suppose some constraints are imposed on resulting approximate circuits (for example, the worst-case error must always be less than 1%, while the mean absolute error is minimized). In that case, all circuits not satisfying these constraints must be excluded from the comparisons.

Most error analysis methods only estimate the exact error as determining the exact error is very time-consuming. The error estimate is obtained by circuit simulation across a reasonably inclusive set of input vectors. The exact error can be obtained by exhaustive simulation, but this approach is not scalable. More scalable approaches are based on formal analysis methods. We will compare the performance of these methods in Sect. 4.7.

Another issue is that many approximate circuits have been developed without assuming any particular data patterns existing in a given application, i.e., the design method assumes that all input vectors will occur with the same probability. This is a common approach if the approximate circuit should be provided as a reusable component. If the designer knows the input data distribution, the approximate circuit can be designed to reflect this knowledge and thus to provide better trade-offs between the error(s) and electrical parameters. This can naturally be accomplished by automated approximation methods [14].

Finally, for a few particular implementations of approximate adders and multipliers, detailed probability error analysis methods were published, e.g., [15]. Knowledge of error probabilities becomes very useful if such a circuit is (re)used in a more complex application, and one needs to perform reasoning about the application-level error based on probability models available at the component level. An obvious disadvantage is that a lot of human effort is required to construct reliable probabilistic models for particular circuits.

4.2.6 *Quality Configurable Circuits*

Approximate circuits having some configuration parameters such as the segment size (i.e., the number of sub-adders working in parallel), the number of fractional bits, or the number of active subcircuits are called *quality configurable circuits*. The setting of these configuration parameters is determined either at the design

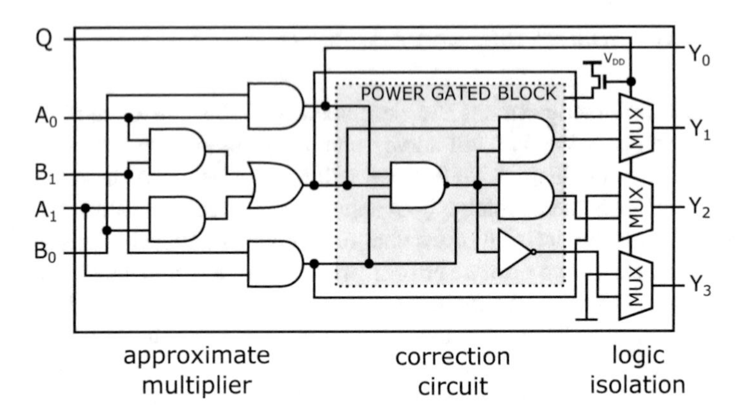

Fig. 4.2 Quality configurable 2-bit multiplier according to [17]

Table 4.1 Parameters of 2-bit and 8-bit exact and configurable approximate multipliers according to [16, 17]. e_{mae} and e_{wce} denote the mean absolute error and the worst-case error

Multiplier	Operating mode	Power [μW]	Delay [ns]	Area [μm^2]	e_{mae} [%]	e_{wce} [%]
2-bit exact	–	3.8	0.15	19	0	0
2-bit QCM	Approximate	2.4	0.09	12[a]	13	13
	Exact	4.7	0.21	23	0	0
8-bit exact	–	428.3	1.25	727	0	0
8-bit QCM	Approximate	483.0	1.51	1197[a]	1.4	22
	Exact	516.4	1.60	1337	0	0

[a] The value is the area of the approximate subcircuit only

time (before the circuit synthesis is conducted) or online, i.e., during the run time, depending on the requested quality of service. This strategy can also be interpreted as an error compensation support or dynamic approximation. For example, these circuits can be utilized in the signal processing applications that can thus benefit from an *in situ* dynamic adaptation of the quality of processing in response to variable requirements on the quality of result and available resources.

We will demonstrate this idea by extending the 2-bit approximate multiplier from Fig. 4.1 to support two modes of operation as introduced in [16, 17], see Fig. 4.2. In the first mode, the quality configurable multiplier (QCM) works exactly as the approximate multiplier from Fig. 4.1, i.e., it generates an incorrect result (7) when both the inputs are 3. In the second mode, a correction circuit is activated, which modifies the output value of the approximate multiplier if it equals 7. Then, the incorrect value (7) is increased by two. The reconfiguration is implemented using the power gating technique. The parameters of the 2-bit QCM synthesized using Synopsys DC with 45 nm FreePDK are given in Table 4.1. Compared to the common multiplier implemented using Verilog star operator, the 2-bit QCM exhibits some overhead when we consider area, power, and delay. In the approximate mode, however, the electrical parameters (i.e., power and delay) are significantly improved. The 2-bit QCM in the approximate mode consumes 36% less power compared to the accurate multiplier. This multiplier can be used as an elementary block to build larger 8-bit and 16-bit quality configurable multipliers.

An important outcome of this brief analysis is that one has to be very careful when properties of common approximate circuits and quality configurable circuits are compared because the quality configurable circuits always exhibit some circuit overhead needed to ensure the reconfiguration.

4.3 Formal Error Analysis

Determining the error of an approximate circuit or deciding whether an approximate circuit satisfies a given error constraint represents not only fundamental theoretical problems but also highly practically relevant problems that must be routinely solved during the design of approximate circuits. This subchapter is focused on the exact error analysis of approximate arithmetic circuits by means of formal methods. But the formal methods can be applied to effectively analyze errors of other combinational circuits as well as sequential systems [9].

Fast and accurate error analysis is especially important in the case automated approximation methods because they usually need to generate and evaluate many candidate designs.

4.3.1 Error Metrics

The functionality of approximate circuits is typically expressed using one or several error metrics. When an arithmetic circuit is approximated, for example, it is necessary to base the error quantification on an arithmetic error metric since the error magnitude could have a significant impact on target application.

Let $f : \mathbb{B}^n \rightarrow \mathbb{B}^m$ be an n-input m-output Boolean function that describes correct functionality (specification) and $\hat{f} : \mathbb{B}^n \rightarrow \mathbb{B}^m$ be an approximation of it, both implemented by two circuits, namely F and $\hat{\text{F}}$. The following paragraphs summarize the error metrics that are relevant for arithmetic circuits.

One of the most popular metrics applied in the context of approximate computing is the *worst-case arithmetic error*, sometimes denoted as *error magnitude* or *error significance*. This metric corresponds with the maximum error the approximation may give and is defined as

$$e_{wce}(f, \hat{f}) = \max_{\forall x \in \mathbb{B}^n} |\text{nat}(f(x)) - \text{nat}(\hat{f}(x))|, \tag{4.1}$$

where nat(x) represents a function nat : $\mathbb{B}^m \rightarrow \mathbb{Z}$ returning a decimal value of the m-bit binary vector x. Typically, a natural binary representation is considered, i.e., nat$(x) = \sum_{i=0}^{m-1} 2^i x_i$. The worst-case error represents the fundamental metric which is typically used as a design constraint and helps to guarantee that the approximate output can differ from the correct output by at most ϵ (i.e., the condition $e_{wce}(f, \hat{f}) \leq \epsilon$ is always satisfied).

The worst-case arithmetic error is closely related to the *relative worst-case error* defined as

$$e_{wcre}(f, \hat{f}) = \max_{\forall x \in \mathbb{B}^n} \frac{|\operatorname{nat}(f(x)) - \operatorname{nat}(\hat{f}(x))|}{\operatorname{nat}(f(x))}. \tag{4.2}$$

This metric can be used to constrain the approximate circuit to differ from the correct one by at most a certain margin. The maximum error magnitude is considered in relation to the correct output value.

There are also statistically oriented error metrics such as the *average-case arithmetic error* or *average-case relative arithmetic error* describing the mean absolute or relative error magnitude. The average-case arithmetic error is defined as the sum of absolute differences in magnitude between the original and approximate circuits, averaged over all inputs:

$$e_{mae}(f, \hat{f}) = \frac{1}{2^n} \sum_{\forall x \in \mathbb{B}^n} |\operatorname{nat}(f(x)) - \operatorname{nat}(\hat{f}(x))|. \tag{4.3}$$

When we replace the expression in the sum by the equation for relative error distance, we can calculate the *mean relative error*:

$$e_{mre}(f, \hat{f}) = \frac{1}{2^n} \sum_{\forall x \in \mathbb{B}^n} \frac{|\operatorname{nat}(f(x)) - \operatorname{nat}(\hat{f}(x))|}{\operatorname{nat}(f(x))}. \tag{4.4}$$

In addition to that, *mean-squared error* corresponding to the average squared error magnitude represents an important metric especially for signal processing applications because it is inversely related to peak signal-to-noise ratio (PSNR). This metric is defined as

$$e_{mse}(f, \hat{f}) = \frac{1}{2^n} \sum_{\forall x \in \mathbb{B}^n} (\operatorname{nat}(f(x)) - \operatorname{nat}(\hat{f}(x)))^2. \tag{4.5}$$

The error metrics mentioned in the previous paragraphs suppose uniform distribution of the input probabilities. There are, however, cases, where we need to consider skewed input distributions. One example is represented by the approximate multiplications carried out in deep neural networks [14]. To evaluate the error metric with respect to a given distribution of input probabilities, *weighted mean error distance* can be introduced as an extension of the conventional mean error [14]:

$$e_{wmae}(f, \hat{f}) = \sum_{\forall x \in \mathbb{B}^n} D(x)|\operatorname{nat}(f(x)) - \operatorname{nat}(\hat{f}(x))|, \tag{4.6}$$

where X corresponds to a discrete random variable representing data at the inputs and D is a probability mass function of X defined as $D(x) = Pr(X = x)$.

In addition to the arithmetic errors, *error rate* referred to as *error probability* can be investigated. The error rate corresponds to the percentage of input vectors for which the output value differs from the original one and is defined as

$$e_{prob}(f, \hat{f}) = \frac{1}{2^n} \sum_{\forall x \in \mathbb{B}^n} [\![f(x) \neq \hat{f}(x)]\!], \tag{4.7}$$

where $[\![f(x) \neq \hat{f}(x)]\!] = 1$ iff the proposition P is satisfied and $[\![P]\!] = 0$ otherwise.

4.3.2 Relaxed Equivalence Checking

Formal verification techniques that are widely adopted in the conventional circuit design flow are often based on *equivalence checking*, i.e., checking whether a mathematical model of a circuit under design meets a given specification. Two main approaches have been developed in this direction—techniques based on Reduced Ordered Binary Decision Diagrams (ROBDDs) and satisfiability (SAT) solvers [18]. In both cases, an auxiliary circuit, the so-called *miter*, is constructed and then analyzed. Figure 4.3a shows that the miter instantiates both the candidate circuit $\hat{F}$ (to be checked) and the golden circuit F and compares their corresponding outputs

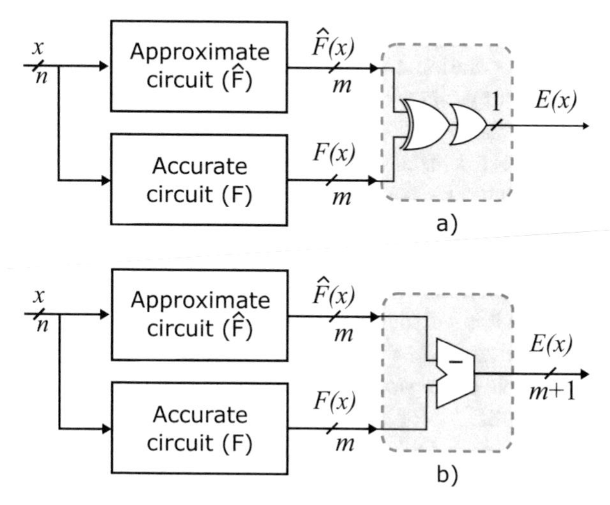

Fig. 4.3 Miter for equivalence checking (**a**) and arithmetic error analysis (**b**). For the equivalence checking, the output E corresponds with $E(x) = [\![f(x) \neq \hat{f}(x)]\!]$. For arithmetic error analysis, the output E equals the error magnitude $E(x) = \operatorname{nat}(f(x)) - \operatorname{nat}(\hat{f}(x))$

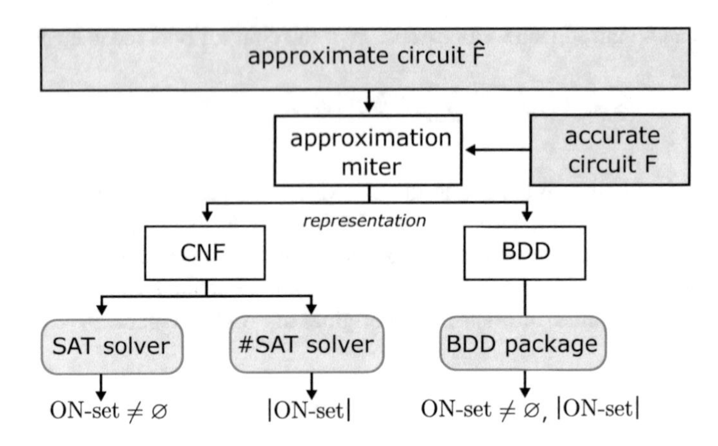

Fig. 4.4 Overview of formal error analysis approaches

to detect a difference in their behavior. In the context of approximate computing, we need to extend this concept to *relaxed equivalence checking*, by stressing the fact that the considered circuits will be checked to be equal up to some bound w.r.t. a suitably chosen distance (error) metric such as the worst-case error or the average error. The (approximation) miter always contains an additional component enabling us to determine the error, see Fig. 4.3b.

Figure 4.4 provides a brief overview of formal error analysis methods for approximate circuits. If the error analysis is performed using ROBDDs, a new ROBDD representing the miter is constructed by a procedure that reads the miter (gate by gate) and adds appropriate nodes to ROBDD. ROBDDs can be directly used for the worst-case as well as the average-case analysis because every library for ROBDD manipulation is equipped with operations enabling us to address questions related to the satisfiability of the miter, namely finding one satisfying assignment and counting the number of satisfying assignments. The first operation provides a single-input assignment x from the ON-set of a Boolean function. The second operation computes the size of the ON-set. As ROBDDs are inefficient in representing classes of circuits for which the number of nodes in BDD is growing exponentially with the number of input variables (e.g., multipliers and dividers), their use in relaxed equivalence checking is typically possible for adders and other less structurally complex functions. Anyway, for example, 128-bit adders can be quickly analyzed in terms of all relevant error metrics [18].

If the error analysis is based on SAT solving, the miter is represented as a logic formula in Conjunctive Normal Form (CNF) for which SAT solver decides whether it is satisfiable or unsatisfiable. The interpretation of this outcome depends on construction of the miter, see Sect. 4.3.3. Common SAT solvers are, in principle, applicable to the worst-case analysis only. However, this approach is more scalable than ROBDDs for the error analysis of multipliers [6]. Specialized SAT solvers (#SAT) are capable of counting the number of satisfiable assignments, but their scalability is very limited, and thus they are currently less practical for the exact error analysis [18].

Even though ROBDDs offer a more flexible approach than SAT considering the possibilities of computation of approximate error metrics, their application is also limited. Not every error metric can directly be calculated using this technique. ROBDDs, for example, do not allow incorporating input probabilities [19]. Moreover, it is not easy to evaluate statistical error metrics involving computation of the relative error due to the presence of division. To address both issues, a more advanced approach needs to be introduced. The only approach allowing us to evaluate the error metrics reflecting the distribution of input probabilities is based on the usage of a more advanced representation known as Algebraic Decision Diagrams (ADDs) [19]. The usage of ADDs is naturally connected with a higher computational cost.

4.3.3 Worst-Case Error Analysis

The worst-case error analysis is typically based on an iterative approach in which a variant of binary search is applied.

For computing the worst-case arithmetic error, for example, the miter given in Fig. 4.3b is used. Algorithm 1 illustrates the principle of determining the worst-case arithmetic error, i.e., calculating the error magnitude at the m-bit output of the miter denoted as E. The principle of this procedure is to iteratively check whether the error is greater than a given threshold (denoted as t in the algorithm). The search procedure gradually narrows down the interval where the exact error value lies. After a finite number of steps, a single value is determined. As the binary search runs in logarithmic time with respect to the range, at most m comparisons are required. The checking can be ensured by means of the magnitude comparator which is used to form a Boolean function whose output is equal to 1 if and only if a given worst-case error $\mathscr{T}$ is violated by the circuit under analysis.

Algorithm 1 Worst-case absolute error computation

Input: n-input approximation miter with m-bit signed output E in the two's complement
Output: maximum absolute arithmetic error (e_{wce})

```
1 l ← 0; r ← 2^m − 1
2 while l ≤ r do
3     t ← ⌈(l + r)/2⌉
4     if WCEGT(E, t) then
5         l ← t + 1
6     else
7         r ← t − 1
8 return l
```

Algorithm 2 Mean absolute error computation

Input: n-input approximation miter with m-bit signed output e in the two's complement, i.e., $E = 2^m e_m - \sum_{i=0}^{m-1} 2^i e_i$

Output: mean absolute arithmetic error (e_{mae})

1 $\varepsilon, c \leftarrow |\text{ON-set}(e_m)|$

for $i \in \{0, 1, \ldots, m-1\}$ **do**

2   **if** $c > 0$ **then**

3         $\varepsilon \leftarrow \varepsilon + 2^i |\text{ON-set}(e_i \oplus e_m)|$

4   **else**

5         $\varepsilon \leftarrow \varepsilon + 2^i |\text{ON-set}(e_i)|$

6 **return** $2^{-n}\varepsilon$

$$\begin{aligned} \text{WCEGT}(E, \mathcal{T}) &= \exists_{x \in \mathbb{B}^n} |E(x)| > \mathcal{T} \\ &= \text{ON-set}\Big([\overline{e_m} \wedge (E > \mathcal{T})] \vee [e_m \wedge (\overline{E} > (\mathcal{T} - 1))] \Big) \neq \varnothing. \end{aligned} \tag{4.8}$$

Then, the satisfiability of this function can be investigated. An incremental SAT solver should be employed to mitigate a potential overhead caused by the necessity of constructing a different comparator in each iteration [18].

4.3.4 Average-Case Error Analysis

Determining the average-case error represents a substantially harder problem because it requires the counting of the number of satisfiable assignments. For computing the average-case arithmetic error, for example, the same miter as in the previous case is used. The mean absolute error can be obtained by determining the error probability per output bit. The obtained counts are then weighted according to the significance of the output bits and summed up. This is illustrated in Algorithm 2.

4.4 Circuit Approximation as a Multi-objective Optimization Problem

The circuit approximation problem can be seen as a *multi-objective optimization problem*, i.e., an optimization problem that involves *multiple objective functions* $g_1(x), g_2(x), \ldots, g_k(x)$, where $g_i : X \to \mathbb{R}$, k is the number of objectives, and $x, x \in X$ is a candidate circuit from the set of feasible circuits X [20]. In the context of approximate circuits, multiple objectives are typically defined to minimize one or several error metric(s), power consumption, area, and delay. The set of feasible solutions consists of all candidate circuits that satisfy the constraints imposed on the

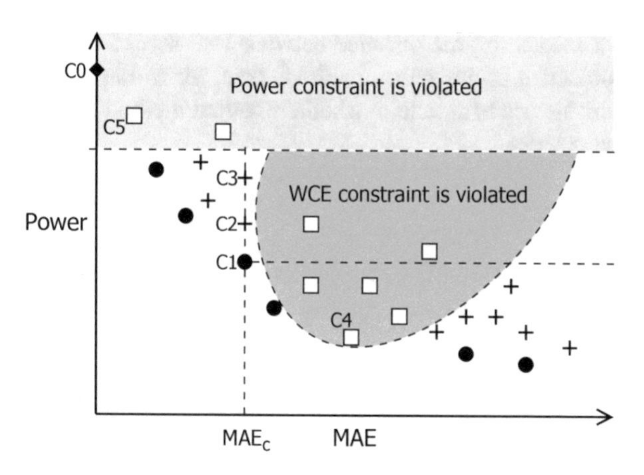

Fig. 4.5 Example of candidate designs in the objective space (power, MAE). Four types of designs are distinguished: non-dominated solutions (circles), dominated solutions (crosses), infeasible solutions (squares), and original solution (diamond)

target circuit. For example, the worst-case error (e_{wce}) has to be less than a given constant, and the power consumption has to be smaller than another constant, as seen in Fig. 4.5.

In the multi-objective optimization, there does not typically exist one feasible solution that minimizes all objective functions simultaneously because the design objectives are conflicting. Hence, rather than one (optimal) solution, the optimization results in a set of solutions, i.e., the solutions that cannot be improved in any of the objectives without degrading at least one of the other objectives. Formally, a feasible solution $x^{(1)} \in X$ is said to *(Pareto) dominate* another solution $x^{(2)} \in X$, if

- $g_i(x^{(1)}) \leq g_i(x^{(2)})$ for all $i \in \{1, 2, \ldots, k\}$ and
- $g_j(x^{(1)}) < g_j(x^{(2)})$ for at least one index $j \in \{1, 2, \ldots, k\}$

and all g_i have to be minimized. A solution $x^* \in X$ is called a *non-dominated solution*, if there does not exist another solution that dominates it. The set of non-dominated solutions is called the *Pareto front*.

Figure 4.5 shows an example of Pareto front containing six non-dominated solutions (circles) and many dominated solutions (crosses) for two objectives to be minimized (e_{mae} and power consumption). The original (accurate) circuit is represented using a black diamond. Figure 4.5 also shows eight infeasible solutions that satisfy the constraints imposed on neither power consumption nor e_{wce}. We say that non-dominated solutions are *Pareto optimal solutions* if all possible candidate solutions are considered during the optimization, and there are not provably better non-dominated solutions in the search space. Claiming that some method finds a Pareto optimal solution without providing a correct proof of it is a clear failure of the author of the method. In practice, we are almost always faced with a situation in which a given method produces suboptimal solutions, i.e., the Pareto front contains the best non-dominated solutions obtained during the experiments conducted with the method. As it is not known "how far" the obtained solutions are from the truly Pareto optimal solutions, a common practice is to introduce a quality metric capable of measuring the distance between two sets of solutions obtained with two multi-objective optimization methods (see, for example, [20]) and compare them under this metric to conclude whether the first method is better than the second method or vice versa.

Another issue is the proper handling of constraints if two approximation methods are compared. For example, solution C4 on Fig. 4.5 would be on the Pareto front if no constraint were imposed on e_{wce}. But in our example, this constraint is specified. A direct comparison with a hypothetical approximation method, which does not specify the same constraint on e_{wce} and claims that C4 is a correct solution, is then meaningless.

In Fig. 4.5, non-dominated solution C1 dominates solutions C2 and C3 (and also some infeasible solutions). A common misinterpretation of this situation is that if one is optimizing for selected criterion (e.g., $e_{mea} = MAE_c$), then solutions C1, C2, and C3 are good candidates and one of them can be selected depending on the available power budget. However, it makes no sense to choose C2 or C3 because C1 is always a strictly better solution under our original assumption that only e_{mae} and power consumption are considered.

4.5 Problem-Specific Approximation Methods

This chapter deals with problem-specific approaches that were developed for obtaining approximate implementations of arithmetic circuits. They are created by experienced engineers who usually start with a common exact implementation. The approximation strategy is dependent on one particular type of circuits (e.g., adders or multipliers). It is not always possible to directly apply these techniques to other types of circuits.

A straightforward approximation technique that can be applied to various arithmetic circuits is *truncation*. In truncation, h-bit (exact) arithmetic circuit is used instead of n-bit circuit ($h < n$) to reduce area, delay, and power consumption. This h-bit circuit is employed to process the most significant h bits of the n-bit operands and the remaining $n-h$ bits are truncated, see also Chap. 2. Because of its simplicity and good results, this technique should be used as a baseline implementation for any comparisons. Moreover, the error profile obtained by truncation is well understood and, hence, the error of complex approximate circuits composed of the truncated circuits can naturally be analyzed.

As stated at the beginning of this chapter, we will primarily deal with approximate adders and multipliers because these circuits are essential in many applications and there is a rich body of literature on this topic.

4.5.1 Approximate Adders

An n-bit (common) adder adds two n-bit operands and produces an $n + 1$ bit result. The most straightforward implementation (the so-called ripple-carry adder) is based on employing n one-bit full adders (FAs) and propagating the carry from the least significant FA to the most significant FA. Although it requires a low amount of logic, its main disadvantage is that the carry chain introduces a long delay increasing linearly with respect to n. In order to reduce this delay, a carry-lookahead adder (CLA) is often employed, which is capable of predicting the input carry to any of the FAs in constant or log time, depending on available additional logic. Other circuit structures that provide some speedup with respect to the ripple-carry adder are carry-select adders and carry-skip adders, but, again, additional logic must be available.

A recent detailed survey of Jiang et al. [3] classified the approximate implementations of n-bit adders into the following classes:

- *Speculative adders*, in which k bits ($k < n$) are used to speculate the carry for each sum bit [21]. This setup leads to a shorter carry chain and thus faster but inexact addition.
- *Segmented adders*, in which the adder is divided into a number of smaller k-bit sub-adders (segments) operating in parallel. Fast addition is obtained as the carry propagation chain is truncated into shorter segments [22–24] and no carry is propagated among the sub-adders.
- *Carry-select adders*, in which the adder is also divided into segments, but the carry input for each sub-adder is selected using different strategies [25–32].
- *Approximate multi-bit full adders*, in which the least significant bits are implemented by approximate FAs that are typically obtained by simplifying the exact FA at the transistor level [4, 33].

Some of these adders are constructed as accuracy configurable circuits, for example, [16, 22, 29].

A detailed analysis conducted in [3] for these approximate adders under several error metrics revealed there is no superior approximation implementation which always provides the best trade-offs. The user has to carefully choose the most suitable implementation for a particular application.

4.5.2 Approximate Multipliers

A typical implementation of the exact unsigned combinational n-bit multiplier is based on generating n n-bit partial products and summing them using $n - 1$ ripple-carry adders organized in an array. In order to reduce delay, the ripple-carry adders are replaced with carry-save adders (the carry and sum signals generated by the adders in a row are passed on to the adders in the next row of the array) and partial results are summed with a structure called Wallace tree, which requires log(n) rows of adders. In these optimized multipliers, a-input/b-output important summing subcircuits (called counters and compressors) can be identified as building blocks. Multiplying of signed binary numbers in the two's complement notation is usually performed with Booth's algorithm, which effectively reduces the number of partial products and their bit-width.

A recent detailed survey of Jiang et al. [3] classified the approximation methods for multipliers into the following classes:

- Approximation in generating *partial products*. Complex multipliers are composed of simplified elementary multipliers (such as the 2-bit approximate multiplier [13] that we discussed in Sect. 4.2 or other smaller approximate multipliers [34]), but the accumulation becomes accurate.
- Approximation in the *partial product tree*, in which some adders or their parts are omitted, for example, because of truncation. Examples include broken array multipliers [33], error-tolerant multipliers [35], and static segment multipliers [36].
- Using *approximate designs of adders, counters, or compressors* to accumulate the partial products, for example, [37–40].
- *Approximate Booth multipliers* [41–46].

Another group of approximation methods does not immediately start with a common multiplier but employs a different approach to obtaining the product. For example, rounding-based approximate (RoBA) multiplier tries to round the operands to the nearest exponent of two to omit the most computationally intensive part of the multiplication [30]. Truncation- and rounding-based scalable approximate multiplier (TOSAM) reduces the number of partial products by truncating each of the input operands based on their leading one-bit position. Hence, the multiplication can be replaced with shift, add, and small fixed-width multiplication operations [47]. In a dynamic range unbiased multiplier (DRUM), an m-bit segment is selected starting from the leading one bit of the input operands and the least significant bit of the truncated values is set to one. The truncated values are multiplied and shifted to the left to generate the final output [48]. Finally, the approximate multiplier can be based on computing an approximate logarithm for both the operands, summing the obtained values and computing antilog [10]. Several schemes for hardware implementation of log and antilog computation exist, but the linear Mitchell approximation techniques are the most area-efficient [49].

With the development of specialized accelerators for deep learning in which it is useful to employ FP number representation, approximate implementations of FP multipliers have been proposed. Some of them are based on converting multiplication to the addition of approximate logarithms of the operands [10]. Another approach is to introduce a specific easy-to-compute function capable of approximating the multiplication [50]. Examples of configurable approximate FP multipliers are [11, 51].

A scalable divide-and-conquer strategy was developed for synthesizing a $2n$-bit approximate multiplier from four n-bit multipliers [52]. The operands are divided into four n-bit chunks (each operand has a lower and higher part) that are independently processed using four multipliers whose outputs are reduced using two adders with one n-bit and one $2n$-bit operand each. The key advantage of this method is that if accurate adders are employed and some of the n-bit multipliers are arbitrary chosen approximate multipliers with known e_{wce}, the upper bound of e_{wce} of the $2n$-bit approximate multiplier can be derived. If only one type of approximate multipliers is used, then e_{wce} can be calculated exactly. Moreover, this construction provides superior trade-offs between the area and error in comparison with many state-of-the-art approximate multipliers [52].

4.6 Automated Approximation and EvoApprox Library

4.6.1 Automated Methods

Automated functional approximation methods start with a common (exact) circuit implementation and define one or several design objectives and constraints. As discussed in Sect. 4.4, the circuit approximation problem can be seen as a multi-objective design problem, where the desired output is a set of non-dominated designs from a Pareto front. As this chapter deals with arithmetic circuits, the approximation is typically conducted at the gate level. The initial circuit is modified by an iterative approximation algorithm to produce an approximate implementation satisfying design objectives and other constraints.

The basic algorithmic approximation techniques are pruning (i.e., removing some parts of the circuit), component replacement (i.e., complex subcircuits are replaced with simpler subcircuits), and approximate resynthesis. If, however, the circuit is provided in a behavioral HDL representation, other more software-oriented techniques (such as loop perforation and memorization, see Chap. 5 for more details on these techniques) can be applied. The automated approximation methods select either randomly or heuristically which parts of the circuit have to be removed, reconnected, or replaced. Table 4.2 gives examples of automated approximation methods, benchmark problems used to evaluate them, and the error evaluation approaches.

One of the automated methods—Cartesian genetic programming (CGP)—is briefly introduced in Fig. 4.6. Based on an original circuit that is supplied by the user, CGP instantiates a population of candidate designs. As CGP is an evolutionary circuit design method, new candidate designs are created by introducing random mutations (i.e., modifications) to the circuit netlist. Candidate designs generated by circuit generator can be constrained in various ways; for example, only circuits having an acceptable number of gates or showing an error below a given threshold are marked as feasible. Candidate designs are evaluated in terms of error (circuit simulation is combined with formal error analysis methods), and the key electrical parameters are quickly estimated. The best-scored circuits then serve as the parents of the new population. This iterative process is repeated for a predefined number

Table 4.2 Selected automated approximation methods, benchmark problems used to evaluate them, and the error evaluation approaches

Method	Ref.	Benchmarks	Error analysis by
ABACUS	[7]	FIR[a], perceptron, block matcher	Simulation
ABM	[53]	6 ISCAS85 benchmark circuits	BDD
ALFANS	[54]	8-bit multipliers, 32-bit adders, MCNC benchmarks	SAT, BDD
ASLAN	[55]	FIR[a], IIR[b], MAC[c], DCT[d], Sobel, and 8-input neuron	Sequential QCC[e] (SAT)
CGP	[34]	2- to 16-bit multipliers, 9-input and 25-input median	Simulation
CGP-BDD	[56]	16 circuits from LGSynth, ITC, and ISCAS	BDD
CGP-SAT	[57]	8- to 32-bit multipliers, 128-bit adders	SAT
SALSA	[5]	Adders, multipliers, FIR[a], IIR[b], DCT[d], etc.	QCC[e] (SAT)
SASIMI	[58]	ISCAS85 benchmarks, multipliers, adders, etc.	Simulation

[a] Finite Impulse Response filter
[b] Infinite Impulse Response filter
[c] Multiply and Accumulate
[d] Discrete Cosine Transform
[e] Quality Constraint Circuit
[f] Fast Fourier Transform

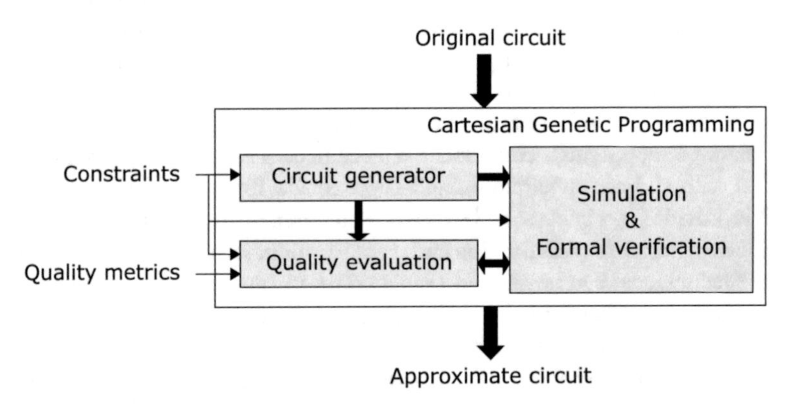

Fig. 4.6 Employing Cartesian genetic programming for automated design of approximate circuits

of iterations. The resulting approximate circuits are fully characterized using professional design tools. The details of the method are presented in [6, 34, 57]. CGP was also employed to evolve efficient implementations of quality configurable circuits [59].

4.6.2 EvoApprox Library

A comprehensive library of approximate arithmetic circuits called EvoApprox8b [60] was introduced in 2017. The idea was to provide well-characterized circuits that can immediately be used in target applications. All circuits were automatically designed by means of CGP. EvoApprox8b contains hundreds of 8-bit approximate adders and multipliers. All circuits were fully characterized in terms of several error metrics and synthesized with Synopsys Design Compiler (45 nm process, V_{dd} =1V) to obtain their area, delay, and power consumption. By means of a simple web user interface, the user can choose the most suitable circuit based on the criteria he/she provides.

In 2019, the library was extended by running additional CGP runs for different objectives and bit-widths. It now contains thousands of various arithmetic circuits, as shown in Table 4.3. In order to simplify the selection of the most suitable circuit for a given application, we identified a subset of circuits and composed EvoApprox8b-Lite. The selection follows the principles of Pareto optimality with respect to several objectives in which power consumption is compared with e_{prob}, e_{mae}, e_{wce}, e_{mse}, and e_{mre} metrics. For each of the five subsets of components, ten circuits evenly distributed along the power axis were included to EvoApprox8b-Lite.

Power vs. e_{mae} trade-offs of thousands of 8-bit approximate multipliers are shown in Fig. 4.7. The black points (corresponding with the EvoApprox8b-Lite) are contrasted with the original circuits of EvoApprox8b (red points) and conventional broken array multipliers (green points), and truncated multipliers (blue points). Note that EvoApprox8b was compared with state-of-the-art approximate circuits in a greater detail [60]. Selected approximate circuits and their various parameters can be downloaded from https://ehw.fit.vutbr.cz/evoapproxlib.

The library provides circuit models in Verilog, Matlab, Python, and C. These models enable the user to integrate the approximate circuits to hardware as well as software projects and design tools. All approximate circuits can thus be simulated in order to obtain their other parameters that are not listed on the website (e.g., the

Table 4.3 The number of approximate implementations of arithmetic circuits in extended EvoApprox library (December 2019)

Circuit	Bit-width	# Approx. implementations
Adder	8	6979
	9	332
	12	4661
	16	1437
	32	916
	64	176
	128	196
Multiplier	8	29,911
	12	3495
	16	35,406
	32	349

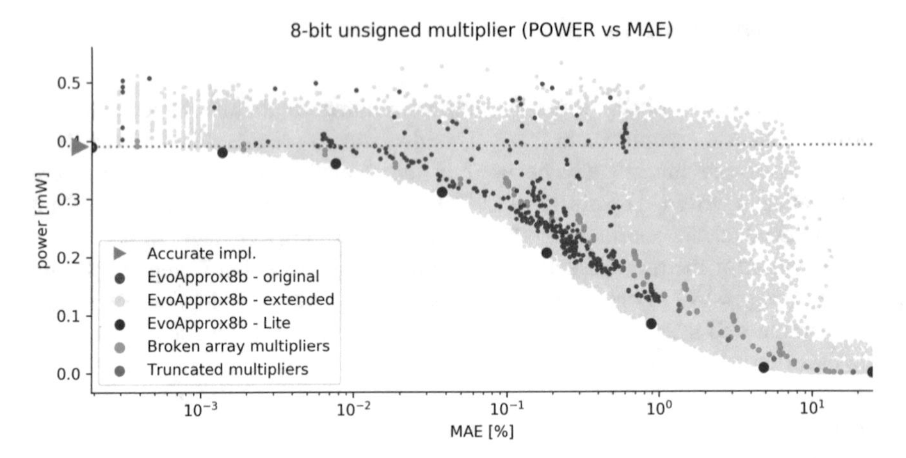

Fig. 4.7 The 8-bit approximate multipliers (black points) that were selected to EvoApprox8b-Lite from all the evolved approximate multipliers (gray points) and compared to the former version of EvoApprox8b library (red points), broken array multipliers (green points) and truncated multipliers (blue points)

errors under different error metrics or power consumption for another fabrication technology).

4.7 Experiments with Error Analysis Methods

This section includes case studies that demonstrate some interesting aspects of the circuit approximation methods, particularly the issues related to the exact error analysis.

4.7.1 Computational Requirements of Error Analysis Methods

A detailed analysis of relaxed equivalence checking algorithms has recently been presented in [18]. The analysis revealed that the computational complexity of the SAT-based methods heavily depends on the actual worst-case error. The computational time increases with a decreasing error, which is noticeable, especially on multipliers. For example, tens of milliseconds are needed to analyze the 12-bit multipliers having an error higher than 2.7%. On the other hand, higher tens of seconds are needed for instances having the error in the range (0.37%, 2.71%], and no result was obtained for multipliers having the worst-case error below 0.05% [18].

Figure 4.8 shows the computational requirements of the WCEGT procedure (i.e., worst-case error checking) for five different thresholds applied to 8-bit multipliers

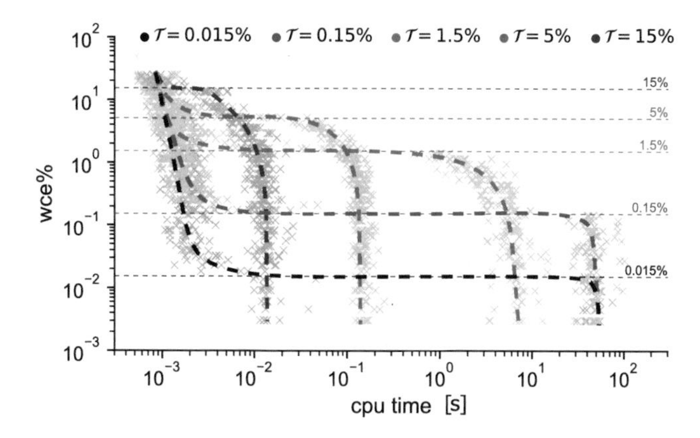

Fig. 4.8 The computational requirements of the WCEGT procedure proving that $e_{wce} > \mathscr{T}$ of 8-bit approximate multipliers taken from EvoApprox library

taken from the EvoApprox library. The worst-case error checking is extremely fast (few milliseconds are required), but only if the actual worst-case error (denoted as wce) is higher than a given threshold $\mathscr{T}$. If this condition is violated, the CPU time may increase by several orders of magnitude. Surprisingly, the difference between the worst-case and the best-case CPU time increases with decreasing the threshold $\mathscr{T}$. Performing WCEGT for thresholds below 1.5% represents the most difficult case. We have to emphasize that the algorithm always terminates for the 8-bit multipliers. Up to 100 seconds are required to analyze the circuit instances whose wce is lower than the chosen threshold.

The same trend was also observed for bigger multipliers. Considering this fact, the design of multiplier-based approximate circuits with low error will be a challenging task because the error analysis will represent a bottleneck of the whole design process.

4.7.2 The Accuracy of Circuit Simulation

For all 8-bit and 16-bit approximate adders and multipliers available in the EvoApprox library, the error was exactly calculated for all relevant error metrics. Knowing the exact errors, we could perfectly analyze the error of the circuit simulation method, which is conducted with a subset of all input vectors. The objective is to determine the minimum number of test vectors that has to be applied to keep the error of circuit simulation below a given threshold.

Let E_{exact} and E_{est} denote the exact error and the error estimated by circuit simulation. Relative difference (RD) [%] between E_{exact} and E_{est} is defined as

$$RD = 100\frac{E_{est} - E_{exact}}{E_{exact}}[\%]. \tag{4.9}$$

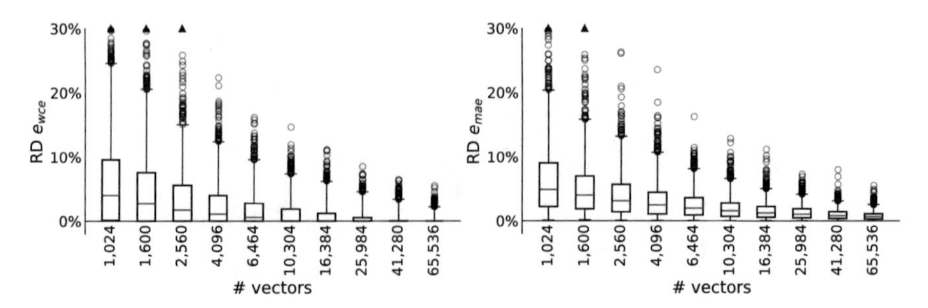

Fig. 4.9 Relative difference between exact and estimated errors for 8-bit approximate multipliers. The whiskers show the 2nd percentile and the 98th percentile. Triangles indicate that there is even higher RD than shown

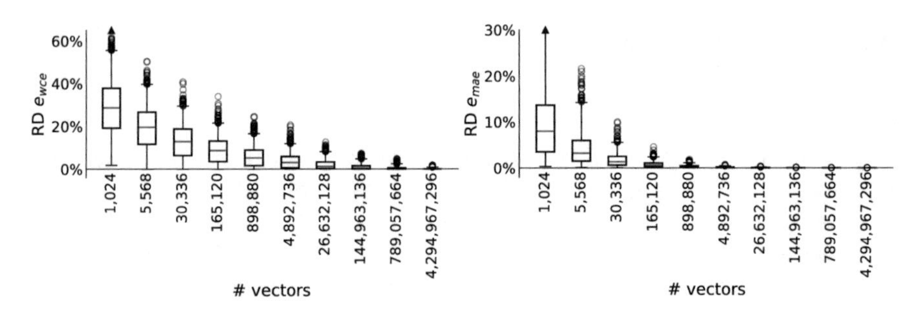

Fig. 4.10 Relative difference between exact and estimated errors for 16-bit approximate multipliers. The whiskers show the 2nd percentile and the 98th percentile. Triangles indicate that there is even higher RD than shown

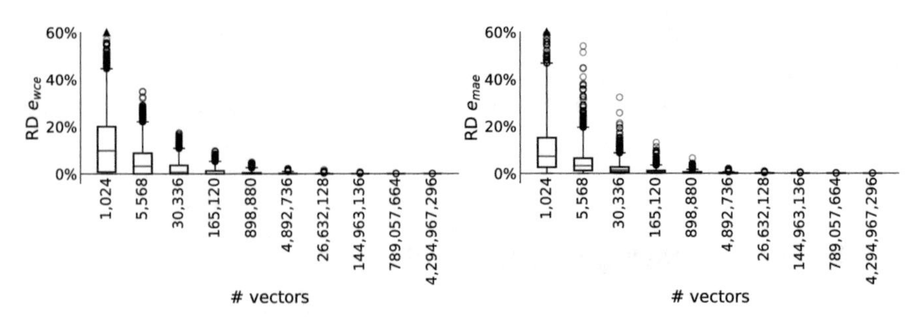

Fig. 4.11 Relative difference between exact and estimated errors for 16-bit approximate adders. The whiskers show the 2nd percentile and the 98th percentile. Triangles indicate that there is even higher RD than shown

Boxplots in Figs. 4.9, 4.10, and 4.11 show how RD depends on the number of input vectors for different circuits. To create one boxplot for e_{wce}, we randomly generated the requested number of input vectors, applied them on 6,275 approximate circuits taken from EvoApprox library and calculated RD. No accurate circuit was considered in the evaluation, i.e., E_{exact} is always greater than zero. The same was done for e_{mae}. A clear consequence of this approach which utilizes randomly generated (but not necessarily unique) vectors is that a non-zero RD is obtained even if the number of generated vectors is identical with the number of all possible input combinations.

In the case of 8-bit approximate multipliers, it makes no sense to use a subset of input vectors during simulation because RD can be higher than 5% even if two-thirds of vectors are used. Moreover, analyzing circuit responses for all $2^{8+8} = 2^{16} = 65,536$ vectors is very fast (few milliseconds on a common CPU [18]). Hence, performing the simulation for all possible input combinations is the best choice.

In the case of 16-bit multipliers, RD for e_{wce} can reach over 10% if $144 \cdot 10^6$ vectors are used. Note that 16-bit approximate multipliers are usually analyzed using only $10 \cdot 10^6$ vectors in some studies [3]. On the other hand, we obtained very reliable error estimates for e_{mae} with less than $5 \cdot 10^6$ vectors. Finally, we analyzed 16-bit adders. A very reliable error characterization in terms of e_{wce} as well as e_{mae} requires a considerably lower number of randomly generated vectors, i.e., less than $5 \cdot 10^6$ vectors as seen in Fig. 4.11.

We can summarize an intuitive fact that estimating e_{mae} with circuit simulation is more reliable than estimating e_{wce} if only a subset of input vectors is used. We encourage the practitioners to provide more statistically relevant error characterizations (e.g., the mean RD and its standard deviation) if the error of approximate circuits is estimated.

4.8 Conclusions

In this chapter, we surveyed various methodological aspects that are relevant for the design of approximate arithmetic circuits. Special attention was given to exact error analysis methods and understanding the circuit approximation problem as a multi-objective optimization problem. We briefly presented problem-specific as well as automated approximation methods developed for adders and multipliers. Unfortunately, misunderstanding of the principles of correct evaluation of approximate circuits and correct benchmarking of circuit approximation methods is still visible in the literature. We believe that this chapter can help in establishing a better practice in this emerging area.

Acknowledgments This work was supported by the Czech science foundation project 19-10137S.

References

1. Melchert, J., Behroozi, S., Li, J., & Kim, Y. (2019). SAADI-EC: a quality-configurable approximate divider for energy efficiency. *IEEE Transactions on Very Large Scale Integration (VLSI) Systems, 27*(11), 2680–2692.
2. Chen, L., Han, J., Liu, W., & Lombardi, F. (2017). Algorithm and design of a fully parallel approximate coordinate rotation digital computer (CORDIC). *IEEE Transactions on Multi-Scale Computing Systems, 3*(3), 139–151.
3. Jiang, H., Liu, L., Lombardi, F., & Han, J. (2019). Approximate arithmetic circuits: design and evaluation. In S. Reda & M. Shafique (Eds.) *Approximate Circuits, Methodologies and CAD* (pp. 67–98) New York: Springer.
4. Gupta, V., Mohapatra, D., Raghunathan, A., & Roy, K. (2013). Low-power digital signal processing using approximate adders. *IEEE Transactions on Computer-Aided Design of Integrated Circuits and Systems, 32*(1), 124–137.
5. Venkataramani, S., Sabne, A., Kozhikkottu, V.J., Roy, K., & Raghunathan, A. (2012). SALSA: systematic logic synthesis of approximate circuits. In *The 49th Design Automation Conference* (pp. 796–801). New York: ACM.
6. Sekanina, L., Vasicek, Z., & Mrazek, V. (2019). Automated search-based functional approximation for digital circuits. In S. Reda & M. Shafique (Eds.) *Approximate Circuits, Methodologies and CAD* (pp. 175–203). New York: Springer.
7. Nepal, K., Hashemi, S., Tann, H., Bahar, R.I., & Reda, S. (2019). Automated high-level generation of low-power approximate computing circuits. *IEEE Transactions on Emerging Topics in Computing, 7*(1), 18–30.
8. Ullah, S., Rehman, S., Prabakaran, B. S., Kriebel, F., Hanif, M. A., Shafique, M., & Kumar, A. (2018). Area-optimized low-latency approximate multipliers for FPGA-based hardware accelerators. In *2018 55th ACM/ESDA/IEEE Design Automation Conference (DAC)*, June 2018 (pp. 1–6)
9. Chandrasekharan, A., Soeken, M., Große, D., & Drechsler, R. (2016). Precise error determination of approximated components in sequential circuits with model checking. In *Proceedings of the DAC'16* (pp. 1–6). New York: ACM.
10. Saadat, H., Bokhari, H., & Parameswaran, S. (2018). Minimally biased multipliers for approximate integer and floating-point multiplication. *IEEE Transactions on Computer-Aided Design of Integrated Circuits and Systems, 37*(11), 2623–2635.
11. Imani, M., Garcia, R., Gupta, S., & Rosing, T. (2018). RMAC: Runtime configurable floating point multiplier for approximate computing. In *Proceedings of the International Symposium on Low Power Electronics and Design*, ISLPED '18, New York, NY (pp. 12:1–12:6). New York: ACM.
12. Gysel, P., Pimentel, J., Motamedi, M., & Ghiasi, S. (2018). Ristretto: A framework for empirical study of resource-efficient inference in convolutional neural networks. *IEEE Transactions on Neural Networks and Learning Systems, 29*(11), 5784–5789.
13. Kulkarni, P., Gupta, P., & Ercegovac, M. (2011). Trading accuracy for power with an underdesigned multiplier architecture. In *2011 24th International Conference on VLSI Design*, Jan 2011 (pp. 346–351)
14. Vasicek, Z., Mrazek, V., & Sekanina, L. (2019). Automated circuit approximation method driven by data distribution. In *Design, Automation and Test in Europe Conference* (pp. 96–101). Leuven: European Design and Automation Association.
15. Mazahir, S., Hasan, O., Hafiz, R., Shafique, M., & Henkel, J. (2017). Probabilistic error modeling for approximate adders. *IEEE Transactions on Computers, 66*(3), 515–530 (2017)
16. Shafique, M., Ahmad, W., Hafiz, R., & Henkel, J. (2015). A low latency generic accuracy configurable adder. In *2015 52nd ACM/EDAC/IEEE Design Automation Conference (DAC)*, June 2015 (pp. 1–6)
17. Shafique, M., Hafiz, R., Rehman, S., et al. (2016). Invited: Cross-layer approximate computing: From logic to architectures. In *DAC'16*.

18. Vasicek, Z. (2019). Formal methods for exact analysis of approximate circuits. *IEEE Access, 7*(1), 177309–177331.
19. Froehlich, S., Große, D., & Drechsler, R. (2019). One method - all error-metrics: A three-stage approach for error-metric evaluation in approximate computing. In *2019 Design, Automation Test in Europe Conference Exhibition (DATE)* (pp. 284–287).
20. Coello Coello, C. A., Gonzalez Brambila, S., Figueroa Gamboa, J., Castillo Tapia, M. G., & Hernandez Gomez, R. (2020). Evolutionary multiobjective optimization: Open research areas and some challenges lying ahead. *Complex & Intelligent Systems, 2020* (pp. 1–16).
21. Verma, A. K., Brisk, P., & Ienne, P. (2008). Variable latency speculative addition: A new paradigm for arithmetic circuit design. In *2008 Design, Automation and Test in Europe* (pp. 1250–1255).
22. Kahng, A. B., & Kang, S. (2012). Accuracy-configurable adder for approximate arithmetic designs. In *DAC Design Automation Conference 2012*, June 2012 (pp. 820–825)
23. Mohapatra, D., Chippa, V. K., Raghunathan, A., & Roy, K. (2011). Design of voltage-scalable meta-functions for approximate computing. In *2011 Design, Automation Test in Europe*, March 2011 (pp. 1–6).
24. Zhu, N., Goh, W. L., & Yeo, K. S. (2009). An enhanced low-power high-speed adder for error-tolerant application. In *Proceedings of the 2009 12th International Symposium on Integrated Circuits*, Dec 2009 (pp. 69–72).
25. Du, K., Varman, P., & Mohanram, K. (2012). High performance reliable variable latency carry select addition. In *2012 Design, Automation Test in Europe Conference Exhibition (DATE)*, March 2012 (pp. 1257–1262).
26. Kim, Y., Zhang, Y., & Li, P. (2013). An energy efficient approximate adder with carry skip for error resilient neuromorphic VLSI systems. In *2013 IEEE/ACM International Conference on Computer-Aided Design (ICCAD)* (pp. 130–137).
27. Lin, I., Yang, Y., & Lin, C. (2015). High-performance low-power carry speculative addition with variable latency. *IEEE Transactions on Very Large Scale Integration (VLSI) Systems, 23*, 1591–1603.
28. Li, L. & Zhou, H. (2014). On error modeling and analysis of approximate adders. In *2014 IEEE/ACM International Conference on Computer-Aided Design (ICCAD)* (pp. 511–518).
29. Hu, J., & Qian, W. (2015). A new approximate adder with low relative error and correct sign calculation. In *2015 Design, Automation Test in Europe Conference Exhibition (DATE)*, 1449–1454.
30. Zendegani, R., Kamal, M., Bahadori, M., Afzali-Kusha, A., & Pedram, M. (2017). ROBA multiplier: A rounding-based approximate multiplier for high-speed yet energy-efficient digital signal processing. *IEEE Transactions on Very Large Scale Integration (VLSI) Systems, 25*, 393–401.
31. Camus, V., Schlachter, J., & Enz, C. (2016). A low-power carry cut-back approximate adder with fixed-point implementation and floating-point precision. In *2016 53rd ACM/EDAC/IEEE Design Automation Conference (DAC)*, June 2016 (pp. 1–6).
32. Ebrahimi-Azandaryani, F., Akbari, O., Kamal, M., Afzali-Kusha, A., & Pedram, M. (2019). Block-based carry speculative approximate adder for energy-efficient applications. *IEEE Transactions on Circuits and Systems II: Express Briefs, 1–1*.
33. Mahdiani, H. R., Ahmadi, A., Fakhraie, S. M., & Lucas, C. (2010). Bio-inspired imprecise computational blocks for efficient VLSI implementation of soft-computing applications. *IEEE Transactions on Circuits and Systems I: Regular Papers, 57*, 850–862.
34. Vasicek, Z. & Sekanina, L. (2015). Evolutionary approach to approximate digital circuits design. *IEEE Transactions on Evolutionary Computation, 19*(3), 432–444.
35. Kyaw, K. Y., Goh, W. L., & Yeo, K. S. (2010). Low-power high-speed multiplier for error-tolerant application. In *2010 IEEE International Conference of Electron Devices and Solid-State Circuits (EDSSC)*, December 2010
36. Narayanamoorthy, S., Moghaddam, H. A., Liu, Z., Park, T., & Kim, N. S. (2015). Energy-efficient approximate multiplication for digital signal processing and classification applications. *IEEE Transactions on Very Large Scale Integration (VLSI) Systems, 23*, 1180–1184.

37. Lin, C., & Lin, I. (2013). High accuracy approximate multiplier with error correction. In *2013 IEEE 31st International Conference on Computer Design (ICCD)*, October 2013 (pp. 33–38)
38. Momeni, A., Han, J., Montuschi, P., & Lombardi, F. (2015). Design and analysis of approximate compressors for multiplication. *IEEE Transactions on Computers, 64*, 984–994.
39. Jiang, H., Liu, C., Lombardi, F., & Han, J. (2019). Low-power approximate unsigned multipliers with configurable error recovery. *IEEE Transactions on Circuits and Systems I: Regular Papers, 66*, 189–202.
40. Ha, M., & Lee, S. (2018). Multipliers with approximate 4–2 compressors and error recovery modules. *IEEE Embedded Systems Letters, 10*, 6–9 (2018)
41. Cho, K.-J., Lee, K.-C., Chung, J.-G., & Parhi, K. K. (2004). Design of low-error fixed-width modified booth multiplier. *IEEE Transactions on Very Large Scale Integration (VLSI) Systems, 12*, 522–531.
42. Song, M.-A., Van, L.-D., & Kuo, S.-Y. (2007). Adaptive low-error fixed-width booth multipliers. *IEICE Transactions on Fundamentals of Electronics, Communications and Computer Sciences, E90-A*, 1180–1187.
43. Wang, J., Kuang, S., & Liang, S. (2011). High-accuracy fixed-width modified booth multipliers for lossy applications. *IEEE Transactions on Very Large Scale Integration (VLSI) Systems, 19*, 52–60.
44. Chen, Y., & Chang, T. (2012). A high-accuracy adaptive conditional-probability estimator for fixed-width booth multipliers. *IEEE Transactions on Circuits and Systems I: Regular Papers, 59*, 594–603.
45. Farshchi, F., Abrishami, M. S., & Fakhraie, S. M. (2013). New approximate multiplier for low power digital signal processing. In *The 17th CSI International Symposium on Computer Architecture Digital Systems (CADS 2013)*, October 2013 (pp. 25–30).
46. Jiang, H., Han, J., Qiao, F., & Lombardi, F. (2016). Approximate radix-8 booth multipliers for low-power and high-performance operation. *IEEE Transactions on Computers, 65*, 2638–2644.
47. Vahdat, S., Kamal, M., Afzali-Kusha, A., & Pedram, M. (2019). TOSAM: An energy-efficient truncation- and rounding-based scalable approximate multiplier. *IEEE Transactions on Very Large Scale Integration (VLSI) Systems, 27*(5), 1161–1173.
48. Hashemi, S., Bahar, R. I., & Reda, S. (2015). Drum: A dynamic range unbiased multiplier for approximate applications. In *2015 IEEE/ACM International Conference on Computer-Aided Design (ICCAD)* (pp. 418–425).
49. Mitchell, J. N. (1962). Computer multiplication and division using binary logarithms. *IRE Transactions on Electronic Computers, EC-11*(4), 512–517.
50. Imani, M., Sokolova, A., Garcia, R., Huang, A., Wu, F., Aksanli, B., & Rosing, T. (2019). ApproxLP: Approximate multiplication with linearization and iterative error control. In *2019 56th ACM/IEEE Design Automation Conference (DAC)*, June 2019 (pp. 1–6).
51. Imani, M., Peroni, D., & Rosing, T. (2017). CFPU: Configurable floating point multiplier for energy-efficient computing. In *2017 54th ACM/EDAC/IEEE Design Automation Conference (DAC)* (pp. 1–6).
52. Mrazek, V., Vasicek, Z., Sekanina, L., Jiang, H., & Han, J. (2018). Scalable construction of approximate multipliers with formally guaranteed worst case error. *IEEE Transactions on Very Large Scale Integration (VLSI) Systems, 26*, 2572–2576.
53. Soeken, M., Grosse, D., Chandrasekharan, A., & Drechsler, R. (2016). BDD minimization for approximate computing. In *21st Asia and South Pacific Design Automation Conference ASP-DAC 2016* (pp. 1–6). New York: IEEE.
54. Wu, Y., & Qian, W. (2019). ALFANS: Multi-level approximate logic synthesis framework by approximate node simplification. *IEEE Transactions on Computer-Aided Design of Integrated Circuits and Systems, 39*, 1–14.
55. Ranjan, A., Raha, A., Venkataramani, S., Roy, K., & Raghunathan, A. (2014). ASLAN: Synthesis of approximate sequential circuits," in *Proceedings of the Conference on Design, Automation and Test in Europe*, DATE'14, EDA Consortium, 2014 (pp. 1–6).
56. Vasicek, Z., & Sekanina, L. (2016). Evolutionary design of complex approximate combinational circuits. *Genetic Programming and Evolvable Machines, 17*(2), 1–24 (2016).

57. Ceska, M., Matyas, J., Mrazek, V., Sekanina, L., Vasicek, Z., & Vojnar, T. (2017). Approximating complex arithmetic circuits with formal error guarantees: 32-bit multipliers accomplished. In *2017 IEEE/ACM International Conference on Computer-Aided Design (ICCAD)* (pp. 416–423).
58. Venkataramani, S., Roy, K., & Raghunathan, A. (2013). Substitute-and-simplify: a unified design paradigm for approximate and quality configurable circuits. In *Design, Automation and Test in Europe, DATE'13*, EDA Consortium, 2013 (pp. 1367–1372).
59. Mrazek, V., Vasicek, Z., & Sekanina, L. (2018). Design of quality-configurable approximate multipliers suitable for dynamic environment. In *Proceedings of the 2018 NASA/ESA Conference on Adaptive Hardware and Systems* (pp. 264–271). New York: IEEE.
60. Mrazek, V., Hrbacek, R., Vasicek, Z., & Sekanina, L. (2017). Evoapprox8b: Library of approximate adders and multipliers for circuit design and benchmarking of approximation methods. In *Design, Automation Test in Europe Conference Exhibition (DATE), 2017* (pp. 258–261).

Chapter 5
Approximate Computing at the Algorithmic Level

Justine Bonnot, Alexandre Mercat, Erwan Nogues, and Daniel Ménard

5.1 Introduction

Data-oriented processing is pervasive in digital industrial and consumer applications in the field of numerous domains like signal and image processing, artificial intelligence, telecommunications. For these applications, to provide new services and to enable innovations, the application complexity continually grows. This complexity rising increases the implementation cost corresponding to the system cost, the energy consumption, the execution time and memory footprint for Software (SW) implementation and the chip area and latency for Hardware (HW) implementation. The concept of approximate computing at the algorithm level aims at reducing the processing complexity to reduce the implementation cost. The implementation cost is reduced by decreasing the number of processing operations and memory exchanges. But, by modifying the original algorithm, the application output is modified and the resulting quality is degraded.

In real-time applications, reducing the processing complexity allows increasing the slack time. As represented in Fig. 5.1, the slack time corresponds to the time span between the completion of a processing and its deadline. Two approaches can be considered to reduce the energy consumption by exploiting the slack time for SW implementation. In the first approach, the slack time is exploited to slow-down the processor by reducing the clock frequency. At the same time, this allows reducing the supply voltage. Given that the supply voltage is a squared term in the dynamic power expression, this allows reducing the power and thus the energy compared to the initial case. This approach corresponds to Dynamic Voltage and Frequency Scaling (DVFS) in which the processor clock frequency and supply voltage are

J. Bonnot · A. Mercat · E. Nogues · D. Ménard (✉)
Univ. Rennes, INSA Rennes, IETR, Rennes, France
e-mail: daniel.menard@insa-rennes.fr

A. Bosio et al. (eds.), *Approximate Computing Techniques*,
https://doi.org/10.1007/978-3-030-94705-7_5

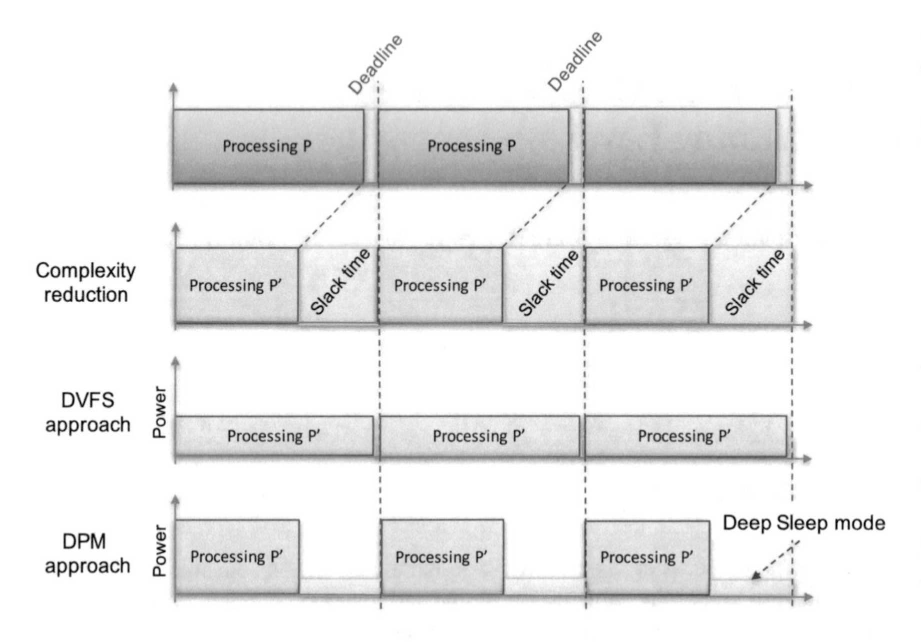

Fig. 5.1 Exploitation of slack time to reduce the energy consumption for real-time processing. (**a**) Initial processing. (**b**) DPM approach. (**c**) DVFS approach

adapted according to the processing load. The second approach corresponds to Dynamic Power Management (DPM) in which the processor enters in a deep sleep mode during the slack time. By exploiting power-gating and clock gating, the power consumption is significantly reduced in the deep sleep mode and the global energy is reduced.

This chapter focus on the approximate computing techniques acting at the algorithm level. In Sect. 5.2, the different available techniques are presented. In Sect. 5.3, the High Efficiency Video Coding (HEVC) video codec is considered as a use-case to illustrate the use of approximate computing techniques at the algorithmic level.

5.2 Techniques for Algorithm-Level Approximate Computing

Different approximate computing techniques acting at the algorithm level have been proposed to reduce the processing complexity of data-oriented applications. These techniques transform the algorithm to enable effective approximation. In this section, the different approaches proposed for approximate computing techniques acting at the algorithm level are presented.

Two directions can be considered to reduce the complexity. The first direction is to skip part of the computation by removing some processing. These approaches are detailed in Sect. 5.2.1. The second direction is based on approximation to replace a part of the computation by a less complex processing. These approaches are detailed in Sect. 5.2.2.

5.2.1 Skip-Based Approaches

Computation skipping consists in not executing parts of the computations to reduce the processing complexity. The limitation of skip-based approaches are exposed in Sect. 5.2.1.1. The selection of the skipped computations can be done at two granularity levels. At fine grained, a part of the computation, corresponding to one up to several expressions, is skipped like in the loop perforation and early termination techniques. These two techniques are described, respectively, in Sects. 5.2.1.2 and 5.2.1.3. At coarse grained a complete part of the computation corresponding to a processing block or a task is skipped as presented in Sect. 5.2.1.4.

5.2.1.1 Scope of Skip-Based Approaches

This approach based on the discarding of a processing block in an application must be used with caution and cannot be applied to any processing. Let us consider a processing block f having x as input and y as output. Skipping f is equivalent to have the relation $y = x$. Thus, f can be skipped if the input x and the output y belong to the same domain, i.e. x and y represent the same kind of data. To illustrate this concept, let us consider the case of a filtering block for which the input and output represent data in the same domain and the case of a transform block for which the input and output belong to different domains. The aim of a filtering block f is to improve the characteristics of the signal x by removing some frequency components which, for example, are not relevant to exploit this signal. If this filtering block f is skipped, the quality of the signal exploitation process will be degraded but this process can be achieved. For the transform block f, like a Fourier transform, the aim is to convert a signal x from the time domain representation to the frequency domain representation. In this case, given that x and y do not represent data in the same domain, eliminating this transform block f will damage the application proper operation and this is not sustainable.

In [45], the fine-grained patterns which suit well for computation skipping are analyzed. For example, *sum* or *argmin* patterns can be considered. In both cases, if some elements of the *sum* or *argmin* are not computed, a result close to the original one can be produced. From this analysis, a list of applications for which skip-based approaches can be used is given and listed below:

1. *Search Space Enumeration:* The application iterates over a search space of items in order to select the best candidate solving the problem. In Sect. 5.2.1.5, the

case of enumerating the search space in the context of discrete optimization is detailed.

2. *Monte-Carlo Simulation:* a set of Monte- Carlo simulations are performed and the skip-based approach aims at reducing the number of performed simulations to reduce the complexity.
3. *Iterative Refinement:* an iterative process is used to improve the accuracy of an algorithm result. The number of iterations is controlled to manage the trade-off between accuracy and complexity. The concept of early termination for iterative refinement applications is discussed in Sect. 5.2.1.3.
4. *Data Structure Update:* a data structure is traversed and updated with computed values. The skip-based approach discards some updates by keeping previous values.

5.2.1.2 Loop Perforation

In a program describing data-oriented applications, loops and especially nested loops represent a significant part of the computation complexity. The loop perforation technique aims at reducing the loop complexity by skipping part of the computation inside the loop. To illustrate the loop perforation concept, a *for-loop* having a loop index i is considered as illustrated in Fig. 5.2a. This loop index is incremented by one at each loop iteration. The loop perforation concept is characterized by a parameter r representing the loop perforation rate, which corresponds to the expected ratio of loop iterations to skip.

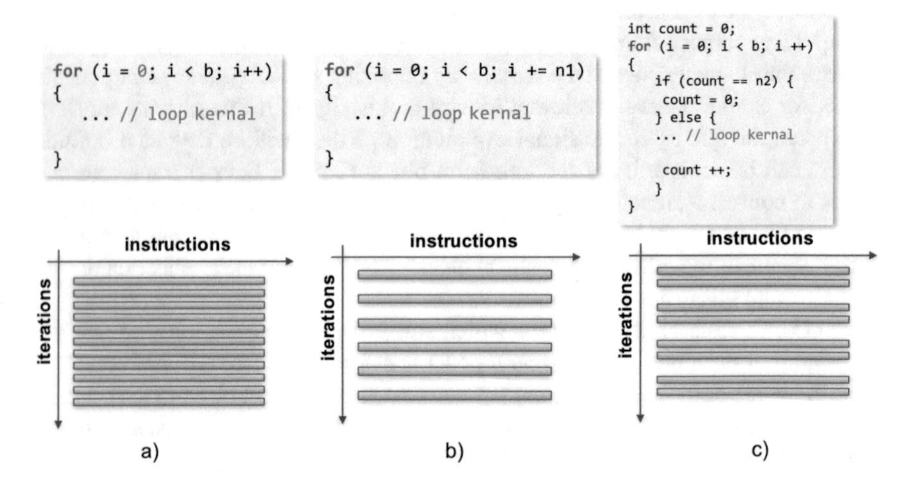

Fig. 5.2 Illustration of the loop perforation concept. (**a**) Initial *for-loop*. (**b**) Loop perforation with the execution of one iteration loop after n_1. (**c**) Loop perforation with the skip of one iteration after n_2

The straightforward approach also named *interleaving perforation* is to increment the loop index by n_1 instead of one as illustrated in Fig. 5.2b. In this case, the loop kernel is performed every n_1^{th} iteration and $n_1 - 1$ iterations are skipped. The perforation rate is equal to

$$r = 1 - \frac{1}{n_1} \quad (5.1)$$

This approach quickly reduces the loop computation complexity. Indeed, half of the iterations are discarded when $n_1 = 2$, the lowest value that can be taken by n_1. The number of discarded iterations increases for higher values of n_1. This high perforation rate can affect significantly the quality of the loop computation result. The alternative is to skip one iteration after n_2 iterations as illustrated in Fig. 5.2c. A conditional structure is inserted in the loop kernel to discard the kernel computation every n_2 iterations. In this case, the perforation rate is equal to

$$r = \frac{1}{n_2} \quad (5.2)$$

The challenges for the loop perforation concept are to detect the best candidate loops and to adjust the perforation rate in order to select the best trade-off between the loop complexity reduction and the degradation of the output. Sidiroglou et al. [45] proposed to identify critical and tunable loops in an application, to reduce the loop complexity. By applying loop perforation on several applications, the authors managed to reduce the execution time by seven while keeping the difference at the output lower than 10%. Besides, the authors have identified several computational kernels that support well perforation, as the computation of a sum, the *argmin* operation. This technique has been applied to applications from the PARSEC 1.0 benchmark suite [7] so as to cover a wide range of application domains as finance, media processing or data mining, for instance.

For the techniques presented above, the decision are taken at design time, but loop perforation can be done at run-time as illustrated in Fig. 5.3a. The number of skipped iterations is adjusted dynamically depending on the targeted output quality or energy requirements. Nevertheless, the selection of the iterations to skip induces a run-time overhead. Selective dynamic loop perforation is proposed in [25], to skip a subset of the instructions inside an iteration, but an overhead appears due to the selection of the iteration. In this case, loops are automatically transformed to be able to skip instructions in chosen iterations. Selective dynamic loop perforation achieves an average speedup of $2\times$ compared to classical loop perforation technique while inducing the same amount of error.

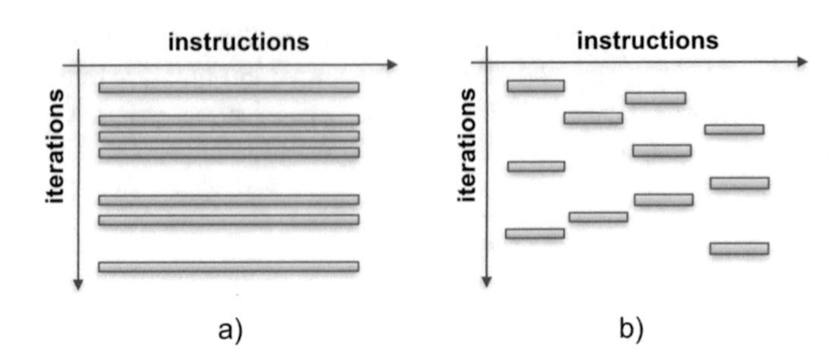

Fig. 5.3 Illustration of the loop perforation concept at run-time. (**a**) Dynamic perforation: the number of skipped iterations is decided at run-time. (**b**) Selective perforation

5.2.1.3 Early Termination

The concept of *Early Termination* ends a computation process before reaching its end. This concept can be widely used in iterative refinement algorithms in which an iterative method is used to improve the output algorithm accuracy at each iteration. To reduce complexity, the iterative process is stopped before it reaches full convergence. Several stopping conditions can be set. In some algorithms, the number of iterations has a direct link with the accuracy and can be computed and fixed in advance. For instance, for the COordinate Rotation DIgital Computer (CORDIC) algorithm [29], the higher the number of iterations, the more accurate the estimation of the trigonometric function result will be.

The stopping condition can be defined when the improvement between several successive iterations falls below a fixed threshold. In convergence-based pruning, the algorithm is stopped when no change or improvement is noticed in the last k iterations. For examples, in the K-means clustering algorithm, the stopping condition can be the ratio of unstable points, i.e. points that changed their memberships in the last iteration.

Nevertheless, to define a stopping criterion directly linked with the output quality, one need to know the reference output which is generally not the case. To answer this problem, the framework ApproxIt [54] has been proposed. It targets only iterative methods and implements a quality estimator to be able to select the best approximation strategy for the next iteration at run-time.

The difference between loop perforation and iterative refinement is illustrated in Fig. 5.4. The different iterations of the loop are figured with i_k. In the accurate implementation of the loop, the N iterations from i_0 to i_{N-1} are successively executed. In the loop perforation version, $r \times N$ iterations of the loop are periodically skipped. For early termination case, from a certain iteration, in the proposed example, i_{stop}, the iterations of the loop do no more need to be executed: the loop stops. In this case, the iterations from i_0 to i_{stop} are executed and allow obtaining a minimal accuracy guaranteed by the early termination stop criterion. On the other hand, for the loop perforation approach, no guarantee on the accuracy is provided.

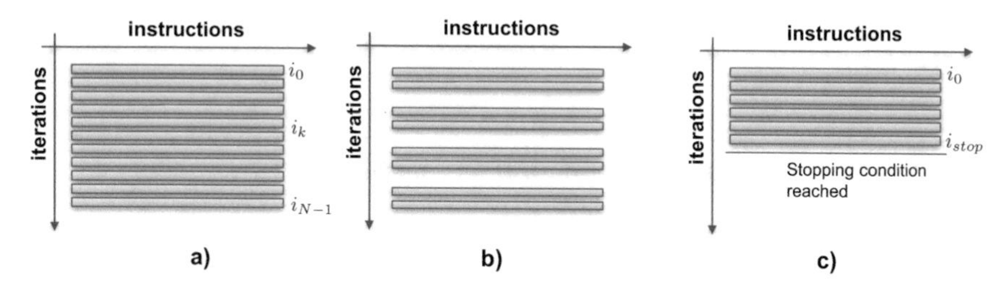

Fig. 5.4 Illustration of the difference between loop perforation and early termination for an iterative refinement algorithm. (**a**) Initial iterative refinement algorithm. (**b**) Loop perforation. (**c**) Early termination. The process is stopped when the condition criterion is reached

5.2.1.4 Task Skipping

The concept of task skipping has been used to discard a task (source code block) when an error or a fault occurs during the task processing [40]. This concept has been exploited in the approximate computing domain to skip a complete task. Compared to the techniques presented before, this skipping approach operates at a higher level of granularity by dropping a block of processing. To explore the trade-off between the implementation cost and the application output quality, a parameter called skip ratio defines the frequency of task skipping. In [51], the skipping ratio is adapted according to the task significance defined by the programmer. In [10] the concept of best effort computing is introduced, task skipping is considered to adapt the processing load to the hardware capabilities. The task skipping technique is used in [16] for *MapReduce* algorithm. Only a subset of randomly chosen map tasks are executed. A skipping ratio is defined by the user and allows adapting the accuracy-complexity trade-off.

Task skipping is considered in the methodology proposed by Nogues et al. in [36] to apply Approximate Computing (AC) at the level of the whole application. The application is modeled with a hierarchical block-based description. The first step selects the signal processing blocks that have the most important potential in terms of energy reduction by applying algorithmic-level approximate computing. The classification step, carried-out by the developer, identifies the class of each block, in order to determine which approximate computing technique can be applied. The profiling step reveals the complexity of each block with respect to global complexity. The second step checks for each signal processing block the potential benefits of approximate computing. Its objective is to explore the trade-offs between energy reduction and quality degradation. Firstly, the reduction of the implementation complexity is evaluated. The considered block is left unmodified if the potential gain appears to be negligible. Then, the quality degradation is evaluated from a set of simulations on a testbench. The aim of the third stage is to develop the optimized algorithm version and to evaluate or measure the real energy consumption reduction. In the proposed approach, task skipping is considered for blocks having inputs and outputs belonging to the same domain, i.e. inputs and outputs representing the same kind of data.

5.2.1.5 Skipping in Discrete Optimization Algorithm

In this section, the focus is put on algorithms that use discrete optimization techniques. When a close-form solution cannot be obtained for the optimization problem, the approach consists in exploring the optimization search space, enumerating the different solutions and selecting the optimal one. These algorithms are referred to as Optimization based on Search Space Exploration (OSSE) algorithms for which the main purpose is to minimize a cost function by exploring a search space. To decrease the implementation cost of OSSE algorithms, the challenge is to reduce the search space by skipping low value-added computation. As illustrated in Fig. 5.5, the studied OSSE algorithms consist of enumerating and testing different candidate solutions S_i ($i \in [1, n]$) to select the optimal one that minimizes the cost function. Numerous applications in the image and signal processing domain integrate OSSE algorithms. In telecommunications, channel decoding and MIMO decoding such as sphere decoding use OSSE algorithms. For example, in [17], authors use the properties of an OSSE algorithm to select the best transmission configuration including the modulation spectral efficiency and ECC code rate that maximizes the quality of a wireless received scalable video over MIMO channels. OSSE algorithms are also used in image processing applications for classification operations such as in the Nearest Neighbor classifier, and in video processing applications, for instance, in motion estimation [48]. In video coding, as depicted in the case-study presented in Sect. 5.3.2, OSSE algorithms are notably used for solving the Rate-Distortion Optimization (RDO) problem for the encoder.

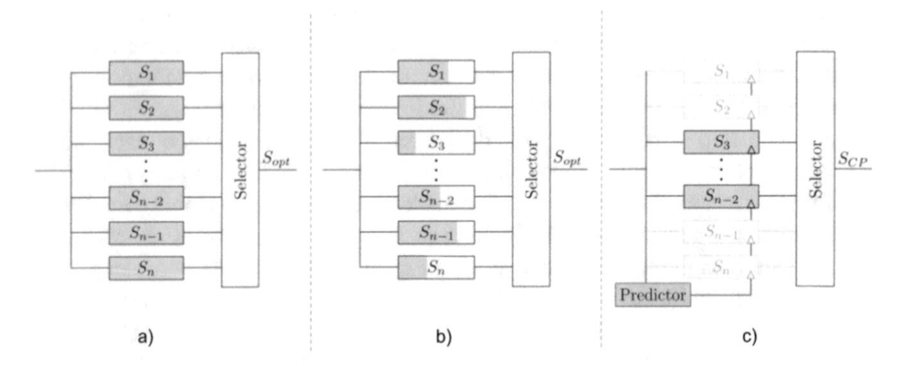

Fig. 5.5 Three techniques to handle OSSE algorithms. (**a**) Exhaustive search: each branch represents a full solution computation. (**b**) Early termination: the exploration of branches which cannot lead to the best solution is stopped before the end. (**c**) SSSR technique: coarse estimation is carried-out and a refinement is applied to the best candidates

An exhaustive search, presented in Fig. 5.5a, is a straightforward approach to process OSSE but it may require a lot of computation, depending on the search space size. A challenge for an OSSE algorithm implementation is to reduce the search space while minimizing the impact on the approximated optimal solution $\tilde{S}_{opt}$ compared to the optimal solution on the full search space S_{opt}. Ideally, the optimal solution must be contained in the reduced search space ($\tilde{S}_{opt} \subseteq_{opt}$). For OSSE algorithms, the search space reduction techniques can be classified into two categories: early termination and Smart Search Space Reduction (SSSR). At the beginning, the entire solution search space is considered. Then, based on intermediate results, the search space can be pruned by excluding the least likely solutions. Hence, parts of the search space which cannot lead to the optimal solution are removed, skipping unavoidable computation. When the discrete optimization problem can be formulated with a tree representing the different solutions S_i, the branch-and-bound technique can be used to efficiently explore the tree and thus reduce the search space. The exploration of the branch is stopped if the minimal cost, which can be obtained for the exploration of this branch, is higher than the best cost which has already been obtained during the exploration of the previous branches. The efficiency of this technique is based on the availability of a heuristic that quickly finds a good solution. This technique can guarantee that the optimal solution S_{opt} is found, even though the search space is pruned. However, the drawback of branch-and-bound techniques is the unpredictability of their execution time [17]. Thus, the gain in terms of energy cannot be predicted or adjusted with parameters controlling the approximation.

The SSSR techniques first select a subset of initial solutions in the search space, based on a coarse estimation of their cost, called prediction. Then a refinement of selected initial solutions is computed to find the best solution among them, as depicted in Fig. 5.5c. An efficient coarse solution predictor providing a good estimation with low computational complexity can improve the quality of such solutions. With SSSR techniques applied to energy minimization, the gain in terms of energy can be controlled by adjusting, at run-time, the search space around the coarse estimation. Nevertheless, optimality cannot be ensured. It depends on the accuracy of the prediction step. This second category of search space reduction techniques is investigated in the case-study presented in Sect. 5.3.2.

The SSSR design method proposed in [33] consists of three steps performed at design time to enable the approximation at run-time. The goal of the first step is to identify the OSSE in the application. Then, these OSSE are classified according to their potential cost gain. The third step consists of designing a predictor to manage the approximation and to reduce the cost of the OSSE according to an acceptable quality degradation. In the fourth step, a run-time management of the approximation is set-up.

5.2.2 Approximation Based Approaches

The alternative to skip a computation block f is to approximate f by replacing it with a less complex computation. The different approximation-based techniques are presented in this Section.

5.2.2.1 Algorithm Selection

The approximate computing technique named *Algorithm Selection* is based on the availability of several version of the same processing block. Each version has its own accuracy and implementation cost. The goal is to switch at run-time to the adequate version. The implementation cost can be decreased when opportunities to reduce the accuracy arise. This adaptation process can be linked to external parameters like, for example, the input data accuracy. More simply, the adaptation can be directly driven by an accuracy or energy target. The techniques presented in the previous sections such as loop perforation, task or operation skipping can be used to generate multiple versions of an application code with different trade-offs between accuracy and cost.

The PetaBricks language and compiler [3] is defined for application in which multiple implementations of multiple algorithms are considered. The programmer defines the different alternatives and PetaBricks automates algorithm selection and autotunes them. In [4], the Bin Packing benchmark has been tested with PetaBricks for different input data sizes and targeted accuracies. To solve the *Bin Packing* problem, 13 algorithm versions are considered. The results show that each of the 13 approximation algorithms perform fastest for some areas of the accuracy and input data size space.

5.2.2.2 Parameter Adjustment

During the design of data-oriented applications, the application parameters are optimized. For each parameter, the value leading to satisfactory complexity-accuracy trade-off is selected. These different parameters have a significant impact on the complexity but also on the quality of the application result. In Ludwig et al. [27], energy consumption is optimized by adjusting the parameter of digital filters. The filter order is dynamically adjusted to reduce the energy consumption which is proportional to the filter order. The filter order is adapted according to the characteristics of the input signal. A lightweight technique is used to measure at run-time the strength of the stopband component (the frequency component to remove) in the input signal. When the stopband component strength is high, the order of the filter is increased to rise the filter stopband attenuation. The filter is designed such as the order increase allows reinforcing the attenuation of the stopband.

In the HEVC decoder use-case, presented in Sect. 5.3.1, the original Motion Compensation (MC) filters are redesigned to provide, for each one, different versions having different number of taps lower than the number of taps of the original filter. The reduction of the number of taps allows reducing the filter complexity and thus its energy consumption. At run-time, the filter having a number of taps compatible with the considered target energy is selected. This approach combines the parameter adjustment and algorithm selection techniques to explore the trade-off between accuracy and complexity.

For digital filters like Finite Impulse Response (FIR), this technique can be considered similar as loop perforation. Indeed, the number of iterations of the loop describing the filter is reduced. But, this reduction comes with a redesign of the filter and not only the discard of some taps. Thus, new coefficients for the filter are defined.

5.2.2.3 Memoization

Memoization is the principle of saving the result of a function execution or computation in a Look-Up Table (LUT) stored in memory, so as to use it for future executions. This technique has first been proposed to reduce the execution time for a computation already done. The main motivation to implement memoization techniques is to remove redundancy due to the repetition of the same input data for a complex computation. The principle of memoization is particularly useful for complex processing for which the search in the LUT of an existing result for the considered input values leads to a significantly lower cost than computing this complex processing. For example, when a function f having x and y as input parameter has to be evaluated, the input operands x and y are used to access the LUT storing the previously executed computation, called the reuse table. If the computation has already been executed with the considered values of x and y, this is a hit, and the result is extracted from the LUT. In the case of a miss. The instruction is executed and its result is stored in the LUT. As Arjun et al. explained in [47], to apply memoization in a program, two conditions have to be satisfied. Firstly, the memoized code has to be transparent to the rest of the code, that is to say that it should not cause any side-effect. Secondly, for the same input, the memoization and original code must produce identical output.

According to these two conditions, no approximation lies in memoization. Besides, the gain brought by memoization may be annihilated by the LUT memory footprint. To further improve the performance in terms of execution time, value locality can be considered. Fuzzy memoization has been proposed and applied for floating-point operations by Alvarez et al. [1]. Indeed, floating-point numbers offer a high dynamic range and imply the need for large LUTs to achieve an acceptable hit rate when accessing the table. Contrary to classical memoization, when implementing fuzzy memoization, before accessing the LUT, N Least Significant Bits (LSBs) of the input operands x and y are dropped. A masking operation is applied before accessing the LUT, which implies that operands with similar Most Significant Bits (MSBs), despite being strictly different, will be affected to the same compartment of the LUT. Fuzzy memoization is particularly well tolerated in multimedia applications, where the end-user generally tolerates errors. The number N of dropped bits can be used to trade-off the output quality and the computation time or energy consumption of the targeted application.

5.2.2.4 Neural Network Approximation

The aim of this approximate computing technique is to replace a complex processing P by a neural network. This technique will be efficient if the cost of the neural network is significantly lower than the cost of P. Neural networks exhibit interesting properties like a high degree of parallelism or resilience to approximation errors. Since a decade, a tremendous amount of research work has been achieved on neural networks, leading to very efficient implementation. Dedicated low power and low cost HW accelerators have been proposed. A wide range of research focus on reducing the neural network complexity with different approaches have been provided. Some come from the approximate computing domains like precision refinement. Neural network structure and the hyperparameters can be tuned to explore the cost/accuracy trade-off. Increasing the number of layers will decrease the approximation error but at the expanse of a cost increase. This research topic is considered in Chap. 15.

In [14], the algorithmic Parrot transformation that allows replacing a code region P of a program by a neural networks is presented. The first step is to detect portion of code that can be approximated by a neural network. Secondly, a training process is carried-out to mimic the behavior of the code region P with the neural network. Thirdly, the neural network is implemented in the neural processing unit (NPU).

In [28], EMEURO, a neural network based emulation and acceleration platform is presented. This approach aims at detecting in an algorithm, portion of code which can be approximated by neural network. The portion of code is restructured in order to have the same data flow as a neural network. Compared to the previous approach, instead of approximating a complex processing with a single neural network, the portion is smaller and a two-layer linear neural network is used. This latter benefits from the availability of libraries including highly optimized versions of these neural networks.

Like in the techniques presented before, the approach proposed in [2] allows approximating a portion of code with neural network. But, the difference is that a analog neural network is used in order to further reduce the energy consumption. On the other hand, the use of analog implementation leads to new challenges like the low precision, limited dynamic range, conversion between analog and digital and temporary result storage.

In [13], neural networks are used to approximate transcendental functions. Multi Layer Perceptron (MLP) neural networks are used to approximate the function on a limited input range. Then, mathematical identities are exploited to evaluate the function with any input value.

5.2.2.5 Mathematical Function Approximation

In this section, the focus is on the approximation of mathematical functions. Data-oriented applications in numerous domains, like signal and image processing, telecommunications, robotics use more and more complex mathematical processing. Especially, these mathematical processing steps integrate complex mathematical functions. The challenge is to implement these mathematical functions with enough accuracy without sacrificing the performances of the application, namely memory usage, execution time and energy consumption. In the context of scientific computation, mathematical libraries like *libm* [24] are available and allow evaluating the different elementary functions. This library provides very accurate approximation of these mathematical functions for floating-point numbers required by scientific applications. Even if computer performances increased over the past decades, this high accuracy is done at the expense of hardware costs, execution time, energy consumption or memory footprint. Such costs can be unsuitable for real-time embedded applications and the proposed accuracy is oversized for most embedded applications. Moreover, this library is generic and the evaluation of a complex function composed of basic functions requires to call the code for each basic function. For embedded applications, the efficient evaluation of a complex function f requires to design a specific source code or HW block dedicated to this function and the considered input range.

Several solutions can be used to compute an approximate value of a mathematical function f over a segment I, according to a maximum error value ϵ. They can be classified in three categories. The first one groups together iterative approaches using shift-and-add algorithms, the second one corresponds to table-based methods, and the third one approximate the function with polynomials.

Iterative Approaches Specific algorithms can be adapted to a particular function [39]. Iterative methods as the shift-and-add BKM algorithm [5] or the CORDIC algorithm [29] are generally easy to implement. For instance, the CORDIC algorithm computes approximate values of trigonometric, logarithmic or hyperbolic functions. To compute the tangency of an angle θ, the principle of the algorithm is to apply successive rotations to a vector $\mathbf{v}$ whose initial coordinates are $(1, 0)$ and final coordinates (X, Y). Indeed, to rotate a vector whose coordinates are (x_{in}, y_{in}) from an angle θ, the operation applied to compute the coordinates of the resulting vector is:

$$\begin{bmatrix} x_o \\ y_o \end{bmatrix} = \cos\theta \begin{bmatrix} 1 & -\tan\theta \\ \tan\theta & 1 \end{bmatrix} \cdot \begin{bmatrix} x_{in} \\ y_{in} \end{bmatrix} \tag{5.3}$$

Nevertheless, to obtain an efficient implementation of the CORDIC algorithm on low cost hardware, the multiplications have to be avoided. To do so, instead of applying a single rotation of θ, several rotations of small angles θ_i are applied, such that $\theta \simeq \sum_{i=0}^{n} \theta_i$. Besides, in the efficient hardware implementation of the CORDIC algorithm, the values of $\tan\theta_i$ are taken equal to 2^{-i} to replace the multiply operations by shifts. Finally, the obtained values of x_o and y_o are equal to $\cos\theta$ and $\sin\theta$, respectively. The accuracy of the CORDIC algorithm is strongly dependent on the number of iterations.

Table-Based Techniques The simplest approach to approximate a function f on an input range I is to tabulate the function in a table T [34]. The n Most Significant Bit (MSB)s of the input x are used to address this table T. This approach is equivalent to segment the input range I in 2^n sub-intervals and to approximate for each sub-intervals I_i the function f by a constant c_i, i.e. a 0-order polynomial. This constant c_i is chosen such as it minimizes the absolute error ϵ with all the values of $f(x)$ for x belonging in the interval I_i. The interval I on which the function has to be evaluated is segmented until the absolute error ϵ is lower than the maximal acceptable error value $\epsilon_{\max}$ on each sub-interval. This type of segmentation has to be uniform: if the error criterion is not fulfilled on a single sub-interval I_s, all the sub-interval of I have to be segmented again. The approximation error depends on the function f characteristics and the table size. Thus, reasonable table size leads to low accuracy. This table-based method consumes the most memory space but is the most efficient in terms of computation time.

To reduce the table size, bi-partite methods have been proposed and generalized for any function in [43] by Schulte. The input value x is decomposed into three groups of bits representing a value x_i. A first-order Taylor decomposition is used to approximate the function f with a linear function [34]. Two tables are used to store the coefficients of this first-order polynomial. De Dinechin and Tisserand [12] detailed improvements of the bi-partite method called multi-partite methods in which several smaller tables are used. The initial values of each segment as well as the values of the offsets to add to these initial values to get whichever value in a segment have to be saved in tables. The size of these tables is then reduced compared to bi-partite table methods exploiting symmetry on each segment. That method allows quick computations and reduced tables to store but is limited to low-precision approximation. This method is efficient for hardware implementation.

Polynomial Approximation Polynomial approximation is a good alternative for function evaluation, especially when several elementary functions are combined. Tools like Sollya [11] provide the polynomial coefficients to approximate a function f on an interval I for a predefined polynomial order. For fixed-point arithmetic, polynomial approximation can give very accurate results for a low implementation cost if the interval I is segmented finely enough. That is to say, a polynomial P_i approximates the function f on each segment I_i of I. The segmentation is required so as to approximate the function f according to a maximum error of approximation $\epsilon_{\max}$. The polynomial order is then a trade-off between the approximation error and the segment size. To obtain a given maximum approximation error, the decrease in the polynomial order implies the reduction of the segment size. This increases the number of polynomials to store in memory. For a given data-path word-length, the increase in polynomial order raises the fixed-point computation errors and annihilates the benefit of lower approximation error obtained by a too high polynomial order. Thus, for fixed-point arithmetic, the polynomial order must be relatively low. Consequently to obtain a low maximal approximation error, the segment size is reduced which is at the expanse of the number of polynomials to store in memory.

Three steps are required to evaluate a function f with this polynomial approximation including a uniform segmentation of the initial interval I. The first steps aims at finding the index of the polynomial p_i associated with the sub-interval I_i in which the value x belongs to. This index is obtained by analyzing the value of the MSB of the input x. This approach allows obtaining easily the polynomial index, but it requires that the interval boundaries values correspond to a power of two. In the second step, the coefficients of the polynomials p_i are loaded from the table storing the different polynomial coefficients. In the third step, the polynomial p_i is evaluated. Most of the time, the Horner scheme is used for the polynomial evaluation in order to limit the computation errors.

To limit the number of polynomials p_i stored in memory, non-uniform segmentation can be considered to adapt each sub-interval size. Thus, the challenge is to find the accurate segmentation of the interval I. Lee et al. have proposed different non-uniform segmentations [23] for hardware function evaluation. On each sub-interval, the function f is approximated by the Remez algorithm. Afterwards, a simple logic circuit is used to find the segment corresponding to an input value x. LUTs are used to store the coefficients of the polynomials. For software function evaluation, Bonnot et al. [8] proposed a non-uniform segmentation technique. The first step of this method consists in finding the optimal non-uniform segmentation for approximating the function f on the interval I. Each sub-interval is divided in two as long as the approximation error criterion ϵ_{max} is not satisfied. The non-uniform segmentation is then stored in a tree structure T. Each node of the tree structure represents a sub-interval on which the approximation error criterion is not satisfied, while the leaves are sub-intervals where the approximation error criterion ϵ_{max} is fulfilled. Consequently, an approximating polynomial p_i is associated with each leaf. The coefficients of the different polynomials as well as required shifts to compute the value $P_i(x) \simeq f(x)$ with input $x \in I$ are stored in tables.

The challenge of the proposed method, is to efficiently access the approximating polynomial p_i depending on the input value x. To be efficient, the MSBs of the input x are analyzed sequentially by group of bits to traverse the tree T and to access to the leaf associated with the considered sub-interval and the approximating polynomial p_i. In this method, the maximum error of approximation is used as an user-defined parameter, and has an impact on the memory footprint of the system as well as the computation time.

Multivariate Function Most of the research works consider univariate functions. For multivariate functions, significantly fewer methods have been proposed. Indeed, for scientific computation, these functions are decomposed as a sequence of univariate standard functions. The evaluation of multivariate functions is used in various domains. As an example, in [30], multivariate function approximation is used in the Direction Cosine Matrix update algorithm to enable inertial navigation. In the context of control system, Instrumental Variable approach integrates bivariate function evaluation to identify some system parameters [19]. In the context of high performance wireless communication like 5G, QR-decomposition are required for channel pre-coding. Bivariate, non-linear functions are used for efficient computation of the Givens-Rotation [42].

In [41], the authors propose a linear approximation of bivariate functions. The function f is approximated by a bivariate polynomial of degree 1. The coefficients of this polynomial are computed through the Minimum Mean Square Error (MMSE) method. A bivariate non-uniform segmentation similar to the univariate case in [8] is performed. If the error criterion is not satisfied, the original set is split into four equal squared subsets. When the recursive process is terminated, a set merging step is applied. A merging between two sets is performed when they are neighbors, i.e. they have a complete side in common, and the approximation created by the mean of their coefficients satisfies the error criterion on the whole set. Similarly in [26] a piecewise-linear approximation is proposed for multivariate continuous non-linear functions. Genetic algorithms are used to decompose the initial set. In [38], polynomial approximation for software implementation of bivariate functions is proposed. This approach is based on the Mean Square Minimization method which is appropriate for the majority of approximation problems. A smart non-uniform segmentation is proposed to better fit irregularities of the approximated function. Furthermore, to improve the performances, the degree of the bivariate approximating polynomial is adapted for each subset which avoid data and time waste. The proposed approach is composed of two elements. The first element is the technique proposed to determine the polynomials used to approximate the desired function f and to decompose the initial set into subsets in order to satisfy the maximal approximation error criterion. The second element is an optimized C source code which implements this approximation technique targeting real-time embedded systems.

5.3 Algorithmic-Level Approximate Computing for Video Codec

In this section, the HEVC video codec is considered as a use-case to illustrate the use of the approximate computing techniques at the algorithmic level presented before. In Sect. 5.3.1, task skipping, algorithm selection and parameter adjustment techniques are exploited to optimize an HEVC decoder. In Sect. 5.3.2, the skipping technique for discrete optimization algorithm is illustrated in the case of HEVC encoder.

5.3.1 HEVC Decoder Use-Case

The standard structure of an HEVC decoder is depicted in Fig. 5.6. The classification presented in [36] is used to decompose the target application into two types of blocks: blocks producing control-oriented data and blocks producing signal-oriented data. A block that generates signal-oriented data can be of two types depending on whether the input and output data are in the same domain, i.e. represent the same type of information (for instance, data or frequency information). If the input and output data are not in the same domain, the signal processing block carries out a domain transformation. This kind of block can be approximated but not skipped. In the opposite case, the block is in charge of enhancing specific characteristics of the signal. This kind of domain conservation block can be skipped, approximated or both.

When receiving compressed data, the entropy decoder first extracts the different syntax elements from the video stream using arithmetic coding. The entropy decoding block is a hard control block, as each decoded value controls the type of processing executed on the data stream. Any approximation on its result would potentially break the downstream execution pipeline because wrong headers for a sequence, picture, or block could be inferred. Then the residual data are dequantized and transformed using an inverse Discrete Cosine Transform (DCT)-like process, resulting in the error of the block intra or inter prediction [46]. The inverse transform block performs a DCT-like transformation and thus is a (signal-oriented) domain transformation block. The prediction of the pixel blocks is then applied and can be either of intra- or inter-frame type depending on input bitstream parameters. Intra-prediction is also a signal-oriented block consisting in replicating pixels in a given direction so as to predict a new block. Intra/inter mode selection is a simple switch that is control-oriented and requires very few resources. In the case of inter-frame prediction, a prediction is computed based on the previously decoded pictures. Relative translation motion vectors are transmitted in the bitstream and these motion vectors have a fractional pixel resolution. Finally the Deblocking Filter (DF) and Sample-Adaptive Offset filter (SAO) filters are applied on the reconstructed data to reduce potential artifacts and increase subjective image quality. In-loop filters and

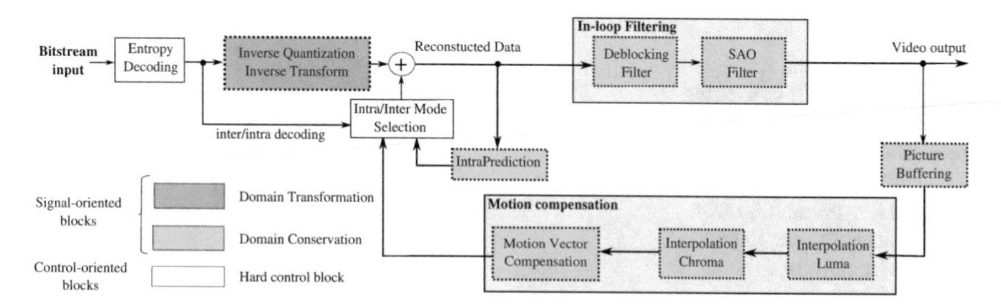

Fig. 5.6 Algorithm classification of an HEVC decoder

motion compensation blocks are all signal-oriented filter blocks. Picture buffering corresponds to a queue of images retained for display and future picture predictions.

5.3.1.1 HEVC Decoder Block Selection

Blocks are considered for approximation in decreasing order of computational complexity. The profiling of the HEVC decoding process is given in Fig. 5.7.

Motion Compensation Filters In video codecs, a motion estimation technique is used by the video encoder to generate a compressed representation of a video by exploiting the temporal redundancy between the frames of the transmitted video sequence. A motion vector is defined for each block in the picture as the relative position of the predicted block with respect to the reference block in a previously decoded picture. However, the true movements of the blocks are not perfect translations and cannot perfectly match the sampling rate of the digitalized video. Therefore, a fractional precision is used for the motion vectors to reduce the prediction residual error and improve the compression performance. If a motion vector has a fractional value, the reference block needs to be interpolated accordingly to generate the prediction.

In the HEVC standard, the fractional motion vector compensation is performed by two separable 1-D interpolation filters for the horizontal and vertical directions [46]. Representing 74% of the decoding effort and being signal-oriented domain conservation blocks, these filters could be skipped [35, 37] and should be tackled first in the algorithmic-level approximate computing method.

In-loop Filters The in-loop filters are composed of two entities: the deblocking filter, already present in the previous JCT-VC H.264/AVC standard [52] and the SAO filter that has been newly adopted in HEVC. The in-loop filters have the particularity of processing decoded frames and can therefore be considered as enhancement functions to improve the decoding quality.

The SAO filter is used to reduce sample distortion by classifying reconstructed samples into categories, obtaining an offset for each category, and then adding that offset to the sample value [15].

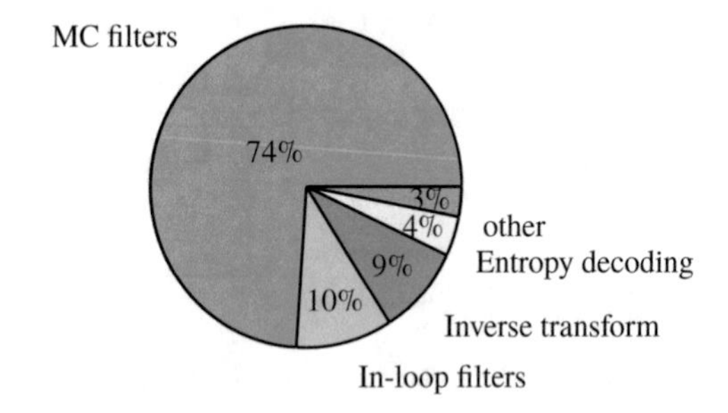

Fig. 5.7 Average relative complexity of data processing blocks in an optimized HEVC decoder on a general purpose processor. Input sequences: Kimono RA 1920x1080 with Quantization Parameter from 22 to 32 by steps of 2

The deblocking filter processes decoded frames to reduce the artifacts generated by inaccurate predictions. A blocking effect may appear between blocks because of the block transform used for residual coding. This blocking effect may be more visible near block boundaries. The deblocking filter enhances the decoded frames global quality by smoothing the transitions between blocks.

In-loop filters represent 10% of the decoding effort and they are signal-oriented domain conservation blocks, these filters need to be addressed second in the algorithmic-level approximate computing method.

The next block in terms of computational complexity is the inverse transform block. It is a domain transformation block, for which approximation can spread over a large set of output data. As the 2 first candidates already represent 83.3% of the global decoding complexity, the next blocks, including inverse transform block, have not been considered for approximation.

5.3.1.2 HEVC Decoder Block Transformations

Block Class Modification of the Motion Compensation Filters The first blocks to approximate are the MC filters as they represent the highest share of the processing load. The first proposal is based on a reduction of the filtering complexity which is a block class modification. The modification is described in Fig. 5.8 where the legacy filters are replaced by approximate filters. An *approximation level control* signal chooses the filter complexity.

The HEVC legacy interpolation filters are analyzed in details in [21]. The HEVC standard uses FIR filters to perform the luminance and chrominance interpolation. Tseng et al. [50] have proposed the design methodology adopted by the HEVC standard. Whereas the actual standard proposes a fixed configuration for the filters, the proposed solution is based on a wide range of filters from the minimum size to the legacy size. In HEVC [46], the standard filter size is fixed to $N_h = 8$ for the luminance component and $N_h = 4$ for the chrominance components. Five categories of *approximate computing* filters are defined, adapting their computational complexity: low, middle, intermediate, high and legacy, respectively, with $N_h = 1, 3, 5, 7, 8$ for luminance and with $N_h = 1, 2, 3, 4, 4$ for chrominance.

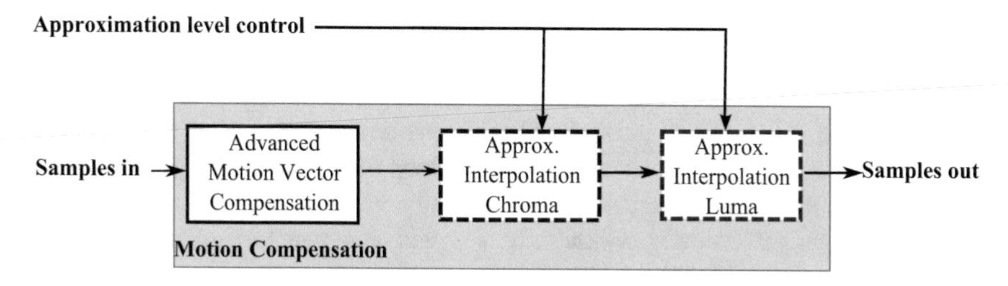

Fig. 5.8 Interpolation filter of the MC block approximation

Table 5.1 Filter size per configuration

Configuration	Chrominance filter size	Luminance filter size
Low	1	1
Middle	2	3
Intermediate	3	7
High	4	7

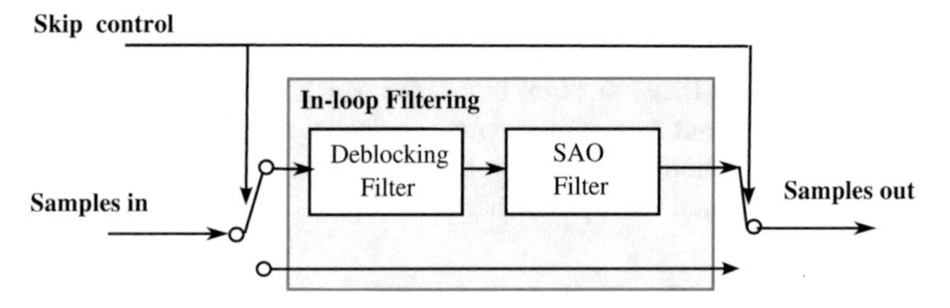

Fig. 5.9 In-loop filter skip

The overall approximation process is controlled by a parameter called *Approximation level control* as described in Fig. 5.8. It sets the number of taps of the MC interpolation filters. The different filter categories: low, middle, intermediate, and high and summarized in Table 5.1.

Computation Skipping of the Motion Compensation Filters and In-loop Filters Another alternative for signal-oriented blocks with domain conservation is to skip processing. Because the error may be high, the block can be either totally skipped or skipped only periodically. A balance must be struck between quality and computational complexity. A parameter called *skip control* is proposed that dynamically activates the processing block. All processing blocks that are classified as *signal-oriented* blocks with domain conservation can be modified with this approximate computing approach, namely the in-loop filters and the MC interpolation filters. Figure 5.9 shows an example of such implementation of the in-loop filters.

The *skip control* parameter provides tuning of the video distortion at the decoder side. A decision is taken at the frame level to activate or not the filters, providing a coarse grain tuning parameter on decoding quality. In an approximate computing system, this tuning capability is used to stay in the acceptable area of the application quality for a given use-case. The frequency of block skipping is set as a percentage of frames where filters are skipped. It leads to a fine quantum of quality distortion. By setting the *skip control* parameter to 0%, the decoder is similar to the legacy HEVC. If the *skip control* is set to 100%, the in-loop filters and the MC interpolation filters of chrominance and luminance are skipped permanently.

Four $skipcontrol$ configurations are selected and categorized as shown in Table 5.2. These configurations are used in the rest of the study.

Table 5.2 Percentage of block skipping per configuration

Configuration	Block skipping percentage
Low	89%
Middle	63%
Intermediate	25%
High	8%

5.3.1.3 Experiments and Results

Power measurements are conducted on an octa-core Exynos 5410 SoC based on the big.LITTLE configuration with four ARM Cortex-A15 cores and four ARM Cortex-A7 cores. The CPU has a maximum clock frequency of 1600 MHz and its frequency can be scaled down to 250 MHz. The software HEVC decoder is *OpenHEVC* [49]. It runs real-time on top of a standard Linux kernel which uses automatic CPU cluster switching in the kernel. The input HEVC bitstreams are taken from the standardization reference one [6, 9].

For the quality metric, the Structural Similarity Index Measure (SSIM) is considered. This metric is more appropriate as its outputs are closer to the human perception than the classical Peak Signal to Noise Ration (PSNR) metric [18]. The aim is to obtain a SSIM as close as possible to the value 1, indicating that the degradation between the decoded frame and the original one is very low.

5.3.1.4 Quality: Energy Consumption Trade-Off

In Figs. 5.10 and 5.11, the energy scalability of QP tuning is compared to the proposed *computation skip* and *computation approximation* methods. The results show that increasing the QP reduces power consumption. The proposed approximate computing methods offer an alternative to reduce power consumption while degrading the quality. In the scenarios where the device can choose the QP, the proposed methods are shown to perform better than modifying the QP. For example, and assuming a reference video at $QP = 27$, changing to $QP = 32$ can offer a reduction of 1 W with a SSIM degraded from 0.84 to 0.8 (Fig. 5.10). The decoder using *computation skip* can also achieve 1 W of power reduction but with an SSIM reduced to only 0.82. With the decoder based on *computation approximation*, the SSIM can be only degraded to 0.83.

The maximum energy gains that are obtained on the test sequences is close to 40%.

Besides, in broadcast scenarios where only one bitrate and QP configuration is usually available, the proposed methods are the only ones that can be envisioned to save power in energy-limited devices. For finer grain tuning, the two techniques can be combined together.

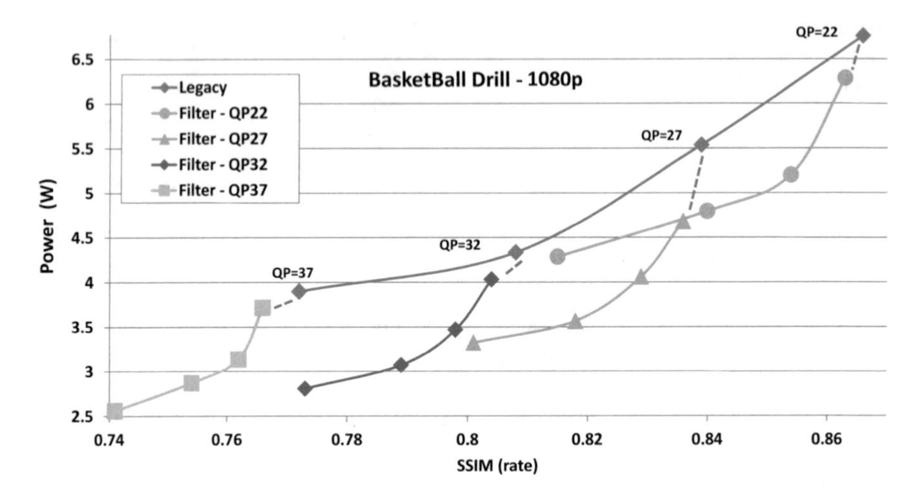

Fig. 5.10 Power consumption of *computation approximation* decoder vs SSIM distortion

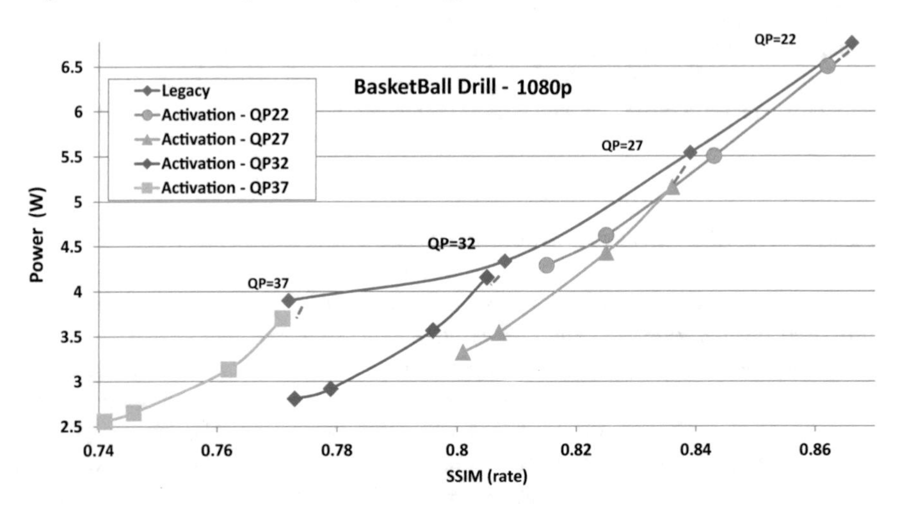

Fig. 5.11 Power consumption of *computation skip* decoder vs SSIM distortion

5.3.2 HEVC Encoder Use-Case

In this section, the skipping technique for discrete optimization algorithm is illustrated in the case of HEVC encoder. An HEVC encoder is classically based on a *hybrid video encoder* structure that combines Inter and Intra-predictions. The solution described in this section focuses only on Intra encoding, Fig. 5.12 illustrates the block diagram of an HEVC intra encoder. While encoding in HEVC, each frame is split into equally sized blocks named Coding Tree Units (CTUs) (Fig. 5.13). Each Coding Tree Unit (CTU) is then divided into Coding Unit (CU), themselves nodes in

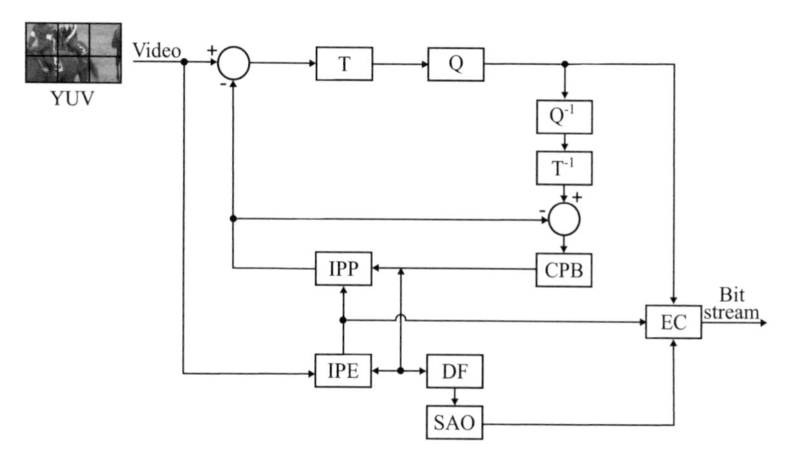

Fig. 5.12 Block diagram of HEVC intra encoder composed by several blocks: Intra Picture Process (IPP), Intra Picture Estimation (IPE), Transform (T), Quantization (Q), Inverse Quantization (Q^{-1}), Inverse Transform (T_{-1}), Current Picture Buffer (CPB), Deblocking Filter (DF), Sample-Adaptive Offset (SAO) and Entropy Coding (EC)

a quad-tree. In HEVC, the size of CU is equal to $2N \times 2N$ with $N \in \{32, 16, 8, 4\}$. The HEVC encoder starts by predicting the blocks from their environment (in time and space). To perform the predictions, CU may be split into Prediction Block (PB) of smaller size.

In intra-prediction mode, PB are square and may take the size of $2N \times 2N$ (or $N \times N$ only when $N = 4$). The HEVC intra-frame prediction is complex and supports a total of 35 modes (illustrated on Fig. 5.14) performed at the level of PB including planar (surface fitting) mode, DC (flat) mode and 33 angular modes [46]. Figure 5.14 shows an example of an intra-prediction with $N \times N$ PB size of 8×8 and the intra-prediction modes. After computing this prediction, the encoder calculates the residuals (prediction error) by subtracting the prediction from the original samples. The residual is then transformed by a linear spatial transform, quantized, and finally entropy coded.

The HEVC encoder also contains a decoder processing loop since the decoded picture is required by the encoder to perform Intra and Inter predictions. This decoder loop is composed of inverse quantization and inverse transform steps that reconstruct the residual information (i.e. the error of the prediction). The residuals are added to the predicted samples to generate a decoded picture (also called reconstructed samples). In the case of Intra encoding, reconstructed samples are stored in the current picture buffer and used for predicting future blocks. Finally, reconstructed samples are post-processed by a deblocking filter and a SAO filter (used for Inter prediction) that generates the parameters of the decoding filter and appends them to the bitstream. To achieve the best Rate-Distortion (RD) performance, the encoder performs an exhaustive search process, named RDO,

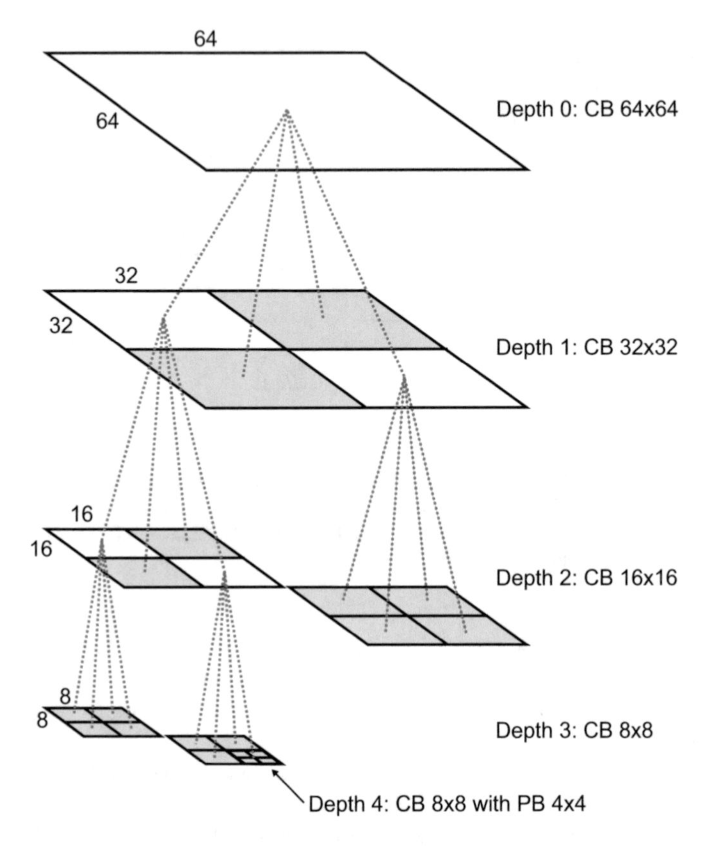

Fig. 5.13 Quad-tree structure of a Coding Tree Block (CTB) divided into Coding Block (CB)

testing every possible combination of partitioning structures combined with the 35 Intra-prediction modes. This exhaustive search constitutes an OSSE algorithm.

In order to decrease the computational complexity of HEVC Intra encoding, a fast intra-mode decision called Rough Mode Decision (RMD) [53, 55] was added in the reference software HEVC test Model (HM) [20]. This technique splits the Intra-prediction process into two successive steps: RMD and RDO as illustrated in Fig. 5.15. RMD consists in constructing a candidate mode list which is then tested in the full RDO process. RMD method computes for each mode m a cost $J_{RMD}(m)$. The N_m modes with the lowest costs $J_{RMD}(m)$ are then evaluated by the full RDO process to select the best among them. N_m depends on the CU size N. The RDO step is much more complex than the RMD step. As the RMD step orders the modes according to their costs, the RDO step can be skipped to limit the encoding complexity. In this work, only the RMD step is applied and the mode $m_{\hat{s}}$ with the smallest cost $J_{RMD}(m)$ is selected.

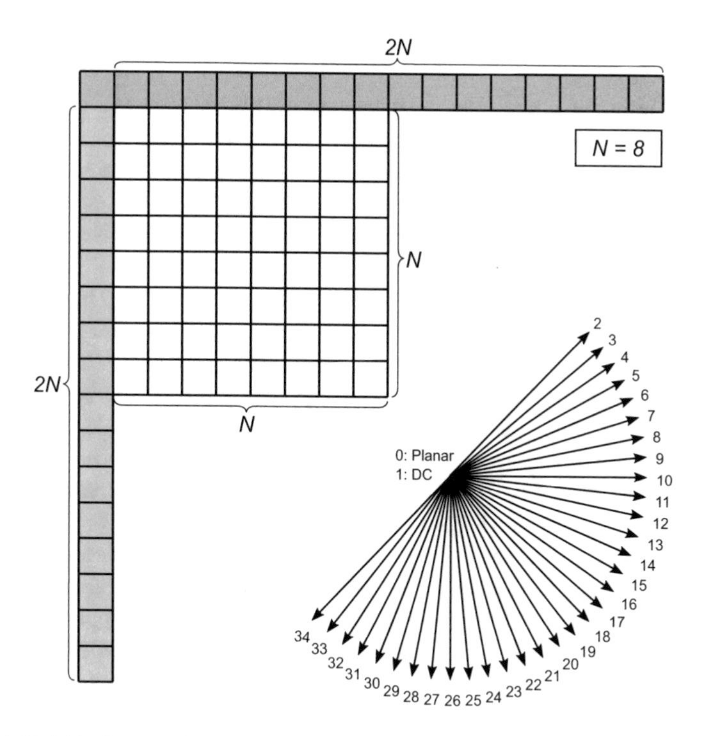

Fig. 5.14 Neighboring samples used for intra-prediction in an $N \times N$ PB with $N = 8$ and intra-prediction modes

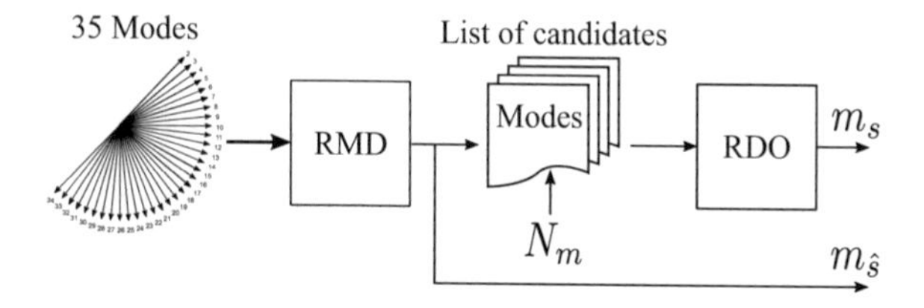

Fig. 5.15 Intra-prediction steps

5.3.2.1 Experimental Set-up

All experimentations are performed on one core of the embedded *EmETXe-i87M0* platform from *Arbor Technologies* based on an Intel Core i5-4402E processor at 1.6 GHz. The energy consumption is used as the optimized cost. To measure the energy consumed by the platform, Intel Running Average Power Limit (RAPL) interfaces are used to get the energy of the CPU package, which includes cores, IOs, DRAM, and integrated graphic chipset. Bjøntegaard Delta Bit Rate (BD-BR) [52] is commonly used in video compression to measure the compression efficiency difference between two encodings. The BD-BR reports the average bit rate difference in percent for two encoding at the same quality: PSNR. The aim is to obtain the lowest BD-BR.

5.3.2.2 OSSE Algorithm Identification and Classification

In HEVC Intra encoding, the selections of RD-wise best PB size and Intra-prediction mode are determined by the RDO process. The RDO process is composed of two nested OSSE: Coding-tree partitioning (CT) and Intra-mode prediction (IM). CT aims at finding the best quad-tree decomposition of a CTU of 64x64 pixels into CU as illustrated in Fig. 5.13. Then, for all Coding Units (CUs), IM aims at finding the best mode to predict blocks from its neighbors.

An energy metric is used to classify and evaluate the OSSE. Theoretical lower bound of the energy consumption are defined and named MCP(CT) and MCP(IM) for the two OSSE of the RDO process: respectively, CT and IM. The MCP is the energy obtained when the encoder is able to perfectly predict the best partitioning solution and thus only the optimal solution is processed to encode the CTU [32]. Therefore, the energy consumption of the search process is reduced to the energy consumption of the solution and the MCP is the minimal energy consumption point that can be achieved for the highest encoding quality.

Table 5.3 summarizes the energy reduction opportunities between optimal (best complexity case) and full search (worst case) solutions at different video resolutions. The results are extracted from [32]. They are obtained by applying the two-pass approach as defined in the OSSE Classification Step presented in [33]. The results show that the search space is similar across all resolutions and the largest energy reduction search space occurs when optimizing the *Coding-tree partitioning*, with up to 76.3% of potential energy reduction while working on the *Intra-mode prediction* offers 27.9% at best. The results lead to the conclusion that the energy problematic can be more efficiently addressed by reducing complexity at the *Coding-tree partitioning*.

Table 5.3 Energy reduction opportunities (in J) [32]

		Energy for Minimal Cost Point (MCP)			Reduction (in %)		
Res.	Energy for exhaustive search			IM	CT	IM	CT
2k	9710	7438	3398	2272	6311	23.4	65.0
1080p	4813	3663	1560	1150	3253	23.9	67.6
720p	2204	1722	911	483	1294	21.9	58.7
480p	1120	833	317	287	803	25.6	71.7
240p	291	209	69	81	222	27.9	76.3
				Average		24.5	67.9

5.3.2.3 Coding-Tree Partitioning OSSE Approximation

Coarse Solution Predictor Design The coarse solution predictor aims at predicting the coding-tree partitioning from video frame content. Authors of [22, 44] show the relationship between CU size and the corresponding block variance of the image. Based on this observation, they propose a variance-aware coding-tree prediction. The energy reduction technique used in this paper follows a similar algorithm. A video sequence is split into equal Groups of Frames (GOF) of size F. The first frame of a Group of Frames (GOF) is encoded with a full RDO process (unconstrained in terms of energy). Then the variance of the selected CU according to their sizes are used to compute variance thresholds *on-the-fly*. For following frames of the GOF, the variance of each CU of each size is recursively compared to the thresholds to choose if the CU has to be split. The coding-tree partitioning is built by this process.

Approximation Management The first parameter that impacts the encoding quality and energy consumption is the number of frames F in the GOF. The second parameter N_d defines the number of depth values tested around the prediction for each constrained CTU [31]. Since applying the RDO process on the predicted depth map is the result of a coarse estimation, it is possible, without compromising too much the complexity, to improve the process by exploring more depths around the predicted optimum.

Since video encoding is time consuming, a fast quality evaluation approach with a restricted parameter set is used to extract the configurations close to the Pareto front.

Quality & Cost Evaluation The Rate-Energy space of all the combinations of parameters for F and N_d has been explored. Results lead to a significant gap in term of BD-BR for $N_d = 1$ and $N_d = 2$. This observation requires to refine N_d and to use non-integer values. To explore non-integer numbers of depths, CTU in a constrained frame are split into two categories [31]: $(N_d - \lfloor N_d \rfloor) \times 100$ per cent of CTU are encoded with $\lceil N_d \rceil$ depths and the rest with $\lfloor N_d \rfloor$ depths.

Figure 5.16 shows the Rate-Energy space for all the combinations of parameters. This figure shows that for normalized energy reduction of up to 60% (below 40% in Fig. 5.16), the points of the Pareto front are generated with a high value of F and a low value of N_d. On the other hand, for normalized energy reductions of less than 40% (higher than 60% in Fig. 5.16) the configurations are obtained with $F = 2$ and a high value of N_d. The encoder has to play on both F and N_d parameters, respectively, the size of the GOF and the number of explored depths to control the energy consumption of the HEVC encoder.

5.3.2.4 Intra-Mode Prediction OSSE Approximation

Coarse Solution Predictor Design The Kvazaar encoder includes a feature that reduces the computational complexity of RMD. This feature reduces the number of

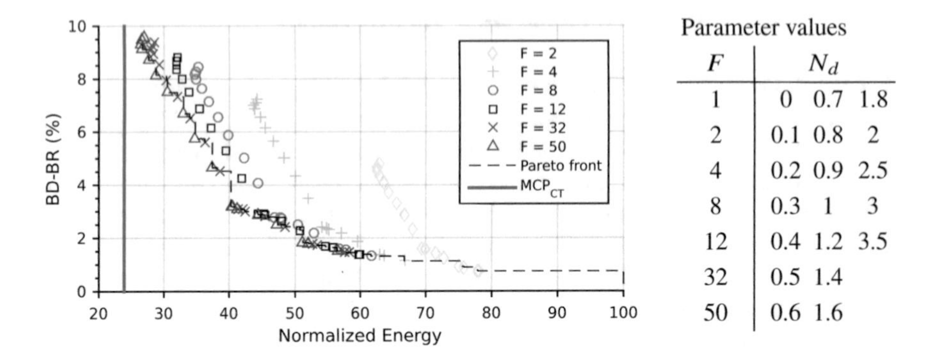

Parameter values

F	N_d		
1	0	0.7	1.8
2	0.1	0.8	2
4	0.2	0.9	2.5
8	0.3	1	3
12	0.4	1.2	3.5
32	0.5	1.4	
50	0.6	1.6	

Fig. 5.16 Pareto in Rate-Energy space from the set of parameters F and N_d

angular prediction modes candidates and is divided in two successive steps. In the first step, for each Prediction Unit (PU) $N \times N$, the number of angular modes tested in RMD is reduced by increasing the angular step-size (θ_N). Let Θ be the set of $(\theta_{32}, \theta_{16}, \theta_8, \theta_4)$. In Kvazaar: Θ is fixed to $(8, 8, 4, 2)$. This *coarse step* always tests DC, Planar and Most Probable Mode (MPM) modes. The goal of the second step is to refine the dominant prediction direction $m_{\hat{s}}$ obtained from the previous step. The angular step-size is reduced by half $\theta'_N = \dfrac{\theta_N}{2}$ and the RMD process computes the cost $J_{RMD}(m_{\hat{s}} \pm \theta'_N)$ of the direction around the prediction mode obtained from the previous step. This step is repeated with the new dominant direction until the angular step-size becomes 1. In the reference configuration, this feature is disabled and all modes are tested in RMD process.

Approximation Management The minimal and maximal number of modes tested by RMD according to θ_N are given, respectively, by Eqs. 5.4 and 5.5. The number of modes tested by RMD depends on whether the MPM is already included in the set of modes. The first and second terms of Eqs. 5.4 and 5.5 correspond to the first and second steps of the RMD algorithm while the third term adds the number of no angular modes plus the MPM.

$$\min_{mode}(\theta_N) = \left\lceil \frac{33}{\theta_N} \right\rceil + \lfloor \log_2(\theta_N) \rfloor + 2 \tag{5.4}$$

$$\max_{mode}(\theta_N) = \left\lceil \frac{33}{\theta_N} \right\rceil + \lfloor \log_2(\theta_N) \rfloor + 5 \tag{5.5}$$

To explore the OSSE linked to the Intra-mode prediction, a set of $\theta_N \in \{2, 4, 8, 12\}$ (corresponding to testing, respectively, 20, 13, 10, and 8 modes) is defined.

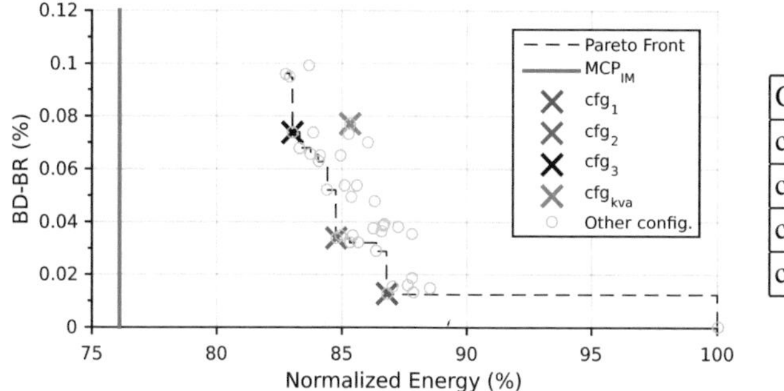

Configuration	Θ
cfg_{kva}	(8, 8, 4, 2)
cfg_1	(8, 4, 2, 2)
cfg_2	(8, 4, 2, 4)
cfg_3	(8, 4, 4, 4)

Fig. 5.17 Pareto in Rate-Energy space generated from all $3 \times 3 \times 2 \times 2 = 36$ combinations of parameter values

Quality & Cost Evaluation For $N \in \{32, 16, 8, 4\}$, 4096 encodings are needed to try all combinations of Θ with $\theta_N \in \{2, 4, 8, 12\}$. The number of experimentations is reduced to study the impact of θ_N for each size N of CU independently. The video sequences are encoded with $\theta_N \in \{2, 4, 8, 12\}$ for a fixed value of $N \in \{32, 16, 8, 4\}$ one at a time. The other angular step-sizes are fixed to the default values of Kvazaar: $\Theta = (8, 8, 4, 2)$.

The impact of θ_N for each value of N has been analyzed independently. Results show that the relation between energy consumption and BD-BR according to θ_N is not linear, and this for all CUs sizes. The configurations with bad trade-off between energy reduction and BD-BR increase has been removed to build a new set of θ_N parameters summarized in Fig. 5.17. Figure 5.17 shows the results of the 36 configurations defined by the table depicted in Fig. 5.17. The difference between the energy reduction opportunities $C_{MCP(IM)}$ and C_{CP} of the Intra-Mode Prediction OSSE is around 5% of energy. Figure 5.17 shows that for the set $(\theta_{32}, \theta_{16}, \theta_8, \theta_4)$, a better configuration than the Kvazaar default one (8, 8, 4, 2) (cfg_{kva} in green in Fig. 5.17) can be used for the same energy reduction.

5.3.2.5 Combination of OSSE

The goal of this section is to study the combination of the two OSSE: CT and IM. The OSSE linked to the CT can be explored with two parameters F and N_d. From results of Fig. 5.16, the configuration of the parameters F and N_d of the Pareto Front are extracted. The OSSE linked to the IM depends on a set of θ_N which is viewed as one parameter to combine the OSSE. In addition to the Kvazaar default configuration (cfg_{kva}), 3 other configurations (cfg_1, cfg_2, cfg_3) are extracted from results of Fig. 5.17 which correspond to significant gap in the Pareto front.

Figure 5.18 shows the results when the configurations described in the associated table are applied on the configurations extracted from the front of the Rate-Energy space of Fig. 5.16. From 100% to 45% of normalized energy consumed, the results

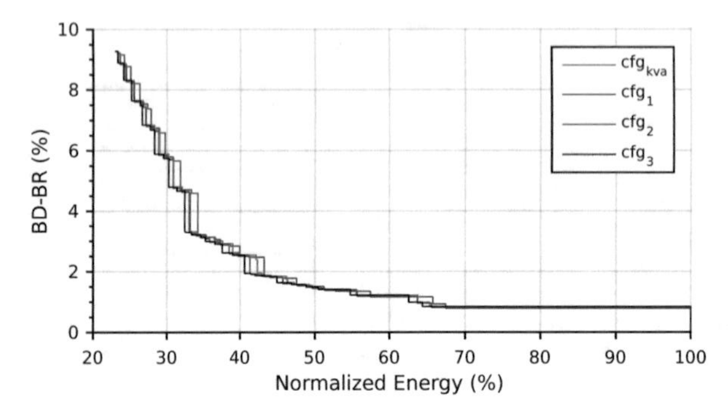

Configuration	Θ
cfg_{kva}	(8, 8, 4, 2)
cfg_1	(8, 4, 2, 2)
cfg_2	(8, 4, 2, 4)
cfg_3	(8, 4, 4, 4)

Fig. 5.18 Pareto in Rate-Energy space from the set of Θ defined in the table

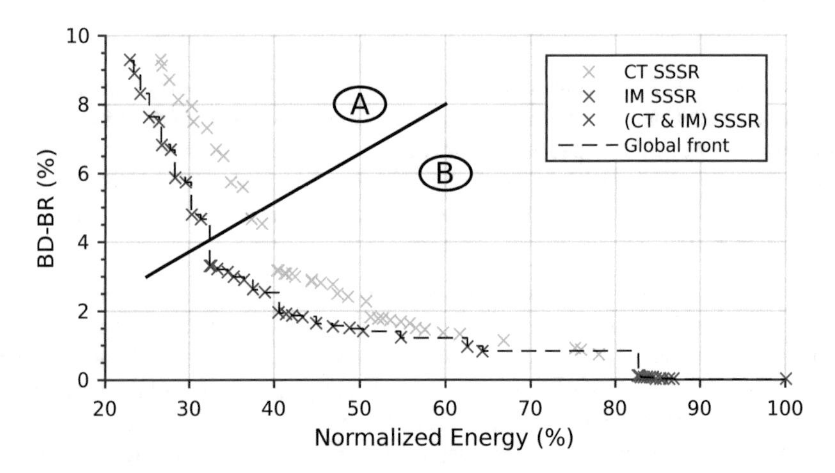

Fig. 5.19 Pareto in Rate-Energy space from the CT OSSE, the IM OSSE and the combination of the two OSSE: CT & IM

of the 4 configurations are intertwined. In the other hand, for an energy consumed less than 45%, the cfg_3 have better results for a major part of the Rate-Energy space. As for the CT OSSE, Fig. 5.18 shows that it is possible to control the energy consumed from 100% to 23%. The results of Fig. 5.17 are finally added in the Rate-Energy space as shown in Fig. 5.19.

Figure 5.19 summarizes the best results (extracted from the Pareto front) of the CT OSSE study of the Sect. 5.3.2.3, the IM OSSE study of the Sect. 5.3.2.4 and the combination of these two OSSE CT & IM, i.e. when the three parameters F, N_d and Θ are used.

Figure 5.19 shows that for all normalized energy target, the combination of the two OSSE (CT & IM) obtains better results than the exploration of the CT OSSE alone. For example, for 32.5% of normalized energy consumed, the combination of the two SSSR compared to the case of CT alone reduces the BD-BR by 4%: from 7.3% to 3.3%. Figure 5.19 shows that the Pareto front has an inflection point (illustrated by the black line in Fig. 5.19). This inflection point splits the Pareto front into two parts (A and B). In part A, a normalized energy reduction of up to 23% of energy consumed has a strong impact on the quality. In the other hand, in part B, the quality degradation is less impacted when the consumed energy is reduced.

To conclude on these results, playing with the two OSSE of the HEVC use-case has been demonstrated to yield better energy efficiency than just using one OSSE, and the SSSR methodology has been shown to give precise answers on the opportunities of gain brought by each OSSE. These results motivate for the SSSR methodology that provides a systematic mechanism to explore and evaluate the approximation opportunities of OSSE-based applications.

On the considered use-case, inflexion points on the Pareto curves guide the designer when choosing the right configuration that does not suffer significantly of quality degradation. This is the case, for instance, in Fig. 5.19 where a designer is advised to target the left-hand side of region B where energy gains are relatively high and BD-BR losses are low.

5.4 Conclusion

In this chapter, the concept of approximate computing techniques at the algorithmic level has been investigated. The aim is to transform the algorithmic description of the application in order to decrease the processing complexity and thus to reduce the implementation cost. Different approximate computing techniques acting at the algorithm level have been proposed and can be classified in two categories. The first category corresponds to approaches skipping parts of the computation by removing some processing. The second category is based on techniques using approximation to replace a part of the computation by a less complex processing. Acting at the algorithm level allows to significantly reduce the implementation cost as illustrated in this chapter with the HEVC video codec use-case. Task skipping, algorithm selection and parameter adjustment techniques are exploited to optimize an HEVC decoder. This technique proposes a Pareto front exploring an energy reduction up to 40% for a reasonable quality degradation of 4 dB for the PSNR. The skipping technique for discrete optimization algorithm has been illustrated in the case of the HEVC encoder. This technique proposes a Pareto front exploring an energy reduction up to a factor 3 for a reasonable quality degradation of 3.3 % for the BD-BR. The use of these techniques is restricted by the lack of tools automating the algorithm transformations. Most of the transformations are carried-out manually by the application designer.

References

1. Alvarez, C., Corbal, J., & Valero, M. (2005). Fuzzy memoization for floating-point multimedia applications. *IEEE Transactions on Computers, 54*(7), 922–927.
2. Amant, R. S., Yazdanbakhsh, A., Park, J., Thwaites, B., Esmaeilzadeh, H., Hassibi, A., Ceze, L., & Burger, D. (2014). General-purpose code acceleration with limited-precision analog computation. In *2014 ACM/IEEE 41st International Symposium on Computer Architecture (ISCA)* (pp. 505–516).
3. Ansel, J., Chan, C., Wong, Y. L. Olszewski, M., Zhao, Q., Edelman, A., & Amarasinghe, S. (2009). Petabricks: A language and compiler for algorithmic choice. *ACM Sigplan Notices, 44*(6), 38–49.
4. Ansel, J. J. A. (2014). *Autotuning Programs with Algorithmic Choice*. PhD thesis, Massachusetts Institute of Technology.
5. Bajard, J.-C., Kla, S., & Muller, J.-M. (1994). BKM: A new hardware algorithm for complex elementary functions. *IEEE Transactions on Computers, 43*(8), 955–963.
6. BBC HEVC bistreams.
7. Bienia, C., Kumar, S., Singh, J. P., & Li, K. (2008). The parsec benchmark suite: Characterization and architectural implications. In *Proceedings of the 17th International Conference on Parallel Architectures and Compilation Techniques* (pp. 72–81). ACM.
8. Bonnot, J., Nogues, E., & Menard, D. (2016). New non-uniform segmentation technique for software function evaluation. In *2016 IEEE 27th International Conference on Application-specific Systems, Architectures and Processors (ASAP)* (pp. 131–138). IEEE.
9. Bossen, F. (2012, February). *Common Conditions and Software Reference Configurations*. Document JCTVC-H1100, Joint Collaborative Team on Video Coding (JCT-VC) of ITU-T SG 16 WP 3 and ISO/IEC JTC 1/SC 29/WG 11, San Jose, CA.
10. Chakradhar, S. T., & Raghunathan, A. (2010). Best-effort computing: Re-thinking parallel software and hardware. In *Design Automation Conference* (pp. 865–870).
11. Chevillard, S., Joldeş, M., & Lauter, C. (2010). Sollya: An environment for the development of numerical codes. In *International congress on mathematical software* (pp. 28–31). Springer.
12. De Dinechin, F., & Tisserand, A. (2005). Multipartite table methods. *IEEE Transactions on Computers, 54*(3), 319–330.
13. Eldridge, S., Raudies, F., Zou, D., & Joshi, A. (2014). Neural network-based accelerators for transcendental function approximation. In *Proceedings of the 24th Edition of the Great Lakes Symposium on VLSI, GLSVLSI '14* (pp. 169–174). New York, NY: Association for Computing Machinery.
14. Esmaeilzadeh, H., Sampson, A., Ceze, L., & Burger, D. (2012). Neural acceleration for general-purpose approximate programs. In *2012 45th Annual IEEE/ACM International Symposium on Microarchitecture* (pp. 449–460). IEEE.
15. Fu, C.-M., Alshina, E., Alshin, A., Huang, Y.-W., Chen, C.-Y., Tsai, C.-Y., Hsu, C.-W., Lei, S.-M., Park, J.-H., & Han, W.-J. (2012, December). Sample adaptive offset in the HEVC standard. *IEEE Transactions on Circuits and Systems for Video Technology, 22*, 1755–1764.
16. Goiri, I., Bianchini, R., Nagarakatte, S., & Nguyen, T. D. (2015). Approxhadoop: Bringing approximations to MapReduce frameworks. In *Proceedings of the Twentieth International Conference on Architectural Support for Programming Languages and Operating Systems* (pp. 383–397).
17. Hamidouche, W., Olivier, C., Pousset, Y., & Perrine, C. (2013, April). Optimal resource allocation for Medium Grain Scalable video transmission over MIMO channels. *Journal of Visual Communication and Image Representation, 24*, 373–387.
18. Hore, A., & Ziou, D. (2010). Image quality metrics: PSNR vs. SSIM. In *2010 20th International Conference on Pattern Recognition (ICPR)* (pp. 2366–2369). IEEE.
19. Janot, A., Vandanjon, P., & Gautier, M. (2014). A generic instrumental variable approach for industrial robot identification. *IEEE Transactions on Control Systems Technology, 22*(1), 132–145.
20. JCT-VC. (2016). HEVC reference software.

21. Kemal, U., Alshin, A., Alshina, E., Bossen, F., Han, W., Park, J., & Lainema, J. (2013). Motion compensated prediction and interpolation filter design in H.265/HEVC. In *IEEE TCSVT*.
22. Khan, M. U. K., Shafique, M., & Henkel, J. (2013). An adaptive complexity reduction scheme with fast prediction unit decision for HEVC intra encoding. In *2013 20th IEEE International Conference on Image Processing (ICIP)* (pp. 1578–1582). IEEE.
23. Lee, D.-U., Cheung, R. C., Luk, W., & Villasenor, J. D. (2009). Hierarchical segmentation for hardware function evaluation. *IEEE Transactions on Very Large Scale Integration (VLSI) Systems, 17*(1), 103–116.
24. Li, R.-C., Markstein, P., Okada, J. P., & Thomas, J. W. (2001). The libm library and floating-point arithmetic for HP-UX on itanium. Technical report, Hewlett-Packard Company.
25. Li, S., Park, S., & Mahlke, S. (2018). Sculptor: Flexible approximation with selective dynamic loop perforation. In *Proceedings of the 2018 International Conference on Supercomputing* (pp. 341–351). ACM.
26. Linaro, D., & Storace, M. (2008, May). A method based on a genetic algorithm to find PWL approximations of multivariate nonlinear functions. In *2008 IEEE International Symposium on Circuits and Systems* (pp. 336–339).
27. Ludwig, J. T., Nawab, S. H., & Chandrakasan, A. P. (1996). Low-power digital filtering using approximate processing. *IEEE Journal of Solid-State Circuits, 31*(3), 395–400.
28. McAfee, L., & Olukotun, K. (2015). Emeuro: A framework for generating multi-purpose accelerators via deep learning. In *2015 IEEE/ACM International Symposium on Code Generation and Optimization (CGO)* (pp. 125–135).
29. Meher, P. K., Valls, J., Juang, T.-B., Sridharan, K., & Maharatna, K. (2009). 50 years of cordic: Algorithms, architectures, and applications. *IEEE Transactions on Circuits and Systems I: Regular Papers, 56*(9), 1893–1907.
30. Meirhaeghe, A., Boutellier, J., & Collin, J. (2019). The direction cosine matrix algorithm in fixed-point: Implementation and analysis. In *ICASSP 2019 - 2019 IEEE International Conference on Acoustics, Speech and Signal Processing (ICASSP)* (pp. 1542–1546).
31. Mercat, A., Arrestier, F., Hamidouche, W., Pelcat, M., & Menard, D. (2017). Constrain the Docile CTUs: An in-frame complexity allocator for HEVC intra encoders. In *2017 IEEE International Conference on Acoustics, Speech and Signal Processing (ICASSP)*. IEEE.
32. Mercat, A., Arrestier, F., Hamidouche, W., Pelcat, M., & Menard, D. (2017). Energy reduction opportunities in an HEVC real-time encoder. In *2017 IEEE International Conference on Acoustics, Speech and Signal Processing (ICASSP)* (pp. 1158–1162). IEEE.
33. Mercat, A., Bonnot, J., Pelcat, M., Desnos, K., Hamidouche, W., & Menard, D. (2017). Smart search space reduction for approximate computing: A low energy hevc encoder case study. *Journal of Systems Architecture: Embedded Software Design, 80*, 56–67.
34. Muller, J.-M. (2020). Elementary functions and approximate computing. *Proceedings of the IEEE, 108*(12), 2136–2149.
35. Nogues, E., Holmbacka, S., Pelcat, M., Menard, D., & Lilius, J. (2014). Power-aware hevc decoding with tunable image quality. In *2014 IEEE Workshop on Signal Processing Systems (SiPS)* (pp. 1–6). IEEE.
36. Nogues, E., Menard, D., & Pelcat, M. (2019). Algorithmic-level approximate computing applied to energy efficient hevc decoding. *IEEE Transactions on Emerging Topics in Computing, 7*(1), 5–17.
37. Nogues, E., Raffin, E., Pelcat, M., & Menard, D. (2015). A modified HEVC decoder for low power decoding. In *Proceedings of the 12th ACM International Conference on Computing Frontiers* (p. 60). ACM.
38. Percelay, M., Bonnot, J., Arrestier, F., & Menard, D. (2020). Polynomial approximation with non-uniform segmentation for bivariate functions. In *2020 IEEE Workshop on Signal Processing Systems (SiPS)* (pp. 1–6).
39. Press, W. H., Teukolsky, S. A., Vetterling, W. T., & Flannery, B. P. (1988). *Numerical recipes in c* (Vol. 1, p. 3). Cambridge University Press.

40. Rinard, M. (2006). Probabilistic accuracy bounds for fault-tolerant computations that discard tasks. In *Proceedings of the 20th Annual International Conference on Supercomputing* (pp. 324–334).
41. Rust, J., & Paul, S. (2016, November). Bivariate function approximation with encoded gradients. In *2016 IEEE Nordic Circuits and Systems Conference (NORCAS)* (pp. 1–6).
42. Rust, J., & Paul, S. (2017). Exploiting special-purpose function approximation for hardware-efficient QR-decomposition. In *Design, Automation Test in Europe Conference Exhibition (DATE), 2017* (pp. 1378–1383).
43. Schulte, M., & Stine, J. (1999). Approximating elementary functions with symmetric bipartite tables. *IEEE Transactions on Computers, 48*(8), 842–847.
44. Shafique, M., & Henkel, J. (2014). Low power design of the next-generation high efficiency video coding. In *Design Automation Conference (ASP-DAC), 2014 19th Asia and South Pacific* (pp. 274–281). IEEE.
45. Sidiroglou-Douskos, S., Misailovic, S., Hoffmann, H., & Rinard, M. (2011). Managing performance vs. accuracy trade-offs with loop perforation. In *Proceedings of the 19th ACM SIGSOFT Symposium and the 13th European Conference on Foundations of Software Engineering* (pp. 124–134). ACM.
46. Sullivan, G. J., Ohm, J.-R., Han, W.-J., & Wiegand, T. (2012, December). Overview of the high efficiency video coding (HEVC) standard. *IEEE Transactions on Circuits and Systems for Video Technology, 22*(12), 1649–1668.
47. Suresh, A., Swamy, B. N., Rohou, E., & Seznec, A. (2015). Intercepting functions for memoization: A case study using transcendental functions. *ACM Transactions on Architecture and Code Optimization (TACO), 12*(2), 18.
48. Sze, V., Budagavi, M., & Sullivan, G. J. (Eds.) (2014). *High efficiency video coding (HEVC)*. Integrated Circuits and Systems. Cham: Springer International Publishing.
49. The Open HEVC - open source project. https://github.com/OpenHEVC/openHEVC
50. Tseng, C.-C., & Lee, S.-L. (2008). Design of fractional delay fir filter using discrete cosine transform. In *IEEE Asia Pacific Conference on Circuits and Systems, 2008. APCCAS 2008* (pp. 858–861). IEEE.
51. Vassiliadis, V., Parasyris, K., Chalios, C., Antonopoulos, C. D., Lalis, S., Bellas, N., Vandierendonck, H., & Nikolopoulos, D. S. (2015). A programming model and runtime system for significance-aware energy-efficient computing. *ACM SIGPLAN Notices, 50*(8), 275–276.
52. Wiegand, T., Sullivan, G. J., Bjontegaard, G., & Luthra, A. (2003, July). Overview of the H. 264/AVC video coding standard. *IEEE Transactions on Circuits and Systems for Video Technology, 13*(7), 560–576.
53. Yinji, P., Junghye, M., & Jiangle, C. (2010). Encoder improvement of unified intra prediction. In *JCTVC-C207* (Vol. 7674, pp. 568–577).
54. Zhang, Q., Yuan, F., Ye, R., & Xu, Q. (2014). Approxit: An approximate computing framework for iterative methods. In *Proceedings of the 51st Annual Design Automation Conference* (pp. 1–6). ACM.
55. Zhao, L., Zhang, L., Ma, S., & Zhao, D. (2011). Fast mode decision algorithm for intra prediction in HEVC. In *Visual Communications and Image Processing (VCIP), 2011 IEEE* (pp. 1–4). IEEE.

Part II
Methods and Tools for Approximate Computing

Chapter 6
Analysis of the Impact of Approximate Computing on the Application Quality

Justine Bonnot, Daniel Ménard, and Karol Desnos

6.1 Introduction

Approximate Computing (AC) is one of the main approaches for post-Moore's Law computing. It exploits the error resilience of numerous applications in order to save energy or accelerate processing. The numerical accuracy of an application is now taken as a new tunable parameter to design more efficient systems. Nevertheless, the numerical accuracy of an application has to stay within an acceptable limit to be usable. For this reason, the impact of the induced errors on the application has to be studied.

Before analyzing the effects of the errors induced by the chosen approximations on the application quality metric, the errors induced by the AC techniques themselves have to be characterized. A thorough characterization of the approximation error allows, during the Design Space Exploration (DSE) phase, choosing the most suitable AC technique with respect to the implementation constraints and quantifying the impact of the approximation on the application quality metric. The impact of the approximation on the application quality metric is generally evaluated numerous times because the DSE requires testing many configurations. Consequently, this evaluation of the approximation error has to be fast so as to limit the DSE time.

AC techniques generate various error profiles. When implementing AC in an application, the objective of error analysis is to derive the impact of the induced approximations on the application quality metric. The evaluation of the impact of the approximation on the application quality metric can be done in three steps as presented in Fig. 6.1. To illustrate the different concepts associated with this figure,

J. Bonnot · D. Ménard (✉) · K. Desnos
Univ. Rennes, INSA Rennes, IETR, Rennes, France
e-mail: justine.bonnot@insa-rennes.fr; daniel.menard@insa-rennes.fr; karol.desnos@insa-rennes.fr

A. Bosio et al. (eds.), *Approximate Computing Techniques*,
https://doi.org/10.1007/978-3-030-94705-7_6

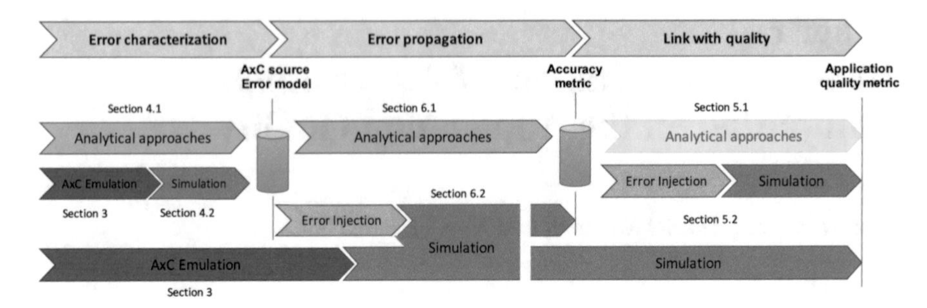

Fig. 6.1 Different steps to analyze the impact of approximation on quality

an example is considered. The application is a digital communication receiver and the AC technique is finite precision with fixed-point arithmetic.

The errors induced by the approximations are first characterized so as to provide an AC source error model according to several error metrics for further error propagation. Several error metrics are used for error characterization. The error metrics can be the mean error amplitude μ, the standard deviation of the error σ, the probability mass function (PMF) of the error, or an interval that contains the error bounds. The error characterization is presented in Sect. 6.4 and can be done with two different techniques: analytical and simulation-based techniques. Analytical techniques are presented in Sect. 6.4.1. They provide a mathematical model of the error. In the case of the considered example, the Widrow model [1] can be used to analyze the AC error due to fixed-point arithmetic. The error resulting of the quantization of a data is modeled by a random variable and the mathematical expressions of the PMF and the error mean and variance are provided. Simulation-based techniques are presented in Sect. 6.4.2 and are more generic but require the emulation of the implemented approximation. The different methods to emulate AC techniques are presented in Sect. 6.3. In the case of the considered example, C++ classes can be used to emulate fixed-point data-types and collect by simulation a set of values to analyze this error.

Then, the errors are propagated through the application. This step described in Sect. 6.6 allows deriving an intermediate accuracy metric. As for the error characterization step, analytical techniques presented in Sect. 6.6.1 can be used. Simulation-based techniques presented in Sect. 6.6.2 can also be used along with emulation or error injection techniques. In the case of the considered example, the power (second-order moment) of the error can be considered as an accuracy metric. For analytical approaches, perturbation theory can be used to propagate the error inside the digital communication receiver and obtained the mathematical expression of the power at the output of the receiver.

Finally, the approximation errors may have to be linked to the output quality as presented in Sect. 6.5. In this case, a few analytical approaches have been proposed and are described in Sect. 6.5.1. Analytical approaches are depicted as shaded in Fig. 6.1 since no general analytical approach has been proposed. They have to be derived specifically for the considered quality metric. Simulation-based techniques are generally preferred and are described in Sect. 6.5.2. They can be implemented along with error injection. In the case of the considered example, the Bit Error Rate (BER) metric can be used to evaluate the quality. This metric evaluates the ratio between the number of corrupted bits and the total number of received bits. Simulation of the digital receiver with fixed-point data-types can be used to evaluate the quality metric (BER) or the accuracy metric (error power). The link between the BER and the error power can be obtained by simulation with error injection or by an analytical approach in this specific case.

6.2 Metrics to Analyze the Impact of AC

Introducing approximations in an application leads to an unavoidable error $\widehat{e_i}$ between the exact value z and the erroneous value $\widehat{z}$ due to approximation. The error of approximation $\widehat{e_i}$ is defined as follows:

$$\widehat{e_i} = \widehat{z} - z \tag{6.1}$$

As defined by Chippa et al. in [2], errors due to AC can be classified into two separate categories denominated *fail small* and *fail rare*. In order to guarantee a functional system, the impact of approximation errors on application quality must be limited. In the *fail small* category, an error is either always present or very frequent, and thus its amplitude must be limited. In the *fail rare* category, an error can have a high amplitude, and thus its probability of occurrence must be low to limit its impact on quality.

The maximum admissible error is plotted in Fig. 6.2 versus its probability of occurrence. The techniques used for AC must lead to errors located in the

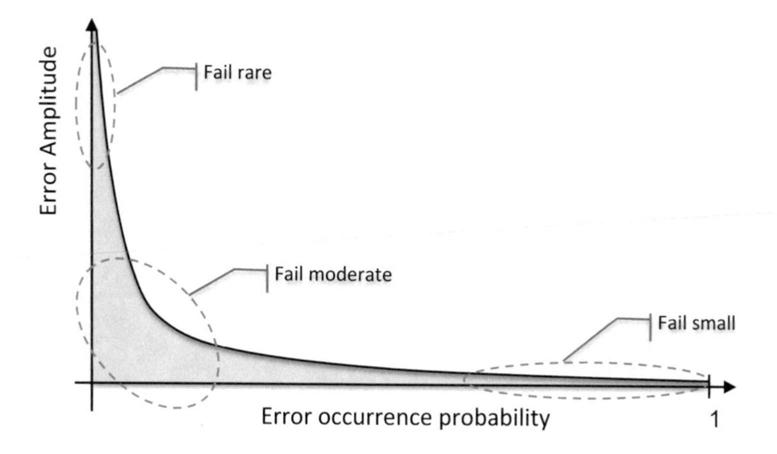

Fig. 6.2 Evolution of maximal error amplitude according to its occurrence probability

gray area. The shape of this gray area defines the quality requirements of the application. A third AC error category named *fail moderate* can be introduced. This category corresponds to errors having both moderate probability of occurrence and moderate amplitude. This characterization of the AC error shows that the rate and the amplitude of the error have to be considered to analyze the effect of AC on application quality.

6.2.1 AC Error and Accuracy Metrics

The introduced errors $\widehat{e_i}$ are characterized and modeled with error metrics to ease the process of linking the induced errors to the quality evaluation function of the application. Numerous error metrics have been proposed, and the choice of the considered error metrics depends on the implemented AC technique as well as on the nature of the output of the application, as presented by Akturk et al. [3]. To quantify the errors induced by an approximation, the deviation between the approximate output and the accurate output has to be measured. Nevertheless, as explained by Akturk et al. [3], the considered error metric has to be robust to several side effects. For instance, when taking a relative error metric, the reference should not introduce any bias on the measured deviation. Besides, averaging the error metric over a certain number of points may hide large deviations on particular points, hence the need for a metric to measure the extreme errors. Another important point in deriving error metrics is to define how to aggregate errors.

When these restrictions have been taken into account, different metrics to compute the deviation exist: statistical, bitwise, and interval-based metrics.

6.2.1.1 Statistical Metrics

According to Chippa et al. [4], the induced errors by an AC technique can be characterized with statistics according to three parameters. The three parameters are all derived from the deviation expressed as the error distance e_i:

$$e_i = |\widehat{z} - z| \tag{6.2}$$

The first parameter to characterize the error is the mean error amplitude, μ_e, which corresponds to the average value of the different error distances e_i

$$\mu_e = \frac{1}{N} \sum_{i \in I} e_i \tag{6.3}$$

with I, the set containing all the values taken by the error e_i. The second parameter f is the Error Rate (ER), which represents the frequency of error occurrence

$$f = \frac{1}{N} \sum_{i \in I} f_{e_i}, \text{ with } f_{e_i} = \begin{cases} 0 \text{ if } e_i = 0 \\ 1 \text{ else} \end{cases} \tag{6.4}$$

The third parameter is the standard deviation of the error e_{rms}, which represents the dispersion around the average error value and is considered as the error predictability by Chippa et al. [4].

$$e_{rms} = \sqrt{\frac{1}{N} \sum_{i \in I} {e_i}^2} \tag{6.5}$$

For a continuous statistical distribution of the error, the different parameters are represented with a distribution. This continuous distribution is called a probability density function (PDF). The ER is represented by the area under the curve representing the distribution, the mean error amplitude is the mean value of the distribution, and the error predictability is the standard deviation of the distribution.

When the distribution of the error is discrete, as for instance with inexact arithmetic operators, the statistical distribution of the error is the probability mass function (PMF). The PMF of the approximation error is the function indicating the probability that the error distance is equal to a particular value. It represents the ER depending on the Error Distance (ED) of the induced errors. The PMF can be represented as a bar chart, and the higher a bar is, the more frequent the considered error occurs. PMFs can be highly asymmetric as presented in Fig. 6.3 for the two inexact operators Almost Correct Adder (ACA) on 16 bits with a carry chain length of 4 and the Approximate Array Multiplier (AAM) on 16 bits, respectively. These PMFs have been obtained with 10,000 uniformly drawn inputs in the input space $[0; 2^{16}]$. The more inputs are drawn, the more accurate the PMF is. To build the PMF of an approximate operator error or statistics on the error, Monte Carlo simulations are generally used [5].

The error can also be characterized in terms of Maximum Error Distance (maximum ED) M_e defined as

$$M_e = \max_{i \in I} e_i \tag{6.6}$$

6.2.1.2 Bitwise Metric

In inexact operators, some bits of the operation output can be erroneous. To evaluate the ratio of erroneous bits in a data, the Bitwise Error Rate (BWER) metric can be considered. In digital systems, a data is encoded with a set of bits. The BWER represents the Bit Error Rate of each bit position in a binary word. This metric is the

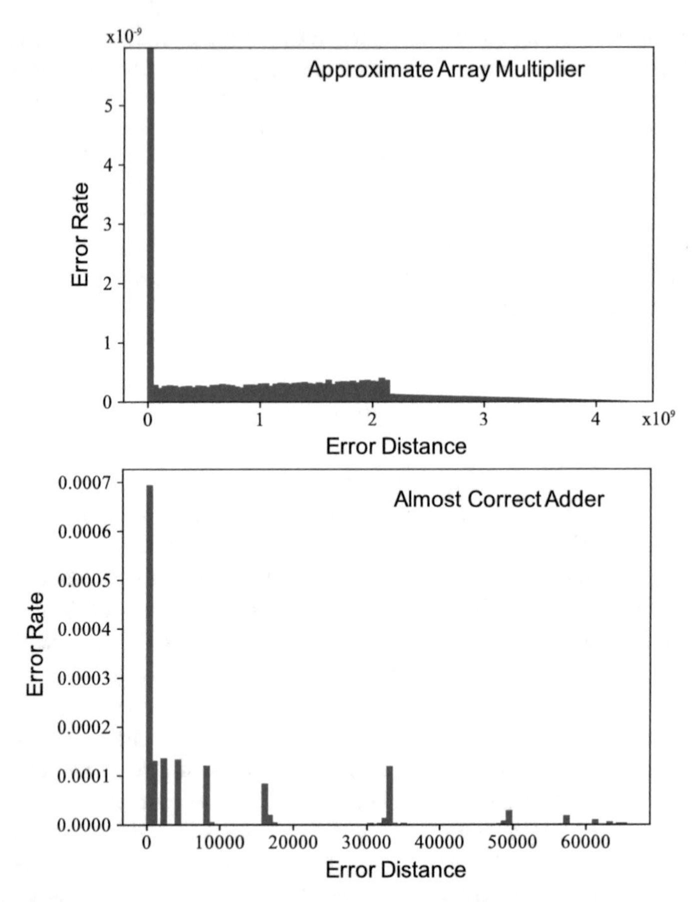

Fig. 6.3 Probability mass function of almost correct adder, $N = 16, C = 4$, and approximate array multiplier, $N = 16$

ratio between the number of erroneous bits on the total number of bits. For instance, if x is an n-bit binary data, the BWER is a vector $\mathbf{b}$ depending on x expressed as $\mathbf{b} = \{p_i\}$ with $i \in [\![0\ ;\ n-1]\!]$, where p_i is the probability of bit i in x to be erroneous. Numerous methods have been proposed to propagate the BWER error metric through an inexact operator and are presented in Sect. 6.6.1.

6.2.1.3 Interval-Based Metric

Error metrics can be represented by intervals and propagated by Interval Arithmetic (IA). IA has first been proposed by Ramon Moore [6] and consists in propagating intervals instead of real numbers in the application. For instance, if variable x lies in $[\underline{x}; \overline{x}]$, the interval will be propagated through the different computations of the application. IA is for instance used to bound the effects of round-off errors in computations and allows guaranteeing the output accuracy. For instance, intervals are used to produce conservative error bounds for the computations in a digital computing machine where the computations are a succession of rounded arithmetic operations whether it be in floating-point or in fixed-point arithmetic. In this case, the produced interval is guaranteed to contain the accurate output of the computation and the radius of the interval is the error bound.

Nevertheless, in the case of errors induced by inexact operators, the error may be unsmooth and the resulting PMF highly asymmetrical. To better render the error induced, Huang et al. [7] proposed an adaptation of IA to inexact circuits called Modified Interval Arithmetic (MIA). In the proposed method, each bar of the PMF of an operator is modeled by an interval.

6.2.2 Application Quality Metrics

The characterization of the error induced by AC allows knowing the impact of the approximation on the application output quality. This impact can be measured with the application quality of service (QoS) or with an intermediate quality metric. The application quality metric, whose nature and measurement depend on the application, quantifies the output quality of the application. For instance, for a signal processing application, the application quality metric can be the signal-to-noise ratio (SNR), whereas for an image processing application, the application quality metric can be the Structural Similarity Index Measure (SSIM). The application quality metric is used to compare the output generated by the approximate version of the algorithm with the output generated by the reference version of the algorithm. Nevertheless, in some cases, an intermediate metric can be easier to compute. An intermediate metric is a generic error metric independent from the QoS metric and that may be linked to the application QoS.

Application quality metrics are generally user-defined quality evaluation functions and have to be provided along with the error tolerance of the application output. For an application, several quality evaluation functions can be used and the impact of the approximation on the different functions may strongly vary. Different quality evaluation functions have been reported in Table 6.1 for various data processing applications.

6.3 Emulation of AC Techniques

Numerous techniques use simulations to evaluate the impact of AC on the application quality metric or on an intermediate accuracy metric. In this case, the approximation mechanism must be emulated so as to reproduce its internal behavior. Emulation techniques have mainly been proposed for inexact arithmetic operators

Table 6.1 Various application quality metrics depending on the nature of the application

Data processing domain	Quality evaluation function
Digital signal processing	Signal to noise ratio
	Mean squared error
	Relative difference
Image processing	Peak signal to noise ratio
	SSIM
	Mean squared error
	Pixel difference
Image segmentation & recognition	Ratio of misclustered points
	Mean centroid distance
	Top-1–top-5 classification
Video coding	Bjøntegaard delta peak signal to noise ratio
	Bjøntegaard delta bit rate
Digital communications	Bit error rate
Web search	Number of correct results in top 25 results

as well as for finite precision arithmetic. In both cases, emulation can be done with functional simulation techniques used to reproduce the behavior of the approximation instead of simulating the approximation behavior on the hardware. Functional simulation techniques for inexact operators and finite precision aim at reproducing the behavior of the approximation at the logic level and at the bit level, respectively.

6.3.1 Inexact Arithmetic Operators

The functional simulation of the behavior of inexact arithmetic operators is used in [8–10] to characterize the induced errors or to analyze the QoS at the output of an application implementing inexact arithmetic operators. In this case, emulation by functional simulation allows studying the behavior of the inexact operator before the hardware implementation. Nevertheless, the emulation of inexact arithmetic operators is complex. To mimic the behavior of inexact arithmetic operators, emulation is done with bit-accurate simulations at the logic level (BALL simulations) that are required to reproduce the internal structure modifications of the operator at the logic level. The complexity of reproducing the internal behavior of the operator leads to long simulation times. For instance, as presented in Fig. 6.4, the BALL simulation time of a 16-bit inexact adder, the ACA, is around 300 times longer than the one of a native accurate processor instruction as floating-point simulation. In the case of an inexact multiplier, the AAM is 4000 times longer.

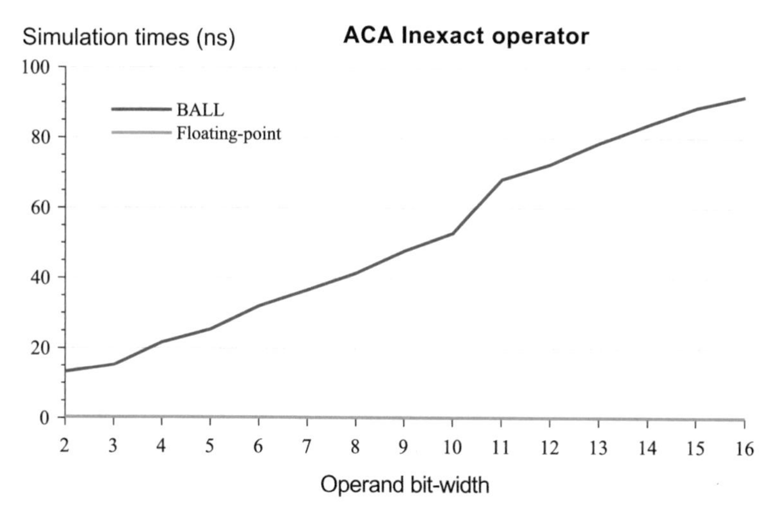

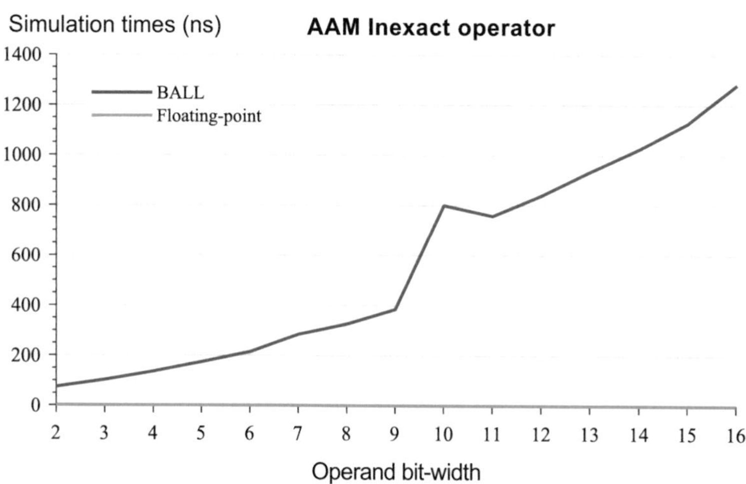

Fig. 6.4 Comparison of the simulation time for the BALL and floating-point simulation of two inexact arithmetic operators

The ratio r between the BALL simulation time and the simulation time for the corresponding accurate floating-point operation of several 32-bit inexact operators is indicated in Table 6.2. The ratio r is evolving in between 31,940 and 73,864,820.

For the emulation of 32-bit inexact arithmetic operators, the required time is very long. Consequently, the simulation of a whole application so as to analyze the impact of 32-bit inexact operators on the QoS at the output of the application becomes prohibitive, if not impossible.

Table 6.2 Ratio r between BALL simulation and simulation of accurate floating-point operation times for 32-bit operators

	Op. name	r
ADD	ETAIV [11]	31,940
	ACA [12]	859,406
	ISA [5]	1,799,519
MPY	AAM [13]	13,375,154
	ABM [14]	73,864,820

6.3.2 Finite Precision Arithmetic

To mimic the finite precision effects, the processing associated with each arithmetic operation can be emulated. Several commercial high-level tools to design digital signal processing applications integrate data-types emulating fixed-point arithmetic, as Signal Processing Worksystem (Cadence), DSP Station (Frontier Design), CoCentric (Synopsis) [15], or C++ classes that have been proposed in SystemC [16–18]. Matlab/Simulink has proposed a fixed-point designer toolbox [19] to emulate the behavior of an application in finite precision. Given the target architecture, the fixed-point simulation of the application can be bit-accurate.

C++-based fixed-point data-types are particularly slow to simulate since they can be two to three orders of magnitude slower than the execution of floating-point data-types. The emulation of fixed-point arithmetic is done on floating-point architectures. The integer word-length, the total word-length, and the quantization and overflow modes can be specified. The quantization mode specifies how to manage a value whose accuracy is greater than the one of the fixed-point variable embedding it, while the overflow mode specifies how to manage a value whose amplitude is larger than the largest that can be encoded on the fixed-point variable. Two different types of fixed-point simulation have been proposed: (1) Constrained data-types also called static fixed-point data-type like `sc_fixed` in SystemC library, with data-type arguments known at compile time. (2) Unconstrained data-types also called dynamic fixed-point data-type like `sc_fixn` SystemC library, with data-type arguments that can be variables and then modified. The static fixed-point data-type simulation is faster than the dynamic one, but the application has to be recompiled each time the data word-lengths are modified. To improve the simulation speed of SystemC fixed-point data-types, a type `_fast` has been proposed for both constrained and unconstrained SystemC data-types, limiting the precision to 53 bits. Other static fixed-point data-type like `ac_fixed` [20] and `ap_fixed` [21] has been proposed. For custom floating-point data-types, C++ classes like `ct_float` [22], `ac_float` [20], `flexfloat` [23], and `floatX` [24] have been proposed.

6.3.3 *Operator Overloading and Approximate Data-Types for Simulation*

For an efficient simulation of an application implementing AC techniques, Sampson et al. proposed EnerJ [25], a Java extension with type qualifiers to indicate which data are approximated and which data have to be accurate. The programmer annotates the code implementing its application and indicates the approximable parts and error-sensitive parts. Approximate storage, for instance, unreliable memory modules as unreliable registers, data caches, or main memory, and computations are emulated to allow quality analysis at the output of the simulation. The inexact arithmetic operators are implemented by overloading the existing accurate operators. Several AC techniques may be emulated through EnerJ, as Dynamic Voltage and Frequency Scaling (DVFS), reduced width in floating-point operations or reduction of the Dynamic Random Access Memory (DRAM) refresh rate.

6.3.4 *Conclusion*

Emulation is an important part of error modeling for AC since it allows avoiding testing the implemented technique within the real system. However, the proposed methods to emulate the impact of inexact operators or finite precision arithmetic for instance lead to long simulation times which impedes the use of exhaustive simulations for characterizing the approximation error and limits the possibilities for the design space exploration. The different abstraction levels for emulation and their associated times are represented in Fig. 6.5. For AC techniques at the computation level, the whole application has to be simulated for emulation with native data-types

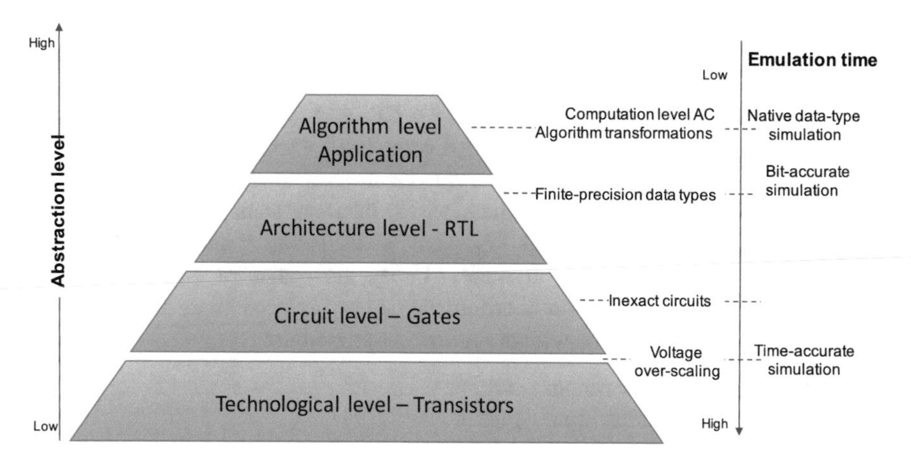

Fig. 6.5 The different abstraction levels and times for emulation

which leads to low emulation time. Finite precision arithmetic has to be emulated at the architecture level and inexact circuits at circuit levels, which leads to moderate emulation time. The longest to emulate is voltage overscaling which has to be emulated at the technological level, that is to say at the transistors level.

6.4 Approximate Computing Error Characterization

AC error characterization aims at developing a model defining the error due to a specific AC technique. Two types of state-of-the-art techniques exist to characterize the error metrics intrinsic to the implemented approximations: analytical and simulation-based techniques. From the different emulation techniques presented in Sect. 6.3, the errors induced by the AC technique are reproduced and can be characterized by simulation. For analytical approach, the aim is to define the mathematical expression of the error metrics.

6.4.1 Analytical Techniques

Analytical techniques have been proposed to provide a mathematical model to evaluate the considered error metric.

6.4.1.1 Finite Precision Arithmetic

Using a statistical representation of the induced error, an error model can be created for various AC techniques. In finite precision arithmetic, signal quantization leads to an unavoidable error. A commonly used model for the continuous-amplitude signal quantization has been proposed in [1] and refined in [26]. The quantization of signal x is modeled by the sum of this signal and a random variable e_x corresponding to the approximation error (quantization noise). This additive noise e_x is a uniformly distributed white noise that is uncorrelated with signal x and any other quantization noise present in the system (due to the quantization of other signals). In Fig. 6.6, $x(n)$ represents the input signal at time n, $x_Q(n)$ the input signal after conversion in fixed-point (Q for quantization), and $e_x(n)$ the statistical error model characterized by its first- and second-order moments. Using the statistical error model, the quantization process in fixed-point arithmetic can be replaced by an additive white noise, as presented in Fig. 6.6 with the following properties:

- Stationary and ergodic random variable
- Non-correlated with the input signal x
- Independent from the other noise sources
- Uniform distribution

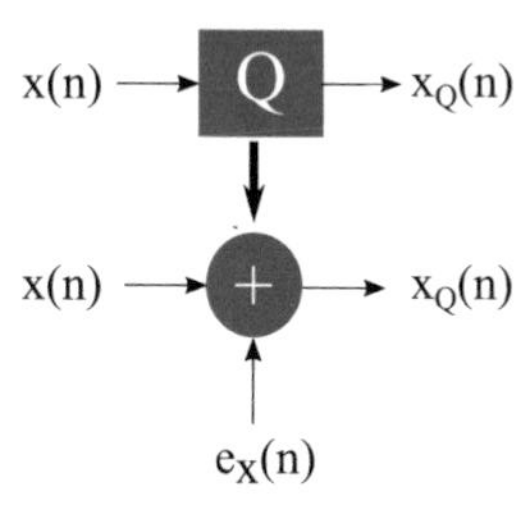

Fig. 6.6 Widrow model for fixed-point quantization noise

The validity conditions of the quantization noise properties have been defined in [26]. These conditions are based on characteristic function of the signal x, which is the Fourier transform of the probability density function (PDF). This model is valid when the dynamic range of signal x is sufficiently greater than the quantum step size and the signal bandwidth is large enough.

This model based on a continuous-amplitude random variable has been extended to include the computation noise in a system resulting from some bit elimination during a fixed-point format conversion. In [27], a model based on a discrete distribution is suggested, and the first- and second-order moments of the quantization noise are given. In this study, the probability value of each eliminated bit to be equal to 0 or 1 is assumed to be $1/2$. The quantization error by truncation is characterized with statistical parameters as the mean error amplitude μ_e and the standard deviation e_{rms} expressed as

$$\mu_e = \frac{q}{2}\left(1 - 2^{-k}\right) \tag{6.7}$$

$$e_{rms}^2 = \frac{q^2}{12}\left(1 - 2^{-2k}\right) \tag{6.8}$$

where k represents the number of eliminated bits and $q = 2^{-n}$, where n is the number of bits to encode the fractional part after quantization. This model has been refined for the different quantization law in [28].

6.4.1.2 Inexact Operators

Probabilistic Analysis To characterize the error metrics of inexact operators, several analytical techniques have been proposed. Liu et al. [29] analytically derive estimated values for the ER and the Mean Error Distance (mean ED) of several block-based inexact adders, namely the ACA, the Error-Tolerant Adder Type II (ETAII), the Equally Segmented Adder (ESA), and the Speculative Carry Selection Adder (SCSA). After having established how to compute the signal propagate p_i, which indicates whether a carry signal is propagated to the ith sum bit, they handle the derivation of error metrics for the different adders separately. The assumption that inputs are uniformly distributed is taken.

To derive the error metrics of the ACA, the authors form the universal error set composed by all the possible error patterns in the inexact operator. With an n-bit ACA, it is possible to derive n disjoint subsets whose union forms the universal error set. Each subset, denoted Π_i, is composed of the error patterns in which the ith bit is erroneous, the upper bits are accurate, and the lower bits are either accurate or not. The total mean ED of the operator is then defined by the sum of the mean ED in each subset. The mean ED in each subset Π_i is approximately equal to $2^i \cdot q_i$, where q_i is the probability to be in the considered subset Π_i. Indeed, the error induced by the ith bit is dominant, while the possible errors on the Least Significant Bits (LSBs) may cancelled each other.

The ER can be derived as $\sum_i q_i$. Through the probabilistic analysis of the inexact operator, the values of q_i are analytically derived.

When it comes to an n-bit ESA divided into $r = \lceil \frac{n}{k} \rceil - 1$ sub-adders, since all the sub-adders have an equal size except the first sub-adder which is exact, the ER is approximately equal to $1 - (\frac{1}{2})^r$. An approximation is then used to compute the mean ED, since the errors in the lower sub-adders can be neglected compared to the one of the higher sub-adders. A similar method is applied for the ETAII giving approximate values of the ER and mean ED. The proposed method relies on a probabilistic analysis of the structures of each considered inexact adder, hence the impossibility to generalize this method to other structure of inexact operators or other AC techniques.

As an improvement of the method proposed in [29], Wu et al. [30] derived a method to compute the exact error profile of block-based inexact adders. Another improvement brought by Wu et al. is to provide a generic method to compute the error statistics of block-based adders. Making the assumption that the inputs are uniformly distributed in $[0; 2^N - 1]$, where N is the size of the adder, the authors compute the probabilities of the signals propagate, generate, and kill the carry, p, g, and k, respectively. Given these probabilities, the computation of the ER is possible. Finally, a result of the inexact arithmetic adder is correct if and only if all the speculated inputs of carries are correct. The authors compute in a recursive way the probability of this event. To derive the error distribution, the binary representation of the ED, named the "error pattern," is analyzed. All the possible error patterns are enumerated and their probability of occurrence is computed, giving the exact PMF of the error induced by the inexact adder.

Mazahir et al. [31, 32] proposed a complete study on a probabilistic evaluation of the exact PMF of inexact adders and inexact recursive multipliers. The targeted class of inexact adders is adders implementing carry chain truncation and carry prediction between successive accurate sub-adders. An error occurs in these adders when the number of bits to predict the carry is not sufficient to predict the accurate carry signal. In this case, an error in a sub-adder can propagate to the upper sub-adders and leads to an output of the adder lower than the accurate adder output. In the end, the method analyzes the probabilities that an error occurs in each sub-adder to derive the accurate PMF. The method is more complex due to its genericity. The method not considers only block-based adders. Nevertheless, this method is particularly long to analyze large bit-width adders. The conditions on the inputs that led to an error are identified and treated as independent events using probabilities.

Matrix-Based Determination of the mean ED
Roy and Dhar [33] extend the method proposed in [31], deriving the accurate value of the mean ED of inexact Least Significant Bits (LSB) adders. This method is based on the structure of these n-bit adders decomposed into an $n - m$-bit-accurate adder on the Most Significant Bits (MSBs) and several inexact sub-adders on the LSBs. The analysis of the mean ED is done by building a 2D memory database of size $(M, 2)$, where $M = 2^{m+1}$ if m LSBs are approximated. To build this database, 4 matrices are used, which consider 4 different carry-out conditions on the mth bit.

Given the truth tables of accurate and inexact adders, the 4 matrices storing the different error amplitudes for each sub-adder are built to finally compute the mean ED. The asymptotic runtime of the proposed matrix-based method for mean ED computation is linear with the number of approximated LSBs, $O(m)$, if the size of the inexact sub-adders is negligible compared to m. The proposed method targets only the estimation of the mean ED.

Empirical Model and Gate-Level Error Characterization Sengupta et al. [34] proposed a gate-level error characterization method to determine the error variance e_{rms} of an adder whose approximation relies on the LSBs. The error variance is characterized as a function of the number of approximated LSBs y. The induced error e can be modeled as a random variable x that lies in $[-(2^y - 1); (2^y - 1)]$ since y LSBs are approximated, and whose probability to be equal to e is p_e. The error variance is then computed as

$$e_{rms}^2(y) = \sum_{x=-(2^y-1)}^{2^y-1} x^2 p_x \tag{6.9}$$

Assuming that x is uniformly distributed in $[-(2^y - 1); (2^y - 1)]$, the error variance can be written as

$$e_{rms}^2(y) = \frac{(2^{y+1} - 1)^2}{12} \tag{6.10}$$

Finally, since the error variance when no LSBs are approximated is 0, an empirical formulation of the error variance is derived as $e_{rms}^2(y) = a \cdot (2^{by} - 1)$. The values (a, b) are constants derived from fitting experimentally obtained error variance values with Monte Carlo simulations to the empirical model.

Hierarchical Analysis Sengupta et al [35] proposed to derive the error PMF at the output of inexact adders or multipliers by first focusing on the characterization of smaller units, for instance, Full Adder (FA). For a signed approximate FA, the error can affect the output of the sum, as well as the sign bit.

The derivation of the error PMF is done in a general case where the input distribution is not uniform. The obtained expressions for the PMF depend on the probabilities of the input signals to be 1, which are known only when the input distribution is known.

6.4.1.3 Voltage-Overscaled Circuits

Liu et al. [36] proposed an analytical analysis of the impact of supply voltage overscaling on arithmetic units, as adders or multipliers. The proposed method estimates the mean ED at the output of an exact operator subject to voltage overscaling. The objective was to generalize a method to derive the mean ED under voltage overscaling since the behavior of three different 16-bit-accurate adders, the Ripple-Carry-Adder (RCA), Carry-Look-Ahead Adder (CLA), and Carry-Select Adder with similar critical supply voltage, was very different under similar voltage overscaling operation.

To compute the mean ED at the output of an arithmetic operator subject to voltage overscaling, for each internal signal of the application, the error significance W_k^e, which depends on the maximum error amplitude as well as on the switching activity, is determined for each internal signal k. If the considered internal signal k impacts the computation of a single output bit, W_k^e is equal to the weight of the output bit (if the output bit is the ith bit, the weight is equal to 2^i). If the considered signal k impacts several output bits, then W_k^e is the minimum of the weights of the output bits. Then, the switching activity is estimated in each node of the circuit and at all discrete time points. The switching activity corresponds to the probability of a transition from a bit a to b, with $(a, b) \in \{0; 1\}$. For a transition $0 \to 1$ of signal k at time $t = T_{CLK}$, the switching activity is denoted P_k^{01} and for a transition $1 \to 0$ P_k^{10}. Given this information, the hypothesis that the different signals independently and additively contribute to the computation of the mean ED gives the following equation for the mean ED computation:

$$\mu_e = \sum_{k \in S} \left(W_k^e \times \left(P_k^{01} + P_k^{10} \right) \right) \tag{6.11}$$

Indeed, with overscaled supply voltage, the critical path delay may be larger than the clock period T_{CLK}. Consequently, at time $t = T_{CLK}$, switching may induce errors. The proposed method has been demonstrated on the error analysis of circuits implementing digital signal processing applications subject to supply voltage overscaling. Giving an accurate analysis of the mean ED at the output of the circuit, it reduces the characterization time by several orders of magnitude compared to classical Monte Carlo simulation techniques. The complexity of the proposed characterization is in $[\mathcal{O}(N); \mathcal{O}(N^m)]$ for the characterization of N-bit operations with m inputs.

6.4.1.4 Conclusion

Numerous methods have been proposed to characterize the errors induced by inexact operators as in [29–31, 33] or to quantify the error induced by finite precision. A few work has been done to analytically evaluate the error induced by DVFS on an arithmetic circuit. Analytical methods proposed for inexact operators are dedicated to specific structures. If the application designer is willing to test inexact operators belonging to different types, the analytical method to compute the error statistics requires a new mathematical derivation.

6.4.2 Simulation-Based Techniques

To characterize the errors induced by AC techniques, simulation-based techniques are massively used. Simulation-based techniques are more and more employed due to their ease of use. Functional simulation techniques run the approximate system on a representative input data set and compute the required statistics for computing the error metrics. To mimic the behavior of the approximation, emulation techniques can be used.

The principle of functional simulation-based techniques for error metric characterization is presented in Fig. 6.7. Functional simulation techniques can be used to link the approximation with an error metric. In this case, the approximate application and its accurate version are run on $N_{Samples}$ points extracted from real input data. The accurate and approximate output values, z and $\hat{z}$, respectively, are used to measure the obtained error according to a chosen metric, for instance, the mean ED μ, the standard deviation of the error σ or the error PMF.

6.4.2.1 Exhaustive Simulations

Exhaustive functional simulations can be used to compute exact statistics of the error induced by an AC technique. In this case, the AC technique is simulated for its whole input data set, and statistics on the induced error are computed.

For instance, an inexact operator can be simulated exhaustively which means simulated for all possible inputs. If the considered inexact operator has two unsigned

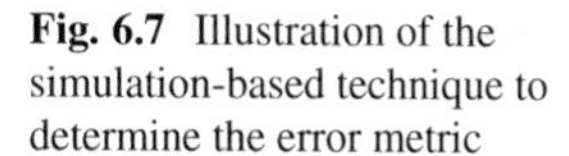
Fig. 6.7 Illustration of the simulation-based technique to determine the error metric

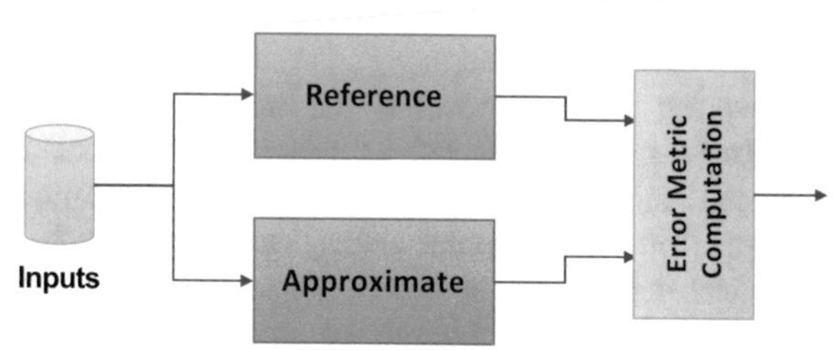

inputs x and y coded on N_x bits and N_y bits, respectively, the exhaustive input set $I = I_x \times I_y$ is composed of 2^{Nx+Ny} values. Consequently, exhaustive functional simulations for high bit-widths inexact operators, and more generally if the input design space of an AC technique is large, are not feasible because of the required simulation time. Besides, as presented in Sect. 6.3.1, the emulation of the approximation mechanisms at the hardware level is complex and long to simulate. To mimic the inexact operator behavior, bit-accurate simulations at the logic-level are required to model the internal structure modifications of the operator. Nevertheless, BALL simulations are two to three orders of magnitude more complex than classical simulations with native data-types for 16-bit operators. This simulation time overhead can reach 7 orders of magnitude for complex 32-bit inexact operators. Thus, exhaustively testing the operator for all the input value combinations is not feasible for high bit-widths because of the required simulation time.

Mazahir et al. [31] have exhaustively simulated the inexact adder proposed in [37] on an Intel Core i7 processor working at 2.4 GHz for various input operands word-length. The evolution of the simulation time depending on the input operands' word-length is exponential. Given this important simulation time overhead, exhaustive simulation is impossible in most cases.

6.4.2.2 Monte Carlo Simulations

Functional simulation is commonly applied on a given number of random inputs, as presented for inexact operators in [8–10] or for DVFS in [36]. Inexact operators are generally simulated with five million random inputs as proposed in [10], which is the typical inexact circuit characterization method. The quality of the statistical characterization obtained from a random sampling is highly dependent on the number of samples taken and on the chosen input distribution. Besides, classical simulation-based analysis does not provide any confidence information on the obtained statistical estimation. Using a great number of samples can be ineffective in terms of simulation time. In this context, a generic simulation-based framework to statistically characterize the error induced by inexact operators has been proposed in [38, 39]. This method is based on inferential statistics and extreme value theory to derive a subset to simulate according to user-defined confidence requirements.

6.4.2.3 Pre-characterization for Analytical Techniques

Noteworthy, a large part of the analytical techniques to characterize error metrics presented in Sect. 6.4.1 relies on a pre-characterization phase. The pre-characterization phase generally relies on simulations to get error information required for the analytical derivation of the error. For instance, before using analytical techniques as the MIA or Modified Affine Arithmetic (MAA) to propagate the errors through an application, Huang et al. [7, 40] launch a characterization phase based on simulations. This characterization phase is required to derive the PMF of the error-free input data, the PMF of the error generated by the inexact operator, and the PMF of the error on the input data if the input is noisy. Once the different PMFs have been derived, they are stored in Look-Up Tables (LUTs).

Similarly, Chan et al. [41] need to characterize the behavior of different error metrics as the ER, mean ED, or maximum ED depending on various input distributions and various hardware configurations. They first simulate the considered inexact operator for various hardware configurations, for instance, different carry-chain length for the ACA. They then record the evolution of different error metrics depending on the standard deviation of the input distribution. The obtained results are saved in LUTs and further used for error composition.

Sengupta et al [34] used a pre-characterization phase with gate-level characterization of inexact adders depending on the number of approximated LSBs. As presented in Sect. 6.4.1.2, the error pre-characterization step for inexact adders consists in fitting the parameters (a, b) in equation $e_{rms}^2(y) = a \cdot (2^{by} - 1)$ giving the standard deviation of the error depending on the number of approximated LSBs y. To do so, the standard deviation of the error is computed for different values of y with exhaustive simulations, and (a, b) are derived by regression.

6.4.2.4 Conclusion

To analyze the errors induced by AC techniques, simulation-based techniques are hardly scalable for large applications, large input data sets, or numerous and different perturbation types. Since exhaustive simulations are not scalable, Monte Carlo simulations are generally used. The number of samples used for the simulation has to be defined carefully to ensure the quality of the obtained statistics.

6.5 Quality Metric Determination

6.5.1 Analytical Techniques

Analytical techniques derive a mathematical expression of the application quality metric. For instance, Liu et al. [29] proposed to analyze the impact of inexact adders on an image processing application and particularly to analytically derive the Peak Signal to Noise Ration (PSNR) metric at the output of an image processing application implementing specific types of inexact arithmetic adders.

The drawback of analytical techniques is their specificity with respect to the considered application as well as to the application quality metric, which makes them difficult to automate. Being more generic and easier to automate, simulation-based techniques are often preferred to analytical techniques to evaluate the quality metric.

6.5.2 Simulation-Based Techniques

6.5.2.1 Direct Quality Metric Determination

Application quality metric can be directly evaluated at the output of the application through simulations. In this case, the application has to be simulated on a representative input data set and the quality metric is measured at the output. Nevertheless, some quality functions are very difficult to evaluate directly. For instance, the BER in digital communication systems is long to directly evaluate because of the required evaluation time. When a BER at receiver output of 10^{-k} is targeted, 10^{2+k} simulations are required for a good quality evaluation of the application. This metric is really long to evaluate when introducing approximation in an application, which leads to the need to separate the quality metric determination process in two steps. First, the impact of the approximation on an intermediate quality metric is evaluated, and then the link between the intermediate accuracy metric and the application quality metric is established.

6.5.2.2 Error Injection

To analyze the impact of an AC technique on the quality metric, the errors have to be emulated within the application. Error emulation can be done at the bit level for finite precision or at the logic level for inexact operators. When error emulation is not directly possible, perturbation-based methods can be implemented. In this case, errors are directly injected in the application, as for instance proposed in the framework REACT [42] with dynamic error injection that extends the ACCEPT framework [43]. To use the REACT framework for error injection, the code has to be annotated with the approximable parts and the injected errors are extracted from a pre-built library of several AC techniques, provided by the user. Errors can be injected at two granularity levels. At the fine grain level, errors are injected in the instructions, while at the coarse grain level, errors are injected at the output of functions.

The error injection technique is widely used to derive the relationship between the intermediate accuracy metric and the application quality metric. Chippa et al. [4] propose the ARC framework to analyze the sensitivity of the different parts of an application in order to identify the error-resilient parts. In this case, the intermediate accuracy metrics are the error amplitude and the error rate. In the context of fixed-point refinement, Parashar et al. [44] proposed to deal with the finite precision conversion of a multi-kernels system by modeling the behavior of each kernel converted in fixed-point by a single noise source located at the output of the kernel.

In this case, the intermediate accuracy metric is the noise power. This technique based on error-injection allows finding the different noise power values at the output of each kernel subject to a quality constraint at the output of the application. An optimization problem is formulated to budget the noise power on each kernel such that the quality constraint at the output of the application is satisfied. A steepest descent greedy algorithm is used to solve this problem. The single source statistical model proposed to mimic the quantization noise at the output of a processing block integrating smooth operations is proposed in [45] and refined in [46].

6.5.3 *Conclusion*

Analytical techniques have been proposed to determine the quality metric at the output of an application but are often giving approximate values of the quality metric at the output of an application. Simulation-based techniques can lead to very accurate estimation of the quality metric at the output of an application, but the accuracy depends on the number of input samples simulated for the quality metric measurement. Besides, simulation-based techniques are more generic for the determination of the quality metric without relying on the type of implemented approximation.

6.6 Accuracy Metric Determination

As presented for the characterization of error metrics, two types of state-of-the-art approaches can be used to evaluate the accuracy metric: analytical and simulation-based approaches. These techniques allow propagating errors within an application.

6.6.1 *Analytical Techniques*

Analytical methods mathematically express the error characteristics at the output of the application.

6.6.1.1 Interval-Based Arithmetic

Interval Arithmetic (IA) and Affine Arithmetic (AA) In IA, an interval is assigned to each internal variable of the application. The interval is then propagated within the different computations according to arithmetic rules. Let us define $[\underline{x}; \overline{x}] = \{x | \underline{x} \leq x \leq \overline{x}\}\}$, with $\underline{x}$ and $\overline{x}$ the minimum and maximum values of a variable x, respectively. IA is particularly suited to simple operations, and the intervals can be propagated from inputs to outputs through basic arithmetic operations as additions, subtractions, multiplications, and divisions represented by the $\diamond$ operator in Eq. 6.12, and if the system is non-recursive. In this case, the non-recursivity means that a variable in the system does not depend on its previous values as it is the case in Infinite Impulse Response (IIR) filters. The general propagation rule is

$$[\underline{x}; \overline{x}]\diamond[\underline{y}; \overline{y}] = [\min(\underline{x}\diamond\underline{y}, \underline{x}\diamond\overline{y}, \overline{x}\diamond\underline{y}, \overline{x}\diamond\overline{y}); \max(\underline{x}\diamond\underline{y}, \underline{x}\diamond\overline{y}, \overline{x}\diamond\underline{y}, \overline{x}\diamond\overline{y})] \quad (6.12)$$

IA allows keeping track of measurement errors, errors caused by the inputs, and errors caused by inexact computations. The asset of using IA is its ease to compute, but the produced error bounds are not tight and generally conservative and pessimistic. Indeed, IA does not take into account any correlation between the variables to be composed and is particularly pessimistic when variables are correlated. Numerous libraries have been proposed to directly compute with intervals, as the C++ library Boost [47].

To improve the estimation of error bounds, AA has been proposed by Stolfi et al. [48] in the 1990s. The variables are no longer modeled with intervals but with affine forms as follows:

$$\hat{x} = x_0 + \sum_{i=1}^{n} x_i \times e_i \quad (6.13)$$

AA improves the tightness of the error bounds by taking into account the first-order correlations between the variables to be composed. In Eq. 6.13, x_0 is the central value of variable x, x_i the partial deviations, and e_i the error terms in $[-1; 1]$. Rules have been proposed to compose different affine forms together and are presented in Eqs. 6.14–6.17, where $\hat{x}$ and $\hat{y}$ are affine forms and c is a constant. As shown for the composition of two affine forms by a multiplication, which is not an affine operation, a residual error symbol is produced as $rad(\hat{x})rad(\hat{y})$, where rad corresponds to the radius of the affine form. As proposed for IA, numerous libraries have been proposed to compute with affine forms, as the C++ library LibAffa [49].

$$c \times \hat{x} = c \times x_0 + c\sum_{i=1}^{n} x_i \times e_i \quad (6.14)$$

$$c \pm \hat{x} = (c \pm x_0) + \sum_{i=1}^{n} x_i \times e_i \quad (6.15)$$

$$\hat{x} \pm \hat{y} = (x_0 \pm y_0) \pm \sum_{i=1}^{n} (x_i \pm y_i) \times e_i \quad (6.16)$$

$$\hat{x} \times \hat{y} = x_0 y_0 + \sum_{i=1}^{n} (x_i y_0 + y_i x_0) \times e_i + rad(\hat{x})rad(\hat{y}) \quad (6.17)$$

AA improves the tightness of the error bounds by considering the first-order correlations of error signals through the sharing of the error terms e_i, but to the detriment of more complex computations. It should be noted though that both types

Table 6.3 Comparison between IA and AA

	Advantages	Drawbacks
IA	Numerically stable	Inefficient for interval products
		Linear convergence
AA	More accurate for interval products keep track of correlation	Less stable higher computational cost

of arithmetic ensure guaranteed error bounds. To sum up, a comparison of both types of arithmetic is proposed in Table 6.3.

Both techniques are well adapted to represent symmetric distributions, for instance, the errors induced by fixed-point arithmetic [50]. Caffarena et al. [51] proposed a method based on AA to estimate the Signal-to-Quantization-Noise Ratio (SQNR) at the output of a digital signal processing application converted in fixed-point and to target nonlinear algorithms.

Highly asymmetric distributions such as the errors produced by inexact arithmetic operators are not well represented by either IA or AA. To better render their asymmetric error distributions, modified interval and affine arithmetics have been proposed.

Modified Interval/Affine Arithmetic Based on the error propagation method proposed with IA or AA, Huang et al. [7, 40] proposed an adaptation of these methods to inexact circuits, more adapted to the highly asymmetric PMFs representing the errors induced by inexact operators. Indeed, IA and AA need to be centered and consequently fail to represent errors of inexact operators. When implementing MIA or MAA, the approximation error metric is the PMF of the inexact operator.

In [7], MIA and MAA are proposed to represent asymmetric distributions. The proposed method allows statically estimating the impact of errors induced by inexact operators on the application quality metric. In MIA or MAA, the whole distribution is decomposed into multiple intervals/affine forms. In the case of MIA, each bar of the PMF of an inexact operator is modeled by an interval as in Eqs. 6.18 and 6.19. In Eq. 6.19, the variable n represents the error magnitude.

$$\text{MIA}(x) = P(a \leq x \leq b), \text{ if } a \leq x \leq b \tag{6.18}$$

$$f_X(n) = \begin{cases} P(2^{\epsilon+n-1} \leq X \leq 2^{\epsilon+n}) & \text{if } n > 0 \\ P(-2^{\epsilon+n+3} \leq X \leq -2^{\epsilon+n+2}) & \text{if } n < -1 \\ P(0 \leq X \leq 2^{\epsilon}) & \text{if } n = 0 \\ P(-2^{\epsilon} \leq X \leq 0) & \text{if } n = -1 \end{cases} \tag{6.19}$$

To compute the total error probability, the function $\text{MIA}(x)$ has to be integrated. The rules to compose different intervals are similar to IA, but in MIA, each bar of the first PMF has to be composed with each bar of the second. MIA can be used to propagate the error induced by inexact circuits through simple blocks. To do so, rules are proposed by Huang et al. [40]. Nevertheless, MIA is still very pessimistic on the error bounds and suffers from range explosion.

In the case of MAA, each bar of the PMF of an inexact operator is modeled by an affine form as in Eq. 6.20.

$$f_X = \begin{cases} P_1 : x_{1,0} + x_{1,1}\alpha_0 + x_{1,2}\beta_0 + \dots \\ P_2 : x_{2,0} + x_{2,1}\alpha_1 + x_{2,2}\beta_1 + \dots \\ P_3 : x_{3,0} + x_{3,1}\alpha_2 + x_{3,2}\beta_2 + \dots \end{cases} \tag{6.20}$$

The problem of range explosion is tackled since MAA takes into account the first-order correlation between variables. However, when dealing with operations such as multiplications or divisions, the output form is approximated to an affine form. Consequently, the output of multiplications or divisions is not guaranteed to be more optimistic than the result obtained with MIA. In addition to this problem of range explosion, MAA can suffer from storage explosion due to the operation of several affine forms which can result in the storage of numerous additional terms. The solution to this problem is to group some terms of the PMF but then reducing the output accuracy.

6.6.1.2 Perturbation Theory for Finite Precision Systems

Existing approaches to compute the analytical expression of the quantization noise power in fixed-point systems are based on perturbation theory, which models finite precision values as the addition of the infinite precision values and a small perturbation. These analyses are based on the Widrow model presented in Sect. 6.4.1.1. At node i, a quantization error signal e_i is generated when some bits are eliminated during a fixed-point format conversion (quantization). This error is assimilated to an additive noise which propagates inside the system. This noise source contributes to the output quantization noise e_y through the gain α_i, as shown in Fig. 6.8.

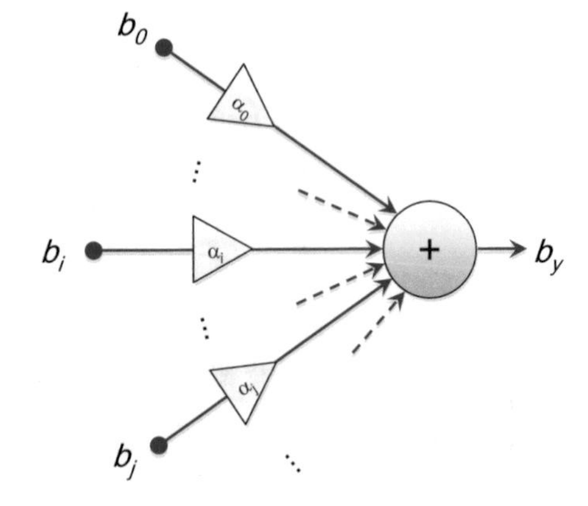

Fig. 6.8 Model for the computation of output quantization error power based on noise sources e_i and gains α_i

The aim of this approach is to define the output noise e_y power expression according to the noise source e_i parameters and the gains α_i between the output and a noise source. This analysis is made up of three phases: (1) *noise generation*, (2) *noise propagation*, and (3) *noise aggregation*. The first phase is presented in Sect. 6.4.1.1 and the two phases are presented in the rest of this section.

Noise Propagation Each noise source e_i propagates to the system output and contributes to the noise b_y at the output. The propagation noise model is based on the assumption that the quantization noise is sufficiently small compared to the signal to consider that the finite precision values can be modeled by using the addition of the infinite precision values and a small perturbation. A first-order Taylor approximation [52, 53] is used to linearize the operation behavior around the infinite precision values. This approach allows obtaining a time-varying linear expression of the output noise according to the input noise [54]. In [55], a second-order Taylor approximation is used directly on the expression of the output quantization noise. In [56] and [57], affine arithmetic is used to model the propagation of the quantization noise inside the system. Affine expression allows obtaining directly a linear expression of the output noise according to the input noises. For non-affine operations, a first-order Taylor approximation is used to obtain a linear behavior. These models, based on the perturbation theory, are only valid for smooth operations. An operation is considered to be smooth if the output is a continuous and differentiable function of its inputs.

Noise Aggregation Finally, the output noise b_y is the sum of all the noise source contributions. The second-order moment of b_y can be expressed as a weighted sum of the statistical parameters of the noise source:

$$E\left(e_y^2\right) = \sum_{i=1}^{N_e} K_i \sigma_{e_i}^2 + \sum_{i=1}^{N_e} \sum_{j=1}^{N_e} L_{ij} \mu_{e_i} \mu_{bj} \tag{6.21}$$

where μ_{e_i} and $\sigma_{e_i}^2$ are, respectively, the mean and the variance of noise source e_i, and N_e is the total number of error sources. These terms depend on the fixed-point formats and are determined during the evaluation of the accuracy analytical expression. The terms K_i and L_{ij} are constant and depend on the computation graph between e_i and the output. Thus, these terms are computed only once for the evaluation of the accuracy analytical expression. These constant terms can be considered as the gain between the noise source and the output.

For the case of linear time-invariant systems, the expressions of K_i and L_{ij} are given in [58]. The coefficient L_{ij} can now be computed by the multiplication of terms L_i and L_j, which can be calculated independently. The coefficients K_i and L_{ij} are determined from the transfer function $H_i(z)$ or the impulse response $h_i(n)$ of the system having e_i as input and b_y as output. In [59, 60], a technique is proposed to compute these coefficients from the SFG (Signal Flow Graph) of the application. The recurrent equation of the output contribution of e_i is computed by traversing the SFG representing the application at the noise level. To support recursive systems, for which the SFG contains cycles, this SFG is transformed into several Directed Acyclic Graphs (DAGs). The recurrent equations associated with each DAG are computed and then merge together after a set of variable substitutions. The different transfer functions are determined from the recurrent equations by applying a Z transform.

In [57], AA is used to keep track of the propagation of every single noise contribution along the datapath, and from this information, the coefficients K_i and L_i are extracted. The method has been proposed for LTI in [56] and for non-LTI systems in [57]. An affine form, defined by a central value and an uncertainty term (error term in this context), is assigned to each noise source. These terms depend on the mean and variance of the noise source. Then, the central value and the uncertainty terms associated with each noise source are propagated inside the system through an affine arithmetic-based simulation. The values of the coefficients K_i and L_{ii} are extracted from the affine form of the output noise. In the case of recursive systems, it is necessary to use a large number of iterations to ensure that the results converge to stable values. In some cases, this may lead to large AA error terms and therefore to long computation time.

In the method proposed in [61], an analytical expression of the coefficients K_i and L_{ij} is determined. For each noise source e_i, the recurrent equation of the output contribution of e_i is determined automatically from the application SFG with the technique presented in [60]. A time-varying impulse response h_i is computed from each recurrent equation. The output quantization noise b_y is the sum of the noise source e_i convolved with its associated time-varying impulse response. The second-order moment of b_y is determined. The expression of the coefficients is proposed in [61]. These coefficients can be computed directly from their expression by approximating an infinite sum, or a linear prediction approach can be used to obtain more quickly the value of these coefficients. The statistical parameters of the signal terms involved in the expression of the coefficients are computed from a single floating-point simulation, leading to reduced computation times. The analysis to compute coefficients K_i and L_{ij} is done on an SFG representing the application and where the control flow has been removed. To avoid loop unrolling which can lead to huge graph, a method based on polyhedral analysis has been proposed in [62].

Different hybrid techniques [52, 55, 63] that combine simulations and analytical expressions have been proposed to compute the coefficients K_i and L_{ij} from a set of simulations. In [55], these $N_e(N_e + 1)$ coefficients are obtained by solving a linear system in which K_i and L_{ij} are the variables. The way to proceed is to carry out several fixed-point simulations where a range of values for σ_{e_i} and μ_{e_i} is covered for each noise source. The fixed-point parameters of the system are set carefully to control each quantizer and to analyze its influence on the output. For each simulation, the statistical parameters of each noise source e_i are known from the fixed-point parameter and the output noise power is measured. At least $N_e(N_e + 1)$ fixed-point simulations are required to be able to solve the system of linear equations. A similar approach is used in [63] to obtain the coefficients by simulation. Each quantizer is perturbed to analyze its influence at the output to determine K_i and L_{ii}. To obtain the coefficients L_{ij} with $i \neq j$, the quantizers are perturbed in pairs. This approach requires again $N_e(N_e + 1)$ simulations to compute the coefficients, which requires long computation times.

During the last fifteen years, numerous works on analytical approaches for quantization error power estimation have been conducted and interesting progresses have been made for the automation of this process. These approaches allow the evaluation of the quantization error power and are very fast compared to simulation-based approaches. Theoretical concepts have been established enabling the development of automatic tools to generate the expression of the quantization error power. The limit of the proposed methods has been identified. Analytical approaches based on perturbation theory are valid for systems made up of only smooth operations.

6.6.2 Simulation-Based Techniques

When using simulation-based techniques, an intermediate accuracy metric is generally evaluated. Indeed, the direct evaluation of the quality metric at the output of an application may require numerous samples to be simulated. For instance, Huang et al. [40] illustrated the example of functional simulation technique on a length-10 dot product with data formatted on 32-bit. To simulate the approximate application covering the whole input space, $32^{20} \leq 1.3 \times 10^{30}$ different input vectors would be simulated.

For the evaluation of an intermediate accuracy metric as the noise power, the metric is generally computed by simulating an arbitrary, and large, number of random inputs N_S [64]. For fixed-point arithmetic, in the literature, the number of samples ranges from 10^5 [64] to 10^{12} [65]. For determining the noise power P induced by finite precision, two different versions of the application are simulated as presented in Fig. 6.7. The distance between the output of the application with infinite precision and the output of the application with finite precision is measured and squared for each simulated sample. The expected value of these distances is then computed to obtain the value of the noise power P at the output of the application. The slow software simulation of fixed-point data-types as well as the high number of samples to simulate makes generally fixed-point conversion a long and tedious task. Sedano et al. [66] proposed to improve the speed of fixed-point conversion, to use inferential statistics to infer the number of inputs to simulate.

The number of significant bits for finite precision arithmetic can be considered as an intermediate accuracy metric. In this context, discrete stochastic arithmetic has been proposed by Jezequel and Chesneaux [67] for floating-point and fixed-point arithmetic. In this method, for each data, the last significant bit is randomly perturbed and the application is run N times using the same input data for each run. The average over these N runs is used as an estimate of the exact value. A t-distribution is used to provide a confidence interval for the estimated value. The number of significant bits is deduced from the number of common bits between the obtained values and the estimation. The main advantage of this approach is the low number of runs N required to obtain a good estimation. In practice, $N = 3$ is used.

Finally, simulation-based techniques are widely employed since they are not limited by the applicability of analytical techniques. However, due to the long emulation time of the approximation techniques, functional simulation techniques do not scale well with large applications.

6.6.3 Conclusion

Analytical techniques have majorly been proposed to characterize errors induced by approximations. These techniques are specific to the considered error metric and the considered approximation technique. Simulation-based techniques have the asset of being generic but do not scale with large systems. Statistical models allow studying the number of simulations to lead so as to get an estimation of the error properties according to user-defined confidence requirements.

6.7 Conclusion

By exploiting the error resilience of numerous applications, AC allows saving energy or reducing the execution time but at the expense of introducing errors in the processing. Error analysis is one of the crucial steps in the integration of AC in the design and implementation process.

The error analysis process can be decomposed in three main steps. The first step corresponds to the AC error characterization which aims at developing a model defining the error due to a specific AC technique. Two types of techniques can be considered to characterize the AC error metrics. Analytical approaches aim at defining a mathematical model of the error metrics. These approaches have been widely used to model the error induced by finite precision like in fixed-point and floating-point arithmetic. For AC techniques for which a mathematical model cannot be obtained, simulation-based techniques can be considered to characterize the error. Emulation techniques are incorporated in the application source code to mimic the AC error behavior.

The second and third steps aim at propagating the error inside the application to determine, respectively, an accuracy metric or directly the quality metric. Determining an intermediate accuracy metric is a good option when determining directly the quality metric is not feasible. Two types of techniques can be considered to determine the accuracy or quality metric. Analytical approaches have the advantage to lead to low evaluation times. Nevertheless, these approaches are limited in terms of supported applications, and they are more adapted for the evaluation of an accuracy metric than an application-specific quality metric. Simulation-based techniques have the advantage to support any kind of systems and can be used to evaluate an accuracy metric or directly the quality metric. The AC technique is emulated or the error model is injected during the simulation. Nevertheless, these techniques lead to high evaluation times especially when this process is included in the AC design space exploration.

References

1. Widrow, B. (1960). Statistical analysis of amplitude quantized sampled-data systems. *Transaction on AIEE, Part. II: Applications and Industry, 79*, 555–568.
2. Chippa, V., Chakradhar, S., Roy, K., & Raghunathan, A. (2013). Analysis and characterization of inherent application resilience for approximate computing. In *50th ACM/IEEE Design Automation Conference (DAC)* (pp. 1–9).
3. Akturk, I., Khatamifard, K., & Karpuzcu, U. R. (2015). On quantification of accuracy loss in approximate computing. In *Workshop on Duplicating, Deconstructing and Debunking (WDDD)* (Vol. 15).
4. Chippa, V. K., Chakradhar, S. T., Roy, K., & Raghunathan, A. (2013). Analysis and characterization of inherent application resilience for approximate computing. In *Proceedings of the 50th Annual Design Automation Conference* (p. 113). ACM.
5. Camus, V., Schlachter, J., & Enz, C. (2015). Energy-efficient inexact speculative adder with high performance and accuracy control. In *2015 IEEE International Symposium on Circuits and Systems (ISCAS)* (pp. 45–48). IEEE.
6. Moore, R. E. (1962). Interval arithmetic and automatic error analysis in digital computing. *Ph. D. Dissertation, Department of Mathematics, Stanford University.*
7. Huang, J., Lach, J., & Robins, G. (2011). Analytic error modeling for imprecise arithmetic circuits. *Proc. SELSE, 51*, 64–69.
8. Du, K., Varman, P., & Mohanram, K. (2012). High performance reliable variable latency carry select addition. In *Design, Automation & Test in Europe Conference & Exhibition (DATE), 2012* (pp. 1257–1262). IEEE.
9. Jiang, H., Liu, C., Liu, L., Lombardi, F., & Han, J. (2017). A review, classification and comparative evaluation of approximate arithmetic circuits. *ACM Journal on Emerging Technologies in Computing Systems (JETC), 13*, 1–34.
10. Camus, V., Cacciotti, M., Schlachter, J., & Enz, C. (2018). Design of approximate circuits by fabrication of false timing paths: The carry cut-back adder. In *IEEE Journal on Emerging and Selected Topics in Circuits and Systems (JETCAS), 8*, 746–757.
11. Zhu, N., Goh, W. L., Zhang, W., Yeo, K. S., & Kong, Z. H. (2010). Design of low-power high-speed truncation-error-tolerant adder and its application in digital signal processing. *IEEE Transactions on Very Large Scale Integration (VLSI) Systems, 18*(8), 1225–1229.
12. Verma, A. K., Brisk, P., & Ienne, P. (2008). Variable latency speculative addition: A new paradigm for arithmetic circuit design. In *Proceedings of the Conference on Design, Automation and Test in Europe* (pp. 1250–1255). ACM.
13. Van, L.-D., Wang, S.-S., & W.-S. Feng (2000). Design of the lower error fixed-width multiplier and its application. *IEEE Transactions on Circuits and Systems II: Analog and Digital Signal Processing, 47*(10), 1112–1118.
14. Juang, T.-B., & Hsiao, S.-F. (2005). Low-error carry-free fixed-width multipliers with low-cost compensation circuits. *IEEE Transactions on Circuits and Systems II: Express Briefs, 52*(6), 299–303.
15. Berens, F., & Naser, N. (2004). Algorithm to system-on-chip design flow that leverages system studio and SystemC 2.0. 1. *Synopsys Inc.*
16. Grötker, T., Liao, S., Martin, G., & Swan, S. (2010). *System design with SystemCTM*. Springer Science & Business Media.

17. Müller, W., Rosenstiel, W., & Ruf, J. (2007). *SystemC: Methodologies and applications*. Springer Science & Business.
18. Akbarpour, B., & Tahar, S. (2003). Modeling SystemC fixed-point arithmetic in HOL. In *International Conference on Formal Engineering Methods* (pp. 206–225). Springer.
19. Bhurat, H., Bryan, T., & Wall, J. (2014). *Best Practices for Converting MATLAB Code to Fixed Point*. Technical Report, MathWorks.
20. SIEMENS EDA. (2022). *Algorithmic C (AC) datatypes reference manual* (Software Version v4.4.2).
21. Feist, T. (2012). Vivado design suite. *White Paper, 5*, 30.
22. Barrois, B., & Sentieys, O. (2017). Customizing fixed-point and floating-point arithmetic - a case study in k-means clustering. In *2017 IEEE International Workshop on Signal Processing Systems (SiPS)* (pp. 1–6).
23. Tagliavini, G., Marongiu, A., & Benini, L. (2018). Flexfloat: A software library for transprecision computing. *IEEE Transactions on Computer-Aided Design of Integrated Circuits and Systems, 39*(1), 145–156.
24. Flegar, G., Scheidegger, F., Novaković, V., Mariani, G., Tom´ s, A. E., Malossi, A. C. I., & Quintana-Ortí, E. S. (2019). Floatx: Ac++ library for customized floating-point arithmetic. *ACM Transactions on Mathematical Software (TOMS), 45*(4), 1–23.
25. Sampson, A., Dietl, W., Fortuna, E., Gnanapragasam, D., Ceze, L., & Grossman, D. (2011). EnerJ: Approximate data types for safe and general low-power computation. *ACM SIGPLAN Notices, 46*, 164–174. ACM.
26. Sripad, A., & Snyder, D. L. (1977). A necessary and sufficient condition for quantization error to be uniform and white. *IEEE Transactions on Acoustics, Speech, and Signal Processing, 25*, 442–448.
27. Constantinides, G. A., Cheung, P., & Luk, W. (1999). Truncation noise in fixed-point SFGs. *IEE Electronics Letters, 35*(23), 2012–2014.
28. Menard, D., Novo, D., Rocher, R., Catthoor, F., & Sentieys, O. (2011). Quantization mode opportunities in fixed-point system design. In *18th European Signal Processing Conference (EUSIPCO-2010) (2010)*, (Aalborg, Denmark) (pp. 542–546). EURASIP.
29. Liu, C., Han, J., & Lombardi, F. (2015). An analytical framework for evaluating the error characteristics of approximate adders. *IEEE Transactions on Computers, 64*(5), 1268–1281.
30. Wu, Y., Li, Y., Ge, X., & Qian, W. (2017). An accurate and efficient method to calculate the error statistics of block-based approximate adders. *Preprint arXiv:1703.03522*.
31. Mazahir, S., Hasan, O., Hafiz, R., Shafique, M., & Henkel, J. (2017). Probabilistic error modeling for approximate adders. *IEEE Transactions on Computers (TC), 66*(3), 515–530.
32. Mazahir, S., Hasan, O., Hafiz, R., & Shafique, M. (2017). Probabilistic error analysis of approximate recursive multipliers. *IEEE Transactions on Computers, 66*(11), 1982–1990.
33. Roy, A. S., & Dhar, A. S. (2018). A novel approach for fast and accurate mean error distance computation in approximate adders. In *2018 IEEE International Symposium on Circuits and Systems (ISCAS)* (pp. 1–5). IEEE.
34. Sengupta, D., Snigdha, F. S., Hu, J., & Sapatnekar, S. S. (2017). Saber: Selection of approximate bits for the design of error tolerant circuits. In *Proceedings of the 54th Annual Design Automation Conference 2017* (p. 72), ACM.
35. Sengupta, D., Snigdha, F. S., Hu, J., & Sapatnekar, S. S. (2018). An analytical approach for error PMF characterization in approximate circuits. *IEEE Transactions on Computer-Aided Design of Integrated Circuits and Systems, 38*, 70–83.
36. Liu, Y., Zhang, T., & Parhi, K. K. (2010). Computation error analysis in digital signal processing systems with overscaled supply voltage. *IEEE Transactions on Very Large Scale Integration (VLSI) Systems, 18*(4), 517–526.
37. Shafique, M., Ahmad, W., Hafiz, R., & Henkel, J. (2015). A low latency generic accuracy configurable adder. In *2015 52nd ACM/EDAC/IEEE Design Automation Conference (DAC)* (pp. 1–6). IEEE.

38. Bonnot, J., Camus, V., Desnos, K., & Menard, D. (2018). Cassis: Characterization with adaptive sample-size inferential statistics applied to inexact circuits. In *2018 26th European Signal Processing Conference (EUSIPCO)* (pp. 677–681). IEEE.
39. Bonnot, J., Camus, V., Desnos, K., & Menard, D. (2019). Adaptive simulation-based framework for error characterization of inexact circuits. *Microelectronics Reliability, 96*, 60–70.
40. Huang, J., Lach, J., & Robins, G. (2012). A methodology for energy-quality tradeoff using imprecise hardware. In *Proceedings of the 49th Annual Design Automation Conference* (pp. 504–509). ACM.
41. Chan, W.-T. J., Kahng, A. B., Kang, S., Kumar, R., & Sartori, J. (2013). Statistical analysis and modeling for error composition in approximate computation circuits. In *2013 IEEE 31st International Conference on Computer Design (ICCD)* (pp. 47–53). IEEE.
42. Wyse, M. (2015). Modeling approximate computing techniques. *Academic paper*.
43. Sampson, A., Baixo, A., Ransford, B., Moreau, T., Yip, J., Ceze, L., & Oskin, M. (2015). Accept: A programmer-guided compiler framework for practical approximate computing. *University of Washington Technical Report UW-CSE-15-01*, Vol. 1, No. 2.
44. Parashar, K., Rocher, R., Menard, D., & Sentieys, O. (2010). A hierarchical methodology for word-length optimization of signal processing systems. In *2010 23rd International Conference on VLSI Design* (pp. 318–323). IEEE.
45. Menard, D., Serizel, R., Rocher, R., & Sentieys, O. (2008). Accuracy constraint determination in fixed-point system design. *EURASIP Journal on Embedded Systems, 2008*, 12.
46. Rocher, R., & Scalart, P. (2012). Noise probability density function in fixed-point systems based on smooth operators. In *Proceedings of the 2012 Conference on Design and Architectures for Signal and Image Processing* (pp. 1–8, 2012).
47. Brönnimann, H., Melquiond, G., & Pion, S. (2006). The design of the boost interval arithmetic library. *Theoretical Computer Science, 351*(1), 111–118.
48. De Figueiredo, L. H., & Stolfi, J. (2004). Affine arithmetic: Concepts and applications. *Numerical Algorithms, 37*(1–4), 147–158.
49. Gay, O., Coeurjolly, D., & Hurst, N. (2006). *Libaffa-C++ affine arithmetic library for GNU/Linux*.
50. Fang, C. F., Rutenbar, R. A., & Chen, T. (2003). Effcient static analysis of fixed-point error in DSP applications via affine arithmetic modeling. In *Proceedings of IEEE International Conference on Computer Aided Design* (pp. 275–282). Citeseer.
51. Caffarena, G., López, J. A., Fernández-Herrero, A., & Carreras, C. (2010). SQNR estimation of non-linear fixed-point algorithms. In *2010 18th European Signal Processing Conference* (pp. 522–526). IEEE.
52. Constantinides, G. A. (2006). Word-length optimization for differentiable nonlinear systems. *ACM Transactions on Design Automation of Electronic Systems, 11*(1), 26–43.
53. Rocher, R., Menard, D., Scalart, P., & Sentieys, O. (2007). Analytical accuracy evaluation of Fixed-Point Systems. In *Proceedings of the European Signal Processing Conference (EUSIPCO)*, (Poznan).
54. Menard, D., Rocher, R., Scalart, P., & Sentieys, O. (2004). SQNR determination in non-linear and non-recursive fixed-point systems. In *European Signal Processing Conference (EUSIPCO)* (pp. 1349–1352).
55. Shi, C., & Brodersen, R. (2004). A Perturbation theory on statistical quantization effects in fixed-point DSP with non-stationary inputs. In *IEEE International Symposium on Circuits and Systems (ISCAS), 3*, 373–376.
56. Lopez, J., Caffarena, G., Carreras, C., & Nieto-Taladriz, O. (2008). Fast and accurate computation of the roundoff noise of linear time-invariant systems. *IET Circuits, Devices and Systems, 2*, 393–408.
57. Caffarena, G., Carreras, C., López, J., & Fernández, A. (2010). SQNR estimation of fixed-point DSP algorithms. *EURASIP Journal on Advances in Signal Processing, 2010*, 1–11.
58. Menard, D., & Sentieys, O. (2002). A methodology for evaluating the precision of fixed-point systems. In *Proceedings of the IEEE International Conference on Acoustics, Speech, and Signal Processing (ICASSP)*, (Orlando).

59. Menard, D., & Sentieys, O. (2002). Automatic evaluation of the accuracy of fixed-point algorithms. In *IEEE/ACM Design, Automation and Test in Europe (DATE)*, (Paris).
60. Menard, D., Rocher, R., & Sentieys, O. (2008). Analytical fixed-point accuracy evaluation in linear time-invariant systems. *IEEE Transactions on Circuits and Systems I: Regular Papers, 55*, 3197–3208.
61. Rocher, R., Menard, D., Scalart, P., & Sentieys, O. (2012). Analytical approach for numerical accuracy estimation of fixed-point systems based on smooth operations. *IEEE Transactions on Circuits and Systems I: Regular Papers, PP*(99), 1–14.
62. Deest, G., Yuki, T., Sentieys, O., & Derrien, S. (2014). Toward scalable source level accuracy analysis for floating-point to fixed-point conversion. In *Proceedings of the 2014 IEEE/ACM International Conference on Computer-Aided Design*, ICCAD '14, (Piscataway, NJ) (pp. 726–733). IEEE Press.
63. Fiore, P. (2008). Efficient approximate wordlength optimization. *IEEE Transactions on Computers, 57*(11), 1561 –1570.
64. Keding, H., Hurtgen, F., Willems, M., & Coors, M. (1998). Transformation of floating-point into fixed-point algorithms by interpolation applying a statistical approach. In *9th International Conference on Signal Processing Applications and Technology (ICSPAT 98)*.
65. Keding, H., Willems, M., Coors, M., & Meyr, H. (1998). Fridge: A fixed-point design and simulation environment. In *Proceedings of the Conference on Design, Automation and Test in Europe* (pp. 429–435). IEEE Computer Society.
66. Sedano, E., López, J. A., & Carreras, C. (2012). Acceleration of Monte-Carlo simulation-based quantization of DSP systems. In *2012 19th International Conference on Systems, Signals and Image Processing (IWSSIP)* (pp. 189–192). IEEE.
67. Jézéquel, F., & Chesneaux, J.-M. (2008). CADNA: A library for estimating round-off error propagation. *Computer Physics Communications, 178*(12), 933–955.

Chapter 7
Accuracy-Aware Compilers

Sasa Misailovic

7.1 Introduction

Users expect their programs to run fast, consume minimum energy, and be resilient to faults. These expectations put major pressure on developers and the tools they use. Approximate computing offers an exciting opportunity to provide a better performance, energy-efficiency, and resilience at the cost of a small amount of an application's quality. Many application domains, e.g., data analytics, machine learning, multimedia processing, robotics, and scientific simulations, are inherently approximate and can gracefully tolerate small amounts of error. Besides algorithmic approximations (which have been the responsibility of algorithm designers), the past two decades have brought up many ways in which systems expose new approximations at different levels of the computing stack. However, unlocking the potential of system-level approximations requires us to rethink the entire system stack—and programming systems in particular—to include accuracy as a first-class concern.

Compilers are programming systems that translate programs from the developer-readable source code to the binary code that executes on hardware. A fundamental property of traditional optimizing compilers is that they generate efficient binary programs that *preserve the semantics* of the original source programs. It means that the binary program should produce *the same result* as the source code that the programmer wrote. Consequently, each traditional program optimization has aimed to improve performance while ensuring that these transformations are *correct*—they preserve the program semantics (proved using static program analysis [1]). The compile-time static analysis uses concepts from discrete mathematics as its

S. Misailovic (✉)
University of Illinois Urbana-Champaign, Champaign, IL, USA
e-mail: misailo@illinois.edu

A. Bosio et al. (eds.), *Approximate Computing Techniques*,
https://doi.org/10.1007/978-3-030-94705-7_7

foundations: the relevant program properties have often been represented as sets or logical formulas over program elements, and the program analysis operates on these mathematical objects.

Despite the prominent role of approximation in applications, architectures, and systems, standard program analysis and compilation systems do not take advantage of approximation opportunities. Traditional optimizations are too rigid to exploit the full optimization potential of the applications. Some compilers have exposed several approximate optimizations as unsafe flags (e.g., relaxing the semantics of floating-point operations). The compilers left a software developer solely responsible for managing all aspects of approximation, which can often result in inflexible approximation choices hard-coded in the program implementations.

Accuracy-aware compilers embrace approximation as a first-class concept in the optimization process. They allow program transformations to automatically apply approximations that *intentionally change the semantics* of programs and expose a *tradeoff* between the performance of the compiled programs and the accuracy of the programs' results. This means that the binary program can produce different acceptable results that may satisfy the user's end-to-end accuracy or latency requirements. Unlike their traditional counterparts, *accuracy-aware transformations* need to reason about the changes they introduce in the program semantics and whether they influence the program safety (e.g., that the approximation does not cause a crash during execution).

Reasoning about accuracy-aware program transformations requires a more flexible framework than the rigid correctness-based analysis framework offered by traditional compilers. First, we need to define analyses for reasoning about two key properties: (1) *accuracy analysis* reasons about the change in the result caused by the transformation and is often based on numerical or probabilistic techniques, and (2) *safety analysis* ensures that applying the transformation does not cause the program to crash or experience other unacceptable errors and is often based on logical reasoning (in an absolute or relative sense [2]). Furthermore, accuracy-aware compilers need to include a search for profitable tradeoffs in their core framework. This search leverages the information that the analyses provide about accuracy and safety to navigate the more promising parts of the tradeoff space.

This chapter will focus on the foundation of accuracy-aware compilers and different techniques that help us to characterize the accuracy–performance tradeoffs and improve confidence about the safety of the approximated programs. We will present the main concepts for the optimization of sequential programs. Through two case studies, we will describe two main directions in developing accuracy-aware compilers: (1) *sensitivity profiling*-based techniques use concrete executions of candidate transformed programs and dynamic program analysis to reason about their accuracy and safety, while (2) *static analysis*-based techniques use probabilistic or relational static analysis to derive formulas that characterize acceptable results and use mathematical optimization to find the best performing program versions, without necessarily running them.

7.2 Accuracy-Aware Compilation

An accuracy-aware compiler searches the space of possible optimizations to find the composition of optimizations that jointly give an acceptably accurate result. We can intuitively formulate the task as

$$\begin{aligned} \textbf{\textit{Maximize}}: &\ \textsc{Performance}(\textit{Configuration}) \\ \textbf{\textit{Constraints}}: &\ \textsc{AccuracyLoss}(\textit{Configuration}) \leq \textit{AccuracyBound} \\ &\ \textsc{Safety}(\textit{Configuration}) = \text{True} \\ \textbf{\textit{Variables}}: &\ \textit{Configuration} \in \textit{SearchSpace} \end{aligned}$$

This optimization formulation maximizes an expression for performance subject to the constraints on accuracy loss and safety. The available accuracy-aware transformations determine the search space, and the *configuration* of an approximate program represents the approximation type and its level at each applicable program location. The compiler determines the configuration representation (e.g., a vector or a mapping) and computes the expressions for performance and accuracy. However, the compiler needs the developer's help to identify the accuracy requirement.

The accuracy specification is the main departure of accuracy-aware compilers from classical compilers. It consists of three components, which are specific to the context of the use of the application, need to be provided by the developer, and help construct the accuracy constraint:

- **Output Abstraction.** The output abstraction is a function that works with the program's output (and optionally its input) to compute a value or a list of values that represent relevant properties of the output. We denote a result of output abstraction as $\mathbf{o} = (o_1, \ldots, o_m)$. Typically, an output abstraction function selects relevant numbers from an output file or files or computes an application-specific measure of the quality of the output. Many approximate computations come with such abstractions already defined and available, for instance, as a part of the program testing effort.
- **Accuracy Metric.** The accuracy metric function computes the distance between the results of the original and transformed programs.[1] The function $Q(\mathbf{o}, \hat{\mathbf{o}})$ takes as input the two numerical vectors computed by the output abstraction function. The abstracted output $\mathbf{o} = (o_1, \ldots, o_m)$ comes from the execution of

[1] Despite the name "accuracy metric" being traditionally used, these functions commonly compute the *accuracy loss* compared to some reference result. When these functions evaluate to a value close to 0, that means that the result of the approximate program is more accurate. They also do not always satisfy all the properties of mathematical metrics. For instance, distortion is not symmetric.

the original program. The abstracted output $\hat{\mathbf{o}} = (\hat{o}_1, \ldots, \hat{o}_m)$ comes from the execution of the transformed program. $Q(\cdot, \cdot)$ is also specified by the developer: some common choices include the absolute or square distance between the output abstraction components $\mathbf{o}$ from the original program and the output abstraction components $\hat{\mathbf{o}}$. Another common example is called *distortion* [3], based on relative difference. It is a weighted mean scaled difference between the program outputs (weights capture the relative importance of the components): $Q_{dist}(\mathbf{o}, \hat{\mathbf{o}}) = \frac{1}{m} \sum_{i=1}^{m} w_i \left| \frac{o_i - \hat{o}_i}{o_i} \right|$. The most common comparison output is the result of the original program (before transformation), but it can also be a reference output obtained some other way.
- **Accuracy Goal.** A developer can specify the accuracy bound b that indicates the maximum acceptable (tolerable) accuracy loss. Specifically, the bound represents the extreme value of the accuracy metric.

The accuracy-aware compiler produces the optimized program that can operate at different points in the accuracy–performance tradeoff space. Most existing compilers cover one of the three workflows:

- **Compile with a fixed approximation.** A compiler of this sort applies the transformations to satisfy an explicit accuracy bound. The approximation is fixed and optimizing for a different bound requires recompilation. Common examples are the techniques that optimize the precision of floating-point numbers.
- **Compile with an approximate knob.** A compiler of this sort applies transformations with a parameter that can change the aggressiveness of the approximation. This parameter is known as an *approximation knob* and can be set at the beginning of the execution or during runtime. To characterize the best configurations of the approximation knobs, the compiler produces a *tradeoff curve*—each point on this curve represents a configuration of the program's approximation knobs with the best accuracy/performance tradeoffs.
- **Compile with runtime adaptation.** In addition to the exposed knob and the tradeoff curve, the compiler inserts a runtime system that dynamically chooses to change the approximation level. A runtime system can monitor performance and adjust accuracy to meet the performance target, or it can monitor accuracy and adjust performance to meet the accuracy target. In some special cases, the runtime system can operate without the compile-time tradeoff curve but instead discovers the optimal tradeoffs solely at runtime.

7.2.1 Accuracy-Aware Transformations

Accuracy-aware transformations automate the optimization of approximate applications. These transformations intentionally change the semantics of programs to trade accuracy for improved performance, energy consumption, or resilience by exploiting the properties of the program's inputs, structure, and execution environ-

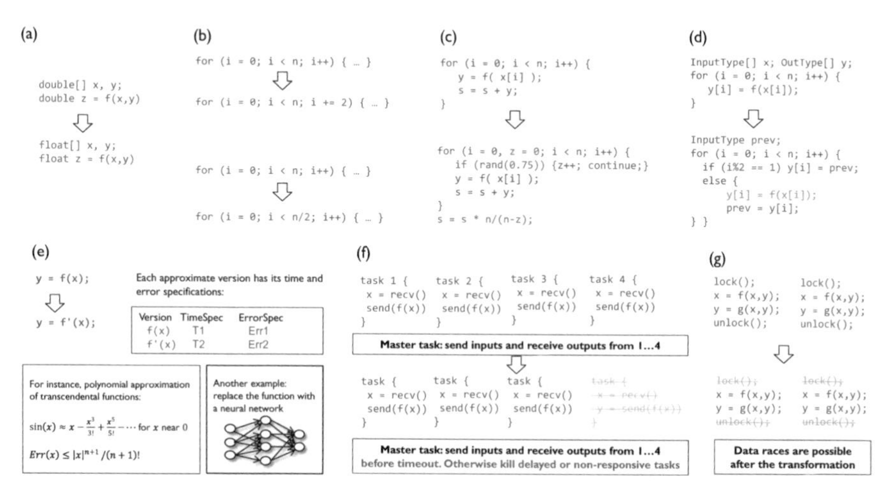

Fig. 7.1 Examples of accuracy-aware optimizations (the approximation code is marked in red). (**a**) Data structure optimization. (**b**) Loop perforation. (**c**) Reduction sampling. (**d**) Approximate tiling. (**e**) Function substitution. (**f**) Dropping tasks. (**g**) Remove locks

ment. Researchers have developed many accuracy-aware software (compiler-level) transformations and approximate hardware components (which can be targeted by a compiler). To achieve their goals, the transformations reduce computation, data representation, and communication. It is common for all of them to create new knobs that expose and control the tradeoff between accuracy and performance/energy. Figure 7.1 presents examples of some popular approximations.

Transformations that reduce computation find instructions in the execution that can be fully skipped or simplified. Examples include task skipping [3–5] (which executes only a subset of tasks), loop perforation [6–8] (which executes only a subset of loop iterations), reduction sampling [9, 10] (which selects only a random subset of inputs for programs that implement reductions like summation or maximization), approximate function substitution [9–12] (which replaces the exact implementation of a function with a less accurate alternative), and dynamic knobs [13] (which select at runtime one of the several approximate versions of the computation based on the application's performance goal).

Transformations that reduce data representation try to reduce the size of data that the application processes. Examples include the selection of floating-point representation (double/float/half-float) [14, 15], quantization [16], or algorithmic data-sketching techniques [17]. Transformations that reduce communication and synchronization are applicable on parallel programs. Examples include early termination of barriers at parallel loops [18] (which results in skipping contributions from interrupted threads) and approximate parallelization with the possibility of data races [19–21].

Finally, for approximate hardware components, we can create abstractions that represent them as computation reductions (e.g., approximate ALU [22] or accelerators [23]), data reductions (e.g., approximate memories [24]), or communication reductions (e.g., approximate communication channels or networks on chip [25]).

7.2.2 Strategies for Exploring the Optimization Search Space

Automatically optimizing approximate programs using accuracy-aware transformations provides new opportunities to reduce engineering effort and resource consumption and increase program resilience. This opportunity, however, comes at a price—the transformations introduce uncertainty into the program's execution, which reflects on the quality of the results it produces. This uncertainty raises many new research questions: (1) how to identify parts of a program that are good candidates for accuracy-aware transformations, (2) how to characterize the effects of a transformation on the program's execution, especially the result's accuracy, and (3) how to automatically discover transformations that provide maximum performance gains for acceptable accuracy losses.

The accuracy-aware optimization techniques fall broadly into two groups—predominantly dynamic approaches doing sensitivity profiling and predominantly static approaches doing program analysis.

Sensitivity profiling-based compilation requires a developer to provide a set of representative inputs and an application-level accuracy metric that quantifies the accuracy of the produced result (e.g., peak signal to noise for video encoders). Then, a sensitivity profiler speculatively applies the accuracy-aware transformations at various points in the program and validates the transformations by testing whether the transformed programs, when executed on the provided inputs, produce results with acceptable accuracy (as calculated by the accuracy metric). These techniques are effective in finding transformed programs with attractive tradeoffs, even in complex programs. However, since this approach relies on representative inputs, its results do not provide guarantees for other inputs.

Analysis-based compilation combines static program analysis (which extracts a set of formulas characterizing the program accuracy) with mathematical optimization techniques to provide a foundation for rigorous program optimization using accuracy-aware transformations. They do not execute the program but leverage accuracy and performance models (which serve as ranking functions to prioritize different optimizations). These techniques operate on time- or energy-consuming subcomputations that we call *approximate kernels*. For each approximate kernel, a developer provides a formal description of the kernel's inputs and the expected output accuracy. Program analysis can ensure that the approximate version of the kernel satisfies the probabilistic output specification for all inputs that adhere to the input specification. The program optimization algorithm can use this analysis to reduce the kernel approximation to a standard mathematical optimization problem. This approach does not require representative inputs, but a developer can optionally use sensitivity profiling to (1) help identify approximate kernel computations and

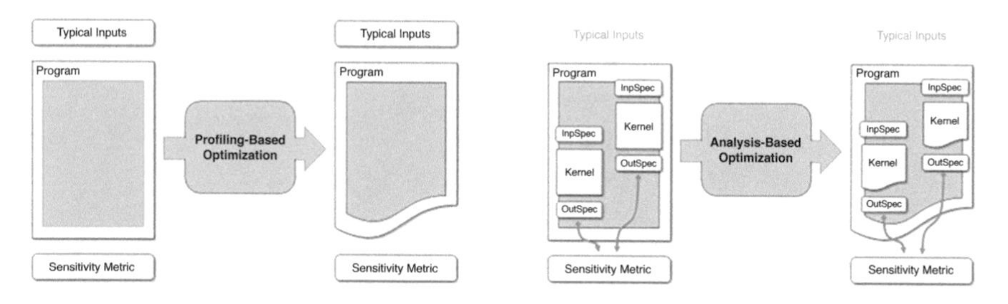

Fig. 7.2 The approaches for accuracy-aware optimization. We start with the original program and produce the approximate program. The analysis-based compiler takes into consideration the annotations on the kernels, while the profiling-based compiler treats the whole program as one entity

(2) derive the kernel-level accuracy specifications that likely satisfy the application-level accuracy metric.

Figure 7.2 illustrates the conceptual difference between the two approaches. The profile-based optimization transforms program subcomputations driven by the inputs and subject to the application-level accuracy metric. While it can often find attractive tradeoffs, it does not provide an intuition for why the transformations work and how we can quantify the effect of the changes (illustrated by a shaded area of the program). In contrast, the analysis-based optimization operates on explicitly exposed approximate kernels with their specifications, which the developer wrote to meet the application-level accuracy requirement (optionally derived using sensitivity profiling). Then, the rigorous analysis and optimization techniques can automatically approximate the kernel functions, while satisfying the developer's specifications. This approach can, therefore, improve a developer's understanding of why accuracy-aware transformations work.

7.2.3 *Approximate Kernels*

Many approximate computations have a specific structure, which can be useful for accuracy-aware optimization. A large portion of a program's work is performed in one or several *approximate kernel* computations. Each execution of a kernel typically processes a part of the application's input and either directly produces a part of the application's output or guides the execution of the application to produce the final output. Kernels are important for understanding the success of program approximations. If approximated kernels have a small accuracy loss, then the full program will likely have a small accuracy loss too.

As an example of an approximate kernel, we show a loop that iterates over a list of elements and aggregates the elements to produce a final sum:

```
float sum = 0;
for (int i = 0; i < n; i++)
     sum = sum + a[i];
```

This loop computes the value of `sum` by aggregating the elements of an array `a` with `n` elements. These approximate kernels have similar structure and functionality and represent instances of computational patterns amenable to approximation. Our previous work divides patterns by their structural properties (e.g., a kernel loop calculates a sum) or functional properties (e.g., a kernel loop's result is used as a distance metric within the application) [7].

We can apply multiple transformations to the summation loop. To run the loop faster, loop perforation can change the induction variable increment from `i++` to `i+=2`. To save energy when executing the loop body on approximate hardware, a compiler can (1) transform the addition operator in the expression `sum = sum + a[i]` to its approximate version `sum = sum +. a[i]`, where `+.` is the approximate sum operator or (2) specify that the array `a` should be stored in unreliable memory using a declaration `float[] a in urel`. Each of these transformations can cause the loop to produce a result `sum` that differs from the result of the original loop.

The success of many approximations rests on the empirical observations that the results of many transformed approximate kernels exhibit *small deviations* from the results of the original kernels *most of the time* [8, 26]. Figure 7.3 illustrates the impact of small and large deviations caused by approximating pixel computation in an image processing application (image scaling). The kernel code is conceptually averaging array elements. To show the impact of errors, on the left, we inject a black pixel with the probability specified under each picture. On the right, we change the absolute value of each pixel's color by the amount specified under each picture. The acceptable results have either many errors small in magnitude or there are infrequent arbitrary deviations. Both optimization approaches leverage these properties of approximate computations—profiling-based approaches do it implicitly, through

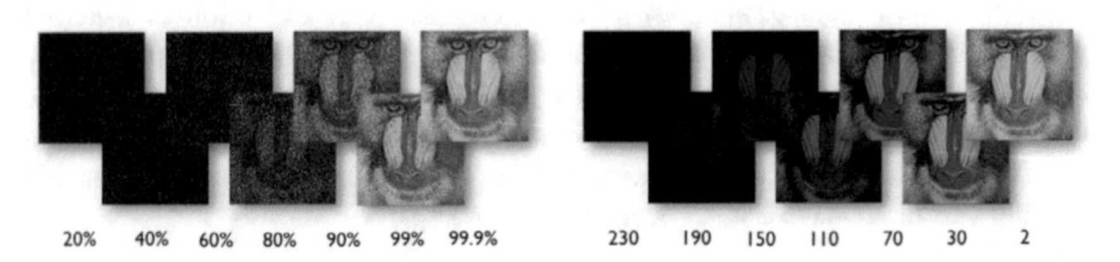

Fig. 7.3 The impact of errors on end-to-end acceptability: the kernel computation calculates the correct pixel value with only specified frequency (left) or computes it with the specified absolute error in each pixel component (right)

performance profiles, and analysis-based approaches do it explicitly, driven by the specifications.

These experiments show the usefulness of probabilistic accuracy specifications of kernels and programs:

- **Kernel Input Specification.** The input specification contains intervals or distributions of the computation's inputs. This specification characterizes the developer's level of knowledge about the inputs. For instance, a developer can specify that the elements of the array `a` in the example sum kernel belong to the interval [0, 10]. Specifications can also be relational—for instance, the elements of the array `a` in the approximate execution have the same value as the elements in the exact execution with probability at least 0.99.
- **Output Accuracy Specification.** The output specification is a probabilistic assertion about the output of the computation. The assertions are relational in that they compare the outputs of the original and approximate executions of the kernel. Their general form is "*Assuming that the inputs have the specified properties, the absolute difference between the results of the original and transformed program is less than or equal to* Δ *with probability at least* p." The developer provides the numerical quantities Δ and p. Two common special cases of these assertions consider only *the magnitude* of numerical error (when $p = 1$) and *the frequency* of producing an incorrect result (when $\Delta = 0$).

The way we defined the specifications for each kernel (i.e., the ranges in the input specifications and thresholds in the output specifications) connects the local error on the kernel level with the global error at the program level. Sensitivity profiling can help derive the specifications. For the example from the data presented in Fig. 7.3, a sensitivity profiler can infer the empirical relationship between the failure rate of the kernel computation and the program's accuracy metric (e.g., PSNR for images). Then, the acceptable reliability of the kernel is chosen to satisfy the end-to-end metric. We discussed how to use sensitivity profiling to derive specifications in [27, Section 2] and [28].

In the rest of this chapter, we present the main components of the two flavors of accuracy-aware optimization. We will show how to do an end-to-end profiling-based compilation that treats programs as a single black box. We will then show how to analyze kernel computations with analytical accuracy and performance models.

7.3 Case Study 1: Profile-Based Optimization

This case study presents a profiling-driven optimization framework SpeedPress [6, 7, 29] that optimizes programs using loop perforation (an accuracy-aware transformation that skips loop iterations). *Sensitivity profiling* is a dynamic program analysis that identifies program locations (in this case loops) where an approximation can be applied. A developer provides to SpeedPress the original application, a set of representative inputs, and an accuracy specification. SpeedPress automates optimizations in three steps:

1. It performs standard performance profiling to find loops in which the program spends the most time. Those loops, if perforated, are most likely to make the program faster.
2. It finds the set of time-consuming loops that can be successfully perforated (i.e., they are faster, have small accuracy loss, and do not exhibit critical errors). Specifically, it generates candidate approximate programs and checks if they satisfy the developer's accuracy specification. SpeedPress uses loop perforation as its transformation—for each loop it can control the fraction of loop iterations that can be skipped.
3. It perforates multiple loops at the same time and constructs a *Pareto-optimal tradeoff curve*, which contains the approximate programs that exhibit the most profitable tradeoffs between performance and accuracy.

The first two steps identify potentially profitable *approximation knobs*—parameters of perforated loops that we can vary to get different accuracy/performance tradeoffs. The last step performs algorithmic autotuning to identify the best configurations of approximation knobs.

7.3.1 Loop Perforation Transformation

SpeedPress implements the loop perforation transformation within the LLVM compiler framework [30]. The perforation pass works with any loop that the existing LLVM loop canonicalization pass, *loop-simplify*, can convert into the following form:

```
for (i = 0; i < M; i++) { ... }
```

In this form, the loop has a unique induction variable (in the code above, `i`) initialized to `0` and incremented by `1` on every iteration, with the loop terminating when the induction variable `i` exceeds the bound (in the code above, `M`). The class of loops that LLVM can convert into this form includes, for example, `for` loops that initialize an induction variable to an arbitrary initial value, increment the induction variable by an arbitrary constant value on each iteration, and terminate when the induction variable exceeds an arbitrary bound.

The loop perforation transformation takes as a parameter a loop perforation rate r. It is the approximation knob that represents the expected percentage of loop iterations to skip. *Interleaving perforation* transforms the loop to perform every `n`th iteration (here the perforation rate is $r = 1 - 1/n$). Conceptually, the perforated computation looks like

```
for (i = 0; i < M; i += n) { ... }
```

In addition to interleaving perforation, SpeedPress can also apply other types of perforation. *Truncation perforation* skips a contiguous sequence of iterations at either the beginning or the end of the loop. For example, it can replace the loop condition `i < M` with `i < M/n`. *Random perforation* randomly skips loop iterations. Perforated computations can also skip only one out of `n` iterations. Our previous work [29] presents a detailed treatment of how to implement various loop perforation strategies.

7.3.2 Sensitivity Profiling Algorithm

The loop perforation space exploration algorithm takes as input an application, an accuracy specification for that application, and a set of training inputs. It can also take the set of allowable perforation rates to bound the search space (e.g., skipping a half, a quarter, or three quarters of iterations). The algorithm produces a set S of loops to perforate. Each element of S contains the speedup, accuracy loss, and the configuration of perforated loops.

7.3.2.1 Sensitivity Profiling for Individual Loops

The exploration algorithm (presented initially in [29] and [7]) starts with a set of candidate loops. The algorithm can be configured to consider only loops that execute for more than a certain percentage of the execution time. In general, perforating a candidate loop may cause the program to crash, generate unacceptable output, produce an infinite loop, or decrease its performance. Algorithm in Fig. 7.4 finds and removes such *critical* loops from the set of candidate loops. The algorithm perforates each loop in turn, using each of the specified perforation rates, and then runs the perforated program on the training inputs.

The sensitivity profiling algorithm filters out a loop if its perforation (1) fails to improve the performance as measured by the speedup s, which is the execution time of the perforated application divided by the execution time of the original unperforated program running on the same input, (2) causes the application to exceed the sensitivity bound b, or (3) introduces memory errors (such as out-of-bounds reads or writes, reads to uninitialized memory, memory leaks, double frees, etc.). If a memory error causes the execution to crash on some input t, its sensitivity a_t is ∞. Since testing for some errors is expensive (and various other checks can be added to the one we used here), therefore, to reduce profiling time, they should be checked only if the loop is determined to be a promising candidate for perforation. The result of sensitivity profiling is the set of perforatable loops S with different tradeoff points. Each loop (with its body) specifies one approximate kernel.

A *tradeoff point* is a triple $(spd, acc, config)$, which specifies the performance and accuracy of the configuration on specified inputs. For sensitivity profiling of individual loops, the configuration $config = \langle l, r \rangle$ specifies the perforation of

Inputs:
A - an application
I - a set of representative inputs
Q - an accuracy metric
b - an accuracy bound
L - a set of time-consuming loops
R - a set of perforation rates
Outputs: S - a set of loops and perforation rates for A that satisfy the developer's accuracy specification.

$S = \emptyset$
for $i \in I$ **do**
  Run A on i, record execution time t_i and output abstraction $\boldsymbol{o}_i$
end for
for $\langle l, r \rangle \in L \times R$ **do**
  Let $A_{\langle l,r \rangle}$ be A with l perforated at rate r
  for $i \in I$ **do**
    Run $A_{\langle l,r \rangle}$ on t, record execution time $\hat{t}_i$, output abstraction $\hat{\boldsymbol{o}}_i$, and execution errors Err_i.
    $acc_i = Q(\boldsymbol{o}_i, \hat{\boldsymbol{o}}_i)$ and $spd_i = t_i / \hat{t}_i$.
  end for
  if $\bigcup_{i \in I} Err_i \neq \emptyset$ **then**
    continue
  end if
  $\overline{spd} = (\sum_{i \in I} spd_i)/|I|$ // other statistics (e.g., max) can be used
  $\overline{acc} = (\sum_{i \in I} acc_i)/|I|$ // other statistics (e.g., max) can be used
  if $\overline{spd} > 1 \wedge \overline{acc} < b$ **then**
    for $i \in I$ **do**
      Run $A_{\langle l,r \rangle}$ using Valgrind to find Err_i (memory errors)
    end for
    if $\bigcup_{i \in I} Err_i = \emptyset$ **then**
      $S = S \cup \{ (\overline{spd}, \overline{acc}, \langle l, r \rangle) \}$
    end if
  end if
end for
return S

Fig. 7.4 Sensitivity profiling finds the set of perforatable loops S in application A given representative inputs I, accuracy metric Q, and accuracy goal b

the loop l at rate r. The accuracy and performance can be numerical values (if the optimization does an empirical evaluation) or symbolic expressions (if the optimization does static analysis, as in the next section).

7.3.2.2 Perforation Space Exploration

To navigate the space of approximate programs, we can use various search algorithms. The optimization problem at hand—maximizing performance subject to an accuracy constraint on a finite set of approximation configurations and representative inputs—is a combinatorial optimization problem reminiscent of the well-known Knapsack problem. The configuration of the full program $config \in L \to R$ is a mapping from loop identifiers ($l \in L$) to their perforation rates ($r \in R$), e.g., $config = \{\langle l_1, r_1\rangle, \ldots, \langle l_n, r_n\rangle\}$. The search space is a set of all such mappings, $L \to R$. If a loop identifier is not in the domain, it will not be perforated. The original program's configuration ($\emptyset$) is the one assumed to have no accuracy loss and no execution errors. For the representative inputs $i_1, \ldots, i_n$, we can describe the optimization problem as follows:

$$\begin{aligned}
\textbf{\textit{Minimize}}: \quad & \tfrac{1}{n}\textstyle\sum_{k=1}^{n} \text{TIME}(A, i_k, config) \\
\textbf{\textit{Constraints}}: \quad & \tfrac{1}{n}\textstyle\sum_{k=1}^{n} Q\big(\text{OUTABSTR}(A, i_k, \emptyset), \text{OUTABSTR}(A, i_k, config)\big) \le b \\
& \textstyle\bigcup_{k=1}^{n} \text{ERRORS}(A, i_k, config) = \emptyset \\
\textbf{\textit{Variables}}: \quad & config \in L \to R
\end{aligned}$$

The first constraint is accuracy constraint. The second constraint is safety constraint. If we want a stricter accuracy test, which ensures the maximum error is below the bound b, we can use the constraint of the form $\bigwedge_k Q(\text{OUTABSTR}(A, i_k, config), \text{OUTABSTR}(A, i_k, config)) \le b$. The set of possible errors in the safety constraint can be obtained from either static or dynamic program analysis. The safety constraint can be relaxed to be relative if the original program has non-critical errors unrelated to approximation, by making the set of the approximate program's observed errors be a subset of the set of errors observed in the original program.

We can implement the solver for this optimization problem by following the pattern of Algorithm in Fig. 7.4. Given a search/autotuning algorithm (which finds the next candidate in the search space), developing the optimizer is straightforward. For each candidate perforation, our optimizer (1) transforms the program—perforates the loops suggested by the search algorithm, (2) runs the sensitivity profiling on the perforated program for all inputs, and (3) if the quality loss is below the threshold, but the performance is improved, and there are no runtime errors, adds the configuration to the set of profitable approximate programs. To save time, like in Algorithm in Fig. 7.4, the safety checks can be split into inexpensive ones, which can be done before or together with running the program (e.g., program's execution status check or a type analysis), expensive ones (e.g., finding latent memory errors) that should run only for otherwise acceptable candidate programs.

Some choices of search algorithm include greedy (explored in [6]), exhaustive (explored in [7]), exhaustive with pruning (which does not explore loops/rates if perforating a subset is already causing unacceptable result), or a composite genetic algorithm supported by a generic autotuner [31] (explored in [32]). Exhaustive approach is feasible for applications that spend the majority of their time in relatively few loops. If the application has enough loops to make exhaustive exploration infeasible, then greedy or hybrid heuristic approaches are possible. In all cases, these algorithms will *execute the candidate program* on representative inputs.

We use the results of exploration to compute the set of Pareto-optimal perforations in the induced performance vs. accuracy tradeoff space. A perforation is *Pareto-optimal* if there is no other perforation that provides both better performance and better accuracy. A user can then select a sensitivity bound b^* to obtain the Pareto-optimal perforated program whose sensitivity is the closest below b^* and gives the maximum speedup.

7.3.2.3 Construction of Pareto Sets

We will now formalize the construction of Pareto sets (also known as Pareto frontiers or Pareto curves in a two-dimensional tradeoff space). All possible tradeoff points for a program comprise the *tradeoff space*. To compare tradeoff points, we define a *dominance* relation ($\preccurlyeq$): a point $s_1 = (spd_1, acc_1, c_1)$ is dominated by a point $s_2 = (spd_2, acc_2, c_2)$ if and only if it has both lower accuracy and worse performance. Formally, $s_1 \preccurlyeq s_2$ iff $spd_1 \leq spd_2$ and $acc_1 \leq acc_2$. We define strict dominance analogously: $s_1 \prec s_2$ iff $spd_1 < spd_2$ and $acc_1 \leq acc_2$, or $spd_1 \leq spd_2$ and $acc_1 < acc_2$, and denote its negation as $\not\prec$. We can describe the optimal points in a tradeoff space via *Pareto sets*. A Pareto set is a subset of a set S that contains only non-dominated points:

$$PS\,(S) = \{\, s \mid s \in S \,\wedge\, \forall s' \in S \,.\, s \not\prec s' \,\} \tag{7.1}$$

One can easily turn this definition into an algorithm (e.g., as the one given in Geilen et al. [33]).

7.3.3 *Perforating a Video Encoder*

The x264 program uses H.264 encoding algorithm to compress raw videos from cameras. Its output is a compressed video. We can define an *output abstraction* ($\boldsymbol{o}$) that characterizes output quality with two components: (1) the peak signal-to-noise ratio (PSNR), which measures the quality of the encoded video relative to the original, unencoded video, as o_1 and (2) the bitrate, which measures the compression achieved by the encoder as o_2. We defined the *accuracy metric* to quantify the quality loss as a distortion function $Q_{dist}(\boldsymbol{o}, \hat{\boldsymbol{o}})$ (Sect. 7.2) that weighs equally the relative difference in PSNR and bitrate of the original and perforated programs. If the reference decoder fails to parse the encoded video during exploration, we record the quality loss of 100% and reject the candidate perforation.

We selected *representative inputs* as follows: since the original benchmark suite from which we obtained the program (PARSEC [34]) contains only a single video, we augmented the input set with standard test videos that are used by developers of software that manipulates video from xiph.org [35]. We selected 16 videos in total: we used four for tuning and the remaining 12 to test the approximation.

To start sensitivity profiling, we considered the loops that contribute at least 1% of the executed instructions (identified by performance profiling). We specified a sensitivity bound b of 0.1 (representing 10%). We instructed SpeedPress to perforate loops with interleaving perforation, which skips every other iteration. We applied four different perforation rates—0.25 (skip a quarter of iterations), 0.5 (skip a half of iterations), 0.75 (skip three quarters of iterations), and execute a single loop iteration (skip all iterations after the first). We performed all of our runs on eight Intel Xeon X5460 3.1 GHz with 8 GB of RAM running Linux.

Figure 7.5 presents the sensitivity profiling results for x264. Out of top 25 loops, SpeedPress identified six that (when individually perforated) improve program performance, while producing acceptably accurate result. The successfully perforated loops are part of *motion estimation*. Motion estimation performs a heuristic search to find similar regions in neighboring frames, so that they can be better compressed. Perforating these loops reduces the quality of the heuristic search, but that reflects only slightly on the final output (the reduced quality was more due to an increased file size than reduced PSNR).

The top two perforated loops in the x264 execution both occur in the `pixel_satd_wxh()`, which computes a similarity metric between two blocks of pixels. Specifically, it takes the difference of two regions of pixels, performs several Hadamard transforms on 4×4 subregions, and then computes the sum of the absolute values of the transform coefficients. Perforation reduces the number of pixels that are being compared.

Figure 7.6 presents the results of exhaustive exploration for x264. The graphs plot a single point for each explored perforation configuration. The Y coordinate of the point is the mean speedup of the perforation over all profiling inputs (higher is better). The X coordinate is the corresponding accuracy metric—the percentage quality loss of the perforation (lower is better). The Pareto set comprises the points connected by the solid green line— these are the best accuracy/performance tradeoffs that one can obtain with applied perforations. We can see that the speedup of the program is above 3x for the bound of 10% quality loss. Running the perforated programs on the held-out inputs shows similar performance improvements, which indicates that the perforation benefits are not overfitting to the profiled inputs.

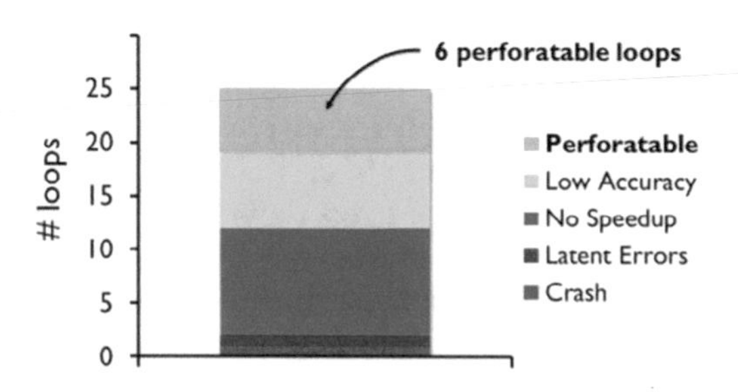

Fig. 7.5 Results of sensitivity profiling of x264

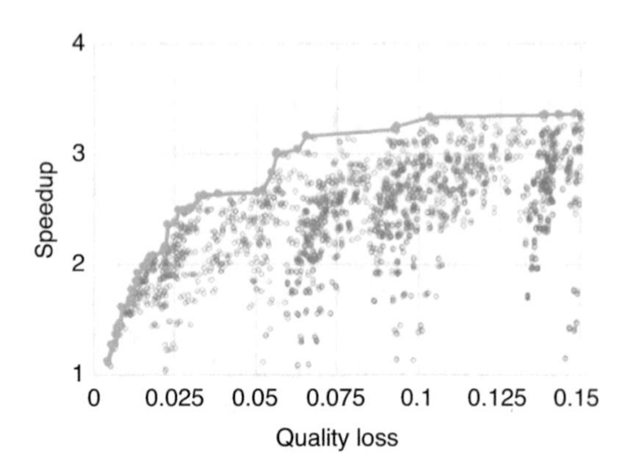

Fig. 7.6 x264 Tradeoff space and Pareto set

7.3.4 Discussion

We attribute the success of loop perforation to transform various programs without a major accuracy loss to three main factors. The first reason is that it targets computational patterns that perform already approximate and heuristic computations—enumerating the search space, computing search metrics (which select desirable elements, filter out undesirable, or direct a search algorithm to stop), running Monte Carlo simulations, and running iterative improvement loops (e.g., searching for a fixed-point solution). We discuss these patterns in greater detail in [7].

The second reason is that most of the errors produced by approximating kernels are small. For instance, in the video encoder, most pixels of the frames produced by the original and perforated codes will be different, but only a small fraction will have a difference greater than 5%. Inspecting the perforatable computations reveals that many compute reductions (sums, averages, or minimums/maximums)—these computations are resilient to small changes to its elements. Chippa et al. [26] present another discussion of general resilience patterns.

The third reason is that some kernel errors can be compensated by the following computation—either by reducing the error of approximation, e.g., by using a negative-feedback control system, or by computational patterns that select only a subset of close approximate values. For instance, when returning the most desirable element, the approximate desirability score for non-top element is irrelevant, and often when the top element is not selected, other selected elements will still have high desirability.

Safety The criticality test we presented here is a dynamic program analysis. It can guarantee the absence of critical errors such as crashes *only on* tested inputs. To prove the absence of errors caused by approximation, one needs to resort to static analysis, e.g., approximate types [22, 36] or relational reasoning [37]. Some of these (e.g., types) can be applied before transformations to reduce the search space a priori. More time-consuming analyses should be applied only if the perforation is deemed profitable, to save the search time.

Accuracy The worst-case analysis of approximations is often too conservative and fails to identify any perforation opportunity. Instead, a more productive way to think about reasoning about the accuracy of approximation is by complementing worst-case with average-case (probabilistic) reasoning. The accuracy estimation provided by the criticality analysis is valid only if the inputs in the program deployment are sampled from the same distribution as in the profiling. Static probabilistic analysis of approximations is an interesting alternative (e.g. see [8] for the analysis of loop perforation).

Extensions Over the years, the researchers proposed various refinements of loop perforation for different domains. For instance, it has been used in computer graphics [38], neural networks [39], GPU kernels [40], and adaptive approximation [41, 42]. For instance, ApproxTuner shows how to extend approximation tuning for heterogeneous systems to three phases—development-time, install-time, and run-time [32].

7.4 Case Study 2: Chisel

Approximate hardware devices provide operations that may produce less accurate or incorrect results to reduce energy consumption (e.g., [43–45]). This hardware is particularly suitable to execute applications whose algorithms are inherently tolerant to inaccuracies in their data and the majority of the computation is performed in several approximate kernels.

This section presents Chisel, an optimization framework that automatically selects approximate instructions and data that may be stored in approximate memory, given the exact kernel computation and the associated reliability and/or accuracy specification [27]. Chisel reduces the effort required to develop efficient approximate computations and enhance their portability. Figure 7.7 presents an overview of Chisel. We next describe the inputs that the developer and hardware designer provide.

Exact Program A Chisel program consists of a kernel function written in the Rely base language [46] (which is a simple imperative language with control-flow constructs and arrays) and code written in an implementation language (such as C) that calls the kernel. The kernel function can compute the return value but may also write computed values into array parameters passed by reference into the kernel from the outer C code. Chisel transforms the kernel function according to the developer's specification.

Kernel's Reliability and Accuracy Specifications Reliability specifications of the form <r*R(x1, ..., xn)> are integrated into the type signature of the kernel. Here r specifies the probability that the kernel (in spite of unreliable hardware operations) computes the value correctly. The term R(x1, ..., xn) is a *joint reliability factor* that specifies the probability that x1,...,xn all have correct values at the start of the kernel. In the following specification, for example,

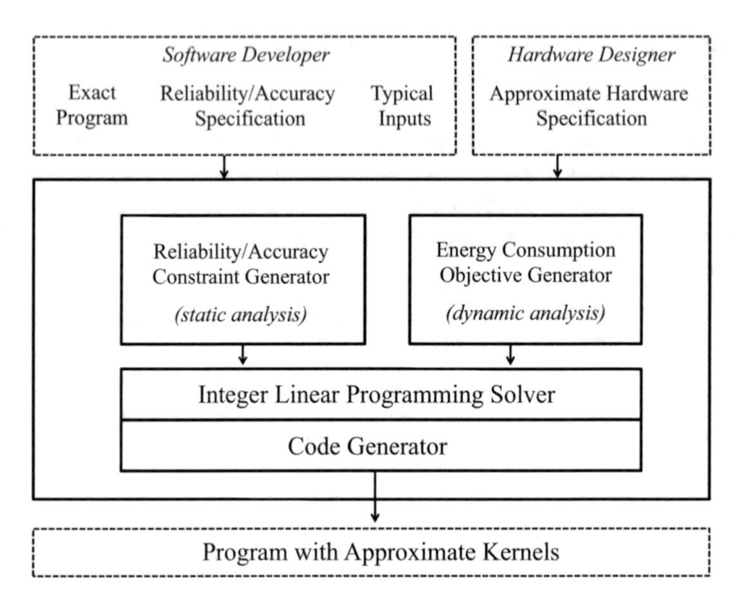

Fig. 7.7 Chisel overview

```
int <0.99 * R(x)> f(int[] <0.98*R(x)> x);
```

the return value has reliability at least `.99` times the reliability of the input `x`; when `f` returns, the probability that all elements in the array `x` (passed by reference into `f`) have the correct value is at least `.98` times the reliability of `x` at the start of `f`.

Chisel also supports combined reliability and accuracy specifications of the following form (these specifications are *relational* in that they specify the combined accuracy and reliability with respect to the fully accurate exact computation):

```
<d >= Δf, r*R(d1 >= Δx1, ..., dn >= Δxn)> f( ... )
```

Here `d` is a maximum acceptable absolute difference between the approximate and exact result values, `r` is the probability that the kernel computes a value (`Δf`) within `d` of the exact value, and the term `R(d1 >= Δx1, ..., dn >= Δxn)` is a joint reliability factor that specifies the probability that each `xi` is within distance `di` of the exact value at the start of the computation. If `r=1`, then the specification is a pure accuracy specification; if `d=0` and all the `di=0`, then it is a pure reliability specification. We discussed accuracy analysis in [27]. The values for all `d` and `r` bounds can be selected to connect with end-to-end accuracy metric by mathematical derivation [27] or sensitivity profiling [27, 28].

Typical Program Inputs A developer provides a set of typical inputs that Chisel uses to estimate the energy savings of approximate computations. In addition, they can help developers derive the accuracy and reliability specification through sensitivity profiling.

Approximate Hardware Specifications Figure 7.8 presents the model of approximate hardware, which consists of approximate ALU, main memory, and cache memory. Chisel works with a hardware specification provided by the designers of the approximate hardware platform [22, 24]. To automatically optimize the implementation of the computation, the optimization algorithm requires a specification of approximate components.

The approximate hardware specification consists of:

- **Operation and Memory Accuracy/Reliability.** The hardware specification identifies (1) approximate arithmetic operations and (2) the approximate regions of the main and cache memories. The specification contains the reliability, ρ_{op}, and the accuracy loss, δ_{op}, of each arithmetic operation. It also contains the probability that read and write operations to approximate main memory and cache complete successfully.
- **Energy Model Parameters.** To compute the savings associated with selecting approximate arithmetic operation, the energy model specifies the expected energy savings of executing an approximate version, α_{op} (as a percentage of the energy of the exact version). To compute the savings associated with allocating data in approximate memory, the energy model specifies the expected energy savings for memory cells, α_{mem}, and cache, α_{cache}. To compute system energy savings, the energy model also provides (1) the specification of the relative portion of the system energy consumed by the CPU versus memory, (2) the relative portion of the CPU energy consumed by the ALU, cache, and other on-chip resources, and (3) the ratio of the average energy consumption of floating-point instructions and other non-arithmetic instructions relative to integer instructions.

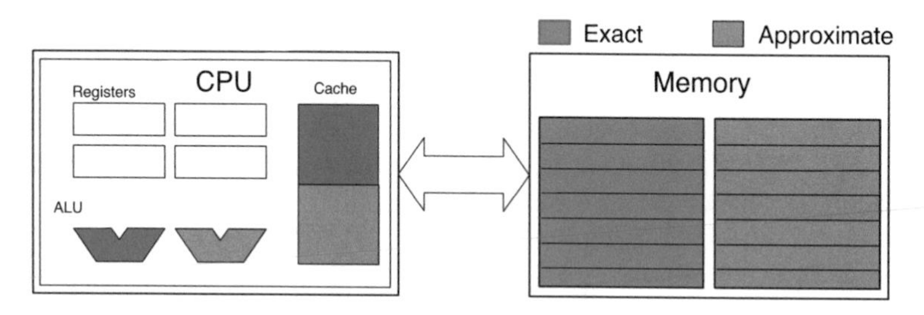

Fig. 7.8 Model of approximate hardware, with exact (blue) and approximate (orange) components

7.4.1 Chisel's Optimization Algorithm

Chisel's optimization algorithm selects approximate instructions and variables allocated in approximate memories. Chisel's optimization algorithm uses program analysis to reduce the problem of selecting approximate operations and variables to an integer linear programming (ILP) problem. Importantly, it does not need to run the approximated programs, which is especially important for fine-grained approximations, that may cause a large tradeoff space. Chisel performs the following main steps:

- **Specify Decision Variables.** Chisel represents the choice whether to approximate each arithmetic operation or a variable in a kernel as one decision variable. Recall that the instructions in Chisel's language have a label (which specifies that an instruction can be either exact or approximate).
- **Compute Reliability and Accuracy Constraints.** Chisel generates reliability and accuracy constraints via a static *precondition generator analysis*. In general, a precondition generator C_ψ operates backward and is often used in program verification. It takes as input a predicate Q_{post} and the program statements $\langle S_1, \ldots, S_n \rangle$. It produces a predicate $Q_{pre} = C_\psi(\langle S_1, \ldots, S_n \rangle, Q_{post})$, such that if Q_{pre} is valid before the execution of the kernel, then Q_{post} will be valid at the end of the execution.

  The predicates Q_{post} for the kernel come from the function specifications, which can state that, e.g., the reliability of the kernel's result should be greater than `0.99*R(x)`. The analysis starts by constructing the corresponding predicate Q_{post} := `0.99*R(x)` $\leq$ `R(result)`. The analysis transforms each such predicate, operating backward by analyzing the statements from S_n to S_1 recursively, i.e., $Q_{pre} := C_\psi(\langle S_1, \ldots, S_{n-1} \rangle, C_\psi(\langle S_n \rangle, Q_{post}))$. Chisel's reliability and accuracy analyses produce predicates Q_{pre}, which are parameterized by the instruction and variable labels. For any particular set of labels, the predicates Q_{pre} are valid for *all* inputs specified in the kernel's input specification.
- **Compute Energy Savings Objective.** Chisel generates the energy savings objective using a dynamic program analysis. Sect. 7.4.6 presents how Chisel constructs an estimate of the savings (as the function of the decision variables) from the traces of the kernel when executed on representative inputs.

Chisel dispatches the generated ILP problem to an off-the-shelf solver, which returns the labels of instructions and variables that maximize savings. Chisel uses these labels to generate the approximate kernel.

7.4.2 Intermediate Language

We will consider a simple intermediate language that operates on numerical data. To enable optimization with approximate instructions and data, we augment our program representation to create an intermediate representation that includes *labels*,

$$
\begin{aligned}
i \in I ::= & \; r = \texttt{init}\ n \mid r = op\ \ r\ \ r \mid r_1 = \texttt{load}\ r_2 \mid \texttt{store}\ r_1\ \ r_2 \mid \\
& r_{val} = \texttt{loada}\ r_{arr}\ \ r_{idx} \mid \texttt{storea}\ r_{arr}\ \ r_{idx}\ \ r_{val} \mid \texttt{return}\ r_{ret} \mid \\
& r = op^{\ell}\ \ r\ \ r \\
s \in S ::= & \; i \mid s\texttt{;}s \mid \texttt{if}\ r_{cond}\ s_{then}\ s_{else}
\end{aligned}
$$

Fig. 7.9 Chisel's intermediate language

where each label $\ell \in \mathcal{L}$ is uniquely associated with an instruction or a program variable. Labels enable Chisel to separately mark each instruction and variable as either exact or approximate.

Syntax Figure 7.9 presents the syntax of the intermediate language. Each instruction `i` $\in$ `I` is either an ALU/FPU arithmetic operation (such as add, multiply, and compare), an initialization that loads a constant, or a load/store from memory for scalars (`load` and `store`) and for arrays (`loada` and `storea`). A statement in this language can be a labeled assembly instruction, a sequence of statements, or an intermediate conditional statement, which, based on the result of the register r_{cond}, continues the execution of the statements in the *then* or *else* branches. The intermediate language has only a structured control-flow and no stray `jmp` instructions.

We augment each arithmetic instruction to have a label $\ell \in \mathcal{L}$ to denote whether an instruction is exact or approximate. The label allows us to specify the configuration. The finite map $\chi \in V \to \mathcal{L}$ maps each variable name in the program to a unique label.

Configurations We define a *configuration* $\theta \in \Theta = \mathcal{L} \to K$ as a finite map from labels to exact or precise kind of each of the program's instructions and variables. The set of configurations denotes all possible optimized programs that Chisel can generate. Any element of this set represents one approximate version of the program. An instruction or a variable is of exact kind if $\theta(\ell) = 0$ or approximate kind if $\theta(\ell) = 1$.

7.4.3 *Reliability Predicates*

Chisel's generated preconditions are *reliability predicates* that characterize the reliability of an approximate program. We adapt the reliability definitions from [46] for Chisel's configurable assembly language kernels. A reliability predicate P has the following form:

$$
\begin{aligned}
P &:= R_f \leq R_f \mid P \wedge P \mid \text{True} \mid \text{False} \\
R_f &:= \rho \mid \rho^{\ell} \mid \mathcal{R}(O) \mid R_f \cdot R_f
\end{aligned}
$$

Specifically, a predicate is either a conjunction of predicates or a comparison between *reliability factors*, R_f. A reliability factor has multiple forms:

- **Constant.** A real number, $\rho \in [0, 1]$, represents the probability that an approximate instruction produces a correct result. The analyses will calculate the constants ρ using the elements of the approximate hardware specification ψ.
- **Labeled reliability.** A term ρ^ℓ represents the reliability of a labeled instruction that can be either exact or approximate. The label ℓ helps encode the choice between the exact and approximate versions of an instruction: if $\theta(\ell) = 0$ (exact), this term will have the value 1 (instructions always produce a correct result); if $\theta(\ell) = 1$ (approximate), the term will have the value ρ. We remark the intentional notational similarity of ρ^ℓ with numerical exponentiation. When we write a reliability of the (syntactic) form ρ^ℓ, we will interpret it as $\rho^{\theta(\ell)}$.
- **Joint Reliability Factor.** A term $\mathcal{R}(O)$ represents probability that all registers and variables in the set $O \subseteq R \cup V$ have the same value in the exact and the approximate executions. This term abstracts the probability that the approximate execution's environment has exact values of the operands from which the remaining execution can produce the exact result.
- **Product of Reliability Factors.** This term combines the probability that the instructions produce correct results and the initial program environments have the exact value of the operands.

This definition of reliability predicates is sufficient for expressing properties about *error frequency*. Before we define the analysis, let us illustrate the intuition behind how we intend to use these predicates:

Example 7.1 (Reliability Factor) Consider the predicate $0.9 \le 0.99^\ell$ with reliability factor. Given a configuration θ, we can represent this predicate as $0.9 \le 1$ (if $\theta(\ell) = 0$) or $0.9 \le 0.99$ (if $\theta(\ell) = 1$). We can succinctly rewrite these two predicates over reals as $0.9 \le 0.99^{\theta(\ell)}$.

Example 7.2 (Joint Reliability Factor) A predicate $0.9 \le \mathcal{R}(\{x\})$ bounds the probability that the variable x in the approximate execution has the same value as in the exact execution (at the same program point). Chisel calculates a lower bound on this probability, which is much easier to compute than the exact probability $\mathcal{R}(\{x\})$.

7.4.4 Reliability Constraint Construction

Chisel's reliability constraint generator operates as a precondition generator. Given a postcondition predicate, Chisel's reliability precondition generator produces a precondition that, when true before the execution of the program, ensures that the postcondition is true after. In other words, the precondition, program, and postcondition satisfy the reliability transformer relation we defined in the previous section. The reliability precondition generator is a function $C \in S \times P \to P$ that takes as inputs an instruction and a postcondition and produces a precondition as output.

7.4.4.1 Initial Postcondition

The analysis starts from the return instruction (the last instruction in Chisel's kernel). Recall that the specification of reliability of the function's output has the form `int <rspec*R(v1,...,vn)> f(int v1, ..., int vn)`, where `rspec` is the numerical constant and `v1,...,vn` are the function's parameters. The analysis represents this specification as the reliability factor $\rho_{spec} \cdot \mathcal{R}(V_{spec})$, where $V_{spec} \subseteq \{v_1, \ldots, v_n\}$ is the set that contains the function inputs in the specification's reliability factor.

The analysis of the return instruction starts from the default initial postcondition $Q_0 = \text{True}$ and constructs the following precondition:

$$C_\psi(\texttt{return}\, r_{ret},\ Q_0) = \rho_{spec} \cdot \mathcal{R}(V_{spec}) \leq \mathcal{R}(\{r_{ret}\})\ \wedge\ Q_0$$

This predicate states that the probability that the return register r_{ret} contains the correct output at the end of the kernel function is greater than the probability that the inputs of the function had correct values at the beginning of the kernel execution, multiplied by the constant reliability degradation ρ_{spec}.

Example 7.3 (Analysis of a Function Returning Constant) We will analyze the function returning a constant written in the Rely language: `int<0.99> three() { return 3; }` The compiler generates the two assembly instructions `r0 = init 3; return r0;`.
The analysis constructs the following precondition for the return instruction: $0.99 \cdot \mathcal{R}(\varnothing) \leq \mathcal{R}(\{r_0\})$. The left side of the inequality comes from the function specification (and $\varnothing$ indicates an empty set, i.e., no variables in the specifications). The right-hand side of the inequality is the result of the rule $C_\psi(\texttt{return}\, r_0, \text{True})$.

7.4.4.2 Precondition Generator for Statements

Initialization and Sequence The following equations present the rules for initializing a register with a constant and a sequence of instructions:

$$C_\psi(r = \texttt{init}\ n,\ Q) = Q\,[\,\mathcal{R}(X)/\mathcal{R}(\{r\} \cup X)\,]$$

$$C_\psi(\texttt{s1; s2},\ Q) = C_\psi(\,\texttt{s1},\ C_\psi(\texttt{s2},\ Q))$$

The initialization rule removes the occurrence of the register r in the joint reliability factor because its previous value is not relevant for the reliability of the kernel's outputs. Specifically, the substitution $Q\,[\,\mathcal{R}(X)/\mathcal{R}(\{r\} \cup X)\,]$ matches all occurrences of the destination register r in a reliability term that occur in the predicate Q and removes them (by leaving only the remainder set X).

The sequence rule first computes the precondition for the second instruction (`s2`) and passes it as the postcondition to the analysis of the first instruction (`s1`).

Example 7.4 (Analysis of a Function Returning Constant) We return to the function from Example 7.3 and analyze the instruction `r1 = init 3;` and the postcondition $0.99 \cdot \mathcal{R}(\varnothing) \leq \mathcal{R}(\{r_1\})$ (that the analysis of the return instruction produced). Then, the precondition generator uses the rules for sequence and initialization to generate the function precondition $0.99 \cdot \mathcal{R}(\varnothing) \leq \mathcal{R}(\varnothing)$.

ALU/FPU The next equation presents the generator rule for ALU/FPU operations ($op \in \{add, sub, mul, div\}$):

$$C_\psi(r = op^\ell \ r_1 \ r_2,\ Q) = Q\,[\,\rho_{op}^\ell \cdot \mathcal{R}(\{r_1, r_2\} \cup X))/\mathcal{R}(\{r\} \cup X)\,]$$

The rule works by substituting the reliability of the destination register r with the reliability of its operands and the reliability of the operation itself. The substitution $Q[\mathcal{R}(\{r_1, r_2\} \cup X)/\mathcal{R}(\{r\} \cup X)]$ matches all occurrences of the destination register r in a reliability factor inside the predicate Q and replaces them with the input registers, r_1 and r_2. The substitution also multiplies in the factor ρ_{op}^ℓ, which is the reliability of the operation ρ_{op} from ψ as a function of its label's configuration.

Example 7.5 (Analysis of Addition) We analyze the statement $r = \mathtt{add}^\ell r_1, r_2$, with the postcondition $Q := 0.99 \leq \mathcal{R}(\{r, z\})$ and the hardware configuration ψ.
First, the analysis obtains the reliability of the addition operator $\rho_{add} = \pi_{\text{op}}(\psi)(add)$. Second, the analysis uses the instruction's label ℓ to represent reliability choice, ρ_{add}^ℓ. Third, the analysis generates the new reliability factor $\mathcal{R}(\{r_1, r_2, z\})$ by substituting r with $\{r_1, r_2\}$. Finally, the analysis substitutes $\mathcal{R}(\{r, z\})$ with $\rho_{add}^\ell \cdot \mathcal{R}(\{r_1, r_2, z\})$ in Q to produce the new precondition $0.99 \leq \rho_{add}^\ell \cdot \mathcal{R}(\{r_1, r_2, z\})$.

Scalar Load/Store The rules for loads and stores from potentially approximate memory are:

$$C_\psi(r_1 = \mathtt{load}\ r_2,\ Q) = Q\,[\,\rho_{ld}^{\chi(\eta(r_2))} \cdot \mathcal{R}(\{\eta(r_2)\} \cup X)/\mathcal{R}(\{r_1\} \cup X)\,]$$
$$C_\psi(\mathtt{store}\ r_1\ r_2,\ Q) = Q\,[\,\rho_{st}^{\chi(\eta(r_1))} \cdot \mathcal{R}(\{r_2\} \cup X)/\mathcal{R}(\{\eta(r_1)\} \cup X)\,]$$

These rules define the semantics of strong updates for scalar program variables. They use the auxiliary *register mapping* generated by the compiler ($\eta \in R \rightarrow V$) that maps the address operand register to the program variable that is read or written. The minimum reliability of a load from a potentially approximate variable, ρ_{ld}, is equal to the probability that the read from memory, the write to a cache location, and the read from that cache location all execute correctly. The reliability of a store to a potentially approximate variable, ρ_{st}, assuming a write-through cache, is equal to the reliability of a memory store.

Example 7.6 (Analysis of Scalar Store) We analyze the statement `store` r_1, r_2 with the postcondition $Q := 0.99 \leq \mathcal{R}(\{x\})$ and the hardware configuration ψ. The statement stores the value of the register r_2 to the location in memory referred by

the register r_1. We consider the case when r_1 holds the location of the variable x, i.e., $\eta(r_1) = x$.
First, the analysis computes the constant ρ from ψ. Second, the analysis identifies that the register r_1 holds the address of x (using the map η) and finds the label ℓ that corresponds to the variable x (using the map χ). Third, the analysis generates the new reliability factor $\mathcal{R}(\{r_2\})$ by substituting x with r_2. Finally, it substitutes $\mathcal{R}(\{x\})$ with $\rho^\ell \cdot \mathcal{R}(\{r_2\})$ in Q to produce the precondition $0.99 \leq \rho^\ell \cdot \mathcal{R}(\{r_2\})$.

Array Load/Store The reliability constraint generation rule for stores to scalar variables provides a semantics for strong updates to memory. Updates to arrays, however, are weak in that a variable refers to multiple memory locations. The following reliability constraint generation rule defines the analysis for arrays:

$$C_\psi(r_{val} = \texttt{loada}\ r_{arr}\ r_{idx},\ Q) = Q\,[\,\rho_{ld}^{\chi(\eta(r_{arr}))} \cdot \mathcal{R}(\{\eta(r_{arr}), r_{idx}\} \cup X)/\mathcal{R}(\{r_{val}\} \cup X)\,]$$

$$C_\psi(\texttt{storea}\ r_{arr}\ r_{idx}\ r_{val},\ Q) = Q\,[\,\rho_{st}^{\chi(\eta(r_{arr}))} \cdot \mathcal{R}(\{r_{idx}, r_{val}\} \cup \{\eta(r_{arr})\} \cup X)\ /\ \mathcal{R}(\{\eta(r_{arr}))\} \cup X)\,]$$

The primary difference between this rule and that for strong updates is that the reliability of the array variable is included in the resulting reliability term (after substitution). Since the function $\eta(r_1)$ points to the same variable name for all elements of the array, this rule effectively treats updates to the potentially different array elements as an update to the single (scalar) variable.

We can further expand these rules with the additional *array safety constraints*—we can enforce that the index variable computations are correct and the array pointer is stored in reliable memory. We do it simply by conjuncting the constraint $\mathcal{R}(\{r_{arr}, r_{idx}\}) = 1$ with the previous constraints.

Conditionals The analysis of conditionals relies on the fact that the Rely base language has structured control flow and therefore the intermediate language keeps the conditional structure. The following reliability constraint generation rule implements the analysis for conditionals:

$$\begin{aligned} &C_\psi(\texttt{if}\ r_c\ s'\ s'', Q) \ = \\ &\quad \textbf{let}\ \{o_1, \ldots, o_k\} = \text{modified}(s') \cup \text{modified}(s'') \\ &\quad \textbf{and}\ Q^* = Q\,[\mathcal{R}(\{r_c, o_1\} \cup X)/\mathcal{R}(\{o_1\} \cup X)\,] \ldots [\mathcal{R}(\{r_c, o_k\} \cup X)/\mathcal{R}(\{o_k\} \cup X)]\ \textbf{in} \\ &\qquad C_\psi(s', Q^*) \wedge C_\psi(s'', Q^*) \end{aligned}$$

This rule uses the helper function *modified* $\in$ *Instr* $\to$ O, which denotes the set of operands (variables and registers) that are modified within a loop branch. To encode the probability that an incorrectly computed conditional may cause an incorrect value in each such operand, it adds the reliability of the conditional expression (r_c) to the joint reliability term containing the operand in the predicate Q. The final predicate is the conjunction of the predicates computed for each of the branches.

Example 7.7 (Analysis of Conditionals) Consider `if` r_c $\{r_2 = \texttt{add}^{\ell_1}\ r_1\ 1\}$ $\{r_3 = \texttt{sub}^{\ell_2}\ r_1\ 1\}$ and the postcondition $Q := 0.9 \leq \mathcal{R}(\{r_2, r_3\})$. The analysis first finds that r_2 and r_3 are the operands modified in the loop. The predicate for the then branch is $Q_{then} := 0.9 \leq \rho_+^{\ell_1} \cdot \mathcal{R}(\{r_c, r_1, r_3\})$ and for the else branch is $Q_{else} := 0.9 \leq \rho_-^{\ell_2} \cdot \mathcal{R}(\{r_c, r_1, r_2\})$. The final predicate is therefore $Q_{then} \wedge Q_{else}$.

Bounded Loops Bounded loops are translated to the intermediate language as a sequence of conditional statements. Then, the analysis uses the rule for conditionals that we previously presented to compute the reliability. Chisel does not handle Rely programs with unbounded loops. We do not support unbounded loops. In principle, one can support these loops in the similar way to Rely, i.e., assign a zero reliability to any variable that is written to within the loop [46].

7.4.4.3 Final Precondition

For a given kernel, our analysis computes a precondition that is a conjunction of predicates of the form

$$\rho_{spec} \cdot \mathcal{R}(V_{spec}) \leq r(\ell_1, \ldots, \ell_n) \cdot \mathcal{R}(V),$$

where $\rho_{spec} \cdot \mathcal{R}(V_{spec})$ is a reliability factor for a developer-provided specification of an output and $r(\ell_1, \ldots, \ell_n) \cdot \mathcal{R}(V)$ is a lower bound on the output's reliability computed by the analysis, parameterized by the labels $\ell_1, \ldots, \ell_n$ of the candidate approximate operations.

Each ρ_{spec} is a real-valued constant and each r is, syntactically, a product of a real-valued constant and labeled reliabilities, i.e.,

$$r(\ell_1, \ldots, \ell_n) = \rho \cdot \Pi_k\ \rho_k^{\ell_k}. \tag{7.2}$$

The product operator iterates over the sequences of instructions that the analysis traversed. If this precondition is valid for a given kind configuration $\theta(\cdot)$, then that configuration satisfies the developer-provided reliability specification.

Example 7.8 (Analysis of a Function) We consider a simple function, `int<0.99 *R(x)> f(int x) { return x+3; }`, for which the compiler generates these assembly instructions:

$$\texttt{r0 = init}\ \langle \texttt{x} \rangle^{\ell_x}\texttt{; r1 = load r0; r2 = init 3;}$$
$$\texttt{r3 = add}^{\ell_+}\ \texttt{r1 r2; r4 = mul}^{\ell_*}\ \texttt{r3 r2 ; return r4;}$$

(the operator $\langle \texttt{x} \rangle$ denotes the stack offset of the variable x). The configuration θ has three elements: ℓ_+ and ℓ_* for the arithmetic operators and ℓ_x for the parameter x

(which can be stored in exact or approximate memory). The analysis constructs the following preconditions for the instructions:

$$Q_1 = C_\psi(\texttt{init}\, r_4,\ \text{True}) := 0.99 \cdot \mathcal{R}(\{x\}) \le \mathcal{R}(\{r_3\})$$

$$Q_2 = C_\psi(r_4 = \texttt{mul}\, r_3\, r_2,\ Q_1) := 0.99 \cdot \mathcal{R}(\{x\}) \le \rho_{mul}^{\ell_*} \cdot \mathcal{R}(\{r_1, r_2\})$$

$$Q_3 = C_\psi(r_3 = \texttt{add}\, r_1\, r_2,\ Q_2) := 0.99 \cdot \mathcal{R}(\{x\}) \le \rho_{add}^{\ell_+} \cdot \rho_{mul}^{\ell_*} \cdot \mathcal{R}(\{r_1, r_2\})$$

$$Q_4 = C_\psi(r_2 = \texttt{init 3},\ Q_3) := 0.99 \cdot \mathcal{R}(\{x\}) \le \rho_{add}^{\ell_+} \cdot \rho_{mul}^{\ell_*} \cdot \rho_{ld}^{\ell_x} \cdot \mathcal{R}(\{r_1\})$$

$$Q_5 = C_\psi(r_1 = \texttt{load}\ r_0,\ Q_4) := 0.99 \cdot \mathcal{R}(\{x\}) \le \rho_{add}^{\ell_+} \cdot \rho_{mul}^{\ell_*} \cdot \rho_{ld}^{\ell_x} \cdot \mathcal{R}(\{x\})$$

$$Q_6 = C_\psi(r_0 = \texttt{init}\ \langle x \rangle,\ Q_5) := 0.99 \cdot \mathcal{R}(\{x\}) \le \rho_{add}^{\ell_+} \cdot \rho_{mul}^{\ell_*} \cdot \rho_{ld}^{\ell_x} \cdot \mathcal{R}(\{x\})$$

Q_6 is the final precondition for the function. This derivation combines the rules for the instructions we previously described. We note that the analysis of load statement immediately inserts the variable x in the joint reliability factor (because of the mapping $\eta(r_0) = x$), and therefore the subsequent analysis of the instruction $r_0 =$ $\mathtt{init}\ \langle x \rangle$ does not modify the predicate.

7.4.4.4 Constraint Simplification

The number of constraints that the generator produces can, in principle, grow exponentially in the number of conditional statements in the program. However, in practice, the number of constraints can be significantly decreased by using a simplification of the constraints after each step of the algorithm [46]. Chisel extends Rely's simplification procedure, which uses the ordering property of the joint reliability factors and the subsumption property of the reliability predicates.

Ordering of Joint Reliability Factors Ordering enables comparing two joint reliability factors by comparing their sets of variables [46, Proposition 1]. This proposition states that for the two sets of variables V and V_{spec},

$$V \subseteq V_{spec} \Rightarrow \mathcal{R}(V_{spec}) \le \mathcal{R}(V). \tag{7.3}$$

Therefore, the reliability of any subset of a set of variables is greater than or equal to the reliability of the set as a whole. It immediately extends from the sets of variables to the sets of operands (registers and variables).

Ordering of Labeled Reliabilities Chisel operates on products of labeled reliabilities, which can also be ordered. Specifically, $\hat{\rho}_1^{\ell_1} \cdot \ldots \cdot \hat{\rho}_n^{\ell_n} \le \rho_1^{\ell_1} \cdot \ldots \cdot \rho_m^{\ell_m}$ if $\{\ell_1, \ldots, \ell_m\} \subseteq \{\ell_1, \ldots, \ell_n\}$, and for each $\ell_i \in \{\ell_1, \ldots, \ell_n\}$, either $\hat{\rho}_i \le \rho_i$ or ρ_i does not show up in the product on the right-hand side (in which case the reliability is by default equal to one).

Subsumption The subsumption property (i.e., sound replacement) defines the condition under which a predicate is trivially satisfied, given another more general predicate [46, Proposition 2]. Specifically, this proposition states that a predicate $\rho_1 \cdot \mathcal{R}(V_1) \leq \rho_2 \cdot \mathcal{R}(V_2)$ subsumes a predicate $r_1' \cdot \mathcal{R}(X_1') \leq \rho_2' \cdot \mathcal{R}(V_2')$ iff $\rho_1' \cdot \mathcal{R}(V_1') \leq \rho_1 \cdot \mathcal{R}(V_1)$ and $\rho_2 \cdot \mathcal{R}(V_2) \leq \rho_2' \cdot \mathcal{R}(V_2')$. This proposition follows immediately from the ordering of joint reliability factors and the ordering of labeled reliabilities.

7.4.5 *Optimization Constraint Construction*

When the configuration $\theta(\cdot)$ is unknown, the final precondition that Chisel's generator produces represents a constraint that lists *all* approximation choices represented by θ. Then, each $\theta(\ell)$ is a variable that can be either 0 (reliable) or 1 (unreliable). The precondition parameterized by $\theta(\cdot)$ therefore represents all approximate versions of the program that satisfy the developer's reliability specification. To generate the constraint for the optimization problem, Chisel analyzes separately the reliability degradation and joint reliability factors in each conjunct: (1) $\rho_{spec} \leq r(\ell_1, \ldots, \ell_n)$ and (2) $\mathcal{R}(V_{spec}) \leq \mathcal{R}(V)$.

Validity Checking To check the validity of this precondition, we use the ordering property, from Eq. 7.3. Therefore, Chisel can soundly ensure the validity of each inequality in the precondition by verifying that (1) $\rho_{spec} \leq r(\ell_1, \ldots, \ell_n)$ and (2) $V \subseteq V_{spec}$. Since V and V_{spec} are not parameterized by the labels ℓ, Chisel can immediately check if these set inclusion constraints are satisfied.

Constraint Construction After checking the validity of the reliability factors, Chisel is left with the inequality

$$\rho_{spec} \leq r(\ell_1, \ldots, \ell_n). \tag{7.4}$$

The denotation of the reliability expression r is $\rho \cdot \Pi_k\ \rho_k^{\theta(\ell_k)}$. The factor ρ is the product of all the constant terms. Recall that the denotation of each ρ^ℓ from Eq. 7.2 under the configuration θ is $\rho^{\theta(\ell)}$, i.e., 1 if $\theta(\ell) = 0$ or ρ if $\theta(\ell) = 1$.

Chisel produces a final optimization constraint by taking the logarithm of both sides of Inequality 7.4:

$$\log(\rho_{spec}) - \log(\rho) \leq \sum_k \theta(\ell_k) \cdot \log(\rho_k). \tag{7.5}$$

The expression on the right side is linear with respect to all labels' kinds $\theta(\ell_k)$. The reliabilities ρ are constants and their logarithms can be immediately computed. Each label's kind can take a value 0 or 1.

7.4.6 Energy Objective Construction

The objective of the optimization is to minimize the energy consumption of the unreliable computation, as a function of the configuration θ. To approximate this optimization objective, we consider a set of traces of the original program. We now define a set of functions that operate on these traces and give an estimate of the energy consumption of the unreliable program executions. The approximate hardware model presents relative savings of operations and memories (e.g., approximate instruction saves 20% of the energy of the exact operation), instead of unknown absolute savings (e.g., approximate instruction consumes 8 pJ instead of 10 pJ). We next show how the analysis computes the expression for the relative energy consumption.

We denote the relative energy savings from hardware specification for each approximate arithmetic operation: α_{int} for integer, α_{fp} for floating-point instructions), α_{mem} for approximate memory, and α_{cache} for cache regions; the specification also contains the relative energy consumption of the system's components, denoted as (μ_{CPU}, μ_{ALU}, and μ_{cache}), and relative instruction class energy rates (w_{fp} and w_{oi}).

7.4.6.1 Absolute Energy Model

Energy of System We model the energy consumed by the system when executing a program under configuration θ with the combined energy used by the CPU and memory: $E_{sys}(\theta) = E_{CPU}(\theta) + E_{mem}(\theta)$.

Energy of CPU We model the energy consumption of the CPU as the combined energy consumed by the ALU, cache, and the other on-chip components: $E_{CPU}(\theta) = E_{ALU}(\theta) + E_{cache}(\theta) + E_{other}$.

Energy of ALU Each instruction in the hardware specification may have a different energy consumption associated with it. However, for the purposes of our model, we let $\mathcal{E}_{int}$, $\mathcal{E}_{fp}$, and $\mathcal{E}_{oi}$ be the average energy consumption (over a set of traces) of an ALU instruction, an FPU instruction, and other non-arithmetic instructions, respectively: $E_{ALU}(\theta) = E_{int}(\theta) + E_{fp}(\theta) + n_{oi} \cdot \mathcal{E}_{oi}$.

Using the instructions from the traces that represent kernel execution on representative inputs, we derive the following sets: *IntInst* is the set of labels of integer arithmetic instructions and *FPInst* is the set of labels of floating-point arithmetic instructions. For each instruction with a label ℓ, we also let n_ℓ denote the number of times the instruction executes for the set of inputs. Finally, let α_{int} and α_{fp} be the average savings (i.e., percentage reduction in energy consumption) from executing integer and floating-point instructions, respectively. Then, the energy consumption of integer and floating-point instructions is $E_{int}(\theta) = \sum_{\ell \in IntInst} n_\ell \cdot (1 - \theta(\ell) \cdot \alpha_{int}) \cdot \mathcal{E}_{int}$ and $E_{fp}(\theta) = \sum_{\ell \in FPInst} n_\ell \cdot (1 - \theta(\ell) \cdot \alpha_{fp}) \cdot \mathcal{E}_{fp}$. This model assumes that the instruction count in the approximate execution is approximately equal to the instruction count in the exact execution.

Memory Energy We model the energy consumption of the system memory (i.e., DRAM) using an estimate of the average energy per second per byte of memory, $\mathcal{E}_{mem}$. Given the execution time of all kernel invocations, t, the savings associated with allocating data in approximate memory, α_{mem}, the size of allocated arrays, S_ℓ, and the configurations of array variables in the exact and approximate memories, $\theta(\ell)$, we model the energy consumption of the memory as $E_{mem}(\theta) = t \cdot \mathcal{E}_{mem} \cdot \sum_{\ell \in ArrParams} S_\ell \cdot (1 - \theta(\ell) \cdot \alpha_{mem})$.

Cache Memory Energy We model the energy consumption of cache cell, $\mathcal{E}_{cache}$, similarly. Let S_c be the size of the cache and α_{cache} the savings of approximate caches. In addition, we need to specify the strategy for determining the size of approximate caches. We analyze the strategy that scales the size of approximate caches proportional to the percentage of the size of the arrays allocated in the approximate main memory. If c_u is the maximum fraction of the approximate cache lines, the energy consumption of the cache is $E_{cache}(\theta) = t \cdot \mathcal{E}_{cache} \cdot S_c \cdot (1 - c_u \cdot \frac{\sum_\ell S_\ell \theta(\ell)}{\sum_\ell S_\ell} \cdot \alpha_{cache})$.

7.4.6.2 Relative Energy Model

However, we can use these equations to derive a numerical optimization problem that instead uses cross-design parameters (such as the relative energy between instruction classes and the average savings for each instruction) to optimize the energy consumption of the program relative to an exact configuration of the program. For each energy consumption modeling function in the previous section, we introduce a corresponding function that implicitly takes the exact configuration as its parameter (e.g., E_{sys}, E_{CPU}, E_{mem}).

System Relative Energy The energy model contains a parameter that specifies the relative portion of energy consumed by the CPU versus memory, μ_{CPU}. Using this parameter, we derive the relative system energy consumption as follows:

$$\mu_{CPU} \cdot \frac{E_{CPU}(\theta)}{E_{CPU}} + (1 - \mu_{CPU}) \cdot \frac{E_{mem}(\theta)}{E_{mem}}.$$

CPU Relative Energy The energy model contains a parameter that specifies the relative portion of energy consumed by the ALU, μ_{ALU}, cache, μ_{cache}, and other components $\mu_{other} = 1 - \mu_{ALU} - \mu_{cache}$. We can then derive the relative CPU energy consumption similarly to that for the whole system: $\frac{E_{CPU}(\theta)}{E_{CPU}} = \mu_{ALU} \cdot \frac{E_{ALU}(\theta)}{E_{ALU}} + \mu_{cache} \cdot \frac{E_{cache}(\theta)}{E_{cache}} + \mu_{other}$.

ALU Relative Energy We apply similar reasoning to derive the relative energy consumption of the ALU: $\frac{E_{ALU}(\theta)}{E_{ALU}} = \mu_{int} \cdot \frac{E_{int}(\theta)}{E_{int}} + \mu_{fp} \cdot \frac{E_{fp}(\theta)}{E_{fp}} + \mu_{oi}$. The coefficients μ_{int}, μ_{fp}, and μ_{oi} are computed from the execution counts of each instruction class (n_{int}, n_{fp}, and n_{oi}) and the relative energy consumption rates of each class with respect to that of integer instructions (w_{fp} and w_{oi}). For example, if we let w_{fp} be the ratio of energy consumption between floating-point instructions and integer instructions (i.e., $w_{fp} = \frac{\mathcal{E}_{fp}}{\mathcal{E}_{int}}$), then $\mu_{fp} = \frac{w_{fp} \cdot n_{fp}}{n_{int} + w_{fp} \cdot n_{fp} + w_{oi} \cdot n_{oi}}$.

Memory and Cache Relative Energy Applying similar reasoning to the memory subsystem yields $\frac{E_{mem}(\theta)}{E_{mem}} = \frac{1}{H} \cdot \frac{t'}{t} \cdot \sum_{\ell \in ArrParams} S_\ell \cdot (1 - \theta(\ell) \cdot \alpha_{mem})$ and $\frac{E_{cache}(\theta)}{E_{cache}} = \frac{1}{H} \cdot \frac{t'}{t} \cdot \sum_{\ell \in ArrParams} S_\ell \cdot (1 - c_u \cdot \theta(\ell) \cdot \alpha_{cache})$, where $H = \sum_\ell S_\ell$ is the total size of heap data. The execution time ratio t'/t denotes possibly different execution time of the approximate program. One can use the results of reliability profiling to estimate this ratio.

Relative Energy for Multiple Inputs The relative energy consumption for multiple inputs is the average of the relative energy consumption $E_{sys}(\theta)/E_{sys}$ for each input. Since this quantity is a sum of relative energy consumption of the components (CPU, ALU operations, and memories), the analysis computes and assigns these average relative energy consumption to each operation and variable label.

7.4.7 Final Optimization Problem Statement

We now state the optimization problem for a kernel computation:

$$\textit{Minimize}: \quad \mu_{CPU} \cdot \frac{E_{CPU}(\theta)}{E_{CPU}} + (1 - \mu_{CPU}) \cdot \frac{E_{mem}(\theta)}{E_{mem}}$$

$$\textit{Constraints}: \log(\rho_{spec,i}) - \log(\rho_i) \le \sum_k \theta(\ell_{k_i}) \cdot \log(\rho_{k_i}) \qquad \forall i \in \{1, \ldots, c\}$$

$$\sum_k \theta(\ell_{k_i}) \cdot \log(\rho_{k_i}) = 0 \qquad \forall i \in \{1, \ldots, c'\}$$

$$\textit{Variables}: \quad \theta(\ell_1), \ldots, \theta(\ell_n) \in \{0, 1\}$$

The decision variables $\theta(\ell_1), \ldots, \theta(\ell_n)$ are the configuration kinds of arithmetic instructions and array variables. Since they are integers, the optimization problem belongs to the class of integer linear programs. The index i iterates over all constraints generated by the reliability and accuracy analyses. The index k iterates over the sequences of the candidate approximate instructions in the constraints. There is a possibility to have multiple accuracy and safety constraints, which we denote with enumerations of i between 1 and c (respectively, c').

Example 7.9 (Analysis of a Function) Let us return to Example 7.8 and repeat the instructions:

$$\texttt{r0 = init}\langle\texttt{x}\rangle\texttt{; r1 = load r0; r2 = init 3;}$$

$$\texttt{r3 = add}^{\ell_+}\texttt{r1 r2; r4 = add}^{\ell_*}\texttt{r3 r2; return r4;}$$

We have three decision variables, $\theta(\ell_+)$, $\theta(\ell_*)$, and $\theta(\ell_x)$. The relative energy savings expression in this case is simple, since only the add and mul instruction can be approximated (because x is on stack, we can immediately set $\theta(\ell_x) = 0$). The expression for the relative energy consumption for these two expressions is $c \cdot (1 - \frac{\alpha_{int}}{2} \cdot (\theta(\ell_+) + \theta(\ell_*)))$. Here, α is the saving of the individual instruction, obtained from the hardware specification and $c = \mu_{cpu}\mu_{alu}\frac{n_{int}}{n_{total}}$ (the instruction counts $n_{int} = 2$ and $n_{total} = 6$ are obtained from the trace). We elided the remaining terms for heap and cache memory, as they remain constant—although they would be necessary for estimating the total system energy consumption.

The reliability constraint computed in the previous step is $0.99 \cdot \mathcal{R}(\{x\}) \leq \rho_{add}^{\theta(\ell_+)} \cdot \rho_{mul}^{\theta(\ell_*)} \cdot \rho_{ld}^{\theta(\ell_x)} \cdot \mathcal{R}(\{x\})$. We first ensure that the reliability factors on the left and right sides match (here, they are both equal $\mathcal{R}(\{x\})$). We then construct the constraint for the optimization problem by taking the logarithm of both sides of the inequality. The final optimization problem is

$$\begin{aligned} &\textit{\textbf{Minimize}}: && c \cdot (1 - \tfrac{\alpha_{int}}{2} \cdot (\theta(\ell_+) + \theta(\ell_*))) \\ &\textit{\textbf{Constraint}}: && \log(0.99) \leq \theta(\ell_+) \cdot \rho_{add} + \theta(\ell_*) \cdot \rho_{mul}. \\ &\textit{\textbf{Variables}}: && \theta(\ell_+),\ \theta(\ell_*) \end{aligned}$$

We can pass this problem to an off-the-shelf ILP solver to get the assignments of the variables $\theta(\ell_+)$ and $\theta(\ell_*)$.

7.4.8 Discussion

Soundness The reliability analysis is sound with respect to the paired execution semantics of the approximate computation. The soundness argument for the reliability predicates follows from the soundness of Rely [46]. When the configuration

$\theta(\cdot)$ for the instructions is known, the analysis can substitute each $\theta(\ell)$ and check whether the final precondition is correct, using the same approach as in Rely.

Safety Constraints Some safety constraints can be easily added to the optimization problem. For instance, if we want to ensure that array indices are always correct, then during the analysis of statement `x[idx]`, we may add the constraint $\mathcal{R}(\{idx\}) = 1$, which would ensure that all the labels of instructions/variables used to compute `idx` are marked as reliable. While we could use ILP solvers for linear constraints, it is an interesting open question how to specify more expressive safety properties. One possible direction is to consider solving the optimization problems using an SMT solver [47] to support general first-order logic constraints.

Extending the Languages and/or Error Models We presented a simple language with a simple error model and a fine instruction-level approximation granularity. The constraint generator can be straightforwardly extended to a granularity of functions, where each function has its output specification $r \cdot \mathcal{R}(\{x_1, \ldots, x_n\})$—the rule for the reliability of function call will be analogous to the one for binary operators (which are just functions with two inputs). The original paper discusses the optimization with functions and some other constructs [27, Section 8].

To reason about more complicated (even arbitrary) error models, one needs a more expressive verification system. Leto [48] shows how to define richer error models and automate the computation of the accuracy/safety constraints. Another interesting direction is support for parallel and distributed applications. Parallelly shows how to lift the predictive analyses for sequential programs to support concurrent message-passing ones [49], but performance models in this case remain an open question. ApproxHPVM [28] shows how to use the kernel-level specifications and multi-step optimization to optimize deep learning applications on heterogeneous systems.

Related Approaches Several techniques used mathematical optimization to optimize approximate computations. As a precursor to Chisel, we presented a framework for coarser-grained optimization of functional map-reduce programs using a combination of linear programming and discretization to get guaranteed near-optimal solutions [9]. Capri [50] solves an optimization problem with deterministic or probabilistic constraints, while modeling errors using Bayesian networks. Existing works also use the constraints to express the error and possible savings of floating-point computations, e.g., [51, 52].

7.5 A Brief Survey of Approximate Program Analysis and Optimization Techniques

Approximate computing has flourished in the past decade. Existing surveys, e.g., [53–55], systematize a large amount of work across the system stack. In this section, we will only mention a subset of techniques historically relevant for software-centric compilation and some recent directions.

Sensitivity Analysis A critical region of program code, when transformed, causes unacceptable program errors (such as crashing, becoming unresponsive, or producing inadequate output). An approximable region of program code, when transformed, only affects the accuracy of the computation.

Testing-based sensitivity analyses can identify critical regions of the program. In general, dynamic sensitivity analyses transform a program's code [6], change program's inputs [56], intermediate data [57–59], or change its execution environment. The analyses reason about the effect of these changes on the program's output and classify the code regions accordingly.

The researchers have also proposed techniques for analyzing the worst-case behavior of numerical computations. The researchers in embedded systems have traditionally used rounding error analyses of numerical programs to derive the worst-case error bounds for reduced bit-width floating- or fixed-point computations [60]. Chaudhuri et al. [61] presented a technique for verifying Lipschitz continuity of approximate computations and bounding the error propagation. Techniques such as Rosa [62] verify the precision of approximate numerical computation and compute the bounds on error propagation through nonlinear computation.

Safety Analysis While dynamic sensitivity profiling techniques can help identify critical parts of the program by finding a single failing execution, they are insufficient to prove the absence of errors or incorrect outputs. Therefore, researchers have developed various techniques that let a developer specify important safety property (such as non-interference of approximate and exact code, pointer safety, or range of values that the computation produces) and verify that the transformations preserve these properties.

EnerJ [22] presents an information flow type system that allows the developer to separate code and data in distinct approximate and exact regions of code. FlexJava [36] automates a part of approximate operation annotation through type inference. Carbin et al. [2] present a general framework for reasoning about arbitrary safety properties (and also worst-case error) of approximate programs. A part of this analysis has been automated within Simdiff [63] using SMT solvers.

Search for Accuracy–Performance Tradeoffs The researchers have presented various techniques for exploring accuracy/performance tradeoff space. These techniques typically discretize the input space, by asking a developer to provide representative inputs, and discretize the configuration space, by trying the approximation with a fixed number of values for each knob. The search algorithms execute the transformed programs for various combinations of the knobs. These combinations are selected using various heuristic search strategies, such as exhaustive search (with optional pruning) [3, 7, 13], greedy search [6, 11, 14], genetic search [12, 31, 64, 65], or stochastic search [15, 66], and leveraging information from sensitivity analysis [67] or static type safety [68]. Alternatively, to improve the accuracy of the results, some techniques perform on-line recalibration, by occasionally running both the exact and approximate versions of the subcomputations [11, 69, 70], machine learning-based models [71–73], or runtime approximation tuning by creating smaller (canary) inputs [42, 74]. Advanced search approaches will be especially important in the context of optimizing multiple software–hardware components in modern heterogeneous systems [23, 32, 64, 68, 73].

Acknowledgments The presentation in this chapter is derived from the author's Ph.D. dissertation [75]. The results from Case Study 1 appeared first in the FSE 2011 paper [7]. The results from Case Study 2 appeared first in the OOPSLA 2014 paper [27]. The author would like to thank all of his co-authors who contributed to bringing up the theory and practice of accuracy-aware program optimization; special gratitude goes to Martin Rinard, Hank Hoffmann, Stelios Sidiroglou, and Michael Carbin. The author would also like to thank Keyur Joshi, Olivier Sentieys, Hashim Sharif, and Yifan Zhao for providing feedback on the earlier versions of this chapter.

References

1. Kildall, G. A. (1973). A unified approach to global program optimization. In *Proceedings of the POPL.*
2. Carbin, M., Kim, D., Misailovic, S., & Rinard, M. (2012). Proving acceptability properties of relaxed nondeterministic approximate programs. In *Proceedings of the PLDI.*
3. Rinard, M. (2006). Probabilistic accuracy bounds for fault-tolerant computations that discard tasks. In *Proceedings of the ICS.*
4. Chakradhar, S., Raghunathan, A., & Meng, J. (2009). Best-effort parallel execution framework for recognition and mining applications. In *International Symposium on Parallel & Distributed Processing (IPDPS).*
5. Meng, J., Raghunathan, A., Chakradhar, S., & Byna, S. (2010). Exploiting the forgiving nature of applications for scalable parallel execution. In *International Symposium on Parallel & Distributed Processing (IPDPS).*
6. Misailovic, S., Sidiroglou, S., Hoffmann, H., & Rinard, M. (2010). Quality of service profiling. In *Proceedings of the ICSE.*
7. Sidiroglou, S., Misailovic, S., Hoffmann, H., & Rinard, M. (2011). Managing performance vs. accuracy trade-offs with loop perforation. In *Proceedings of the FSE.*
8. Misailovic, S., Roy, D., & Rinard, M. (2011). Probabilistically accurate program transformations. In *Proceedings of the SAS.*
9. Zhu, Z., Misailovic, S., Kelner, J., & Rinard, M. (2012). Randomized accuracy-aware program transformations for efficient approximate computations. In *Proceedings of the POPL.*
10. Samadi, M., Jamshidi, D., Lee, J., & Mahlke, S. (2014). Paraprox: Pattern-based approximation for data parallel applications. In *Proceedings of the ASPLOS.*
11. Baek, W., & Chilimbi, T. M. (2010). Green: A framework for supporting energy-conscious programming using controlled approximation. In *Proceedings of the PLDI.*
12. Ansel, J., Wong, Y., Chan, C., Olszewski, M., Edelman, A., & Amarasinghe, S. (2011). Language and compiler support for auto-tuning variable-accuracy algorithms. In *International Symposium on CGO.*
13. Hoffmann, H., Sidiroglou, S., Carbin, M., Misailovic, S., Agarwal, A., & Rinard, M. (2011). Dynamic knobs for responsive power-aware computing. In *Proceedings of the ASPLOS.*
14. Rubio-González, C., Nguyen, C., Nguyen, H., Demmel, J., Kahan, W., Sen, K., Bailey, D., Iancu, C., & Hough, D. (2013). Precimonious: Tuning assistant for floating-point precision. In *Proceedings of the SC.*
15. Schkufza, E., Sharma, R., & Aiken, A. (2014). Stochastic optimization of floating-point programs with tunable precision. In *Proceedings of the PLDI.*
16. Han, S., Mao, H., & Dally, W. (2016). Deep compression: Compressing deep neural networks with pruning, trained quantization and Huffman coding. In *International Conference on ICLR.*

17. Cormode, G., & Garofalakis, M. (2007). Sketching probabilistic data streams. In *Proceedings of the SIGMOD/PODS*.
18. Rinard, M. (2007). Using early phase termination to eliminate load imbalances at barrier synchronization points. In *Proceedings of the OOPSLA*.
19. Misailovic, S., Sidiroglou, S., & Rinard, M. (2012). Dancing with uncertainty. In *Proceedings of the RACES*.
20. Misailovic, S., Kim, D., & Rinard, M. (2013). Parallelizing sequential programs with statistical accuracy tests. *ACM Transactions Embedded Computing System Special Issue on Probabilistic Embedded Computing, 12*(2s), 1–26.
21. Campanoni, S., Holloway, G., Wei, G.-Y., & Brooks, D. (2015). Helix-up: Relaxing program semantics to unleash parallelization. In *International Symposium on CGO*.
22. Sampson, A., Dietl, W., Fortuna, E., Gnanapragasam, D., Ceze, L., & Grossman, D. (2011). EnerJ: Approximate data types for safe and general low-power computation. In *Proceedings of the PLDI*.
23. Esmaeilzadeh, H., Sampson, A., Ceze, L., & Burger, D. (2012). Neural acceleration for general-purpose approximate programs. In *International Symposium on MICRO*.
24. Liu, S., Pattabiraman, K., Moscibroda, T., & Zorn, B. (2011). Flikker: Saving DRAM refresh-power through critical data partitioning. In *Proceedings of the ASPLOS*.
25. Fernando, V., Franques, A., Abadal, S., Misailovic, S., & Torrellas, J. (2019). Replica: A wireless manycore for communication-intensive and approximate data. In *Proceedings of the ASPLOS*.
26. Chippa, V., Chakradhar, S., Roy, K., & Raghunathan, A. (2013). Analysis and characterization of inherent application resilience for approximate computing. In *Proceedings of the DAC*.
27. Misailovic, S., Carbin, M., Achour, S., Qi, Z., & Rinard, M. (2014). Chisel: Reliability- and accuracy-aware optimization of approximate computational kernels. In *Proceedings of the OOPSLA*.
28. Sharif, H., Srivastava, P., Huzaifa, M., Kotsifakou, M., Joshi, K., Sarita, Y., Zhao, N., Adve, V., Misailovic, S., & Adve, S. (2019). ApproxHPVM: A portable compiler IR for accuracy-aware optimizations. In *Proceedings of ACM on Programming Languages, 3*(OOPSLA).
29. Hoffmann, H., Misailovic, S., Sidiroglou, S., Agarwal, A., & Rinard, M. (2009, September). Using code perforation to improve performance, reduce energy consumption, and respond to failures. Tech. Rep. MIT-CSAIL-TR-2009-042, MIT.
30. Lattner, C., & Adve, V. (2004). LLVM: A compilation framework for lifelong program analysis & transformation. In *International Symposium on CGO*.
31. Ansel, J., Kamil, S., Veeramachaneni, K., Ragan-Kelley, J., Bosboom, J., O'Reilly, U. M., & Amarasinghe, S. (2014). OpenTuner: An extensible framework for program autotuning. In *International Conference on PACT*.
32. Sharif, H., Zhao, Y., Kotsifakou, M., Kothari, A., Schreiber, B., Wang, E., Sarita, Y., Zhao, N., Joshi, K., Adve, S., Misailovic, S., & Adve, S. (2021). ApproxTuner: A compiler and runtime system for adaptive approximations. In *Proceedings of the PPoPP* (pp. 262–277).
33. Geilen, M., & Basten, T. (2007). A calculator for pareto points. In *Proceedings of the DATE*.
34. P. B. Suite. http://parsec.cs.princeton.edu/
35. X. V. T. Media. http://media.xiph.org/video/derf
36. Park, J., Esmaeilzadeh, H., Zhang, X., Naik, M., & Harris, W. (2015). FlexJava: Language support for safe and modular approximate programming. In *Proceedings of the FSE*.
37. Carbin, M., Kim, D., Misailovic, S., & Rinard, M. (2013). Verified integrity properties for safe approximate program transformations. In *Proceedings of the PEPM*.
38. Lou, L., Nguyen, P., Lawrence, J., & Barnes, C. (2016). Image perforation: Automatically accelerating image pipelines by intelligently skipping samples. *ACM Transactions on Graphics (TOG), 35*(5), 1–14.
39. Figurnov, M., Ibraimova, A., Vetrov, D., & Kohli, P. (2016). PerforatedCNNs: Acceleration through elimination of redundant convolutions.
40. Maier, D., Cosenza, B., & Juurlink, B. (2018). Local memory-aware kernel perforation. In *International Symposium on CGO*.

41. Li, S., Park, S., & Mahlke, S. (2018). Sculptor: Flexible approximation with selective dynamic loop perforation. In *Proceedings of the ISC* (pp. 341–351).
42. Xu, R., Koo, J., Kumar, R., Bai, P., Mitra, S., Misailovic, S., & Bagchi, S. (2018). VideoChef: Efficient approximation for streaming video processing pipelines. In *Proceedings of the USENIX ATC*.
43. Palem, K. (2005). Energy aware computing through probabilistic switching: A study of limits. *IEEE Transactions on Computers, 54*(9), 1123–1137.
44. Leem, L., Cho, H., Bau, J., Jacobson, Q., & Mitra, S. (2010). ERSA: Error resilient system architecture for probabilistic applications. In *Proceedings of the DATE*.
45. Lee, K., Shrivastava, A., Issenin, I., Dutt, N., & Venkatasubramanian, N. (2006). Mitigating soft error failures for multimedia applications by selective data protection. In *Proceedings of the CASES*.
46. Carbin, M., Misailovic, S., & Rinard, M. (2013). Verifying quantitative reliability for programs that execute on unreliable hardware. In *Proceedings of the OOPSLA*.
47. Bjørner, N., Phan, A., & Fleckenstein, L. (2015). νZ-an optimizing SMT solver. In *Internation Conference on TACAS*.
48. Boston, B., Gong, Z., & Carbin, M. (2018). Leto: Verifying application-specific hardware fault tolerance with programmable execution models. *Proceedings of the ACM on Programming Languages, 2*(OOPSLA), 1–30.
49. Fernando, V., Joshi, K., & Misailovic, S. (2019). Verifying safety and accuracy of approximate parallel programs via canonical sequentialization. *Proceedings of the ACM on Programming Languages, 3*(OOPSLA), 1–29.
50. Sui, X., Lenharth, A., Fussell, D., & Pingali, K. (2016). Proactive control of approximate programs. In *Proceedings of the ASPLOS*.
51. Chiang, W., Baranowski, M., Briggs, I., Solovyev, A., Gopalakrishnan, G., & Rakamaric, Z. (2017). Rigorous floating-point mixed-precision tuning. In *Proceedings of the POPL*.
52. Izycheva, A., Darulova, E., & Seidl, H. (2019). Synthesizing efficient low-precision kernels. In *International Symposium on ATVA*.
53. Han, J., & Orshansky, M. (2013). Approximate computing: An emerging paradigm for energy-efficient design. In *Proceedings of the ETS*.
54. Xu, Q., Mytkowicz, T., & Kim, N.-S. (2016). Approximate computing: A survey. *IEEE Design & Test, 33*(1), 8–22.
55. Stanley-Marbell, P., Alaghi, A., Carbin, M., Darulova, E., Dolecek, L., Gerstlauer, A., Gillani, G., Jevdjic, D., Moreau, T., Cacciotti, M., Daglis, A., Enright-Jerger, N., Falsafi, B., Misailovic, S., Sampson, A., & Zufferey, D. (2020). Exploiting errors for efficiency: A survey from circuits to applications. *ACM Computing Surveys, 53*(3), 51:1–51:39.
56. Carbin, M., & Rinard, M. (2010). Automatically identifying critical input regions and code in applications. In *Proceedings of the ISSTA*.
57. Roy, P., Ray, R., Wang, C., & Wong, W. (2014). ASAC: Automatic sensitivity analysis for approximate computing. In *Proceedings of the LCTES*.
58. Venkatagiri, R., Mahmoud, A., Hari, S., & Adve, S. (2016). Approxilyzer: Towards a systematic framework for instruction-level approximate computing and its application to hardware resiliency. In *International Symposium on MICRO*.
59. Nongpoh, B., Ray, R., Dutta, S., & Banerjee, A. (2017). Autosense: A framework for automated sensitivity analysis of program data. *IEEE Transactions on Software Engineering, 43*(12), 1110–1124.
60. Gaffar, A., Mencer, O., Luk, W., Cheung, P., & Shirazi, N. (2002). Floating-point bitwidth analysis via automatic differentiation. In *International Conference on FPT*.
61. Chaudhuri, S., Gulwani, S., Lublinerman, R., & Navidpour, S. (2011). Proving programs robust. In *Proceedings of the FSE*.
62. Darulova, E., & Kuncak, V. (2014). Sound compilation of reals. In *Proceedings of the POPL*.
63. He, S., Lahiri, S., & Rakamaric, Z. (2018). Verifying relative safety, accuracy, and termination for program approximations. *Journal of Automated Reasoning, 60*(1), 23–42.

64. Barone, S., Traiola, M., Barbareschi, M., & Bosio, A. (2021). Multi-objective application-driven approximate design method. *IEEE Access*.
65. Dorn, J., Lacomis, J., Weimer, W., & Forrest, S. (2019). Automatically exploring tradeoffs between software output fidelity and energy costs. *IEEE Transactions on Software Engineering, 45*(3), 219–236.
66. Ha, V., & Sentieys, O. (2021). Leveraging Bayesian optimization to speed up automatic precision tuning. In *Proceedings of the DATE*.
67. Vassiliadis, V., Riehme, J., Deussen, J., Parasyris, K., Antonopoulos, C., Bellas, N., Lalis, S., & Naumann, U. (2016). Towards automatic significance analysis for approximate computing. In *International Symposium on CGO*.
68. Sampson, A., Baixo, A., Ransford, B., Moreau, T., Yip, J., Ceze, L., & Oskin, M. (2015). Accept: A programmer-guided compiler framework for practical approximate computing. Tech. Rep. UW-CSE-15-01, University of Washington.
69. Samadi, M., Lee, J., Jamshidi, D., Hormati, A., & Mahlke, S. (2013). Sage: Self-tuning approximation for graphics engines. *International Symposium on MICRO*.
70. Goiri, I., Bianchini, R., Nagarakatte, S., & Nguyen, T. (2015). Approxhadoop: Bringing approximations to MapReduce frameworks. In *Proceedings of the ASPLOS*.
71. Ding, Y., Ansel, J., Veeramachaneni, K., Shen, X., O'Reilly, U. M., & Amarasinghe, S. (2015). Autotuning algorithmic choice for input sensitivity. In *Proceedings of the PLDI*.
72. Barati, S., Bartha, F., Biswas, S., Cartwright, R., Duracz, A., Fussell, D., Hoffmann, H., Imes, C., Miller, J., Mishra, N., Arvind, Nguyen, D., Palem, K., Pei, Y., Pingali, K., Sai, R., Wright, A., Yang, Y., & Zhang, S. (2019). Proteus: Language and runtime support for self-adaptive software development. *IEEE Software, 36*(2), 73–82.
73. Mahajan, D., Yazdanbakhsh, A., Park, J., Thwaites, B., & Esmaeilzadeh, H. (2016). Towards statistical guarantees in controlling quality tradeoffs for approximate acceleration. In *International Symposium on Computer Architecture (ISCA)*.
74. Laurenzano, M., Hill, P., Samadi, M., Mahlke, S., Mars, J., & Tang, L. (2016). Input responsiveness: Using canary inputs to dynamically steer approximation. In *Proceedings of the PLDI*.
75. Misailovic, S. (2015). *Accuracy-Aware Optimization of Approximate Programs*. PhD thesis, Massachusetts Institute of Technology, Cambridge, MA.

Chapter 8
Design Space Exploration Tools

Mario Barbareschi, Salvatore Barone, Nicola Mazzocca, and Alberto Moriconi

8.1 Introduction

Nowadays, the amount of information needed to be processed by computer systems is quickly becoming unsustainable, since the latter are experimenting an unprecedented growth of data to be processed. Indeed, while, on the one hand, these systems increasingly interact with the physical world and, on the other hand, they process the large amount of data samples coming from all the various sensing sources. This is the root cause of the tremendous growth of power consumption of computing systems. Growth that is increasing year by year so that it is estimated that energy consumption will exceed the energy production capabilities before 2040 [1].

Therefore, power and energy reduction are critical requirements in the design of computing systems, especially in pervasive embedded and mobile electronic devices, where the battery capacity is a limiting factor. Additionally, computationally intensive tasks, such as machine learning applications, have found their way into these power-limited devices, increasing the need for efficient electronics. In this perspective, current technologies and design approaches are bound to become quite soon inadequate to offer suitable solutions to these application requirements; hence, novel design approaches have to be considered.

One of the most promising solutions is the Approximate Computing (AxC) design paradigm. It is based on the observation that while performing exact computations, or maintaining peak-level service performance, may require too many resources, reduction in energy consumption and performance enhancements can be achieved by selectively relaxing correctness requirements, hence exposing a certain

M. Barbareschi · S. Barone (✉) · N. Mazzocca · A. Moriconi
DIETI – Department of Electrical Engineering and Information Technologies, University of Naples Federico II, Napoli, Italy
e-mail: mario.barbareschi@unina.it; salvatore.barone@unina.it; nicola.mazzocca@unina.it; alberto.moriconi@unina.it

A. Bosio et al. (eds.), *Approximate Computing Techniques*,
https://doi.org/10.1007/978-3-030-94705-7_8

degree of approximation [2]. The scientific literature demonstrated the effectiveness of imprecise computation for error-resilient applications, for both software and hardware components implementing inexact algorithms [3, 4].

The AxC leverages the presence of error-tolerant data and algorithms, and the perceptual limitations of the end-user, to carefully trade accuracy for performance gains or energy savings. In other words, it exploits the existing gap between the accuracy level provided by computer systems and the accuracy level effectively needed by the considered application, or by the end-users, with the latter being usually far lower than the former.

8.1.1 Fields of Application

A large variety of applications could potentially benefit from AxC: its use is unavoidable in many scenarios, while the opportunity of using it arises inherently in many others. Notable examples are floating-point applications, which involve a certain degree of approximation due to representation errors. Indeed, to produce acceptable results, computations rarely need to be performed at the maximum available precision.

The perceptual limitations of humans can be exploited to reduce the storage requirements [5] or to improve performances in multimedia and signal processing applications [6, 7]. Furthermore, for several iterative refinement algorithms, running iterations with reduced precision at intermediate computation can improve performances with little or even no effects on the quality of results [8–10]. Moreover, early loop termination [8, 11], memory accesses skipping [12], or tasks skipping [13] can alleviate performance bottlenecks.

8.1.2 A Brief Overview of Approximate Computing Techniques

Since a naive approximation approach, such as uniform approximation, is unlikely to be efficient, different Approximate Computing Techniques (AxCTs) have been proposed in the scientific literature. Some examples are bit-width optimization, also known as precision-scaling, Loop Perforation (LP), memoization, and Load Value Approximation (LVA).

Precision-scaling for input data and intermediate operands has been proposed to improve efficiency for floating-point, fixed-point, and even integer computations in many scientific applications [14–18].

The effectiveness of the LP technique, which works by skipping some iterations of a loop to reduce computational overhead, has been shown by Sidiroglou et al. [11] for iterative refinement algorithms, search space enumeration, and Monte Carlo simulations.

The memoization technique works by storing the results of functions for later reuse. Keramidas et al. [19] proposed the combined use of AxC and memoization, to increase the amount of successfully reused values.

The LVA technique has been proposed to mitigate the latency induced by cache-misses, leveraging the inner nature of applications to estimate values to be loaded, allowing CPUs and Graphic Processing Units (GPUs) to progress without stalling [12, 20].

8.1.3 Issues and Open Challenges

Exploiting AxC requires coping with

1. The designation of parts of the considered software or hardware component which are suitable to be approximate
2. The approach to introduce actual approximation
3. The selection of appropriate error metrics, which generally depend on the particular application
4. The actual error assessment procedure, to guarantee output quality constraints are met [3], and finally
5. The Design-Space Exploration (DSE), to select the best approximate configurations among those generated by a certain approximation technique.

As for the first two of the aforementioned issues, pinpointing approximable code or data portions may require the designer to have deep insights into the application. Moreover, since a naive approximation approach—such as the uniform one—is unlikely to be efficient, and since no approach can be universally applied to all approximable applications, the approximation approach needs to be determined on a per-application basis by the designer.

As for error assessment, it typically requires the simulation of both exact and approximate applications, nevertheless Bayesian inference [21, 22] or machine learning-based approaches [23] have been proposed in the scientific literature.

Finally, concerning DSE, initial approaches either combine multiple design objectives in a single-objective optimization problem or optimize a single parameter while keeping the others fixed. Recently published works address the circuit design problem by using MOP to search for Pareto-optimal approximate circuit implementations [24]. Unfortunately, such approaches did not focus on complex systems, rather on arithmetic components, such as adders and multipliers, since they are building blocks for more complex designs.

In the remainder of this chapter, we discuss the state of the art for AxC automatic tools. In particular, Sect. 8.2.1 details automatic tools targeting the approximation of digital circuits, while Sect. 8.2.2 focuses on tools for the approximation of software applications. Note that we do not aim at a complete survey of the literature, but at highlighting thought-provoking knowledge in the field. Finally, Sect. 8.3 details the $\mathbb{IDEA}$, a unified framework that allows automatically exploring the impact of different approximation techniques on either hardware or software applications, while facing with the DSE problem as a MOP.

8.2 Automatic Tools for Approximate Computing

8.2.1 Approximation Tools Targeting Digital Circuits

In this section, AxC automation tools targeting digital circuits, both combinational and sequential, are discussed, highlighting their main features and the innovative contributions brought to the scientific literature. Table 8.1 summarizes the discussion, reporting for each different tool, its main characteristics, such as the type of circuit to which the tool refers, the model with which the circuit is represented, how the error introduced by the approximation is kept under control, the design space exploration technique used to find the best configurations, and the target technology are reported.

8.2.1.1 Power-Aware and Branch-Aware Word-Length Optimization

One of the first attempts to define an automatic methodology for digital circuits approximation is PowerCutter [17], which focuses on minimizing power consumption by making use of word-length optimization.

The tool needs three inputs: a C/C++ model of the design to be approximated, a set of error constraints defined on the value of output variables, and a set of ranges for input variables. Alternatively, an input dataset can be provided by the user so that input ranges are computed by the tool.

The first stage of the proposed design flow is a static analysis of the given model, which, in turn, consists of three steps. First, a range analysis is performed: arithmetic operations are performed on ranges, instead of single values, in order to gather information about the variability range of intermediate and output variables. Range information is passed to a precision analyzer, which determines the optimal number of fractional bits required by each variable. Leveraging range and precision information, a floating-point to fixed-point conversion is performed. The third step is cost analysis: operations are ranked based on the amount of time they require and the number of times they are performed.

Since results provided by static analysis may be too conservative, an offline dynamic analysis is also performed, in order to further optimize the design. Dynamic analysis consists of three steps: dynamic range analysis, automatic differentiation, and branch analysis. The dynamic range analysis is performed in order to determine whether the floating-point representation may be more effective than the fixed-point one when dealing with very small values. By making use of user-provided input ranges or input dataset, the variability range of all variables is tracked to decide whether to keep the fixed-point representation—adopting a shift to reduce the required bits—or whether to restore the original floating-point representation. Dynamic range analysis is also able to detect input patterns, which can be used for variable-to-constant conversion. Several designs make use of complex functions, such as trigonometric functions. In order to implement such functions in hardware,

Table 8.1 Summary of tools for digital circuits approximation

Tool	Circuit type	Input model	Error control	Search method	AxCT	Quality assurance	Output	Output model	Target technology
PowerCutter [17] (see Sect.8.2.1.1)	Combinational	C/C++	Error bound	–	Precision-scaling	Testing	Annotated C/C++	Set of range for variables	ASIC/FPGA
SALSA [25] (see Sect. 8.2.1.2)	Combinational	Netlist	QCC	–	ADC	–	Approx. circuit	Netlist	ASIC
SASIMI [26] (see Sect. 8.2.1.3)	Combinational	Netlist	Error bound	Hil climbing	Substitute & simplify	Testing	Approx. circuit	Netlist	ASIC
ASLAN [27] (see Sect. 8.2.1.4)	Combinational/ sequential	Structural HDL	SQCC	Hill climbing	Precision-scaling	Formal verification	Approx. circuit	Structural HDL	ASIC
ABACUS [28] (see Sect. 8.2.1.5)	Combinational/ sequential	Behavioral HDL	Iterative refinement	Greedy	AST transformation	Testing	Pareto front	HDL	ASIC
SCALS [29] (see Sect. 8.2.1.6)	Combinational	Netlist	Error bound	Iterative refinement	Functional approximation	Testing	Approx. Circuit	Netlist	FPGA
Češka et al.[30] (see Sect. 8.2.1.7)	Combinational	Netlist	QCC	GA	AIG-rewriting	Formal verification	Approx. circuit	Netlist	ASIC
BLASYS [31] (see Sect. 8.2.1.8)	Combinational	Truth table or Netlist	Error bound	-	BMF	Testing	Approx. circuit	Netlist	ASIC
CIRCA [32] (see Sec. 8.2.1.9)	Combinational/ sequential	Annotated HDL, blif	SQCC	Hill climbing/Simulated-annealing/Monte Carlo tree search	Precision-scaling/AIG-rewriting	Testing/Formal verification	Pareto front	HDL/blif	ASIC/FPGA

specific IP-cores are typically used. This may make error analysis cumbersome because the actual implementation of such IP-cores may be not known. In order to tackle with this issue, the automatic differentiation technique [33] is used to compute the sensitivity of inputs and determine their optimal word-length. To further reduce area requirements, branch analysis is performed: the whole design is split into basic blocks, each of which with a single entry point and a single exit point. Blocks are ranked based on their execution frequency, since blocks executed more frequently may have greater contribution to the error, and precision of variables along less frequently executed blocks is reduced in order to save area.

The tool produces two outputs: the first is a C/C++ source code in which each variable is annotated with range and optimal precision, while the second is a database of statistics that could be used to adapt word-length at runtime using, for example, FPGA re-configuration features or the clock gating for the ASIC technology.

In order to evaluate the proposed approach, different designs, such as Discrete Cosine Transform (DCT) computation blocks, ray tracing applications, and Finite Impulse Response (FIR) filters, have been considered. The authors claimed power savings up to 32%.

8.2.1.2 Systematic Methodology for Automatic Logic Synthesis of Approximate circuits (SALSA)

In [25], a Systematic methodology for the Automatic Logic Synthesis of functionally Approximate circuits (SALSA) is proposed. In order to obtain an approximate version of a given circuit, the proposed methodology starts from its functional Register-Transfer Level (RTL) description. The type and the amount of error allowed by the considered application are encoded in one or mode Boolean logic functions.

Figure 8.1 depicts the QCC used by SALSA to formulate and solve the synthesis problem. It consists of three blocks: the exact logic circuit, the approximate circuit,

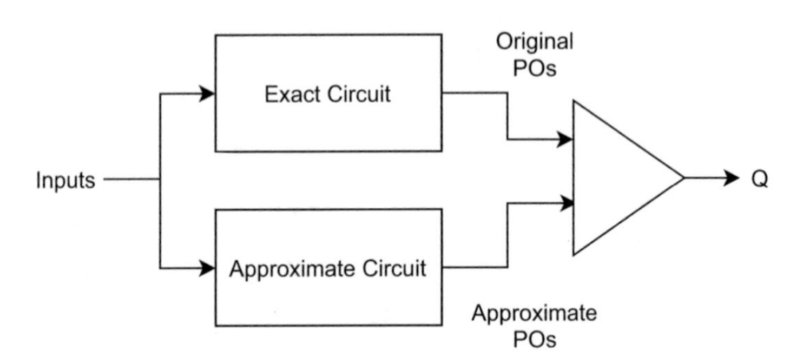

Fig. 8.1 Quality constraint circuit [25]

and the quality function (Q-function) circuit. The first is a structural description of the circuit to be approximate, while the Q-function defines error constraints to be satisfied. During the synthesis process, the approximate circuit is iteratively evolved leveraging the Observability Don't Care (ODC) concept, while preserving the $Q = 1$ invariant.

The ODC set of a node in a logic circuit is the set of input values for which the Primary Outputs (POs) of the circuit are insensitive to the output of the considered node [34]. Since the ODCs of the approximate circuit's POs w.r.t. the output Q do not affect the value of the Q-function, these input combinations can be used to simplify the nodes. These ODCs are called ADCs and they are used to simplify the circuit using standard *don't care*-based synthesis techniques.

In order to find the ADCs set, the tool first co-factorizes the Q-function, using Boole's theorem; considering the ith PO,

$$\begin{aligned} Q &= f(PO_0, \cdots, PO_i, \cdots, PO_N) \\ &= \overline{PO_i \cdot f(PO_0, \cdots, 1, \cdots, PO_N) + \overline{PO_i} \cdot f(PO_0, \cdots, 0, \cdots, PO_N)} \end{aligned} \tag{8.1}$$

Then, the set of inputs for which both negative and positive co-factors have the same value is searched. In the first step, the sensitivity of the Q-function to POs is computed; then, in the second step, the ADCs of Q are expressed in terms of Primary Inputs (PIs) of the circuit, and the approximation takes place. The synthesis process is iterated until quality constraints are met.

SALSA has been tested on a number of logic and arithmetic circuits, from simple ripple-carry adders to DCT computation blocks. Results show a 40–60% reduction for area requirements and a 20–40% reduction of power consumption.

8.2.1.3 Substitute-And-SIMplIfy (SASIMI)

In many applications, the degree of resiliency to error may depend on the dataset being processed, on the working conditions or on the specific context. In these scenarios, quality configurable circuits, which are capable of reconfiguring themself at run-time in order to adapt their accuracy, are needed.

In the work by Venkataramani et al. [26], Substitute-And-SIMplIfy (SASIMI), a new automatic approach targeting the generation of quality configurable circuits, is proposed. The idea is to identify near-identical pairs of signals—signals showing the same value with very high probability—and substitute one in place of another, introducing functional approximation. The signal being replaced is called Target Signal (TS), while its substitution is called Substitute Signal (SS).

Signal substitution has both direct and indirect effects. Well-chosen substitutions can lead to circuit simplification, due to the elimination of the logic computing TSs. Moreover, the logic in the transitive fan-out of TSs can be downsized, since substitutions introduce timing slack, and the logic computing the fan-in of TSs can be sized regardless of TSs themself.

The approximate circuit is obtained using an iterative algorithm. In each iteration, the best candidate signal pair is identified, substitution is performed, the circuit is simplified, and the error is estimated. The algorithm proceeds to the next iteration only if error constraints are met; otherwise, the last suitable version of the approximate circuit is returned. The selection of candidate signal pairs is performed by taking both the size of the logic being deleted and the size of the transitive fan-out into account. The latter is combined with the arrival time of the SS, i.e., the maximum slack that can be introduced, to assess the maximum potential logic downsizing. The score of a signal pair is, then, computed as the normalized sum of logic being deleted and downsized, weighted against the introduced error, which ensures that possible approximations are not exhausted in a few iterations.

In order to generate quality configurable circuits, the tool actually performs signal substitution, but the logic is retained and the difference between TSs and SSs is constantly monitored. Since additional area and power are consumed by the substitution, selection and clock-extension circuits, and since no logic is deleted, logic downsizing becomes crucial. Generated circuits can operate in *approximate* or *accurate* mode: additional logic is introduced in order to selectively choose which output has to be used. In *approximate* mode, any difference between TSs and SSs is simply ignored, as the output meets the error constraints by design. In *accurate* mode, on the other hand, the difference is monitored; if a TS and the relative SS take the same value, the circuit operates in a single clock cycle; otherwise, the result is recomputed from the substitution point with a single clock cycle penalty.

The proposed methodology has been evaluated using a wide range of circuits, including arithmetic circuits, and complex data paths, such as adders, multipliers, FIR filters, DCT computation blocks, and the ISCAS85 benchmark. By making use of two different quality metrics, error rate and average error, and targeting the 45nm CMOS technology, the authors claimed significant gains in terms of area requirements and power consumption.

8.2.1.4 Automatic Methodology for Sequential Logic ApproximatioN (ASLAN)

In [27], the authors focused on the approximation of sequential circuits. With sequential circuits, two key challenges have to be addressed. The first is the estimation of the impact of approximation on the output quality, observed after multiple cycle of operations. In this kind of circuit, in fact, error due to approximation is propagated at each computation cycle, and different cycles may have different significance in error propagation. The selection of an approximation, given certain quality constraints, is the second challenge to be addressed.

In order to formulate the problem of sequential logic approximation, the authors of [27] make use of an SQCC, which is used to characterize the impact of approximation on the POs of a given sequential circuit. Figure 8.2 depicts the SQCC, which consists of three components: the original sequential circuit, the approximate circuit, and the Quality Evaluation Circuit (QEC). The QEC encodes

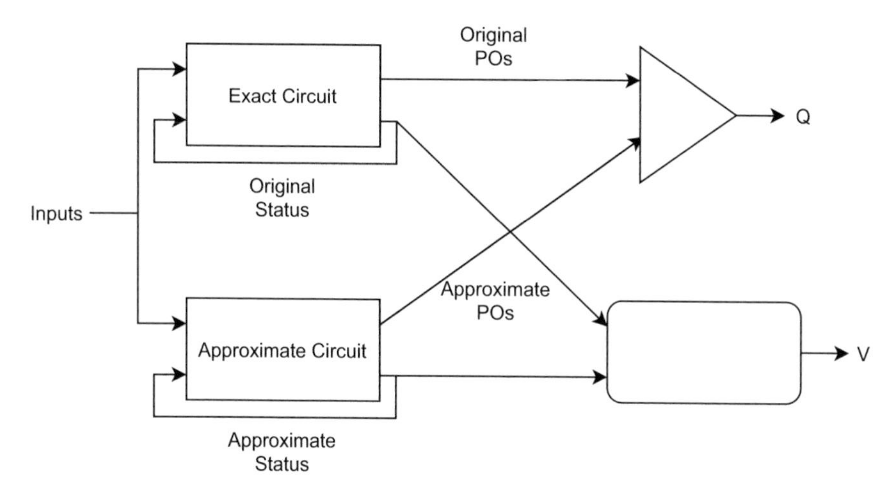

Fig. 8.2 Sequential quality constraint circuit [27]

quality constraints to be met as a multi-output logic function: it monitors POs and status registers in order to indicate whether quality constraints are met through the *quality* Q bit and the *valid* V bit. The latter indicates that the operation performed by the approximate circuit has been completed, and therefore its POs are ready to be evaluated for quality, while the former is set only if quality constraints are satisfied.

Starting from an RTL description of the sequential circuit to be approximated, in order to maximize energy savings, the tool first identifies combinational blocks—such as adders, multipliers, etc. A gradient-descent heuristic is then used to search for the optimal quality-energy operating point for each of these blocks; the ranking is based on a figure of merit computed by taking into account the proportion of energy required by the block w.r.t. the whole circuit, the energy saving obtained by approximating the block, and the error introduced due to the approximation. The best configuration is then selected, and the quality constraints are checked. The process is repeated until no block can be further approximated without violating constraints.

In order to guarantee output quality, the tool performs formal verification by making use of the following properties:

- Safety: in all possible states of the SQCC, if V is true, then Q should be true.
- Liveness: V eventually becomes true along all possible paths through the space state of the SQCC.

The first property ensures that whenever both the original sequential circuit and the approximate one have produced their outputs, i.e., V is high, the latter should satisfy quality constraints, i.e., Q is high. The second property states that both the original and the approximate circuits should eventually produce their respective outputs.

Formal verification is performed by making use of the time expansion technique: the SQCC is iteratively unrolled until the V signal is high, and then quality constraints are checked.

ASLAN has been tested on a number of different circuits, such as FIR filters, vector product blocks, DCT computation blocks, MPEG encoders, and k-means clustering algorithm. Experimental results showed a decrease of up to 70% in required area and a reduction in energy consumption of up to 34%.

8.2.1.5 Automated Behavioral Synthesis of Approximate Computing Systems (ABACUS)

The approach proposed in Nepal et al. [28] operates on the behavioral description of a circuit, in order to generate an approximate version that meets specific quality requirements. The approach is implemented in the Automated Behavioral Synthesis of Approximate Computing Systems (ABACUS) tool.

Given a behavioral or RTL description of a certain circuit, coded in the Verilog Hardware Description Language (HDL), the tool operates on the AST of the whole design in order to generate an RTL description of the approximate circuit, also coded in Verilog. One of the advantages is that it does not need the designer to have in-depth knowledge of semantic and functionality of the considered design.

Five different kinds of AST transformation are available: whenever any of these is invoked, the tool traverses the AST and searches for nodes where changes can be applied. *Data-type* transformations consist in intermediate signal truncation, setting the last significant bits to zero in binary arithmetic operations, for instance. *Operation* transformations substitute standard arithmetic operators, such as adders and multipliers, using approximate operators, which require less silicon area and power. *Expression* and *variable-to-constant* transformations are very clever: they simplify computations by sharing common, or similar, operands and replace variables having small variance—at most 10%—with their constant mean value. The tool is also able to skip some iterations during loop unrolling of behavioral descriptions, replacing the outcomes of skipped iterations using prior ones.

Since the AST obtained applying a certain transformation can be used as input for a different transformation, and since each transformation can be applied several times, the size of the design space grows quickly. To overcome this issue, ABACUS makes use of an iterative stochastic greedy algorithm to identify transformed ASTs on the Pareto front, i.e., transformed ASTs that provide optimal trade-off between accuracy and gains. ABACUS, in fact, embeds a simulation and synthesis engine in order to evaluate accuracy and design metrics, such as area requirements, power consumption, and timing information.

The greedy algorithm goes through several iterations, each of which applies multiple randomly picked transformations on the original design. The accuracy is then evaluated using a user-provided input dataset; if error constraints are met, the design is considered valid and passed to the synthesis engine, in order to assess its requirements. The set of Pareto-optimal variants is then sorted in terms of accuracy, area, and power savings.

The tool has been tested using FIR filters, machine learning classifiers, and image compression blocks. Results show area and power savings in the 15–38% and 10–33% range, respectively, with only 2–8% accuracy loss.

8.2.1.6 Statistically Certified Approximate Logic Synthesis (SCALS)

Liu et al. [29] pointed out that existing approaches for approximate logic synthesis usually simplify a logic network in a technology independent manner. According to the authors, this means that the impact on the quality of results of logic simplification due to technology mapping is not taken into account during the approximation process. In addition, using a uniform distribution to generate test vectors can lead to incorrect conclusions on error, since realistic datasets are unlikely to follow such distribution.

In order to overcome this issues, a new statistically based approximate logic synthesis framework, called Statistically Certified Approximate Logic Synthesis (SCALS), is proposed. SCALS is an extension of the PIMap tool [35], which is a logic synthesis and technology mapping tool targeting Look-Up Table (LUT)-based FPGA. The core of the tool is an iterative improvement algorithm consisting in three steps. First, a *transformation move*, such as balancing and AIG rewriting, is proposed. Then, the quality of the move is evaluated through technology mapping and area requirement estimation. Information gathered during the mapping step is used to determine whether to accept or reject the proposed move.

Given a combinational logic network composed by technology independent gates, SCALS aims at area and/or delay minimization, taking into account the target technology. Both LUT-based FPGAs and ASIC standard cell libraries are supported. The tool makes use of the input distribution for error estimation, using the error rate or the mean relative error magnitude as the error metric. The input distribution can be either user-provided or computed from an input dataset. The user is also required to provide a confidence level to be used during hypothesis testing.

As briefly introduced before, SCALS extends the PIMap flow to approximate logic synthesis. Starting from a gate-level representation of the logic network to be approximated, the tool performs a mapping to the target technology, and then it splits the netlist into a number of sub-netlists, each containing a predefined number of standard cells or LUTs. Each sub-netlist is optimized in isolation by making use of a collection of transformation moves consisting of three exact transformations—i.e., transformations that do not introduce any error—such as depth balancing and gate count reduction, and three approximate transformations. The *reduce* approximate transformation move randomly selects a logic gate and removes a randomly selected fan-in at that gate. If the selected logic gate has only one fan-in, the gate itself is removed from the netlist and its fan-in is directly connected with its fan-out. The *flip* approximate transformation move randomly selects a logic gate and randomly inverts one of its fan-ins. The *add* approximate transformation move adds a two-input logic gate, with randomly selected logic function and randomly selected fan-ins and a fan-out, to the network. After applying the selected transformation moves, the tool performs technology mapping again, in order to measure area requirements. The introduced error is then simulated by performing logic simulations. In order to speed up the simulation process, the tool first simulates the whole exact circuit, obtaining a set of test vectors for each one of the sub-netlists. These test vectors are used to stimulate only the transformed sub-netlist, so that the global error can be inferred from sub-netlists without the need of simulating the whole circuit every time. The Markov Chain Monte Carlo method [36] is used to determine whether to accept or reject the move. In particular, the Metropolis–Hastings [37] method is used to compute the acceptance probability.

SCALS is implemented as an extension of the Berkeley ABC tool [38]. In order to demonstrate their proposal, the author used the tool on the EPFL and the MCNC benchmarks, targeting 4-LUT and 6-LUT technologies. Circuits are split into 16 sub-netlists and error evaluation is performed by making use of 10K randomly generated test vectors. For arithmetic circuits, experimental results showed area savings ranging between 10% and 70%, with an error rate being less than 1%. By using the same error threshold, control circuits showed area savings ranging between 10% and 40%. However, with high error-tolerant circuits, the authors claimed area savings up to 90%.

8.2.1.7 Approximating Complex Arithmetic Circuits with Formal Error Guarantees

In [30], the authors proposed a new method targeting approximate arithmetic circuits.

The proposed method is based on a verifiability-driven search strategy: given a circuit to be approximated, an error metric, and an error threshold, the search strategy searches for an approximate version of the given circuit having minimum area while satisfying error constraints. The search space is explored by making use of GA, and, instead of using simulations to estimate error due to approximation, formal guarantees for error bounds are provided by making use of a SAT solver.

The circuit is represented by making use of a two-dimensional array of nodes, where each node is encoded using three integer numbers: the first two denote input signals, while the third is the logic function performed at the considered node. Bearing in mind the GA terminology, each node is a chromosome and the integer numbers used for its encoding are its genes. New approximate candidates are generated by performing mutations in chromosomes; fitness functions are, then, evaluated, and the best candidate is selected for further evolution. Two fitness functions are computed: the error w.r.t. the original circuit and the area requirements. In order to reduce the amount of time spent during the space exploration, a resource budget is used: candidate circuits that require more resources than a user-defined threshold are thrown away, and new candidates are generated. There are two main disadvantages in this approach: no solution may exist for a given budget, and good solutions are possibly thrown away even if they require a negligible amount of additional resources.

Since formal verification is still not possible for some error metrics, such as average error, the proposed method makes use of the absolute worst-case error. A miter, consisting of the original circuit, the approximate circuit, a subtractor, and a comparator, is used to evaluate it. Outputs of the original and approximate circuit are subtracted, and the result is fed to the comparator, which checks that the error is smaller than a given threshold. Typically, this is done by computing the absolute value of the subtractor output, but, unfortunately, the circuit performing such task consists in long chains of XOR gates, which are known to cause poor performance in SAT solvers [39]. For this reason, the miter makes use of two different comparators: one for the positive and one for the negative threshold. Moreover, since the threshold is fixed, the standard comparator can be replaced with a specialized Boolean logic network, reducing the size of the circuit and the time required to complete the task.

For evaluation purposes, the proposed method has been implemented as part of the Berkeley ABC tool [38], expressing chromosomes as AIG-nodes and leveraging the *iprove* engine for equivalence checking. In order to allow resource management, the engine provides several tunable knobs, such as the number of simulations performed to prove non-equivalence. In order to define a limit for the amount of time needed by the equivalence checking procedure, the number of conflicts in which a single AIG-node is involved during the backtracking process is used.

In order to demonstrate the proposal, complex arithmetic circuits, such as multipliers with inputs up to 32 bits and adders with inputs up to 128 bits, are considered. For purposes of fitness functions evaluation, the size of the circuits computed by the ABC tool, targeting a 45nm standard cell technology library, is used. The power-delay product is also computed, by making use of the Synopsis Design Compiler tool, but it is not involved in the fitness functions computation.

In one of the experiments, the approximation of a 16 bits multiplier is performed, with error ranging between 0.1% and 20%. Three different resource limits have been considered: unlimited resources, 160K and 20K conflicts. Despite the fact that some potentially good candidates are quickly rejected, the aggressive resource limit allowed to generate and evaluate a significantly higher number of candidates, obtaining up to 75% reduction in area requirements. The scalability of the proposed approach has been demonstrated using complex multipliers, with inputs up to 32 bits in length, and various adders with inputs up to 128 bits in length.

8.2.1.8 Approximate Logic Synthesis Using Boolean Matrix Factorization

Hashemi et al. [31] proposed BLASYS, a novel approach for combinational circuits approximation based on BMF. In the proposed approach, a multi-output logic function having k inputs and m outputs is firstly analyzed in terms of its truth-table: being nothing more than a matrix, the truth-table is passed to a BMF algorithm, along with the factorization factor f, in order to produce two sub-matrices, namely B and C, which correspond to the truth-table of a *compressor* and a *decompressor* circuit. Using this technique, any arbitrary circuit can be forced to compress as much information as possible in f intermediate signals, using an AND network. Such information can be decompressed using the decompression circuit, which can be implemented as an XOR network. Approximation is introduced by not preserving the equality between the starting truth-table M and the product of the matrices B and C.

Being BMF an NP-hard problem, the authors of [31] selected the ASSO heuristic [40, 41] to estimate the B and C matrices. In order to measure factorization error, the chosen heuristic makes use of the Euclidean distance L_2, which translates to the Hamming distance in case of Boolean matrices. However, if a signal has to be interpreted as a binary coded number, the Hamming distance is not really an accurate representation of inaccuracy, as mismatches in different bits contribute differently on the actual error. Therefore, the authors of BLASYS selected the weighted Hamming distance $d = |(M - B \cdot C) \cdot w|$, where w is a weight vector.

Since the complexity of BMF grows exponentially with the size of the matrix to be factorized, i.e., the size of the truth-table, the latter is split into a number of smaller sub-circuits, and, for each one of these, BMF is performed in isolation. The decomposition process is very similar to FPGA technology mapping, but the fundamental difference is the aim: circuit decomposition is performed only to address computational complexity. Typical technology mapping algorithms make use of k-feasible cuts enumeration algorithms. Conversely, BLASYS makes use of the KL-cuts algorithm presented in [42] to identify sub-circuits having at most k inputs and m outputs, with k and m being design choices mostly determined by computational budgets.

Decomposing a circuit in smaller sub-circuits does not mean sub-circuits can be tested for error in isolation, because even small error in sub-circuit outputs can propagate and cause larger error. Thus, BLASYS performs error evaluation taking the whole circuit into account. Since an exhaustive simulation is infeasible, a Monte Carlo simulation using one million randomly generated test cases is performed to evaluate circuit accuracy. The approximation process is completed only if the error is greater than a user-defined threshold.

In order to demonstrate the proposal, the authors implemented the BLASYS methodology as part of the Yosys synthesis tool [43], and, by making use of the Synopsys Design Compiler, they evaluated area and power requirements targeting a 65nm standard cell library. The tool has been used on six different kinds of circuits—an adder and several multipliers and FIR filters—claiming an area reduction ranging between 8% and 47% with an error of only 5%. The authors also compared results, in terms of area requirements, obtained by using the weighted and the unweighted error metric. Experimental results prove that, for the same area, using a weighted metric allows for greater accuracy. They also compared their work with the SALSA tool, which has been discussed in Sect. 8.2.1.2, claiming their methodology allows to achieve further area savings up to 20%.

8.2.1.9 CIRCA: Toward a Modular and Extensible Framework for Approximate Computing

In the work by Witschen et al. [32], CIRCA, a framework for approximate computing, is proposed. The framework is developed on the basis of several already existing approximate framework and tools, trying to fill their gaps. It focuses on the design space exploration, which is, according to the authors, the most demanding part of the whole design flow. The authors of CIRCA developed it to be as generic as possible, modular, and extensible: the framework is not restricted to a particular type of circuit, to a particular error metric, to a specific AxCT, to a specific space exploration algorithm, or to a specific technology. It is extensible and its input and output are compatible with all the well-known existing tools, such as Berkeley ABC [38] or the Yosys [43] tool.

The framework consists of three stages: the *input* stage, the QUAES (QUality assurance, Approximation, Estimation, and search space exploration) stage, and the *output* stage.

The input stage manages two main tasks: it preprocesses the input design and ensures compatibility with external tools. The preprocessing task aims at the identification of a set of sub-circuits within the original design which are amenable to approximation. This set is denoted as candidates-set, and its elements can be identified by automatic methods or by manual annotation performed by the user. The input stage also reads user-provided test vectors for testing-based quality assurance.

In the QUAES stage, candidates are subjected to approximation: different *variants* of a candidate and different *configurations* of a certain variant are generated. The approximation flow is split into four blocks: the *quality assurance*, the *approximation*, the *estimation*, and the *exploration* block. The exploration block acts as central control block and implements three procedures: *select*, *expand*, and *evaluate*. The evaluate procedure takes a set of circuit configurations and provides an estimation for error, area requirements, delay, and power consumption. The select procedure takes a set of circuit configurations and selects a configuration to be further considered, based on the previous evaluations. It sends the selected configuration, called CUT (Circuit Under Test), to a quality assurance procedure, which checks if quality constraints are met, using either formal verification or test-based techniques. If the CUT does not meet such constraints, the search algorithm can abort the search procedure or pick up the next configuration, depending on the specific search algorithm. In the latter case, the search procedure terminates if there is no more configuration to be evaluated. If the CUT meets the quality constraints, it is passed to the expand procedure, which further approximates the circuit.

The output stage performs post-processing on valid configurations, in order to select the best ones and to provide a set of Pareto-optimal configurations. The output stage also connects the QUAES stage with external synthesis tools, for actual circuit implementation.

The first implementation of the CIRCA framework is provided with two AxCTs, precision-scaling and AIG-rewriting, and three different design space exploration algorithms, which are hill-climbing, simulated-annealing, and Monte Carlo tree search. In order to perform the quality assurance procedure, as previously anticipated, the framework can employ testing-based or formal verification techniques. For the test-based technique, a set of input vectors must be provided by the user via the input stage. Formal verification is performed using an approximation miter similar to the one used by the ASLAN tool [27]: the outputs coming from the original and approximate circuits are compared using the absolute worst-case error as error metrics.

In order to prove the effectiveness of their proposal, the authors of the framework selected a benchmark of seven circuits—the butterfly data path of the Fast Fourier Transform (FFT), some FIR filters, adder-trees, and an RGB to YCbCr converter—and manually annotated adder and multiplier components. They obtained 55% area savings by making use of the precision-scaling AxCT and up to 33% of area savings by making use of the AIG-rewriting technique.

8.2.2 Approximation Tools Targeting Software Applications

In this section, AxC automation tools targeting software component are discussed. For each of them, the main features and the innovative contributions brought to the scientific literature are highlighted. Table 8.2 summarizes the discussion: for each different tool, its main characteristics are reported.

8.2.2.1 EnerJ

At the software level, a key challenge in AxC is the isolation of parts of the program that have to be precise from those that can be approximated. To this end, Sampson et al. introduce EnerJ [44], an extension to the Java language with type qualifiers that distinguish between *approximate* and *precise* data types.

Values of precise types have the typical correctness guarantees of conventional computing, while values of approximate types have none; moreover, overloaded operators and methods, with approximate versions of algorithms, can be applied to them. Assignment of approximate values to precise variables is illegal; an explicit *endorsement* operation is needed to treat approximate data as precise, allowing assignments. An example of the annotation usage is shown in Listing 8.1.

Table 8.2 Summary of tools for software approximation

Tool	Input	Error control	Search method	AxCT	Quality assurance
EnerJ [44] (see Sect. 8.2.2.1)	Annotated Java source code	Error bound	-	Precision-scaling	Simulation
PetaBricks [45] (see Sect. 8.2.2.2)	PetaBricks source code	Error bound	Heuristic (Genetic Algorithm)	Source code transformations	Configurable (statistical or runtime check)
Precimonious [18] (see Sect. 8.2.2.3)	C source code	Error bound	Iterative (Delta Debugging Algorithm)	Precision-scaling	Simulation
SAGE [7] (see Sect. 8.2.2.4)	C source code	Error bound	Greedy Algorithm	Instructions skipping, precision-scaling, and thread fusion	Simulation
ACCEPT [46] (see Sect. 8.2.2.5)/REACT [47] (see Sect. 8.2.2.6)	C source code	Error bound	Binary Knapsack	Loop perforation, synchronization elision, and hardware acceleration	Simulation
ASAC [48] (see Sect. 8.2.2.7)		Error bound	Latin Hypercube Sampling	Precision-scaling	Kolmogorov–Smirnov hypothesis test
iACT [49] (see Sect. 8.2.2.8)	C source code	Error bound	-	Precision-scaling, fault injection, and memoization	Simulation

```
@Approx int a = ...;
int p;             // precise by default

p = a;             // illegal
p = endorse(a); // legal
```

Listing 8.1 Example of EnerJ annotations usage

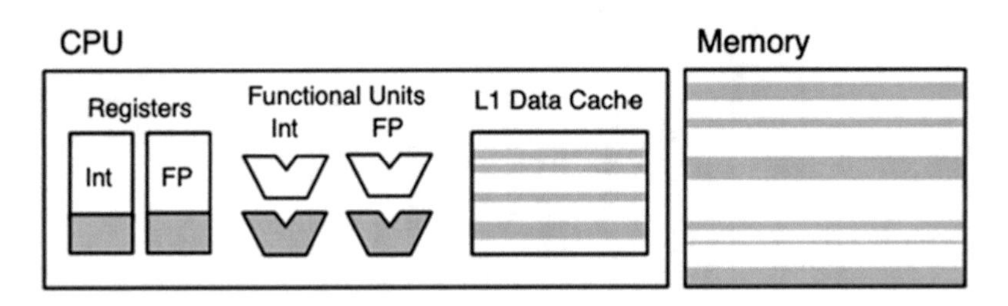

Fig. 8.3 Hardware model used for simulation in [44]

The EnerJ extensions provide the means to implement type-based information-flow tracking [50], but they do not define the specific approximation techniques to be used: approximate algorithms have to be provided by the programmer, while approximate storage and operations require some form of architectural support, in the form of ISA extensions and memory devices.

In order to evaluate the potential savings of this approach, a number of existing Java applications have been manually annotated with type qualifiers. The approximate versions of the programs are executed on a simulator that implements approximate integer and floating-point operations and approximate storage at the register, cache, and memory level; an example of such an architecture is shown in Fig. 8.3. Energy consumption is evaluated with a simplified model that considers three components: instruction execution, SRAM storage (for CPU registers and cache), and DRAM storage. Potential energy savings are shown to be in the 10–50% range, depending on the considered application.

8.2.2.2 Language and Compiler Support for Variable-Accuracy Algorithms

The PetaBricks framework [45] is one of the first attempts to embed the accuracy concept into programming languages and compilers. As the authors pointed out, traditional programming languages work under the assumption that programs always require a fixed and strongly defined behavior. However, for certain classes of algorithms, such as NP-hard problems or programs with tight computational timing constraints, an accurate solution might be unfeasible to compute. Conversely, in many other cases, the programmers could trade accuracy off to achieve better performances.

A key challenge when writing variable-accuracy software arises from maintaining the abstraction boundary between the designer and the user of such software. The former understands the algorithm and all its tunable parameters affecting the accuracy but does not know anything about the requirements of a certain application. On the other hand, the user actually knows the requirements of a specific application but may not know how exposed tunable parameters affect accuracy level. Yet another challenge arises when software is built by composing multiple variable-accuracy modules: manually determining the accuracy level to be used can be extremely cumbersome because of the interdependencies between accuracy-related choices.

The framework proposed by Ansel et al. [45] provides a way to describe multiple manners to solve a problem. The programmer is required to specify all the "transformations" to compute outputs from inputs—i.e., different algorithms to solve a given problem—, the parameters that affect the accuracy, and which error metric has to be used. Accuracy-related parameters are automatically set by the tool in order to explore the solution space. Then, an autotuner automatically determines the configuration that offers the best trade-off between accuracy and performance gains. The autotuner follows a GA approach to search through the solution space, collecting a population of candidate algorithms which is expanded by using mutators. The PetaBricks compiler and autotuner represent different candidate algorithms by making use of configuration files in which a value is assigned to each accuracy-related variable. Mutator functions are automatically generated by the framework, starting from a static analysis of each different algorithm that could be used to solve a given problem, in order to generate new algorithm configurations taking training inputs into account. There are three different mutator categories:

- Decision-tree manipulators, which act on the specific algorithm to be used to solve a given problem
- Log-normal scaling mutators, which scale a configuration parameter by taking a random number from a log-normal distribution of scale 1
- Uniform random mutators, which replace a configuration parameter with a new value taken from a discrete uniform distribution containing all legal values for the considered parameter

The framework supports three different types of accuracy guarantees:

- Statistical guarantees, computed by performing offline testing on a set of training inputs, in order to determine statistical bounds within a user-defined confidence level
- Runtime checking, which assesses accuracy at runtime and provides stronger accuracy guarantees
- Domain-specific guarantees, which require the programmer to provide a lower bound on accuracy

In order to prove the effectiveness of the proposed framework, the authors of [45] tested it by making use of a number of different benchmarks, such as image compression, bin-packing problems, k-means clustering, and Helmholtz and

Poisson equations. For k-means clustering, the authors claim speed-ups ranging from 1.1 to 9.6. For algorithms showing even more error resiliency, such as image compression or Helmholtz and Poisson equation solvers, speed-ups ranging from 1.3 to 34.6 are achieved. Performances dramatically increase, as the authors claim, for the bin-packing problem: speed-ups ranging from 1832 to 13789 are achieved because of the algorithmic changes made by the autotuner, which lower the complexity to $O(n^2)$ or even $O(n)$ when a significant loss of accuracy is allowed.

8.2.2.3 Precimonious

Floating-point computations involve by nature a certain degree of approximation; ideally, the precision of floating-point data should be carefully tailored to the nature of the application, but this can be a difficult task for programmers without specific background or even impossible for bigger programs.

PRECIMONIOUS, introduced in [18], is a tuning assistant for floating-point precision; it aims at finding, if possible, a set of program variables that can have their precision lowered without violating a user-specified error constraint and providing a performance improvement, measured in execution time, relative to the original program.

The tool receives three inputs:

- A C program
- A test suite, in the form of a representative set of inputs
- An accuracy requirement

It then operates in four phases: first, a *search file* that describes the search space is created; the search file contains an entry for each variable whose precision can be tuned, and the candidate types to be explored.

The tool then produces candidate configurations using a modified version of the delta-debugging [51] algorithm; each iteration considers, for each variable, the highest and the second-highest available precision and determines the subset that has to be allocated at the highest precision. The algorithm first divides the search space into two disjoint subsets and checks if the accuracy requirement can be satisfied by only using the highest precision for one subset. If such a configuration exists, it further partition the subsets, in order to minimize their size. If at any step the algorithm cannot find a valid configuration, the number of subsets in the partition is doubled. The algorithm halts when the partition granularity can no longer be increased. An example of the algorithm operation is shown in Fig. 8.4.

The set of program transformations that corresponds to the identified variable type assignment is applied to an LLVM intermediate representation of the original program, which is then compiled and checked for correctness and performance. The result of this phase serves as feedback to the delta-debugging algorithm. A high-level view of the tool flow is shown in Fig. 8.5.

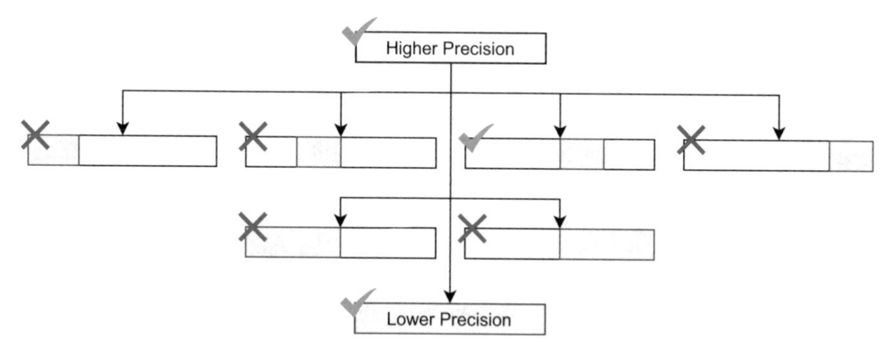

Fig. 8.4 The delta-debugging algorithm used in [18]

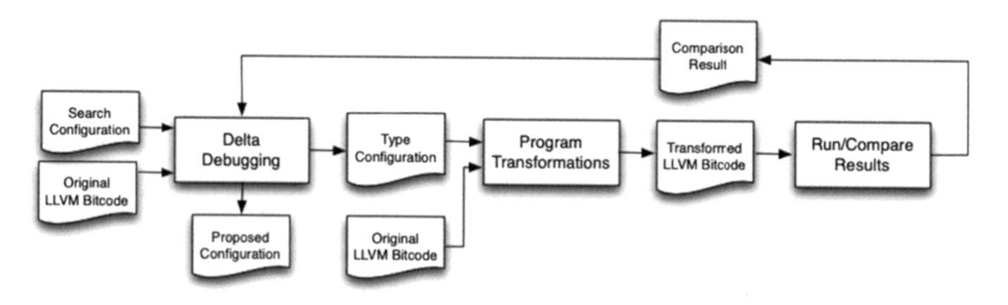

Fig. 8.5 The PRECIMONIOUS flow [18]

The tools have been tested on selected functions from the GNU Scientific Library, two NAS Parallel Benchmarks, and three other numerical programs, showing improvements up to 41% in execution time at 10^{-10} error threshold.

8.2.2.4 Self-tuning Approximation for Graphic Engines

As the authors of [7] claim, several common bottlenecks in GPUs could be alleviated by means of AxC. Some examples are serialization of data accesses and memory bandwidth limitations. In the context of GPU programming, AxCTs have two main limitations: first, the programmer must implement and tune most aspects of the approximation; moreover, he/she is often unaware of the hardware upon which it runs.

Samadi et al. [7] proposed SAGE, a framework for performing runtime approximation on GPUs. It enables the programmer to write the program only once; thus, it trades accuracy off for performance gains, taking user-defined error metrics and thresholds into account.

SAGE consists in two main phases: an offline compilation and a runtime kernel management. The former investigates the input source code, in order to automatically recognize approximable code portions. Then, it automatically generates multiple approximate versions of the original kernel. The runtime kernel management, on the other hand, dynamically selects the best approximate kernel by taking error metrics and quality constraint into account.

During the offline compilation phase, SAGE automatically detects all those operations known as expensive to perform on a GPU, and, then, three different optimizations are performed. Contentions caused by atomic operations have a significant impact on performance, so SAGE improves performances by skipping atomic instructions showing high contention rate. Considering the large amount of cores available on GPUs and bearing in mind that, in order to achieve high throughputs, cores must access data very quickly, input data is packed sacrificing precision, reducing the amount of bits for its representation and traffic on Network On Chips (NOCs). The third optimization performed by SAGE is thread fusion: it computes the output for one of the threads of the original kernel and broadcasts it to its adjacent threads.

During the runtime kernel management phase, in order to reduce timing overhead, SAGE makes use of an online greedy tree algorithm to find reasonable approximation parameters. All the different approximate versions of the original kernel are arranged in a binary tree in which each child node has always a lower output quality and higher performances than its parent. Starting from the root of the tree, which is the original kernel with no approximation, the best version is searched using the steepest-ascent hill-climbing heuristic.

In addition, since the behavior of a program can change at runtime, accuracy and performances are constantly monitored: after a number of invocations of the kernel, a calibration step is performed. If quality constraints are not met, SAGE switches to a less aggressive approximate version of the original kernel. At the beginning of the execution, when there is low confidence on accuracy, the calibration is performed more frequently, in order to converge to a stable solution quickly. As confidence grows, the interval between two calibrations is gradually increased, so the overhead is reduced.

In order to evaluate the impact of optimizations performed by the proposed framework, the authors considered ten different applications from the machine learning and image processing domains, including k-means clustering and several classifiers. The authors claim an average speed-up of 2.5 with less than 10% accuracy loss.

8.2.2.5 Accept

The ACCEPT framework is introduced in [46] as a trade-off between fully manual and fully automated program modification techniques.

It is composed of:

- C type qualifiers for annotating approximate data
- A compile-time analyzer to identify approximable code

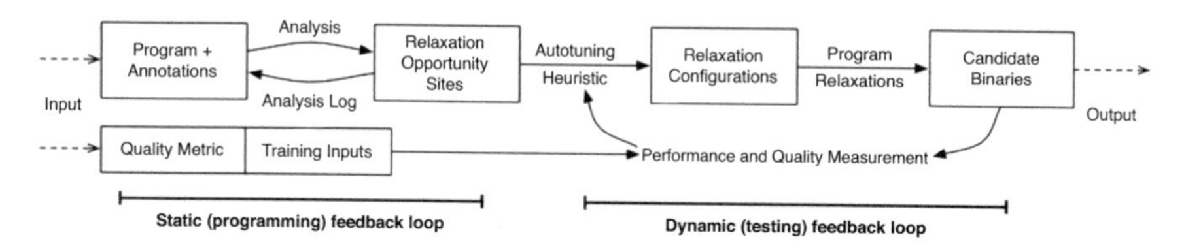

Fig. 8.6 The ACCEPT compiler flow [46]

- A feedback system that guides the programmer in order to improve code annotation
- An autotuning system that chooses the best approximations to be applied to the code

The compiler workflow is shown in Fig. 8.6.

The extension to the C type system is directly derived by the one proposed in EnerJ and introduced in Sect. 8.2.2.1: all types are qualified as *approximate* or *precise*, and noninterference guarantees apply between approximate and precise data; an explicit *endorsement* expression acts as a cast from approximate type to its precise equivalent.

```
1  APPROX int a = ...;
2  int p;            // precise by default
3
4  p = a;            // illegal
5  p = ENDORSE(a);  // legal
```

Listing 8.2 Example of ACCEPT annotations usage

The analysis of the annotated program determines a set of transformations, called *relaxations*, that can be applied to the code in a way that affects only data qualified as approximate; coarse-grained code regions, such as loop and function bodies, are checked, and a relaxation can be applied to a region iff it is *precise-pure*, i.e., iff it:

- Contains no store to precise variables that may be read outside of the region
- Does not call any functions that are not precise-pure
- Does not include an unbalanced synchronization statement

The framework also identifies candidate regions for approximate hardware acceleration, which need to be precise-pure, with single entry and single exit and with identifiable live-ins and live-outs.

The analyzer produces a log that provides feedback to the programmer w.r.t. the relaxations that have been applied and the ones that are not safe, identifying the statements that prevent them, called *blockers*.

An autotuner heuristically finds Pareto-optimal sets of relaxations, solving a binary knapsack problem under the simplifying assumption that error and performance improvement compose linearly.

The framework has been implemented as an extension to the LLVM compiler infrastructure; in order to test its effectiveness, three approximation strategies have been implemented:

- Loop perforation: some loop iterations are skipped
- Synchronization elision: reduction or removal of synchronization in the program
- Neural acceleration: selected code regions are off-loaded to a hardware accelerator that approximates them using a previously trained neural network

A number of benchmark applications, targeting an x86 server, a mobile SoC, and a low-power embedded device, have been manually annotated; the average speed-up on the three platforms is of 2.3×, 4.8×, and 1.5×, respectively, while the error varies widely from practically zero to 26.7%.

8.2.2.6 REACT

The REACT modeling framework is introduced in [47] to enable the exploration of the efficiency-accuracy trade-off of AxC techniques.

It consists of:

- An application profiler
- An energy model
- A quality model

The profiler is implemented using Intel's Pin, a dynamic binary instrumentation framework; it groups ×86 instructions into the *compute* and *memory* categories that are then used by the energy model. Dynamic and static architectural costs are captured in REACT with a simplified linear energy model; a precise baseline cost of an application execution is evaluated as the sum of the energy required for its phases:

$$Energy_{phase} = Static_{compute} + Dynamic_{compute} + Static_{memory} + Dynamic_{memory} \tag{8.2}$$

$$Static = Power \times CPI \times Instructions \tag{8.3}$$

$$Dynamic = Energy_{event} \times Count_{event} \tag{8.4}$$

The $Power$, $Energy$, CPI, and $Instructions$ terms are architectural parameters that define the cost of operations and structures.

This baseline can then be compared with a number of approximate energy totals relative to a variety of AxC techniques. Fine-grained techniques operate at the event level, modifying one or more architectural parameters, while coarse-grained techniques operate at the phase level, modifying one or more of the terms in Eq. (8.2).

Quality modeling is performed extending the ACCEPT framework, presented in Sect. 8.2.2.5, to inject errors at the instruction and function granularities.

8.2.2.7 ASAC

ASAC, introduced in [48], is an automatic tool for sensitivity analysis in AxC; it aims to automatically generate code annotations for distinguishing variables that can be approximated from the ones that have to be precise, by extracting information about the sensitivity of the output to the program data.

The tool operates in three phases, named *discovery*, *probe*, and *testing*. In the discovery stage, the program variables are extracted and a dataflow analysis, using techniques introduced in [52], is performed to determine the range of the variables. The Cartesian product of the range intervals for each variable produces a *hypercube*, where each point is a variable assignment for the program data.

In the probe phase, the edges of the hypercube, which represent the variable ranges, are discretized, thus yielding a number of smaller hypercubes; the *Latin Hypercube Sampling* [53] technique is used to select a bias-free sample with a good coverage of the sample space, as shown in Fig. 8.7a. A number of uniformly random points are then selected from each of the sampled hypercubes; these constitute perturbation vectors, which are used to alter the values of the variables at selected points in the program execution, dynamically injecting them with an instrumentation tool.

The results of each probe run are compared to a Quality of Service (QoS) threshold for the program and are classified accordingly as *good* or *bad*. From the classified samples, a cumulative distribution function for the two classes is constructed for each hypercube dimension, i.e., for each program variable, as shown in Fig. 8.7b.

In the test phase, the sensitivity score for each variable is evaluated as the maximum distance between the two curves, obtained applying the *Kolmogorov–Smirnov* hypothesis test; intuitively, the more the two curves are "near," the less the

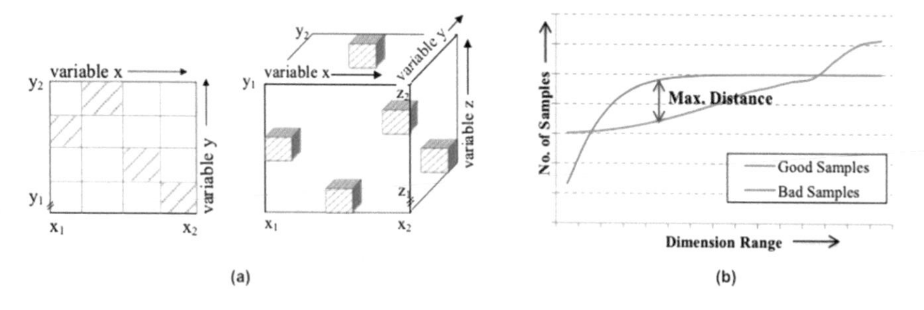

Fig. 8.7 Example of hypercubes (**a**) and CDFs (**b**) in the ASAC tool [48]

program is sensitive to the variable. The sensitivity scores are then used to rank the variables according to the program sensitivity.

The resulting program annotations are then compared with those manually applied to benchmarks from SciMark2, MiBench, and SPEC2006, achieving an average accuracy of 86%. The automatically annotated benchmarks are also evaluated for error according to selected relevant metrics, by randomly bit-flipping variables marked as approximable, obtaining a maximum error of 6%.

8.2.2.8 iACT

Mishra et al. developed the Intel's Approximate Computing Toolkit (iACT) [49] as a mean to analyze the impact of approximation techniques in software applications. The toolkit provides a compiler extension based on the LLVM pragma annotation framework that allows the programmer to specify the approximation techniques to apply; it also provides runtime support for approximate memoization and a hardware simulator based on the Intel Pin dynamic binary instrumentation framework, which handles the required architectural support to the chosen techniques.

Three approximation techniques have been implemented, targeting C programs: automated precision reduction, noisy ALU computations, and approximate memoization. Techniques are applied at a function level or at a loop level. The *axc* pragma specifies that the annotated C function runs on hardware that implements noisy arithmetic and floating-point operations and noisy loads and stores, simulated with different parametrized noise models. The *axc_precision_reduce* pragma down-converts all floating-point values in the function to a 16-bit data type. The *axc_memoize* pragma, applied at function call site, invokes the approximate memoization runtime support; as shown in Listing 8.3, error tolerance percentages are specified for the function arguments, and a global table of the function mapping is populated during program execution; if error tolerances are satisfied on a subsequent call, the memoized value is returned without executing the function.

```
1  void foo(float x, float y, float &z) {
2    z = x + y;
3  }
4
5  [...]
6
7  #pragma axc_memoize [(0:5), (1:10)]{2}
8  foo(x, y, &ret);
```

Listing 8.3 Example of iACT memoization pragma

The iACT toolkit has been tested on three different applications. A bodytracking algorithm has been annotated for precision reduction, obtaining 22% dynamic energy reduction with quality degradation less than 4% compared to precise execution. The memoization technique has been applied to Sobel filtering, with dynamic energy reduction up to 22% with less than 10% of pixels deviating from their expected values. A random bit failure error model, representative of timing failure characteristic of low-voltage memory operation, has been applied to a classification algorithm; even at moderately high bit failure probability, around 50%, output mismatches remain under the 5% threshold.

8.3 The 𝕀DE𝔸 Tool Suite

As stated in Sect. 8.1, one of the challenges that prevent the spread of AxC is the lack of generic automation tools. Existing tools, which have been discussed in previous sections, are not fully automatic or they simply provide a guided approach for approximation. Furthermore, most of such tools are too tied to a specific AxCT or tightly coupled with a certain application. Furthermore, approximate variants have to be found with quicker evaluations, in order to define a systematic and fully automatic methodology for designers.

To accomplish these goals, the authors of [54] presented the 𝕀DE𝔸 (𝕀DE𝔸 is a Design space Exploration tool for Approximate Algorithms) tool suite, which consists in two different tools: the first is Clang-Chimera, an automatic source-to-source mutation engine, while the second tool is Bellerophon, an automatic design space exploration tool. Although developed as completely independent projects, these two tools can be used together to generate different approximate versions of a given algorithm and to find an estimate of the Pareto front. They need a C/C++ implementation of the algorithm to be approximated and alter the original source code using user-defined approximation methods. Each approximate configuration is then compiled and executed in order to evaluate the quality of results.

Figure 8.8 sketches the overall flow of 𝕀DE𝔸. A set of C/C++ source files containing the algorithm to be approximated are placed in input to the Clang-Chimera tool, in order to perform the source code mutation and to generate a set of mutated source files, called *mutants*. Then, Bellerophon evaluates each different mutant, and, when the space exploration phase is completed, the set of dominant configurations can be used to reimplement the algorithm in its approximated version, in hardware or in software.

Both tools are described in Sects. 8.3.1 and 8.3.3, respectively, while Sect. 8.3.5 is a full walk-through.

8.3.1 Clang-Chimera

Named after the Chimera, the famous mythological monstrous animal, Clang-Chimera is a source-to-source mutation engine for the C/C++ programming language: given a C/C++ source code, the tool analyzes the AST in order to apply user-defined modifications.

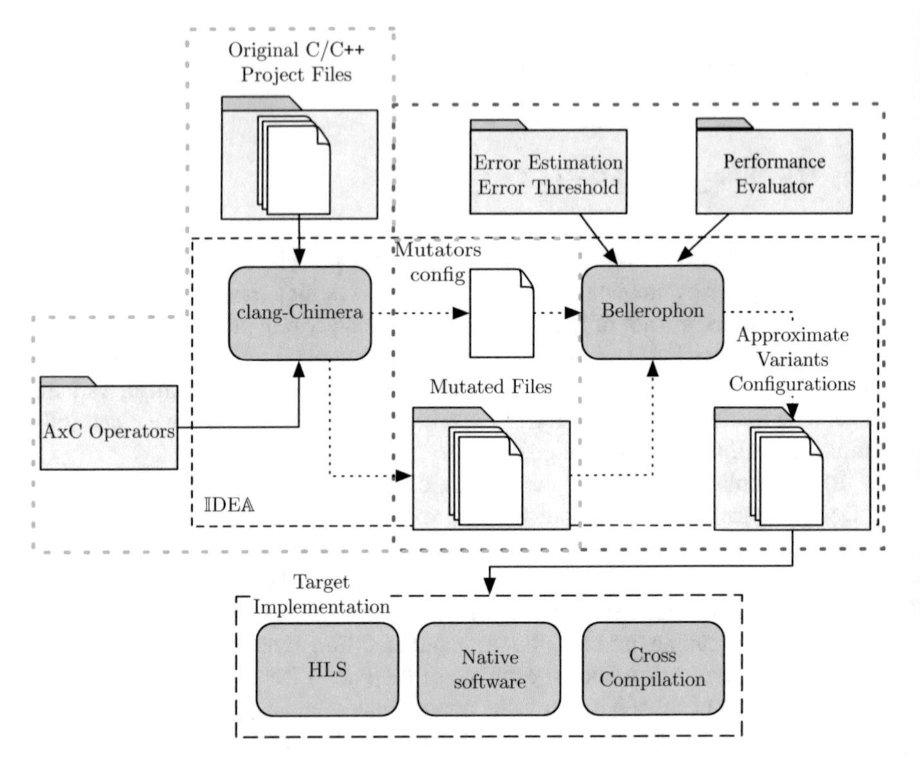

Fig. 8.8 The IDEA flow, which includes Clang-Chimera and Bellerophon tools

```
int main (void) {
  for (int i = 0; i < N; i++) {
      for (int j = 0; j < M; j++) {
          body;
      }
  }
}
```

Listing 8.4 Precise code

The AST is a tree representation of a source code written in a certain programming language. Each node of the AST is an abstract representation of a language construct that appears in the source code. Figure 8.9 depicts the AST representation of the code from Listing 8.4. Each *for* loop is represented using a node having four children nodes: the initialization node, the loop-termination condition node, the loop-modifier node, and loop-body node.

An AST pattern is a set of nodes that follow certain rules, which are specified in terms of properties of nodes and relations between nodes. Properties of a node include its type and its value, while relations express how nodes are connected together.

Clang-Chimera introduces the concepts of *mutator* and *mutation operator*, borrowing terms from the mutation testing field. A mutator is an entity that matches

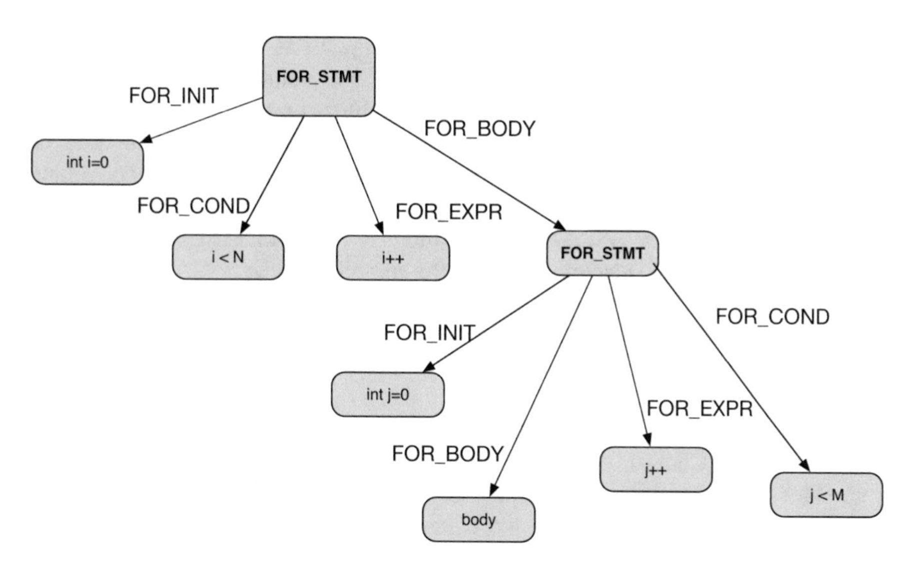

Fig. 8.9 AST representation of a *for* loop

and modifies an AST pattern. It is composed of *matching rules* and *mutation rules*. Matching rules define the AST pattern on which the mutator operates, while mutation rules specify how the matched pattern has to be modified. When a mutator has a set of mutation rules that are slightly different from each other, the *mutation type* properties of the mutator can be exploited to apply a specific mutation. For instance, let us consider a sum operator; it can be modified in three different ways: instead of defining three different mutators, a single mutator having three different mutation types can be defined.

Mutators can be First Order Mutator (FOM) or High Order Mutator (HOM). FOM is the default type and applies a single mutation at a time; if it has multiple mutation types, it generates a different mutated version of the source code, i.e., a mutant, for each different mutation type. Sometimes, multiple mutations have to be applied at the same time; a HOM having multiple types generates a single mutated version of the given source code, containing all the mutations specified by the mutation rules.

A mutation operator, or simply *operator*, is an entity that groups several mutators. It can be used to narrow down the scope of mutators, defining which part of the source code has to be modified. Mutation operator can also be of FOM or HOM type. A FOM operator can be seen as a simple mutator container: mutators it contains can be FOM or HOM. A HOM operator is used when global mutations have to be applied and it must contain only HOM mutators. HOM mutators of a FOM operator act on a per-function basis, so at most a mutant can have mutations accumulated on single functions, while a HOM operator accumulates the transformations on all the functions.

The Clang-Chimera tool makes use of the Clang-LLVM compiler suite. Given a syntactically correct C/C++ source code, a set of user-defined mutation operators, and a configuration file that states the list of available operators for a given function or a given language construct, the tool allows generating mutated versions of the algorithm, called *mutants*. The tool also generates a report file containing information such as the location and the type of each performed mutation. Every action performed by the tool is checked in order to ensure that the modified source code is syntactically correct and ready for the compilation step.

Clang-LLVM allows an easy gathering of all syntactic and semantic information from a C/C++ source code. In particular, Clang-Chimera makes use of the *LibTooling* interface, which gives access to all the capabilities provided by the Clang front-end, such as AST manipulation.

As briefly introduced before, the AST is the internal representation of a C/C++ source code that Clang-LLVM produces to manage the compilation step. The AST resembles the structure of the source code: its root node is the root of the whole translation unit, which represents the whole source file, while other elements of the tree are the Clang internal type definitions, function declarations, parameter declarations, function body elements, variable declarations, binary operators, return statements, etc. The Clang AST nodes are not instances of the same class, and they have not a common base class: there are multiple classes, such as *Decl* for declarations and *Stmt* for statements. Therefore, AST traversing and manipulation are not trivial tasks.

Clang-LLVM provides different utilities for AST traversal and manipulation. The *ASTConsumer* and *RecursiveASTVisitor* utilities are tightly coupled and are used to traverse an AST. While the former manages high-level nodes, such as translation units, the latter copes with lower level constructs, such as function declarations and statements. They traverse the AST and call a handler when a particular node is traversed. Since using these entities can be cumbersome, Clang-LLVM provides also the *ASTMatcher* utility, which allows defining AST pattern matching rules in an easier way. The utility provides three matchers, which can be combined: the *NodeMatcher*, which matches a specific node, the *NarrowingMatcher*, which matches attributes of a node, and the *TraversalMatcher*, which allows to express structural relations between nodes. A matcher can match multiple times: when a match occurs, a callback is called to apply user-defined actions. Clang-LLVM provides the *Rewriter* utility in order to allow AST modifications.

8.3.2 Code Mutation: An Example

In order to allow the reader to better understand how Clang-Chimera works and how it can be used in AxC, we provide the following example.

Let us consider a function implementing an approximate version of the sum operator and suppose it works by setting the N least significant bits of the sum to zero, with N being configurable by the user. The body of such a function is reported in Listing 8.5, where *add1* and *add2* are the addends and *nab* is the number of bits of the sum to be set to zero. Please note that the exact nature of the approximation being made by this function is hidden in the function itself, and, from the Clang-Chimera point of view, it does not matter.

```
int ax_sum(int add1, int add2, int nab)
{
  return (add1 + add2) & (!((1<<nab)-1));
}
```

Listing 8.5 Interface of the approximate sum

Let us consider the code in Listing 8.6, which has to be mutated replacing every exact sum with the approximate sum described above. The user has to configure a mutator (i.e., the AxC Operator in Fig. 8.8) that defines the matching and mutation rules. The Clang-Chimera tool mutates the code in Listing 8.6, and it generates the code in Listing 8.7. The amount of error introduced by the approximation depends on the value assigned to the *nab* parameters. Indeed, the main problem is to find an appropriate value for these parameters, in order to achieve the best trade-off between performance gains and accuracy losses.

```
...
y = x + 2;
z = 2 * x + 3 * y + 2;
...
```

Listing 8.6 Code to be mutated

```
int nab_0 = 0;
int nab_1 = 0;
int nab_2 = 0;
...
y = ax_sum(x, 2, nab_0);
z = ax_sum(ax_sum(2 * x, 3 * y, nab_1), 2, nab_2);
```

Listing 8.7 Mutated code

8.3.3 *Bellerophon*

Named after the famous mythological hero who killed the Chimera, Bellerophon is a design space exploration tool designed to solve MOPs, computing different fitness functions.

Bellerophon makes use of the Non-dominated Sorted Genetic Algoritm (NSGA)-II algorithm and evaluates solutions in terms of three different fitness functions: an *error fitness function*, a *reward fitness function*, and a *penalty fitness function*. The first expresses the amount of error measured with respect to a reference solution, the reward fitness function rewards certain characteristics of one solution with respect to the others, while the penalty fitness function penalizes infeasible solutions by reducing their fitness values in proportion to the degree of constraint violation. Error and penalty functions have to be minimized, while reward has to be maximized. The definition of such fitness functions is up to the user, but only the error fitness function is mandatory to implement.

Bellerophon leverages the Clang/LLVM Just-In-Time compiler (LLVM-JIT) in order to speed up the space exploration phase. The tool consists of two main components: the *evaluation context*, which manages all the information needed to perform NSGA-II and makes use of the LLVM-JIT, and the *evaluator*, which manages all the information needed to compute and evaluate fitness functions.

The LLVM-JIT makes possible to access the code in execution and allows to read and modify any code or data portion of it. This means that every time a mutant needs to be altered and the fitness functions evaluated, Bellerophon can skip the whole code compilation, linking, and loading. Just the code portion that needs to be altered will be recompiled and linked to the already loaded program image in memory, reducing the amount of time required to solve the MOP.

Using a configuration file, Bellerophon is able to generate a random population and then to evolve it in candidate solutions. At the end of the process, a report file containing the set of non-dominated solutions, their genes, and fitness function values is provided to the user.

Since the implementation of a GA is not a trivial task, Bellerophon is based on the ParadisEO framework [55], which is a template-based evolutionary computation library written in C/C++. ParadisEO is an extended version of the Evolving Objects (EOs) framework and it includes various improvements such as local search methods, multi-threading, and grid-computing support.

Although designed and developed as a standalone project, Bellerophon can be used to evaluate different configurations of a certain mutant generated by making use of the Clang-Chimera tool, in order to state which configuration is better. Configurations differ from one another by the value assigned to each different configuration parameter, i.e., the value of each gene.

8.3.4 Space Exploration Example

Bearing in mind the code mutation example given in Sect. 8.3.2, an example of the Bellerophon workflow is provided in this section.

In order to perform the space exploration, the tool needs to be configured. The Clang-Chimera tool produces a configuration file meant to be used for Bellerophon once the code mutation phase is done. This file contains the total amount of configurable parameters, their name, and the range in which they can vary. Each parameter becomes a gene in the Bellerophon context.

In the approximated sum function from Listing 8.5, the addends are 32-bit long integer numbers, and therefore each *nab* parameter can vary in the [0, 32] range. Setting the value to zero means no approximation has to be made, while setting it to 32 means that the approximation has to be made on every bit of the sum.

Using information from the configuration file, Bellerophon is able to generate an initial population of chromosomes. In this case, chromosomes have three genes, one for each *nab* parameter. The value of each genes governs the amount of approximation introduced in each sum. For the sake of clarity, consider the $\{1, 3, 2\}$ chromosome: the *nab_0* parameter will assume the value of 1, while the *nab_1* parameter will assume the value of 3. This means that the approximation will affect only the last significant bit of the first sum and the three last significant bits of the second sum.

Regarding the definition of fitness functions, the difference between the exact sum and the approximate one is a good example of error function. In order to define a meaningful reward function, the target application must be taken into account. Suppose you want to implement an approximate circuit in hardware that realizes the code reported in Listing 8.7, and suppose that silicon area can be saved by increasing the number of approximate bits in the sums, or, in other words, the value assumed by each *nab* parameter. The sum of the values assumed by each of these parameters represents a good example of reward function because the higher this value, the higher the area savings will be. A C/C++ code example of such functions is reported in Listings 8.8 and 8.9. Starting from the initial population of candidate solutions, the tool will search for a solution having low error and high reward value.

```
double BELLERO_getError(){
  $\cdots$
  y = x + 2
  z = 2 * x + 3 * y + 2;
  y_a = ax_sum(x, 2, nab_0);
  z_a = ax_sum(ax_sum(2 * x, 3 * y_a, nab_1),  2, nab_2);
  return abs(z - z_a);
}
```

Listing 8.8 Example of error fitness function

```
double BELLERO_getReward(){
  $\cdots$
  return nab_0 + nab_1 + nab_2;
}
```

Listing 8.9 Example of reward fitness function

8.3.5 *A Walk-Through with the* $\mathbb{I}$DE$\mathbb{A}$ *Tool Suite*

This section provides a full walk-through of the $\mathbb{I}$DE$\mathbb{A}$ tool-suite. A Docker image with the $\mathbb{I}$DE$\mathbb{A}$ tools preinstalled is available from [56].

Since the code mutation example given in Sect. 8.3.2 is too simple, because it is only meant to allow the reader to better understand what code mutation is, this walk-through tackles the approximation of a DCT computation algorithm. The DCT is used in many applications, including JPEG and MPEG compression and digital signal analysis.

In the literature, a significant amount of research papers is focused on applying approximate computing to DCT algorithms, avoiding complex functions and multiplications [57]. They all leverage the DCT coefficients, which can be scaled and approximated by integers so that floating-point multiplications can be replaced by integer multiplications. The resulting algorithms are significantly faster than the original versions and they are extensively used in practical applications. However, integer multiplication is still a complex operation, and therefore many multiplier-less algorithms have been proposed for fast DCT computation, such as BAS08 [58], BAS09 [59], BAS11 [60], BC12 [61], CB11 [62], PEA12 [63], and PEA14 [64]. All algorithms work in the same way: instead of calculating the DCT according to its definition, they operate using the matrix calculation. In addition, by making several transformations on the matrices involved, they express the DCT as a set of linear equations. The example in Eq. (8.5) refers to the set of equations used by the BAS12 [61] algorithm.

$$\begin{aligned}
f_0 &= x_0 + x_1 + x_2 + x_3 + x_4 + x_5 + x_6 + x_7 \\
f_1 &= x_0 - x_7 \\
f_2 &= x_0 - x_3 - x_4 + x_7 \\
f_3 &= x_5 - x_2 \\
f_4 &= x_0 - x_1 - x_2 + x_3 + x_4 - x_5 - x_6 + x_7 \\
f_5 &= x_6 - x_1 \\
f_6 &= x_2 - x_1 + x_5 - x_6 \\
f_7 &= x_4 - x_3
\end{aligned} \tag{8.5}$$

In order to further reduce the resource requirements of multiplier-less algorithms, Almurib et al. [6] noted that inexact adders can be used to compute the less significant bits of each sum.

Consider an inexact adder that works exactly like the inexact sum defined in Listing 8.5, i.e., it sets the value of the least significant N bits to zero, and assume that the goal is to find which algorithm configuration offers the best trade-off between error and performance gains, using Clang-Chimera and Bellerophon.

The walk-through follows a bottom-up approach: Sect. 8.3.5.1 shows how to define a Clang-Chimera mutator, Sect. 8.3.5.2 shows how to create a mutation operator and how to register it, and Sect. 8.3.5.3 explains how to configure the tool in order to generate mutants.

8.3.5.1 Defining a Clang-Chimera Mutator

As introduced in the previous sections, the AST pattern matching logic and the mutation logic are both embedded into the Clang-Chimera mutators. New mutators can be defined by sub-classing the `chimera::mutator::Mutator` class, which is an abstract class defined in the `include/Core/Mutator.h` header file. This class defines a set of pure virtual and non-pure virtual functions that must be implemented or overridden. In particular, the AST pattern matching rules must be defined using one of the `getXXXMatcher` functions, which define coarse-grained matching rules, while fine-grained matching rules are defined by the `match` function. Mutation rules can be defined implementing the `mutate` function.

In Fig. 8.10, a Unified Modeling Language (UML) class diagram showing detailed description of the `Mutator` class is shown. Bearing in mind the code mutation example given in Sect. 8.3.2, Fig. 8.10 also shows the `AdderMutator`

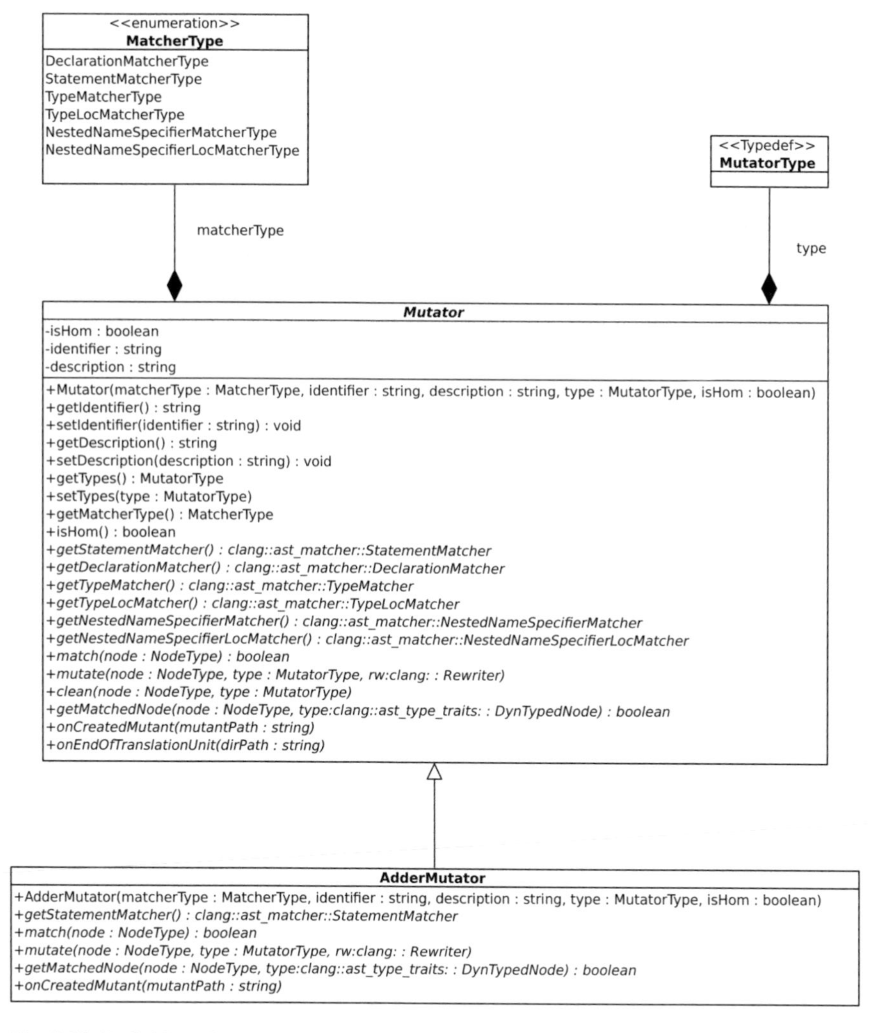

Fig. 8.10 Definition of a new mutator class

class, which is inherited from the Mutator class and is used to define matching and mutation rules to replace each sum operation with a function call, e.g., to mutate the code in Listing 8.6 and to generate the code in Listing 8.7.

In order to perform such mutation, a statement matcher searching for all the additions involving integer variables needs to be defined. An example of such matcher is reported in Listing 8.10: it matches all the sum and subtraction operations (lines 3 and 4) performed between two integer numbers (lines 5 and 6). Moreover, the matcher binds the matched nodes with an identifier (line 7), in order to make the gathering of such nodes easier for the fine-grained matching rules and mutation rules methods.

```
1  StatementMatcher chimera::adder::MutatorAdder::getStatementMatcher() {
2    return stmt(
3    binaryOperator(
4    anyOf(hasOperatorName("+"), hasOperatorName("-")),
5    hasRHS(XHS_MATCHER("int", "rhs")),
6    hasLHS(XHS_MATCHER("int", "lhs"))
7    ).bind("adder_op"));
8  }
```

Listing 8.10 Example of getStatementMatcher function

The matcher in Listing 8.10 returns all AST nodes corresponding to sums and subtractions between integers, including those involving constant values or those performed in order to access array elements. The particular application we are considering needs only sums and subtractions between integer *variables* to be replaced; in addition, sums and subtractions performed while accessing array elements and those involving constants and those performed on other kind of expressions must be left unmodified. Expressing these constraints in a statement matcher is cumbersome, so a fine-grained matcher has to be used. An example of fine-grained matcher is reported in Listing 8.11. This matcher checks that the node refers to a binary operator matched by the statement matcher in Listing 8.10 (lines 2 to 4). Then, the matcher extracts left and right operands (lines 5 and 6) and verifies that they are not binary operators (lines 9 to 16).

```
1  bool chimera::adder::MutatorAdder::match(const NodeType &node){
2    const BinaryOperator *bop = node.Nodes.getNodeAs<BinaryOperator>("adder_op"
     );
3    if (!bop)
4    return false;
5    const Expr *lhs = bop->getLHS()->IgnoreCasts()->IgnoreParens();
6    const Expr *rhs = bop->getRHS()->IgnoreCasts()->IgnoreParens();
7    bool isLhsBinaryOp = ::llvm::isa<BinaryOperator>(lhs);
8    bool isRhsBinaryOp = ::llvm::isa<BinaryOperator>(rhs);
9    if (isLhsBinaryOp){
10     if( (((BinaryOperator*)lhs)->getOpcodeStr() == "+") ||
11     ((BinaryOperator*)lhs)->getOpcodeStr() == "-") return false;
12   }
13   if (isRhsBinaryOp){
14     if( (((BinaryOperator*)rhs)->getOpcodeStr() == "+") ||
15     ((BinaryOperator*)lhs)->getOpcodeStr() == "-") return false;
16   }
17   return true;
18 }
```

Listing 8.11 Example of fine-grained matcher function

The `match` function returns a subset of the nodes selected by the `getStatementMatcher` function: the mutation will take place on all these nodes, and therefore the selection has to be as accurate as possible. Listing 8.12 shows a partial example of `mutate` function, which defines and implements mutation rules. Source code mutation is performed making use of a `::clang::Rewriter` object, which allows source code modifications in C/C++ language. In FOMs, a different
`::clang::Rewriter` object is passed to each call to the `mutate` function, and therefore the modifications are isolated. In HOMs, the `::clang::Rewriter` object passed to each call to the `mutate` function is always the same. First, a reference to the AST node corresponding to the sum operation to be replaced is obtained (line 4), identifying its operands (lines 13 to 15). Then, this operation is replaced with the call to the function `ax_sum` (lines from 26 to 34), defined in Sect. 8.5. Note that the declaration of the variable *nab* is also inserted, to allow the user to choose the degree of approximation to be introduced in the sum (line 24).

```
1   Rewriter & chimera::adder::MutatorAdder::mutate(const NodeType &node,
     MutatorType type, Rewriter &rw) {
2
3     // Retrieve a pointer to function declaration to insert global variables
      before it
4     const FunctionDecl *funDecl = node.Nodes.getNodeAs<FunctionDecl>("
      functionDecl");
5
6     // Set the operation number
7     unsigned int bopNum = this->nabCounter++;
8
9     // Local rewriter to hold the original code
10    Rewriter oriRw(*(node.SourceManager), node.Context->getLangOpts());
11
12    // Retrieve binary operation, left and right hand side
13    BinaryOperator *bop    = (BinaryOperator*) node.Nodes.getNodeAs<
      BinaryOperator>("adder_op");
14    Expr *internalLhs      = (Expr*)           node.Nodes.getNodeAs<Expr>("lhs")
      ;
15    Expr *internalRhs      = (Expr*)           node.Nodes.getNodeAs<Expr>("rhs")
      ;
16
17    Expr *lhs = (Expr*) bop->getLHS()->IgnoreCasts();
18    Expr *rhs = (Expr*) bop->getRHS()->IgnoreCasts();
19    bool isLhsBinaryOp = ::llvm::isa<BinaryOperator>(lhs);
20    bool isRhsBinaryOp = ::llvm::isa<BinaryOperator>(rhs);
21
22    // Create a global var before the function
23    ::std::string nabId = "nab_" + ::std::to_string(bopNum);
24    rw.InsertTextBefore(funDecl->getSourceRange().getBegin(), "int " + nabId +
      " = 0;\n");
25
26    // Retrieve the name of the operands
27    ::std::string lhsString = rw.getRewrittenText(lhs->getSourceRange());
28    ::std::string rhsString = rw.getRewrittenText(rhs->getSourceRange());
29
30    // Form the replacing string
31    ::std::string bopReplacement = "ax_sum((" + lhsString + ", " + rhsString +
      ", " + nabId + ")";
32
33    // Replace all the text of the binary operator with a function call
34    rw.ReplaceText(bop->getSourceRange(), bopReplacement);
35
36    ...
```

```
37
38     return rw;
39 }
```

Listing 8.12 Example of `mutate` function

Testing a mutator means to test its matching and mutation rules. Testing of the matching rules is straightforward due to the following assumption: given a source code, the beginning of any syntax construct has a unique location, which can be identified with a line and column number. On the other hand, testing mutation rules can be cumbersome because there is no simple way to provide an oracle, i.e., an expected mutated code to compare with. So, while it is possible to create an oracle to automatically check matching rules, the mutation rules have to be manually checked.

8.3.5.2 Defining and Registering a Clang-Chimera Mutation Operator

A mutation operator can be seen as a collection of mutators; therefore, from a C/C++ language point of view, it is an object that is built, and, then, mutators are added to it. A simple way to define an operator is to inherit from the `MutationOperator` class, defined in `include/Core/MutationOperator.hpp`. Figure 8.11 shows the `AdderOperator` class, which inherits from the `MutationOperator` class and makes use of the `MutatorAdder` class. In order to make Clang-Chimera aware of new operators, they have to be registered using the `registerMutationOperator` function, provided by the `ClangTool` class. The `AdderOperator` class provides the `getAdderOperator` function, which

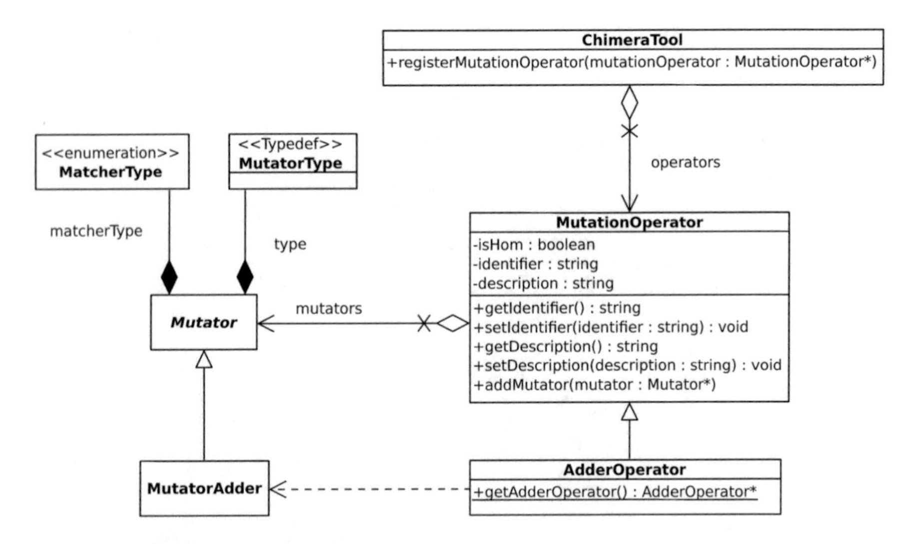

Fig. 8.11 Defining and registering of operators

builds a new `AdderOperator` object and adds a `MutatorAdder` mutator in it and then returns a pointer to the newly created `AdderOperator` instance, in order to allow its registration.

8.3.5.3 Configuring and Running Clang-Chimera

In order to be able to proceed with the code mutation, the Chimera tool must be configured using a CSV file that indicates on which portion of code—i.e., which function—it is necessary to act and which operator must be used to perform the mutations. Suppose you want to make the mutation of the DCT calculation algorithm in Listing 8.13, the configuration file might look like the one in Listing 8.14. In addition, when you launch the tool, you must specify the set of source files on which you want to perform the mutations and any other options for compiling.

When Chimera finishes its execution, it produces the mutated source and a report containing all the mutations that have been performed. An example is reported in Listing 8.15

```
1  void BC12_dct1d(const int * const input, int * const output)
2  {
3    int x0b = input[0] + input[7];
4    int x1b = input[1] + input[6];
5    int x2b = input[2] + input[5];
6    int x3b = input[3] + input[4];
7    int x4b = input[3] - input[4];
8    int x5b = input[2] - input[5];
9    int x6b = input[1] - input[6];
10   int x7b = input[0] - input[7];
11
12   int x0c = x0b + x3b;
13   int x1c = x1b + x2b;
14   int x2c = x1b - x2b;
15   int x3c = x0b - x3b;
16   int x4c = 0 - x4b;
17   int x5c = 0 - x5b;
18   int x6c = 0 - x6b;
19   int x7c = x7b;
20
21   output[0] = x0c + x1c;
22   output[4] = x0c - x1c;
23   output[6] = 0 - x2c;
24   output[2] = x3c;
25   output[7] = x4c;
26   output[3] = x5c;
27   output[5] = x6c;
28   output[1] = x7c;
29 }
```

Listing 8.13 DCT computation using the BC12 algorithm

```
1  BC12_dct1d,MutatorAdder
```

Listing 8.14 The Chimera CSV configuration file

```
int nab_0 = 0;
...
int nab_17 = 0;
...
...
void BC12_dct1d(const int * const input, int * const output)
{
  int x0b = ax_sum(input[0], input[7], nab_0);
  int x1b = ax_sum(input[1], input[6], nab_1);
  int x2b = ax_sum(input[2], input[5], nab_2);
  int x3b = ax_sum(input[3], input[4], nab_3);
  int x4b = ax_sum(input[3], - input[4], nab_4);
  int x5b = ax_sum(input[2], - input[5], nab_5);
  int x6b = ax_sum(input[1], - input[6], nab_6);
  int x7b = ax_sum(input[0], - input[7], nab_7);

  int x0c = ax_sum(x0b, x3b, nab_8);
  int x1c = ax_sum(x1b, x2b, nab_9);
  int x2c = ax_sum(x1b, - x2b, nab_10);
  int x3c = ax_sum(x0b, - x3b, nab_11);
  int x4c = ax_sum(0, - x4b, nab_12);
  int x5c = ax_sum(0, - x5b, nab_13);
  int x6c = ax_sum(0, - x6b, nab_14);
  int x7c = x7b;

  output[0] = ax_sum(x0c, x1c, nab_15);
  output[4] = ax_sum(x0c, - x1c, nab_16);
  output[6] = ax_sum(0, - x2c, nab_17);
  output[2] = x3c;
  output[7] = x4c;
  output[3] = x5c;
  output[5] = x6c;
  output[1] = x7c;
}
```

Listing 8.15 Mutated code

8.3.5.4 Configuring and Running Bellerophon

Since it has a lot of tunable parameters, Bellerophon comes with a set of scripts to make it easy to configure and execute. In addition to the configuration parameters of the genetic algorithm itself, Bellerophon needs to be configured so that it is able to compile and link the mutated code, generated by Chimera, every time the value of a gene is modified.

At the end of its execution, Bellerophon will provide the designer with a set of Pareto-optimal solutions: Fig. 8.12, for instance, depicts the Pareto front for the DSE for Listing 8.15. The error function used during the exploration phase calculates the error as the difference between the DCT calculated using the BC12 algorithm and the one calculated using the different configurations of the approximate BC12 algorithm. The reward function, instead, is calculated simply as the sum of the value assumed by each "nab" parameter. The tool also provides the designer with full details concerning configurations belonging to the Pareto front, allowing him/her to suitably configure the final target application.

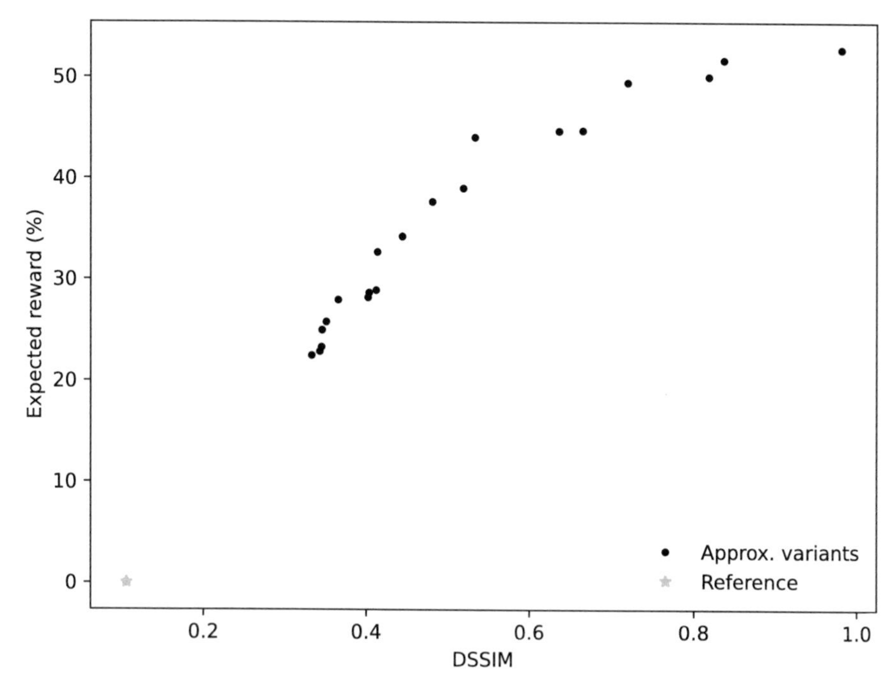

Fig. 8.12 Pareto front estimation provided by Bellerophon

8.4 Conclusion

In the first part of this chapter, we briefly introduced AxC, focusing on its fields of application, the main techniques, and the open issues and challenges that may slow down its adoption.

As we believe that the availability of automatic tools for AxC is an important factor for a wider adoption, we then presented the state of the art for AxC tools targeting digital circuits and software applications, highlighting the main innovations of the discussed tools.

We concluded the chapter presenting $\mathbb{IDEA}$, an extendible tool suite that allows to describe AxCTs, apply them to C/C++ code, and explore the design space of the obtained approximate variants to find an estimate of the Pareto front. A detailed walkthrough is provided, which guides the reader through all the passes required to obtain and evaluate the approximate variants of an original algorithm; the results can then be used directly in software or as a basis to guide a hardware implementation.

References

1. dcadmin, *Rebooting the IT Revolution: A Call to Action* (2015). https://www.semiconductors.org/resources/rebooting-the-it-revolution-a-call-to-action-2/
2. Mittal, S. (2016). A survey of techniques for approximate computing. ACM Comput. Surv. **48**, 1–33 (2016).
3. Chippa, V. K., Chakradhar, S. T., Roy, K., & Raghunathan, A. (2013). Analysis and characterization of inherent application resilience for approximate computing. In *Proceedings of the 50th Annual Design Automation Conference on - DAC '13*, Austin, TX (p. 1) New York: ACM Press.
4. Venkataramani, S., Chakradhar, S. T., Roy, K., & Raghunathan, A. (2015). Approximate computing and the quest for computing efficiency. In *2015 52nd ACM/EDAC/IEEE Design Automation Conference (DAC)*, June 2015 (pp. 1–6).
5. Rahimi, A., Ghofrani, A., Cheng, K.-T., Benini, L., & Gupta, R. K. (2015). Approximate associative memristive memory for energy-efficient GPUs. In *2015 Design, Automation Test in Europe Conference Exhibition (DATE)*, March 2015 (pp. 1497–1502).
6. Almurib, H. A., Kumar, T. N., & Lombardi, F. (2018). Approximate DCT image compression using inexact computing. *IEEE Transactions on Computers, 67*, 149–159 (2018).
7. Samadi, M., Lee, J., Jamshidi, D. A., Hormati, A., & Mahlke, S. (2013). SAGE: Self-tuning approximation for graphics engines. In *Proceedings of the 46th Annual IEEE/ACM International Symposium on Microarchitecture - MICRO-46*, Davis, CA (pp. 13–24) New York: ACM Press.
8. Chippa, V. K., Mohapatra, D., Roy, K., Chakradhar, S. T., & Raghunathan, A. (2014). Scalable effort hardware design. *IEEE Transactions on Very Large Scale Integration (VLSI) Systems, 22*, 2004–2016 (2014).
9. Raha, A., Venkataramani, S., Raghunathan, V., & Raghunathan, A. (2015). Quality configurable reduce-and-rank for energy efficient approximate computing. In *2015 Design, Automation Test in Europe Conference Exhibition (DATE)*, March 2015 (pp. 665–670).
10. Sartori, J., & Kumar, R. (2013). Branch and data herding: Reducing control and memory divergence for error-tolerant GPU applications. *IEEE Transactions on Multimedia, 15*, 279–290 (2013).
11. Sidiroglou-Douskos, S., Misailovic, S., Hoffmann, H., & Rinard, M. (2011). Managing performance vs. accuracy trade-offs with loop perforation. In *Proceedings of the 19th ACM SIGSOFT Symposium and the 13th European Conference on Foundations of Software Engineering - SIGSOFT/FSE '11*, Szeged, Hungary (p. 124). New York: ACM Press.
12. Yazdanbakhsh, A., Pekhimenko, G., Thwaites, B., Esmaeilzadeh, H., Mutlu, O., & Mowry, T. C. (2016). RFVP: Rollback-free value prediction with safe-to-approximate loads. *ACM Transactions on Architecture and Code Optimization, 12*, 1–26 (2016).
13. Samadi, M., Jamshidi, D. A., Lee, J., & Mahlke, S. (2014). Paraprox: Pattern-based approximation for data parallel applications. In *Proceedings of the 19th International Conference on Architectural Support for Programming Languages and Operating Systems*, Salt Lake City, UT (pp. 35–50). New York: ACM.
14. Tong, J. Y. F., Nagle, D., & Rutenbar, R. A. (2000). Reducing power by optimizing the necessary precision/range of floating-point arithmetic. *IEEE Transactions on Very Large Scale Integration (VLSI) Systems, 8*, 273–286 (2000).
15. Fang, F., Chen, T., & Rutenbar, R. A. (2002). Floating-point bit-width optimization for low-power signal processing applications. In *2002 IEEE International Conference on Acoustics, Speech, and Signal Processing*, May 2002 (Vol. 3, pp. III–3208–III–3211).
16. Yeh, T., Faloutsos, P., Ercegovac, M., Patel, S., & Reinman, G. (2007). The art of deception: Adaptive precision reduction for area efficient physics acceleration. In *40th Annual IEEE/ACM International Symposium on Microarchitecture (MICRO 2007)*, December (pp. 394–406).

17. Osborne, W., Coutinho, J., Luk, W., & Mencer, O. (2008). Power-aware and branch-aware word-length optimization. In *2008 16th International Symposium on Field-Programmable Custom Computing Machines*, April 2008 (pp. 129–138)
18. Rubio-González, C., Nguyen, C., Nguyen, H. D., Demmel, J., Kahan, W., Sen, K., Bailey, D. H., Iancu, C., & Hough, D. (2013). Precimonious: Tuning assistant for floating-point precision. In *Proceedings of the International Conference on High Performance Computing, Networking, Storage and Analysis*, Denver, CO, November 2013 (pp. 1–12). New York: ACM.
19. Keramidas, G., Kokkala, C., & Stamoulis, I. (2015). Clumsy value cache: An approximate memoization technique for mobile GPU fragment shaders. In *Workshop on Approximate Computing (WAPCO'15)* (p. 6).
20. Sutherland, M., Miguel, J. S., & Jerger, N. E. (2015). Texture cache approximation on GPUs. In *Workshop on Approximate Computing Across the Stack* (p. 3).
21. Traiola, M., Savino, A., Barbareschi, M., Carlo, S. D., & Bosio, A. (2018). Predicting the impact of functional approximation: From component- to application-level. In *2018 IEEE 24th International Symposium on On-Line Testing And Robust System Design (IOLTS)*, July 2018 (pp. 61–64).
22. Traiola, M., Savino, A., & Di Carlo, S. (2019). Probabilistic estimation of the application-level impact of precision scaling in approximate computing applications. *Microelectronics Reliability, 102*, 113309 (2019).
23. Mrazek, V., Hanif, M. A., Vasicek, Z., Sekanina, L., & Shafique, M. (2019). autoAx: An automatic design space exploration and circuit building methodology utilizing libraries of approximate components. In *2019 56th ACM/IEEE Design Automation Conference (DAC)*, June 2019 (pp. 1–6).
24. Sekanina, L., Vasicek, Z., & Mrazek, V. (2019). Automated search-based functional approximation for digital circuits. In S. Reda & M. Shafique (Eds.) *Approximate Circuits* (pp. 175–203). Cham: Springer International Publishing.
25. Venkataramani, S., Sabne, A., Kozhikkottu, V., Roy, K., & Raghunathan, A. (2012). SALSA: Systematic logic synthesis of approximate circuits. In *DAC Design Automation Conference 2012*, June 2012 (pp. 796–801).
26. Venkataramani, S., Roy, K., & Raghunathan, A. (2013). Substitute-and-simplify: A unified design paradigm for approximate and quality configurable circuits. In *2013 Design, Automation Test in Europe Conference Exhibition (DATE)*, March 2013 (pp. 1367–1372).
27. Ranjan, A., Raha, A., Venkataramani, S., Roy, K., & Raghunathan, A. (2014). ASLAN: Synthesis of approximate sequential circuits. In *2014 Design, Automation Test in Europe Conference Exhibition (DATE)*, March 2014 (pp. 1–6).
28. Nepal, K., Li, Y., Bahar, R. I., & Reda, S. (2014). ABACUS: A technique for automated behavioral synthesis of approximate computing circuits. In *2014 Design, Automation Test in Europe Conference Exhibition (DATE)*, March 2014 (pp. 1–6).
29. Liu, G., & Zhang, Z. (2017). Statistically certified approximate logic synthesis. In *2017 IEEE/ACM International Conference on Computer-Aided Design (ICCAD)*, November 2017 (pp. 344–351).
30. Češka, M., Matyaš, J., Mrazek, V., Sekanina, L., Vasicek, Z., & Vojnar, T. (2017). Approximating complex arithmetic circuits with formal error guarantees: 32-bit multipliers accomplished. In *2017 IEEE/ACM International Conference on Computer-Aided Design (ICCAD)*, November 2017 (pp. 416–423).
31. Hashemi, S., Tann, H., & Reda, S. (2018). BLASYS: Approximate logic synthesis using Boolean matrix factorization. In *Proceedings of the 55th Annual Design Automation Conference*, San Francisco, CA, June 2018 (pp. 1–6). New York: ACM.
32. Witschen, L., Awais, M., Ghasemzadeh Mohammadi, H., Wiersema, T., & Platzner, M. (2019). CIRCA: Towards a modular and extensible framework for approximate circuit generation. *Microelectronics Reliability, 99*, 277–290 (2019).
33. Gaffar, A., Mencer, O., Luk, W., Cheung, P., & Shirazi, N. (2002). Floating-point bit width analysis via automatic differentiation. In *2002 IEEE International Conference on Field-Programmable Technology, 2002. (FPT). Proceedings*, December 2002 (pp. 158–165).

34. De Micheli, G. (1994). *Synthesis and optimization of digital circuits*. New York: McGraw Hill.
35. Liu, G., & Zhang, Z. (2017). A parallelized iterative improvement approach to area optimization for LUT-based technology mapping. In *Proceedings of the 2017 ACM/SIGDA International Symposium on Field-Programmable Gate Arrays*, Monterey, CA, February 2017 (pp. 147–156). New York: ACM.
36. Geyer, C. J. (1992). Practical Markov chain Monte Carlo. *Statistical Science, 7*(4), 473–483 (1992)
37. Hastings, W. K. (1970). Monte Carlo sampling methods using Markov chains and their applications. *Biometrika, 57*(1), 97–109 (1970)
38. Brayton, R., & Mishchenko, A. (2010). ABC: An academic industrial-strength verification tool. In *International Conference on Computer Aided Verification* (pp. 24–40). New York: Springer.
39. Han, C.-S., & Jiang,J.-H. R. (2012). When Boolean satisfiability meets Gaussian elimination in a simplex way. In D. Hutchison, T. Kanade, J. Kittler, J. M. Kleinberg, F. Mattern, J. C. Mitchell, M. Naor, O. Nierstrasz, C. Pandu Rangan, B. Steffen, M. Sudan, D. Terzopoulos, D. Tygar, M. Y. Vardi, G. Weikum, P. Madhusudan, & S. A. Seshia (Eds.), *Computer Aided Verification* (Vol. 7358, pp. 410–426). , Berlin, Heidelberg: Springer.
40. Miettinen, P., & Vreeken, J. (2011). Model order selection for Boolean matrix factorization. In *Proceedings of the 17th ACM SIGKDD International Conference on Knowledge Discovery and Data Mining* (pp. 51–59).
41. Miettinen, P., & Vreeken, J. (2014). MDL4BMF: Minimum description length for Boolean matrix factorization. *ACM Transactions on Knowledge Discovery from Data (TKDD), 8*(4), 1–31.
42. Martinello, O., Marques, F. S., Ribas, R. P., & Reis, A. I. (2010). KL-cuts: A new approach for logic synthesis targeting multiple output blocks. In *2010 Design, Automation & Test in Europe Conference & Exhibition (DATE 2010)* (pp. 777–782). New York: IEEE.
43. Wolf, C., & Glaser, J. (2013). Yosys - A free verilog synthesis suite. In *Proceedings of the 21st Austrian Workshop on Microelectronics (Austrochip)* (p. 6).
44. Sampson, A., Dietl, W., Fortuna, E., Gnanapragasam, D., Ceze, L., & Grossman, D. (2011). EnerJ: Approximate data types for safe and general low-power computation. *ACM SIGPLAN Notices, 46*(5), 11.
45. Ansel, J., Wong, Y. L., Chan, C., Olszewski, M., Edelman, A., & Amarasinghe, S. (2011). Language and compiler support for auto-tuning variable-accuracy algorithms. In *International Symposium on Code Generation and Optimization (CGO 2011)*, April 2011 (pp. 85–96).
46. Sampson, A., Baixo, A., Ransford, B., Moreau, T., Yip, J., Ceze, L., & Oskin, M. (2015). *ACCEPT: A Programmer-Guided Compiler Framework for Practical Approximate Computing*. University of Washington Technical Report UW-CSE-15-01 1.2, p. 14.
47. Wyse, M., Baixo, A., Moreau, T., Zorn, B., Bornholt, J., Sampson, A., Ceze, L., & Oskin, M. (2015). REACT: A framework for rapid exploration of approximate computing techniques. In *Workshop on Approximate Computing Across the Stack (WAX w/PLDI)* (p. 3).
48. Roy, P., Ray, R., Wang, C., & Wong, W. F. (2014). ASAC: Automatic sensitivity analysis for approximate computing. In *Proceedings of the 2014 SIGPLAN/SIGBED Conference on Languages, Compilers and Tools for Embedded Systems - LCTES '14*, Edinburgh (pp. 95–104). New York: ACM Press.
49. Mishra, A. K., Barik, R., & Paul, S. (2014). iACT: A software-hardware framework for understanding the scope of approximate computing. In *Workshop on Approximate Computing Across the System Stack (WACAS)* (p. 6).
50. Sabelfeld, A., & Myers, A. (2003). Language-based information-flow security. *IEEE Journal on Selected Areas in Communications, 21*, 5–19.
51. Zeller, A., & Hildebrandt, R. (2002). Simplifying and isolating failure-inducing input. *IEEE Transactions on Software Engineering, 28*(2), 183–200.
52. Rodrigues, R. E., Campos, V. H. S., & Pereira, F. M. Q. (2013). A fast and low-overhead technique to secure programs against integer overflows. In *Proceedings of the 2013 IEEE/ACM International Symposium on Code Generation and Optimization (CGO)* (pp. 1–11). New york: IEEE.

53. McKay, M. D., Beckman, R. J., & Conover, W. J. (2000). A comparison of three methods for selecting values of input variables in the analysis of output from a computer code. *Technometrics, 42*(1), 55–61.
54. Barone, S., Traiola, M., Barbareschi, M., & Bosio, A. (2021). Multi-objective application-driven approximate design method. *IEEE Access, 9*, 86975–86993.
55. Liefooghe, A., Basseur, M., Jourdan, L., & Talbi, E.-G. (2007). ParadisEO-MOEO: A framework for evolutionary multi-objective optimization. In S. Obayashi, K. Deb, C. Poloni, T. Hiroyasu, & T. Murata (Eds.) *Evolutionary Multi-Criterion Optimization* (Vol. 4403, pp. 386–400). Berlin, Heidelberg: Springer.
56. Barone, S. (2021). *Iideaa-docker*, September 2021.
57. Loeffler, C., Ligtenberg, A., & Moschytz, G. (1989). Practical fast 1-D DCT algorithms with 11 multiplications. In *International Conference on Acoustics, Speech, and Signal Processing*, May 1989 (Vol. 2, pp. 988–991).
58. Bouguezel, S., Ahmad, M. O., & Swamy, M. N. S. (2008). Low-complexity 8$\times$ 8 transform for image compression. *Electronics Letters, 44*(21), 1249–1250.
59. Bouguezel, S., Ahmad, M. O., & Swamy, M. N. S. (2009). A fast 8×8 transform for image compression. In *2009 International Conference on Microelectronics - ICM*, December 2009 (pp. 74–77).
60. Bouguezel, S., Ahmad, M. O., & Swamy, M. (2011). A low-complexity parametric transform for image compression. In *2011 IEEE International Symposium of Circuits and Systems (ISCAS)*, May 2011 (pp. 2145–2148).
61. Bayer, F. M., & Cintra, R. J. (2012). DCT-like transform for image compression requires 14 additions only. *Electronics Letters, 48*(15), 919.
62. Cintra, R. J., & Bayer, F. M. (2011). A DCT approximation for image compression. *IEEE Signal Processing Letters, 18*, 579–582 (2011)
63. Potluri, U. S., Madanayake, A., Cintra, R. J., Bayer, F. M., & Rajapaksha, N. (2012). Multiplier-free DCT approximations for RF multi-beam digital aperture-array space imaging and directional sensing. *Measurement Science and Technology, 23*, 114003.
64. Potluri, U. S., Madanayake, A., Cintra, R. J., Bayer, F. M., Kulasekera, S., & Edirisuriya, A. (2014). Improved 8-point approximate DCT for image and video compression requiring only 14 additions. *IEEE Transactions on Circuits and Systems I: Regular Papers, 61*, 1727–1740.
65. Deb, K., Pratap, A., Agarwal, S., & Meyarivan, T. (2002). A fast and elitist multiobjective genetic algorithm: NSGA-II. *IEEE Transactions on Evolutionary Computation* 6(2), 182–197. https://doi.org/10.1109/4235.996017
66. Lee, D. D., & Seung, H. S. (1999). Learning the parts of objects by non-negative matrix factorization. *Nature, 401*(6755), 788–791. https://doi.org/10.1038/44565

Chapter 9
Wordlength Optimization of Fixed-Point Algorithms

Gabriel Caffarena

9.1 Introduction

Computing systems possess an inherent limitation when it comes to represent numbers and this imposes a tight constraint in the mathematical precision of feasible algorithms. A deep understanding of the way that limited precision affects the quality of applications, as well as the cost of systems (i.e., power consumption) is essential in order to satisfactorily implement algorithms. One of the most interesting aspects of approximate computing is its ability to trade-off accuracy with energy efficiency. Fixed-point refinement [1–3] is a branch of approximate computing [4] where the signals size (e.g., wordlengths) is reduced, trying to minimize the produced errors, as well as the cost. So, the operations are not approximated, but their results are. An important stage of fixed-point refinement is wordlength optimization (WLO), a task devoted to finding the fractional part of signals while optimizing implementation cost. This chapter presents a review on the different available approaches.

This section contains an introduction to the main concepts involved in WLO. The remaining sections cover the following aspects: most important WLO techniques, how to deal with complexity, the different cost functions used to steer optimization, fusion with architectural explorations, a comparison between techniques, and finally, some conclusions.

G. Caffarena (✉)
University CEU-San Pablo, Boadilla del Monte, Spain
e-mail: gabriel.caffarena@ceu.es

A. Bosio et al. (eds.), *Approximate Computing Techniques*,
https://doi.org/10.1007/978-3-030-94705-7_9

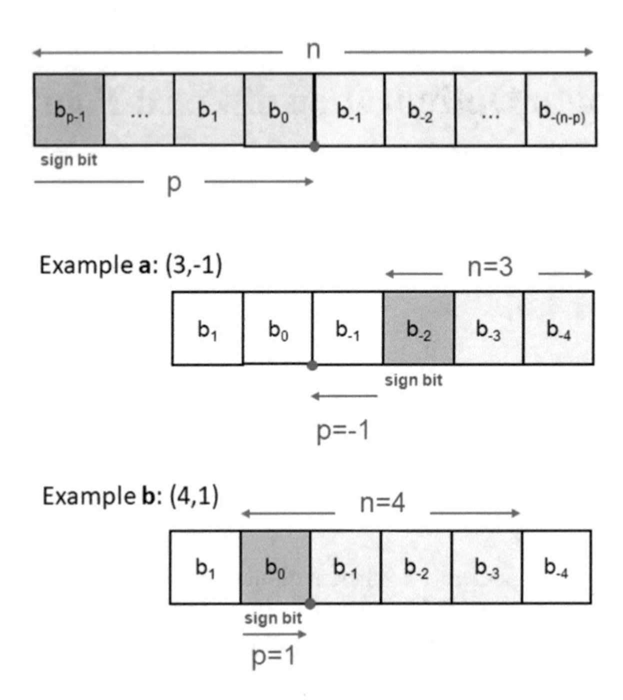

Fig. 9.1 Fixed-point representation. Example *a* shows a number with no integer part. Example *b* shows a number with both integer and fractional parts

9.1.1 Fixed-Point Refinement Overview

Let us briefly introduce fixed-point refinement.[1]

Fixed-point arithmetic is in many occasions preferred to floating-point arithmetic due to its energy efficiency and operation speed [5]. The fixed-point hardware operators are simpler than the floating-point counterparts, but, one of their main drawbacks is that they are not suitable for dealing with large dynamic range signals. However, the benefits are clear. For example, a 32-bit fixed-point addition and a 16-bit fixed-point multiplication require 0.5 pJ and 2.2 pJ, respectively, while a 64-bit floating-point unit consumes 50 pJ [6, 7].

First, we are going to assume that signals are in two's complement fixed-point format. As displayed in Fig. 9.1, a fixed-point number is defined by an n-bit integer mantissa, that includes the sign bit, and the location p of the fractional point from the sign bit. Thus, the value of a fixed-point number x, given its mantissa x_m and the couple (n, p) is:

$$x_{fxp} = x_m \cdot 2^{n-p} \tag{9.1}$$

[1] Also known as *quantization*.

For a given algorithm variable (or circuit signal), the couple (n, p) is selected trying to fulfill two main objectives: (i) the complete dynamic range can be represented (or at least, a high percentage of it); and, (ii) the fractional bits are enough to produce a tolerable error that do not hinder the application performance.

Selecting the right size per variable could have a significant impact in the final cost of the system. Power savings close to 30% [8] and area savings up to 45% [1] are reported. For instance, if the inputs of a multiplier are truncated, this could lead to a quadratic silicon area reduction of the operator. Reducing the wordlength of a multiplier's output does not have any impact on the area of the operator itself, but it definitively has an impact on posterior operations. The decision about which signals must have a reduction in their mathematical precision is not only a function of cost (i.e., area) but also on the quality degradation produced by such changes. Thus, fixed-point refinement is necessary to orchestrate the selection of (n, p) couples for all signals, so that cost is reduced while accuracy is kept within the required bound.

Fixed-point refinement is the procedure that selects the fixed-point format (n, p) for each variable or signal in order to optimize a particular design cost, while complying with a given output error constraint. It is common to perform two main stages: *Dynamic Range Analysis*[2] and *Wordlength Optimization*. The range analysis stage fixes the position of the most significant bit (MSB) of the signal, choosing the position of the binary point (i.e., p). The wordlength optimization stage finds the proper set of signal wordlengths (i.e., n) – the least significant bits (LSB)— which minimizes cost. Dynamic range can be obtained by means of interval-based methods [9–11], statistical [12–15] and stochastic methods [2, 16, 17]. For linear time-invariant (LTI) systems the use of impulse-response evaluation techniques stands as the most efficient option [2, 16, 18].

Wordlength optimization is normally an intensive iterative process in contrast with dynamic range analysis, which can be solved analytically or with several simulations. In the next subsection we elaborate on the particularities of wordlength optimization.

9.1.2 Wordlength Optimization

The ultimate goal of fixed-point refinement is to enable the implementation of mathematical algorithms, bounding the cost to a reasonable figure. Ideally, the designer seeks the minimization of the cost ($C(\boldsymbol{W})$) while keeping a maximum allowed degradation (D_{max}). $\boldsymbol{W}$ is a vector that holds all couples (n, p).

$$\begin{aligned} &\min C(\boldsymbol{W}) \\ &D(\boldsymbol{W}) \leq D_{max} \end{aligned} \tag{9.2}$$

[2] Also called *Scaling* or *Integer Part Analysis*.

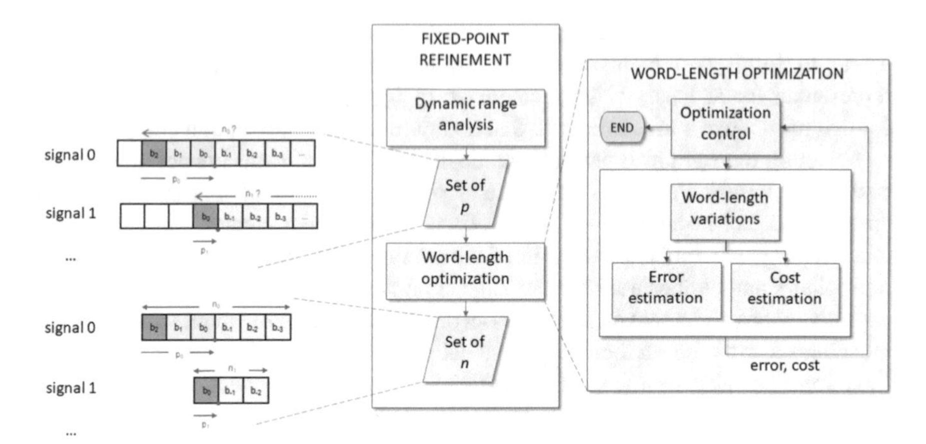

Fig. 9.2 Fixed-point refinement diagram. The wordlength optimization process is detailed

The right side of Fig. 9.2 displays a diagram of the basic blocks forming WLO. Basically, the process iterates trying different wordlengths (i.e., n) combinations until the cost is minimized. This involves an error estimation (or application quality estimation) and, ideally, a cost estimation. The cost estimation is not always performed and in many occasions the optimization is performed to the error. The estimation of the error might require bit-true simulations [14, 19] or the use of analytical expressions [2, 3, 5]. The latter requires an initial stage, not shown in the diagram, to obtain the analytical model. Also, for these kind of estimations is essential to perform a wordlength propagation prior to the error estimation (see Sect. 9.3 for a detailed explanation). The optimizer control block selects the size of the wordlengths (set of n) using the values of the previous error and cost estimations and decides when the optimization procedure has finished.

WLO is a NP-hard problem [20], suffering from an exponential growth in complexity as long as the number of variables is increased. Thus, it is common to oversimplify the process or to resort to heuristics to bound the optimization time. For instance, the quantization of an IIR biquad filter with 8-bit coefficients with a SQNR (signal to quantization noise ratio) of 30 dB requires approximately 300 noise estimates using a gradient descent optimization algorithm. That means carrying out 300 error estimations. Oversimplification comes from schemes such as the uniform wordlength (UWL) approach, which assigns a global parameter (n, p) to all signal. The optimization process is trivial and the number of error estimations are scarce. As expected, the results are far from optimal and the multiple wordlength (MWL) approach [1, 21, 22] is preferred, yet it requires long design times. According to [23], WLO may take up to 25–50% of the whole design time.

9.2 Optimization Techniques

There is a wide variety of optimization techniques applied to WLO. In this section, the most common techniques are presented, focusing on the design of fixed-point circuits. In Sect. 9.6, a comparison between these techniques is presented.

The reader may find very useful the reviews and comparison performed in papers such as [14, 24–28].

The optimization methods presented have been divided into five categories:

- *Optimal*: They provide an optimal solution. Computation times could be prohibitive.
- *Constrained search*: Thorough but not as exhaustive as the optimal approaches. Computation times can be elevated.
- *Local search*: heuristics aiming at producing fast optimization. Computation times are standard but there is a risk or getting stuck in a local minimum.
- *Non-integer wordlengths*: Wordlengths are considered real, so faster optimization methods can be applied. Computation time are standard but results are suboptimal.
- *Stochastic*: Randomness is introduced in the optimization process. Computation times can be high but local minima are avoided.

Figure 9.3 presents the classification used as well as the different published works.

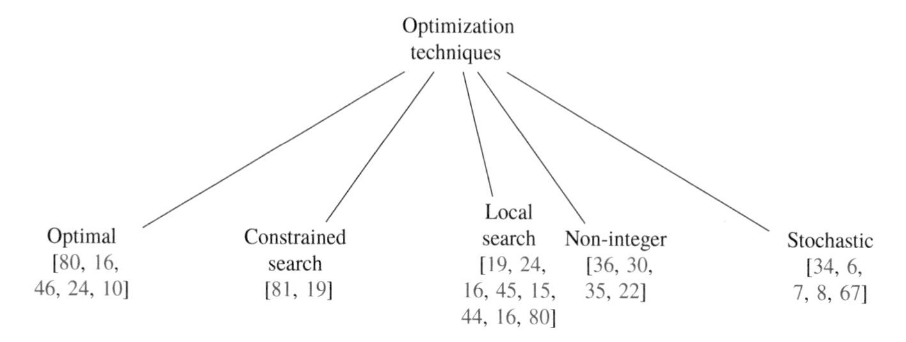

Fig. 9.3 WLO techniques

9.2.1 Optimal Approaches

Complete search[3] [14, 27, 29] applies brute force to test all possible wordlength combinations within a wide space. The number of wordlengths combinations that must be tested is computed in [27]. The best solution found, the one that minimizes cost, is selected. In order to reduce the design space, lower and upper bounds for all wordlengths are computed. Obtaining such bounds is not trivial and the proposals, such as *minimum uniform wordlength*[4] as an upper bound [32] and the *minimum wordlength set*[5] as a lower bound [29, 32] might not be good estimates. The main reason is that, as proven in [1], the quantization error at the output of a system is non-monotonic and non-linear with respect to the wordlengths. In summary, there is a risk that the global minimum is not found within these bounds.

A Mixed Integer Linear Programming (MILP) [42] formulation of the WLO problem for LTI algorithms is presented in [1]. MILP requires to formulate the optimization problem by means of equations and inequalities that are fed to a solver. The formulation is based on the use of an analytical model of error propagation that relates wordlengths with the variance of the error at the output [1]. There might be an explosion in the number of variables and MILP inequalities for medium size problems since the formulation considers: implementation cost (e.g., area of adders, delays, and constant multipliers), inequalities expressing the error function, wordlength propagation rules with possible signal branching (i.e., forks), and the error constraint. Thus, such an approach is used to assess the quality of a faster heuristic proposed. For medium size algorithms, the reported solving times are in the order of several days. This work was extended to consider resource sharing in [30].

9.2.2 Constrained Search

Constrained search is referred to techniques that perform a thorough, but not complete, design space exploration [29].

The *Min+b* [25] starts with a *minimum wordlength set* [29] which is refined iteratively until the error constraint is met. The optimization is based on distributing b bits among several signals. All combinations are assessed, starting with $b = 1$, increasing if the error constraint is not fulfilled, and stopping when the output error is satisfactory. Stopping when the first solution is found might lead to sub-optimal results, since many possibilities are still to be explored. We can find the algorithm

[3] Some authors use the term *exhaustive search* [29].

[4] Minimum global wordlength that complies with the error constraint.

[5] For each variable, the minimum wordlength that complies with the error constraint considering that the wordlength for the rest of signals is infinite, is computed. The set is composed with all the individual minimum wordlengths.

applied to minimizing the error[6] and the cost[7] [27], both referred in this text as $Min{+}b_{err}$ and $Min{+}b_{cost}$.

A *branch and bound* technique is proposed in [32]. This time, the algorithm starts with a suitable uniform solution and it starts decreasing the wordlengths until reaching a minimum wordlength bound (i.e., the minimum wordlength set [29]). Initially, there is only a set of wordlengths and at each iteration new sets are generated by reducing one unit a single signal. All the wordlength sets are then tested and the ones not complying with the error limit are discarded. In the next iteration, a new collection of wordlength sets is created following a similar fashion. The process finishes when it is not possible to further reduce the error. The solution with the smallest cost is selected. Note, that when a wordlength set is discarded, it is assumed that future modifications will never comply with the error constraint. This is not true, given the non-monotonic nature of quantization.

9.2.3 *Local Search*

Many optimization techniques are based on gradient algorithms, since they are easy to implement and do not require an excessive number of iterations.

A gradient descent optimization, that we referred to as $Max\text{-}1_{cost}$ is proposed in [29]. The algorithm starts with an upper bound of the wordlengths and it starts decreasing the wordlengths until reaching a minimum wordlength bound. The effect of reducing one unit for each wordlength individually is assessed. If the error limit is not surpassed, then the cost is evaluated. The signal that produced the maximum cost reduction is then reduced permanently and the process starts again. In [14], a similar algorithm ($Max\text{-}1_{err}$) is presented, though the error, not the cost, is used as the optimization function.

$Max\text{-}1^{enh}_{cost}$ [1] is an enhanced version of the $Max\text{-}1_{cost}$ algorithm. In this method, the wordlengths are reduced individually until the maximum error constraint is violated and the signal that produced the maximum cost reduction is decreased by one bit permanently. The idea is to make gradient descent less dependent on local information, thus, reducing the chances of falling into a local minimum.

In [14, 26], a gradient-ascent algorithm is used instead (*Min+1*). The algorithm starts with the minimum wordlength solution and, then, it tests all possibilities of increasing signals by 1 bit. The signal that produced the maximum error decrease is chosen and its wordlength is increased by one unit permanently.

An approach that considers simultaneously cost and error gradients to guide the local search is presented in [27]: the $Min{+}1^{cost}_{err}$.[8] This idea was previously presented, but only briefly mentioned, in [43]. *Min+1 bit* is performed as explained

[6] Called *sequential search* in [27].

[7] Called *local search* in [27].

[8] *CDM search* (complexity and distortion measure search) in the paper [27].

before, but the decision of the candidate to be reduced is based on the minimum gradient of $\alpha_c \cdot cost + \alpha_d \cdot error$, with $\alpha_c + \alpha_d = 1$. The authors claim that the optimization time is reduced in comparison to considering only cost or only error gradients.

A similar procedure to *Max-1*$^{enh}_{cost}$ [1] is the *evolutive* procedure of [14]. Initially, all signals have infinite precision (i.e., floating-point). Then, a signal is selected and its wordlength is decreased until the error criterion is not met. The signal then is permanently assigned to the minimum wordlength plus two units. Then, the same procedure is applied to another signal. The method stops when all signals are quantized. An extension of this method is proposed in [25] with a final extra phase that applies *Max-1*$_{cost}$ to the obtained solution.

The *All+1* method[9] is presented in [29]. The initial point is the minimum wordlength solution and all signal wordlengths are increased one bit until the specifications are met.

The *preplanned* search in [26] starts with an initial phase that computes the sensitivity of each signal wordlength. The sensitivity information enables to assign priorities to the signals and determine the order of wordlength reduction. The procedure stops when the error is below the permitted value.

The *hybrid* procedure of [25] first applies *Min+1* and it feeds the obtained solution to the *Max-1*$_{err}$. Another method that combines two optimization techniques is the *heuristic* of [25] that first applies *All+1* and then *Max-1*$_{cost}$.

9.2.4 Optimization with Non-integer Wordlengths

There are also approaches where the optimal techniques are applied to solve the quantization problem assuming *real* wordlengths. The obtained solution is then refined looking for a feasible realization with integer wordlengths. Optimality is not guaranteed, but the authors claim that the time of the optimization process has been reduced.

This idea was first proposed in [34]. Three simplifications are performed:

- The wordlengths are real numbers.
- The cost is a linear combination of the wordlengths.
- The quantization noise of a signal only depends on its final wordlength.

The last assumption implies that more refined noise models such as [44] are neglected. The optimization is then based on an approximate solution to the problem using the Lagrangian multipliers that results in a linear expression of the cost that is a function of the wordlength and the so-called noise gains. The noise gains relates the wordlengths to the error. A similar approach is taken by [45]

[9] Called *constrained search* in [29] and *heuristic* in [31].

In [35] the error estimation is performed by means of interval arithmetic and the worst-case error contribution of each signal to the outputs is computed (as in [21]). Also, the objective function is linear, since they use the summation of wordlengths as an implementation complexity measure. Thus, the optimization problem can be solved by Linear Programming, reducing the computation time.

Fiore [36] presents a non-linear iterative algorithm that is applied to find the *fractional* wordlengths minimizing area. The area cost is estimated using a quadratic function for multipliers and a linear function for adders. The convergence of the optimization method is analyzed and bounds on the wordlengths are given. The error estimation is based on the pre-computation of the contribution to the SQNR of each signal quantization. The method allows the generation of pareto-optimal curves—cost vs. error—that helps the designer to choose the proper set of wordlengths that complies with a certain error and cost constraint.

A quadratic objective function is addressed in [37] by means of sequential quadratic programming. The main contributions of this work in comparison to the other continuous-domain approaches are that it considers the discrete noise model from [44] and also the noise correlation in signal forks. Though, forks behavior is simplified to ensure the convexity of the optimization function.

Parashar et al. [46] presents a method that relaxes the integer constraints on wordlengths. It uses the power of the quantization noise as the error function and a cost function based on the energy dissipation of the operators. The authors proves that both the cost and error functions are convex when the wordlength are real and rounding modes with zero mean are used. The operators are analyzed and a continuous convex function with the pareto front is obtained.

9.2.5 Stochastic Optimization

Stochastic methods introduce random steps within the optimization procedure in order to avoid local minima.

Simulated annealing (SA) is an iterative optimization method that is based in changing the values of the problem variables by means of the so-called movements. Each time a movement is performed, the resulting cost as well as the output error of the current solution are computed. If the noise constraint is met and the cost is smaller than the current minimum one, the movement is accepted. If the cost is greater than the minimum achieved so far, the movement is accepted with a certain probability that decays with time (by means of a decreasing *temperature*). If the movements and the SA parameters (i.e., initial solution, temperature annealing function, etc.) are wisely selected the method can reach an optimal solution.

Fiore [38] presented a SA procedure where the movements are based on the use of a Markov-chain. In [24] SA is used to perform WLO and comparison results with $Max - 1$ and $Max - 1^{enh}$ are provided. There are only two possible movements applied to randomly picked signals: increase the wordlength by one unit and decrease it by one unit. The increment had half the probability of happening. Also, in [39, 40], SA is used to perform the simultaneous wordlength optimization and architectural synthesis, and some comparison results with the heuristics in [1, 32] are provided. Here, SA is used to handle the complexity of combining two optimization processes.

The algorithm GRASP (Greedy Randomized Adaptive Search Procedure) is presented in [41]. The algorithm applies two phases iteratively: *construction* and *local search*. The first phase, called *construction*, is based on algorithm such as $Max - 1$ and $Min + 1$ with the particularity that instead of selecting the signal with the best performance, there are T_{RCL} candidates (RCL stands for restricted candidate list) and one is chosen randomly to get the permanent wordlength variation. In the second phase, a TABU search [47] is applied over the previous solution. The TABU search allows both increments and decrements of wordlengths and uses a short buffer to store wordlength sets that did not comply with the error constraint. The whole procedure is repeated several times and the best solution is selected. Two versions are tested; one that uses only error as the optimization function and another that uses the ratio between error decrease and cost increase. The latter follows a similar idea to [33].

9.3 Dealing with Complexity

Quantization is an NP-hard problem [20], thus, optimal or quasi-optimal approaches imply computationally intensive exhaustive analysis in which all possible outcomes need to be tested. Moreover, if a simulation-based approach [13, 48] is followed to evaluate the quantization error of every outcome, then, the optimization times are impractical even for small designs.

Some solutions proposed to deal with complex systems are:

- Make use of parallel computation architectures
- Use fast estimators for the error/quality computation
- Cluster signals to reduce the optimization variables
- Perform multi-objective optimization
- Relax the integer constraints on wordlengths
- Use a divide-and-conquer approach which partitions the system in sub-blocks

Parallel Hardware

Making the best of the parallelization capabilities of microprocessors and graphics processing units (GPU) is a solution to reducing time-to-market constraints. Algorithms that require several starts, such as [41], can exploit parallelism by means of issuing several simultaneous optimization processes or threads in a computer or in a cluster of computers. Also, when bit-true simulations are required, it is possible to run several independent simulations in multi-core or distributed-memory systems.

The massive parallelization of GPUs was first proposed to be exploited for bit-true simulations in [49]. The benchmarks were synthetic algorithms with a total number of operations ranging from 100 to 2000. Accelerations with respect to a single-thread CPU computation are in the approximate range from ×40 to ×60 when the number of operations is no less than 1000 with input vectors with 10^4 samples. In [50], this idea is extended to interval-based error computation. The speedups range from 50 to 1000 in comparison to a 16-thread CPU execution. However, the benchmarks used have less than 25 operations, so the proposal, yet interesting, is not suitable for realistic problems.

Fast Error Estimation

Fast estimators have been applied to LTI systems and some linear systems with mild non-linearities (the so-called *smooth operators* [51]). The basic idea is to perform an analysis of the algorithm in order to obtain an *error model* that accounts for the way that errors propagates to the output. During WLO, the model can be used to quickly assess the quality of the system at each optimization step without requiring to perform costly bit-true simulations. Figure 9.4 depicts the new tasks required to apply fast estimation. The *error analysis* task obtains an error propagation model. This is a function that relates the local errors to the overall error at the output. During the optimization loop, it is common to have a *wordlength propagation* task—not to be confused with the error propagation model—that considers the wordlength changes proposed in the previous *wordlength variations* task to condition all

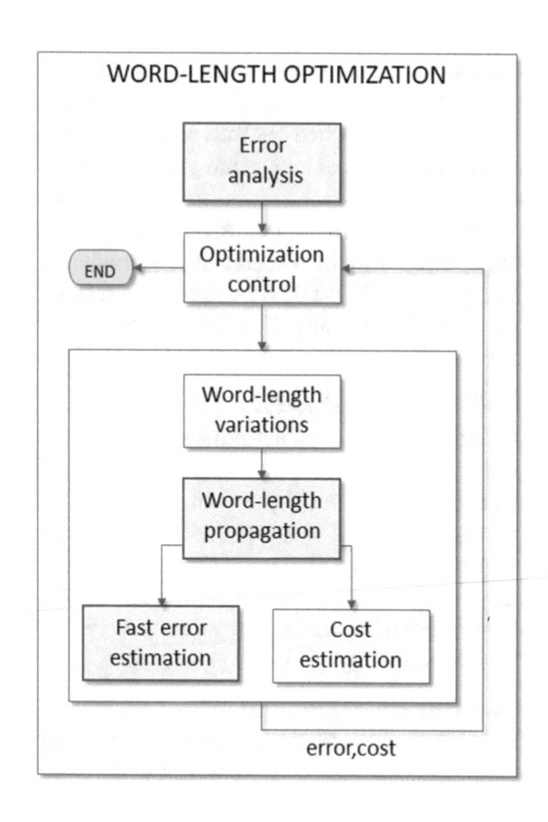

Fig. 9.4 WLO with fast estimation of the error

wordlengths in the algorithms. For instance, if the wordlengths of the inputs of an adder are reduced, it is likely that the output's wordlength can be also reduced without losing precision (the precision loss happened at the output). This new change, in turn, could also affect other wordlengths. Noise models such as [44] depends on having the exact wordlengths for all signals. For other error models, this task might not be necessary. Finally, the *fast error estimation* task makes use of the set of couples (n, p) to obtain an estimation of the output's error. In [3], optimization time reductions of up to three orders of magnitude are reported when comparing fast estimation with bit-true fixed-point simulation.

We comment here only the works that use an additive noise model [44, 52] and use the noise power/variance at the output as a quality metric [2, 3, 18, 51, 53, 54]. The *error model* is based in computing a *gain* that relates the local errors with their contributions to the output. Also, there are other parameters required to express the correlation between errors. For LTI systems, there are works that base the computation of the gains on the impulse response [2] and the use of affine arithmetic [3, 18]. For non-linear systems with smooth operations, a pseudo impulse response is applied in [51], affine arithmetic in [3] and a hybrid method combining simulations and analytical expressions in [53, 54].

Signal Grouping

A way to reduce optimization time is to decrease the number of variables of the optimization problem by grouping together signals [31, 43]. Instead of optimizing the wordlength of each single variable, these are clustered to reduce complexity. The grouping is carried out following different criteria. For instance, in [29, 31, 55], inputs and outputs of adders are grouped together. The same applies to multiplexers or delays.

Also, signal grouping can be made considering architectural issues, such as temporal mobility in resource-shared implementations, as in [56]. When resource sharing is applied it might be sensible to assign the same wordlength to operations that are executed using the same arithmetic resource. This approach tries to match resource binding with wordlength grouping in an interactive fashion (see Sect. 9.5).

Non-integer Wordlengths

Optimization based on non-integer wordlengths was addressed in Sect. 9.2. Some approaches report that applying an optimization with relaxed constraints on the wordlengths, followed by a final stage of rounding up their values leads to better optimization time if compared with algorithms such as $Min - 1$ [35–37, 46] as well as reductions in the final cost [37, 46]. Doi et al. [35] reports time reduction of three orders of magnitude, although the benchmarks are of low complexity. Fiore [36] reports a reduction of optimization time of two order of magnitude with complex algorithms. Finally, [37] also reports up to three orders of magnitude speedups.

Multi-Objective Optimization

The aforementioned gradient-based search approaches from [27] achieve fast quantization by means of steering the optimization using a linear combination of cost and error gradients simultaneously. The number of iterations can be reduced down to a third and the results are quasi-optimal for the tested benchmarks. In [41], the ratio between the error gradient and the cost gradient is used and it is reported that the number of simulations can be reduced a 50% in comparison to using only the error as an optimization metric.

System Partitioning
Finally, it is interesting to mention works where industrial-size designs are addressed in practical time. The key to do so is to use hierarchical methods, where the quantization problem is subdivided in several independent problems. In [57] quantization goes as follows. First, the system is divided in top-level entities—which basically are the main processing blocks (i.e., FFT, channel equalizer, etc.). Then, the top-level signals—the inputs and outputs of top-level entities—are quantized complying with the error constraint and following a cost estimation reduction criteria. Finally, each top-level entity is quantized independently given the error constraints imposed by the quantization of the top-level signals. The method provides a practical solution to the complexity explosion of wordlength optimization.

Also, in [58], a mixed analytical and simulation-based hierarchical approach is presented where significant design time reductions are reported (three orders of magnitude). The technique is based on dividing the algorithm in smooth and non-smooth blocks and applying analytical approaches whenever possible. The partition is manual, but there are recent techniques to automatize this process [59].

In [60], a hierarchical approach applied to a communication system is presented aiming at reducing the number of bit-error-rate (BER) simulations and favoring noise variance simulations, since these require smaller input vectors. First, the blocks' outputs are injected a zero-mean normal noise and the variance that meets a specified maximum BER (β) is found. Secondly, each group is individually optimized applying the $Max - 1$ algorithm complying with the output noise variance. Then quantization is applied even further until the β is met. Finally, a global refinement is applied with the $Max - 1$ algorithm using BER simulations. A speedup of $\times 9$ is reported.

9.4 Optimization Function

It is not common to find works where direct design objectives (such as area, delay, energy consumption) are optimized though quantization. Most of works rely on the naive idea that reducing wordlength lead to reducing cost, but this is a high level cost function which normally is poorly correlated with the real design optimization objectives. Many approaches that minimize the summation of wordlengths as a way to produce cost-efficient designs, can be found [5, 12, 14, 25, 61, 62]. In [5], it is shown how the area and latency costs are benefited from an error-oriented wordlength optimization. However, as pointed out in [27], the direct use of cost as the minimization function clearly outperforms the results obtained by error-oriented approaches.

Pioneer work [32] deals with speed, area and power optimization. They provide objective functions for the three design costs and show that, for some examples (i.e., FIR filter), speed optimization leads to a quasi-uniform approach, while the MWL approach is always the preferred for area and power consumption minimization. Also, the solutions for area and power minimization are quite similar. However, this results cannot be generalized.

Area has been considered in [1, 14, 21, 22, 24–27, 30–32, 36, 39, 53, 55, 63, 64], and area savings of more than 50% with respect to the UWL approach are reported. Some of these works also show that, as a bonus, power or speed are typically improved [1, 65].

There are some approaches which expose the effect of quantization on power consumption or energy dissipation [33, 65–67], and these costs are directly considered as optimization functions in [37, 43, 46, 68–70]. Power reductions up to 20–35% compared to UWL are provided. [69] shows that area optimal solutions can even have worse power performance than UWL solutions.

Finally, it is interesting to highlight that apart from [32] there are not many works that directly consider clock frequency as the minimization function.

9.5 Combined Wordlength Optimization and Architectural Synthesis

Traditionally, WLO has been decoupled from architectural design to manage complexity [8]. However, even though the quantization process is already an NP-hard problem, there have been attempts to extend the optimization problem to consider also architectural issues (i.e., resource sharing, fine grain-coarse grain operator trade-off, etc.).

One of the pioneer works in closing the gap between WLO and architectural synthesis is [21], where the wordlength optimization is carried out by minimizing a lower bound on the area cost of a resource sharing architecture. Once the wordlengths are selected, scheduling, resource allocation and resource binding are performed. The interesting point of this work is that quantization makes use of an estimation of the total area based on both the algorithm execution latency and the wordlengths. The latency of resources is considered variable, although a very simple model is used. The results are compared to an UWL approach, but there is no comparison to the traditional two-step approach, where WLO is performed prior to architectural synthesis.

In [55] WLO and architectural synthesis are interleaved. First, the system is quantized using a noise constraint a little bit more relaxed than the required in the specification. Then, a datapath is created by applying scheduling, resource allocation and resource binding. Finally, wordlengths are refined by increasing the operators' wordlengths until the noise constraint is met. The work is valuable, however, the approach is too simplistic: a single optimization iteration, 1-cycle latency operators, etc. Again, the authors do not provide any comparison with a traditional two-step approach.

In [30] the combined problem is explored by means of MILP. Due to the very long computation times required by MILP, the problem complexity is reduced through some simplifications: 1-cycle latency operators, multiplexers and registers neglected, etc. The results prove the validity of the combined approach and serves as a starting point to develop heuristics. This work has been extended in [39], where a simulated annealing approach is used to intertwine wordlength and architectural optimization, targeting FPGA implementations. Area reductions up to 21% in comparison with traditional approaches are reported. In [40] the same method is applied to FPGAs with embedded DSP blocks and a method to balance logic blocks and DSP blocks is provided with similar results reported.

The work in [56] also presents an iterative approach. Here, the operators costs (area, latency and power consumption) are wordlength dependent. Basically, WLO and architectural synthesis are introduced within the optimization loop. Operations are grouped after architectural synthesis, based on mobility and wordlength information. The wordlength of each group is optimized by means of quantization to reduce cost. Every change in the wordlengths of operations produces a new datapath which leads to new groups of operations. The method iterates several times and records the best solution.

Another interesting point of view is that of exploiting the information about precision sensitivity to wisely distribute operations among fine-grain processing blocks (i.e., logic fabric of FPGAs) and coarse-grain blocks (i.e., DSP blocks of FPGAs) [71] presents the interleaved application of quantization and resource binding (binding between arithmetic operations and FPGA resources) for datapaths with no resource sharing. Here, the operations which require the greatest wordlengths are quantized and are assigned to resources first. In fact, they are assigned to FPGA embedded multipliers, which are suitable for high wordlength operations. By doing that, it is possible to implement operations with small wordlengths in the remaining LUT-based resources, therefore, making a better use of the overall FPGA resources.

9.6 Comparison Between Techniques

Comparing techniques is a difficult task since there is no common framework. In this section a summary of different papers dealing with the comparison of several WLO techniques is presented. The conclusions extracted from these paper may help the reader to choose the appropriate technique. However, it must be bore in mind that each work uses a different set of benchmarks and, also, that the complexity of the benchmarks is low. For instance, system partitioning approaches or non-integer optimization are not considered in the summarized papers.

Table 9.1 All compared techniques. Checkmarks indicate if a technique is considered in one comparison work

Papers	M.A. Cantin 2002 [25]	G. Constantinides 2003 [1]	K.H. Han 2006 [27]	G. Caffarena 2008 [24]
Branch and bound [32]	✓		✓	
Complete search [14, 27, 29]			✓	
Evolutive search [14]	✓		✓	
Exhaustive search [31]	✓		✓	
Heuristic [25]	✓			
Hybrid [25]	✓			
Max-1_{cost} [29]		✓		✓[a]
Max-1^{enh}_{cost} [1]		✓		✓[b]
Max-1_{err} [14]	✓		✓	
Min+1^{err}_{cost} [14]	✓		✓[c]	
Min+b_{cost} [27]			✓[d]	
Min+b_{err} [25]	✓		✓[e]	
MILP [1]		✓		
Preplanned [26]	✓		✓	
Simulated annealing [39]	✓			✓

[a] *GRAD* in [24]
[b] $GRAD_2$ in [24]
[c] *Complexity-distortion measure* (CDM) in [27]
[d] *Local search* in [27] (a gradient-ascent version of the proposal in [32])
[e] *Sequential search* in [27]

The papers summarized are [14, 24, 25, 27]. Table 9.1 shows the techniques covered by each of the four works.

Comparison by M.A. Cantin, 2002 [25]
A total of 12 DSP benchmarks with a number of operations ranging from 3 to 36 were used. The details are not specified. The average number of bits of all the signals and the number of optimization iterations are used as comparison metric.

The main conclusions are:

- All methods reach optimal solutions for benchmarks with a small number of operations.
- In general, methods which do not impose the minimum WL solution as a lower bound obtain improved results. This reinforces the idea that the original upper and lower wordlength bounds has a deep impact in the final results.
- The algorithms which required longer computation times were *evolutive*, *SA*, *exhaustive* and *branch and bound*.
- Combining two optimization methods, such as *hybrid* and *heuristic*, lead to improved results in terms of minimization.
- It seems that the lower bound limitation makes the *exhaustive* method to perform poorly.

- the number of iterations of *SA* is limited to 10 times the number of operations of the algorithm, thus, the result are not necessarily optimal (see below for more results on simulated annealing).
- the fastest for the benchmarks tested is *preplanned.*

These results must be interpreted with care, since the set of benchmarks chosen is limited, mainly formed by small-size examples, and the metric, the average number of bits, is tangentially related to cost.

Comparison by G. Constantinides, 2003 [1]
In this work, the benchmarks used are LTI algorithms with small and medium size (the largest is a 126-tap FIR filter). The optimization function is the area.

The main conclusions are:

- The authors state that *MILP* is not practical for medium size problems, and they point out that for some cases the solver used cannot reach a solution due to stability problems.
- $Max\text{-}1^{enh}_{cost}$ reduces the area cost up to 6% when compared to $Max\text{-}1_{cost}$, proving that local minima are somehow avoided.

Comparison by K. Han, 2006 [27]
The testbench are a CDM demodulator with 5 quantized signals, an OFDM demodulator with 4 quantized signals, an IIR filter with 7 quantized signals and a noise cancellation filter with 5 quantized signals. For the latter, several optimal, constrained and local methods. For the rest of benchmarks, the followed procedure for testing the performance of the algorithms is to use a lower bound for the $Min+b$ algorithms (including $b = 1$) and an upper bound for the $Max - 1$ algorithm. The lower bound is obtained by means of finding the minimum wordlength for each signal that complies with a given quality constraint while the rest of signal has infinite precision. The upper bound is the global minimum value that complies with the constraints. The tested algorithms are used to find a feasible solution that is finally refined with an optimal search.

The main conclusions are:

- The *complete* search requires a number of iterations 4 to 5 orders of magnitude bigger than the rest of options.
- $Min\text{+}b_{err}$ and *preplanned* reduce the number of iterations 65% and 95%, respectively, against *exhaustive*.
- *Preplanned* emerges as the fastest approach, although the results are not quasi-optimal and they are highly dependent on the input vectors used for simulation.
- $Min\text{+}b_{err}$ has fast convergence, but in some cases the results are far from optimal.
- $Min\text{+}b_{cost}$ has a slow convergence, and for some experiments the obtained costs are high in comparison with the optimal.
- $Min\text{+}1^{err}_{cost}$ displays fast convergence and produces minimal solutions if the contribution of the error and the cost is balanced.

In summary, the use of constrained or local search techniques produce results much faster than optimal approaches and also the combination of error and cost in the gradient computation seems to have beneficial effects in both convergence and optimization. However, the conclusions, yet interesting, are not definitive due to the scarcity and simplicity of the benchmarks.

Comparison by G. Caffarena, 2008 [24]
Small-size LTI are used as benchmarks with 21 to 38 quantized signals and the optimization function is the area.
The main conclusions are:

- Regarding computation time:
  - The fastest algorithm is *Max-1*$_{cost}$.
  - *Max-1*$^{enh}_{cost}$ takes times 1 order of magnitude bigger.
  - *SA* times are 2–3 order of magnitude bigger.
- In terms of cost minimization:
  - The average reduction of *Max-1*$^{enh}_{err}$ vs. *Max-1*$_{err}$ is around 5%, while there are a few cases where *Max-1*$_{err}$ performs better.
  - Comparing *SA* and *Max-1*$_{err}$, the latter always performs better and the average cost reduction is around 10% for the benchmarks tested.

9.7 Conclusions

This chapter presented the most common techniques applied to wordlength optimization, a design stage that enables efficient fixed-point implementations. In this last section the authors would like to provide some overall conclusions, not only to help the reader to choose the appropriate techniques for WLO when designing efficient hardware, but also to expose the current gaps that might be overcome in future research works.

Hierarchical techniques (i.e., [46, 57–60]) stand as the key to the optimization of complex systems. The partitioning itself provides a reduction in complexity, since each block can be optimized independently, and time-consuming simulations are deferred to a final stage with less optimization iterations. However, a wise selection of the partitions may allow the partial application of fast techniques (i.e., LTI fast estimators) that will reduce design time even further. Automatic partition can be a clear asset in this context. Also, there are many fast estimators dealing with the variance or the power of the quantization noise, so adapting these metrics to other application performance metrics at a system level (i.e., BER) might lead to significant improvements [60, 72]. In the same line, the application of techniques to reduce number of the required bit-true simulations or the computation time they need, can be essential [7, 58, 73, 74].

Regarding the particular optimization techniques, a wide range of these has been covered in the chapter. The optimal approaches take too long to be considered practical and they seem to be relegated to assessing the quality of heuristics with small-size benchmarks [1, 24, 30]. However, optimal approaches in the context of real wordlengths [34, 34, 35, 37, 45, 46] provide quasi-optimal results in standard optimization times. These techniques require a final *quantization* step that assign a proper integer wordlength to each signal. Many works reported that the results are more optimized than those of heuristics which directly use integer wordlengths. However, there is not a thorough analysis on the implications of that final stage, where a *bit* budgeting must be applied with the risk of moving the solution far from an optimal or quasi-optimal state. Regarding the heuristics presented in works like [1, 14, 24, 25, 32], etc., many conclusions can be gathered. There is a trade-off between optimization time and optimality. Fast heuristics, such as *Min-1* or *Max-1* tend to get stuck in local minima. However, they can be used to obtain an initial solution to more complex optimization processes. The combination of several heuristics seems to provide good results [25, 41]. Also, heuristics with the possibility of increasing and decreasing the wordlength at each iteration seem to avoid local minima, with a penalty in computation time. The recent introduction of TABU search in [41] arises as an interesting option to be further investigated.

As for the cost function, it is clear that using directly cost as the optimization function is the sensible thing to do, although there is again no thorough analysis about the effect on the final cost of using other—sometimes easier to obtain—metrics (i.e., error, summation of wordlengths, etc.) in the optimization. It would be interesting to see the benefits and drawbacks (i.e., time vs cost reduction) of using simple models such as in [25, 34]. Moreover, the simultaneous use of error and cost gradients accelerates the optimization process and avoid local minima [27, 41, 73]. Getting to the opposite side, using an accurate cost model that not only considers the cost of each operation but also the specific architectural details of the final system has proved to provide cost reduction up to 20% [21, 30, 39, 40, 55, 56, 71]. The combination of system-level partitioning with the combined WLO and architectural synthesis poses as an interesting research line.

As a final remark, this chapter ends providing the reader with some references about floating-point WLO. The decision about when using floating-point rather than fixed-point is not always easy to make [5, 75, 76]. The use of non-standard precisions for floating-point arithmetic implementations has been researched in the last couple of decades [5, 68, 77–80] in order to achieve efficient implementations in applications that require a wide dynamic range. WLO optimization has been applied in several works [17, 81–83] aiming at applying customized wordlengths for the exponent and mantissa of each signal within the algorithms. Moreover, there are also works where fixed-point and floating-point arithmetic are combined in the same designs [77, 84–87]. In general, the techniques presented in this chapter can also be applied to floating-point WLO, but there are still many open topics, such as the elaboration of accurate error models, techniques to efficiently combine fixed-point and floating-point, etc.

Acknowledgments This research has been partially funded by the Spanish Ministry of Science, Innovation and Universities through the project RTI2018-095324-B-I00.

References

1. Constantinides, G. A., Cheung, P. Y. K., & Luk, W. (2003). Wordlength optimization for linear digital signal processing. *IEEE Transactions on Computer-Aided Design of Integrated Circuits and Systems, 22*(10), 1432–1442 (2003).
2. Menard, D., Rocher, R., & Sentieys, O. (2008). Analytical fixed-point accuracy evaluation in linear time-invariant systems. *IEEE Transactions on Circuits and Systems I: Regular Papers, 55*(10), 3197–3208.
3. Caffarena, G., Carreras, C., López, J., & Fernández, A. (2010). SQNR estimation of fixed-point DSP algorithms. *International Journal of Advances in Signal Processing, 2010*, 1–11.
4. Mittal, S. (2016). A survey of techniques for approximate computing. *ACM Computing Surveys, 48*, 61.
5. Gaffar, A., Mencer, O., & Luk, W. (2004). Unifying bit-width optimisation for fixed-point and floating-point designs. In *IEEE Symposium on Field-Programmable Custom Computing Machines* (pp. 79–88).
6. Balfour, J. (2010). *Efficient Embedded Computing*. PhD thesis, Stanford University.
7. Rocher, R., Menard, D., Scalart, P., Nehmeh, O. S. R., Menard, D., Nogues, E., Banciu, A., Michel, T., & Rocher, R. (2015). Fast integer word-length optimization for fixed-point systems. *Journal of Signal Processing Systems, 84*, 113–128.
8. Gal, B., & Casseau, E. (2011). Word-length aware DSP hardware design flow based on high-level synthesis. *Journal of Signal Processing Systems, 62*(3), 341–357.
9. Keding, H., Willems, M., Coors, M., & Meyr, H. (1998). FRIDGE: A fixed-point design and simulation environment. In *Design, Automation and Test in Europe*, Paris (pp. 429–435).
10. Lee, D.-U., Gaffar, A., Cheung, R., Mencer, W., Luk, O., & Constantinides, G. (2006). Accuracy-guaranteed bit-width optimization. *IEEE Transactions on Computer-Aided Design of Integrated Circuits and Systems, 25*(10), 1990–2000.
11. Lopez, J., Carreras, C., Caffarena, G., & Nieto-Taladriz, O. (2003). Fast characterization of the noise bounds derived from coefficient and signal quantization. In *IEEE International Symposium on Circuits and Systems* (Vol. 4, pp. 309–312).
12. Cmar, R., Rijnders, L., Schaumont, P., Vernalde, S., & Bolsens, I. (1999). A methodology and design environment for DSP ASIC fixed point refinement. In *Proceedings of the conference on Design, automation and test in Europe, DATE '99* (p. 56).
13. Kim, S., Kum, K.-I., & Sung, W. (1998). Fixed-point optimization utility for C and C++ based digital signal processing programs. *IEEE Transactions on Circuits and Systems II: Analog and Digital Signal Processing, 45*, 1455–1464.
14. Cantin, M.-A., Savaria, Y., Prodanos, D., & Lavoie, P. (2001). An automatic word length determination method. In *IEEE International Symposium on Circuits and Systems*, Sydney (Vol. 5, pp. 53–56)
15. Singhee, A., Fang, C. R., Ma, J. D., & Rutenbar, R. A. (2006). Probabilistic interval-valued computation: Toward a practical surrogate for statistics inside cad tools. In *2006 43rd ACM/IEEE Design Automation Conference* (pp. 167–172).
16. Sarbishei, O., Radecka, K., & Zilic, Z. (2012). Analytical optimization of bit-widths in fixed-point LTI systems. *IEEE Transactions on Computer-Aided Design of Integrated Circuits and Systems, 31*(3), 343–355.
17. Boland, D., & Constantinides, G. A. (2013). A scalable precision analysis framework. *IEEE Transactions on Multimedia, 15*(2), 242–256.
18. Lopez, J., Caffarena, G., Carreras, C., & Nieto-Taladriz, O. (2008). Fast and accurate computation of the roundoff noise of linear time-invariant systems. *IET Circuits, Devices & Systems*, *2*(4), 393–408.
19. Kim, S., Kum, K.-I., & Sung, W. (1995). Fixed-point simulation utility for C and C++ based digital signal processing programs. In *IEEE Workshop on VLSI Signal Processing* (pp. 197–206).

20. Constantinides, G., & Woeginger, G. (2002). The complexity of multiple wordlength assignment. *Applied Mathematics Letters, 15*, 137–140.
21. Wadekar, S., & Parker, A. (1998). Accuracy sensitive word-length selection for algorithm optimization. In *International Conference on Computer Design* (pp. 54–61).
22. Caffarena, G., López, J. A., Leyva, G., Carreras, C., & Nieto-Taladriz, O. (2009). Architectural synthesis of fixed-point DSP datapaths using FPGAs. *International Journal of Reconfigurable Computing, 2009*, 703267:1–703267:14.
23. Clark, M., Mulligan, M., Jackson, D., & Linebarger, D. (2005). Accelerating fixed-point design for MB-OFDM UWB systems. In *CommsDesign* (Vol. 3).
24. Caffarena, G. (2008). *Combined Word-Length Allocation and High-Level Synthesis of Digital Signal Processing Circuits*. PhD thesis, Universidad Politécnica de Madrid, 2008.
25. Cantin, M.-A., Savaria, Y., & Lavoie, P. (2002). A comparison of automatic word length optimization procedures. In *IEEE International Symposium on Circuits and Systems* (Vol. 2, pp. 612–615).
26. Han, K., Eo, I., Kim, K., & Cho, H. (2001). Numerical word-length optimization for CDMA demodulator. In *The 2001 IEEE International Symposium on Circuits and Systems, 2001. ISCAS 2001* (Vol. 4, pp. 290–293).
27. Han, K., & Evans, B. (2006). Optimum wordlength search using sensitivity information. *EURASIP Journal of Applied Signal Processing, 2006*, 1–14. https://doi.org/10.1155/ASP/2006/92849.
28. Holzer, M., Knerr, B., Belanovic, P., & Rupp, M. (2006). Efficient design methods for embedded communication systems. *EURASIP Journal on Embedded Systems, 2006*, 64913.
29. Sung, W., & Kum, K.-I. (1994). Word-length determination and scaling software for a signal flow block diagram. In *1994 IEEE International Conference on Acoustics, Speech, and Signal Processing, 1994. ICASSP-94* (Vol. ii, pp. II/457 –II/460, Vol. 2, 19–22).
30. Caffarena, G., Constantinides, G., Cheung, P., Carreras, C., & Nieto-Taladriz, O. (2006). Optimal combined word-length allocation and architectural synthesis of digital signal processing circuits. *IEEE Transactions on Circuits and Systems II: Express Briefs, 53*(5), 339–343.
31. Sung, W., & Kum, K.-I. (1995). Simulation-based word-length optimization method for fixed-point digital signal processing systems. *IEEE Transactions on Circuits and Systems II: Express Briefs, 43*(12), 3087–3090.
32. Choi, H., & Burleson, W. (1994). Search-based wordlength optimization for VLSI/DSP synthesis. In *IEEE Workshop VLSI Signal Processing* (pp. 198–207).
33. Han, K., Evans, B., & Swartzlander, E. (2004). Data wordlength reduction for low-power signal processing software. In *IEEE Workshop on Signal Processing Systems* (pp. 343–348).
34. Fiore, P., & Lee, L. (1999). Closed-form and real-time wordlength adaptation. In *1999 IEEE International Conference on Acoustics, Speech, and Signal Processing, 1999. ICASSP '99. Proceedings* (Vol. 4, pp. 1897–1900, Vol.4, 15–19).
35. Doi, N., Horiyama, T., Nakanishi, M., & Kimura, S. (2004). Minimization of fractional wordlength on fixed-point conversion for high-level synthesis. In *Design Automation Conference, 2004. Proceedings of the ASP-DAC 2004. Asia and South Pacific* (pp. 80–85, 27–30)
36. Fiore, P. (2008). Efficient approximate wordlength optimization. *IEEE Transactions on Computers, 57*, 1561–1570.
37. Clarke, J. A., Constantinides, G. A., & Cheung, P. Y. K. (2009). Word-length selection for power minimization via nonlinear optimization. *ACM Transactions on Design Automation of Electronic Systems, 14*(3), 1–28.
38. Fiore, P. (2000). *A Custom Computing Framework for Orientation and Photogrammetry*. PhD thesis, Massachusetts Inst. of Technology, 2000.
39. Caffarena, G., & Carreras, J. (2010). Precision-aware architectural synthesis of DSP circuits. In *European Signal Processing Conference*, Aalborg.
40. Caffarena, G., & Carreras, C. (2010). Architectural synthesis of DSP circuits under simultaneous error and time constraints. In *2010 18th IEEE/IFIP International Conference on VLSI and System-on-Chip* (pp. 322–327).

41. Nguyen, H., Menard, D., & Sentieys, O. (2011). Novel algorithms for word-length optimization. In *2011 19th European Signal Processing Conference* (pp. 1944–1948).
42. Garfinkel, R. S., & Nemhauser, G. L. (1972). *Integer programming*. New York: Wiley.
43. Fang, F., Chen, T., & Rutenbar, R. (2002). Floating-point bit-width optimization for low-power signal processing applications. In *IEEE International Conference Acoustics, Speech, and Signal Processing* (Vol. 3, pp. 3208–3211).
44. Constantinides, G., Cheung, P., & Luk, W. (1999). Truncation noise in fixed-point SFGs. *IEE Electronics Letters, 35*(23), 2012–2014.
45. Chan, S. C., & Tsui, K. M. (2005). Wordlength determination algorithms for hardware implementation of linear time invariant systems with prescribed output accuracy. In *2005 IEEE International Symposium on Circuits and Systems* (Vol. 3, pp. 2607–2610).
46. Parashar, K. N., Menard, D., & Sentieys, O. (2013). A polynomial time algorithm for solving the word-length optimization problem. In *2013 IEEE/ACM International Conference on Computer-Aided Design (ICCAD)* (pp. 638–645).
47. Glover, F. (1989). Tabu search — Part I. *INFORMS Journal of Computing, 1*(3), 190–206 (1989)
48. Keding, H., Hürtgen, F., Willems, M., & Coors, M. (1998). Transformation of floating-point into fixed-point algorithms by interpolation applying a statistical approach. In *International Conference Signal Processing, Applications and Technology*.
49. Caffarena, G., & Menard, D. (2012). Many-core parallelization of fixed-point optimization of VLSI circuits through GPU devices. In *Proceedings of the 2012 Conference on Design and Architectures for Signal and Image Processing* (pp. 1–8).
50. Kapre, N., & D. Ye (2016). GPU-accelerated high-level synthesis for bitwidth optimization of FPGA datapaths. In *Proceedings of the 2016 ACM/SIGDA International Symposium on Field-Programmable Gate Arrays*, FPGA '16, New York, NY (pp. 185–194). New York: Association for Computing Machinery.
51. Rocher, R., Menard, D., Scalart, P., & Sentieys, O. (2012). Analytical approach for numerical accuracy estimation of fixed-point systems based on smooth operations. *IEEE Transactions on Circuits and Systems I: Regular Papers, 59*(10), 2326–2339.
52. Widrow, B., & Kollár, I. (2008). *Quantization noise: Roundoff error in digital computation, signal processing, control, and communications*. Cambridge: Cambridge University Press.
53. Shi, C., & Brodersen, R. (2004). A perturbation theory on statistical quantization effects in fixed-point DSP with non-stationary inputs. In *IEEE International Conference on Circuits and Systems, 3*, 373–376.
54. Constantinides, G. A. (2006). Word-length optimization for differentiable nonlinear systems. *ACM Transactions on Design Automation of Electronic Systems, 11*, 26–43.
55. Kum, K. I., & Sung, W. (2001). Combined word-length optimization and high-level synthesis of digital signal processing systems. *IEEE Transactions on Computer-Aided Design of Integrated Circuits and Systems, 20*, 921–930.
56. Herve, N., Menard, D., & Sentieys, O. (2007). About the importance of operation grouping procedures for multiple word-length architecture optimizations. In *International Workshop on Applied Reconfigurable Computing* (pp. 107–122).
57. Weijers, J., Derudder, V., Janssens, S., Petré, F., & Bourdoux, A. (2006). From MIMO-OFDM algorithms to a real-time wireless prototype: A systematic Matlab-to-Hardware design flow. *EURASIP Journal on Applied Signal Processing, 2006*, 1–12.
58. Parashar, K. N., Menard, D., & Sentieys, O. (2014). Accelerated performance evaluation of fixed-point systems with un-smooth operations. *IEEE Transactions on Computer-Aided Design of Integrated Circuits and Systems, 33*(4), 599–612.
59. Sedano, E., Menard, D., & López, J. A. (2014). Automated data flow graph partitioning for a hierarchical approach to wordlength optimization. In Goehringer, D., Santambrogio, M. D., Cardoso, J. M. P., & Bertels, K. (Eds.) *Reconfigurable Computing: Architectures, Tools, and Applications* (pp. 133–143). Cham: Springer International Publishing.
60. Novo, D., Tzimi, I., Ahmad, U., Ienne, P., & Catthoor, F. (2013). Cracking the complexity of fixed-point refinement in complex wireless systems. In *SiPS 2013 Proceedings* (pp. 18–23).

61. Willems, M., Keding, H., Grötker, T., & Meyr, H. (1997). FRIDGE: An Interactive Fixed-Point Code Generation Environment for Hw/Sw-Codesign. In *International Conference on Acoustics, Speed, Signal Processing*, Munich.
62. Caffarena, G., Fernandez, A., Carreras, C., & Nieto-Taladriz, O. (2004). Fixed-point refinement of OFDM-based adaptive equalizers: A heuristic approach. In *EUSIPCO* (pp. 1353–1356).
63. Chang, M., & Hauck, S. (2002). Precis: A design-time precision analysis tool. In *IEEE Symposium on Field-Programmable Custom Computing Machines* (pp. 229–238).
64. Rocher, R., Menard, D., Herve, N., & Sentieys, O. (2006). Fixed-point configurable hardware components. *EURASIP Journal on Embedded Systems, 2006*, 13 pp. Article ID 23197. https://doi.org/10.1155/ES/2006/23197
65. Constantinides, G. (2003). Perturbation analysis for word-length optimization. In *IEEE Symposium on Field-Programmable Custom Computing Machines* (pp. 81–90).
66. Jevtic, R., Carreras, C., & Caffarena, G. (2008). Fast and accurate power estimation of FPGA DSP components based on high-Level switching activity models. *International Journal of Electronics, 95*(7), 653–668.
67. Clarke, J., Gaffar, A., & Constantinides, G. (2005). Parameterized logic power consumption models for FPGA-based arithmetic. In *Proceedings of International Conference on Field Programmable Logic and Applications* (pp. 626–629).
68. Tong, J., Nagle, D., & R. Rutenbar (2000). Reducing power by optimizing the necessary precision/range of floating-point arithmetic. *IEEE Transactions on Very Large Scale Integration (VLSI) Systems, 8*, 273–286.
69. Gaffar, A., Clarke, J., & Constantinides, G. (2006). PowerBit - Power aware arithmetic bit-width optimization. In *IEEE International Conference on Field Programmable Technology* (pp. 289–292).
70. Mallik, A., Sinha, D., Banerjee, P., & Zhou, H. (2007). Low-power optimization by smart bit-width allocation in a systemC-based ASIC design environment. *IEEE Transactions on Computer-Aided Design of Integrated Circuits and Systems, 26*, 447–455.
71. Bouganis, C.-S., Constantinides, G., & Cheung, P. (2005). A novel 2D filter design methodology for heterogeneous devices. In *IEEE Symposium on Field-Programmable Custom Computing Machines*.
72. Menard, D., Serizel, R., Rocher, R., & Sentieys, O. (2008). Accuracy constraint determination in fixed-point system design. *EURASIP Journal on Embedded Systems, 2008*, 1–12.
73. Parashar, K., Rocher, R., Menard, D., & Sentieys, O. (2010). A hierarchical methodology for word-length optimization of signal processing systems. In *23rd International Conference on VLSI Design, 2010. VLSID '10* (pp. 318–323).
74. Sedano, E., (2016). *Automated Wordlength Optimization Framework for Multi-source Statistical Interval-Based Analysis of Nonlinear Systems with Control-Flow Structures*. PhD thesis, Universidad Politécnica de Madrid.
75. Janhunen, J., Pitkanen, T., Silven, O., & Juntti, M. (2011). Fixed- and floating-point processor comparison for MIMO-OFDM detector. *IEEE Journal of Selected Topics in Signal Processing, 5*(8), 1588–1598.
76. El-Ghazawi, T., El-Araby, E., Huang, M., Gaj, K., Kindratenko, V., & Buell, D. (2008). The promise of high-performance reconfigurable computing. *Computer, 41*, 69–76.
77. Belanovic, P., & Leeser, M. (2002). A library of parameterized floating-point modules and their use. In *Field Programmable Logic and Applications* (pp. 657–666).
78. Leyva, G., Caffarena, G., Carreras, C., & Nieto-Taladriz, O. (2004). A generator of high-speed floating-point modules. In *12th Annual IEEE Symposium on Field-Programmable Custom Computing Machines, 2004. FCCM 2004* (pp. 306–307).
79. Fousse, L., Hanrot, G., Lefèvre, V., Pélissier, P., & Zimmermann, P. (2007). MPFR: A multiple-precision binary floating-point library with correct rounding. *ACM Transactions on Mathematical Software, 33*(2), 13–es. https://doi.org/10.1145/1236463.1236468
80. Caffarena, G., & Menard, D. (2016). Quantization noise power estimation for floating-point DSP circuits. *IEEE Transactions on Circuits and Systems II: Express Briefs, 63*(6), 593–597 (2016)

81. Ho, C., Leong, M., Leong, P., Becker, J., & Glesner, M. (2002). Rapid prototyping of fpga based floating point dsp systems. In *13th IEEE International Workshop on Rapid System Prototyping, 2002. Proceedings* (pp. 19–24).
82. Fang, C., Chen, T., & Rutenbar, R. (2002). Lightweight floating-point arithmetic: Case study of inverse discrete cosine transform. *EURASIP Journal on Applied Signal Processing, 2002*(2002), 879–892.
83. Gaffar, A., Mencer, O., Luk, W., Cheung, P., & Shirazi, N. (2002). Floating-point bitwidth analysis via automatic differentiation. In *IEEE International Conference on Field-Programmable Technology* (pp. 158–165).
84. Tsoi, K., Ho, C., Yeung, H., & Leong, P. (2004). An arithmetic library and its application to the N-body problem. In *IEEE Symposium Field-Programmable Custom Computing Machines* (pp. 68–78).
85. Dido, J., Geraudie, N., Loiseau, L., Payeur, O., Savaria, Y., & Poirier, D. (2002). A flexible floating-point format for optimizing data-paths and operators in FPGA based DSPs. In *ACM/SIGDA International Symposium Field-programmable Gate Arrays* (pp. 50–55).
86. Andraka, R. (2006). Hybrid floating point technique yields 1.2 Gigasample per second 32 to 2048 point floating point FFT in a single FPGA. In *High-Performance Embedded Computing*.
87. Lenart, T., & Owall, V. (2006). Architectures for dynamic data scaling in 2/4/8K pipeline FFT cores. *IEEE Transactions on Very Large Scale Integration (VLSI) Systems* (Vol. 14, pp. 1286–1290).

Part III
Approximate Computing Applied to Real-Life Applications

Chapter 10
Exploiting Approximations in Real-Time Scheduling

Kamyar Mirzazad Barijough, Lin Huang, I-Hong Hou, Sachin S. Sapatnekar, Jiang Hu, and Andreas Gerstlauer

10.1 Introduction

In real-time systems, reaction time and latency guarantees are of critical importance. In such systems, tasks have constraints in the form of deadlines that must be met during system execution. Real-time analysis methods can provide deadline guarantees, but they rely on knowledge about upper bounds for individual task execution times [1]. Furthermore, when there are dependencies between tasks, upper bounds are also needed for communication times. These bounds are generally chosen with pessimism to account for all scenarios and, therefore, are significantly larger than average computation and communication times. This pessimism often leads to overdesign of real-time systems, which ultimately reduces schedule admissibility and resource utilization.

In case of independent tasks, as an alternative to pessimistic upper bounds, one can use tighter bounds for execution times and schedule tasks according to the optimistic bounds. While on average such bounds may be satisfied, task deadlines might be violated occasionally, e.g., when the system is under stress or task execution times grow large for certain inputs. In the general case, deadline violations will lead to task results being dropped. In mixed-criticality settings [2], tasks can thereby be partitioned into different criticalities based on their overall contribution to quality,

K. M. Barijough · A. Gerstlauer (✉)
Electrical and Computer Engineering, University of Texas at Austin, Austin, TX, USA
e-mail: kammirzazad@utexas.edu; gerstl@ece.utexas.edu

L. Huang · I-H. Hou · J. Hu
Electrical and Computer Engineering, Texas A&M University, College Station, TX, USA
e-mail: liushui0820@tamu.edu; ihou@tamu.edu; jianghu@tamu.edu

S. S. Sapatnekar
Electrical and Computer Engineering, University of Minnesota, Minneapolis, MN, USA
e-mail: sachin@umn.edu

A. Bosio et al. (eds.), *Approximate Computing Techniques*,
https://doi.org/10.1007/978-3-030-94705-7_10

such that only low-criticality tasks with minor quality impact will be dropped. Nevertheless, this provides only a very coarse notion of quality management, e.g., on a complete frame basis in audio/video applications. By contrast, many such applications support a more graceful tradeoff between execution time and quality degradation. For example, instead of completely skipping a task, its precision in computation can be lowered in exchange for earlier completion. The inherent error tolerance of these applications can be generalized and exploited through approximate or, as it is often also called, imprecise computing where deadline violations are translated into reduced quality. In such scenarios, scheduling can be formulated as a quality optimization problem in terms of task execution time budgets under total scheduling constraints.

In case of dependent tasks, in addition to task execution times, there is the added problem of accounting for communication latency in allocating task budgets. Oftentimes, such as in single-device systems, tightly coupled systems or distributed systems with closed networks, communication times are either highly predictable or insignificant compared to computation times. Traditional distributed real-time frameworks [3, 4] rely on such assumptions to provide overall guarantees. However, many systems, such as the Internet of Things (IoT), utilize inherently open and often wireless networks for communication, which can have losses and unpredictable or potentially unbounded delays. This requires assignment of timeouts to bound communication delays. Real-time network protocols [5] follow such approaches to tradeoff latency for quality, e.g., in video or audio streaming, but such protocols are intended for end-to-end communication and not distributed computation. In distributed real-time systems, scheduling algorithms need to allocate communication budgets as well as computation budgets. Similar to computation budgets, communication budgets can be enforced by skipping partial communication, dropping intermediate data and thus reducing quality. Therefore, in the general case, scheduling of real-time systems is a quality optimization problem in terms of both task computation and communication budgets.

In this chapter, we consider reducing pessimism in scheduling of real-time systems through exploiting approximate computation or communication while optimizing overall quality. Towards this end, we first consider the case of independent tasks executing in a single multiprocessor device. Then, we turn our attention to dependent tasks executing locally or in a distributed fashion. Finally, we conclude with a summary.

10.2 Scheduling Independent Tasks

A common application scenario for real-time scheduling is for computing tasks without inter-dependencies among them, i.e., there is no constraint specifying that a task must proceed another. We introduce the application of approximate computing for a recently popular independent task scheduling model—mixed-criticality systems, with an emphasis on multiprocessor scheduling.

10.2.1 Mixed-Criticality (MC) System Model

A mixed-criticality (MC) system [2] has a set of independent sporadic tasks $\mathcal{T} = \{\tau_1, \tau_2, \ldots\}$ and each task $\tau_i \in \mathcal{T}$ consists of an infinite sequence of jobs $\{J_i^1, J_i^2, \ldots\}$. A basic and perhaps the most common type of mixed-criticality system is dual-criticality system, where only two criticality levels exist. Without loss of generality, the system model description here is based on dual-criticality systems. As such, task set $\mathcal{T}$ is composed by two disjoint subsets of low-criticality tasks $\mathcal{T}_L$ and high-criticality tasks $\mathcal{T}_H$. The system may operate in either low-criticality mode or high-criticality mode. We use subscript in $\{L, H\}$ to indicate task criticality and superscript in $\{lo, hi\}$ to differentiate low-criticality and high-criticality modes.

Each task $\tau_i \in \mathcal{T}$ is characterized by $(T_i, \chi_i, R_i^{lo}, R_i^{hi})$, where T_i is the minimal time interval between two consecutive jobs of task τ_i, $\chi_i \in \{L, H\}$ indicates its criticality level, and R_i^{lo} and R_i^{hi} are estimated job execution time in low-criticality and high-criticality mode, respectively. If job J_i^j is released at time a_i^j, its deadline is $a_i^j + T_i$. Therefore, T_i implicitly specifies job deadline and is also called period for convenience.

A system starts with low-criticality mode, which is also the ordinary operation mode. In order to reduce the pessimism of WCET (Worst-Case Execution Time) estimate in conventional real-time scheduling, execution time estimate R_i^{lo} is not very conservative for high-criticality tasks. Consequently, there is a low but non-zero probability that an actual execution time exceeds R_i^{lo}. In the classic MC model [2], as long as any high-criticality job J_i^j has actual execution time exceeding R_i^{lo}, the system switches to high-criticality mode. In high-criticality mode, all high-criticality tasks are scheduled with very pessimistic R_i^{hi}, i.e., $R_i^{hi} > R_i^{lo}$, such that there exists a guarantee for meeting deadlines of all high-criticality tasks. Meanwhile, all low-criticality tasks are discarded and no longer executed in high-criticality mode.

10.2.2 Scheduling of Mixed-Criticality Systems

The goal of scheduling is to decide when to execute each job. For a multiprocessor system, the scheduling also allocates jobs to certain processors. We consider typical preemptive scheduling, where a low-priority job can be preempted by a high-priority job during its execution. For each task τ_i, its utilizations in low and high-criticality modes are defined as

$$u_i^{lo} = \frac{R_i^{lo}}{T_i}, \text{ and } u_i^{hi} = \frac{R_i^{hi}}{T_i},$$

respectively. Then, the total utilizations of all low-criticality tasks are given by

$$U_L^{lo} = \sum_{\tau_i \in \mathcal{T}_L} u_i^{lo} \text{ and } U_L^{hi} = \sum_{\tau_i \in \mathcal{T}_L} u_i^{hi}.$$

Likewise, the total utilizations for high-criticality tasks are defined as

$$U_H^{lo} = \sum_{\tau_j \in \mathcal{T}_H} u_j^{lo} \text{ and } U_H^{hi} = \sum_{\tau_j \in \mathcal{T}_H} u_j^{hi}.$$

Please note in classic MC model, $U_L^{hi} = 0$ since all low-criticality tasks are discarded in high-criticality mode.

10.2.2.1 Uniprocessor Scheduling

One well-known technique for MC scheduling on uniprocessors is EDF-VD (Earliest Deadline First with Virtual Deadlines) [6]. EDF-VD is similar to EDF [7] except that the implicit deadlines of all high-criticality tasks are scaled by a factor $x \in (0, 1)$ in low-criticality mode. That is, for each high-criticality task τ_i, its virtual implicit deadline is $\hat{T}_i = x \cdot T_i$ in low-criticality mode while its implicit deadline in high-criticality mode remains to be T_i. The sufficient schedulability condition in low-criticality mode is given by the following theorem.

Theorem 1 (Theorem 1 in [6]) *If the following condition is satisfied, sporadic task set $\mathcal{T}$ is schedulable with EDF-VD method on uniprocessor in low-criticality mode.*

$$U_L^{lo} + \frac{U_H^{lo}}{x} \le 1. \tag{10.1}$$

The scaling factor x can be obtained as $x = U_H^{lo}/(1 - U_L^{lo})$.

10.2.2.2 Partitioned Scheduling on Multiprocessors

In this approach [8], $n = |\mathcal{T}|$ tasks are partitioned onto m unit-speed processors. After the partitioning, each task is never changed to another processor. As such, there is a fixed subset of tasks on each processor. Then, uniprocessor scheduling methods can be applied to each processor individually. In [8], a 2-phase task partitioning algorithm is introduced. In phase 1, high-criticality tasks are assigned to each processor one by one as long as the high-criticality utilization U_H^{hi} for each processor does not exceed $\frac{3}{4}$. In phase 2, low-criticality tasks are further assigned to processors one by one following the condition that the low-criticality utilization $U_L^{lo} + U_H^{lo}$ is no greater than $\frac{3}{4}$.

10.2.2.3 Global Scheduling on Multiprocessors

Global scheduling allows the jobs of a task to be allocated onto different processors. The method of fpEDF (fixed-priority EDF) [9] is a state-of-art global scheduling approach for multiprocessor in traditional real-time systems without mixed-criticality. Fixed-priority means the priority of one job cannot be changed during its execution. For a task set $\bigcup_{\tau_i \in \mathcal{T}}(T_i, R_i)$ to be scheduled on m identical processors, fpEDF first chooses a subset $\mathcal{T}_{hp} \subset \mathcal{T}$ of at most $m - 1$ tasks, each with utilization greater than $\frac{1}{2}$, and assigns them on m_{hp} processors with the highest priority. The priorities of remaining tasks $\mathcal{T}_{lp} \subset \mathcal{T}$ are lower and scheduled according to EDF (Earliest Deadline First) principle on the other $m_{lp} = m - m_{hp}$ processors. The schedulability condition for fpEDF is given in Lemma 1 [9].

Lemma 1 *Consider a task set* $\bigcup_{\tau_i \in \mathcal{T}}(T_i, R_i)$ *to be scheduled on m identical processors. Let* U_{lp}^{total} *be the total utilization of the tasks in* $\mathcal{T}_{lp}$*, and* U_{lp}^{max} *be the maximum utilization of tasks in* $\mathcal{T}_{lp}$*. If* $U_{lp}^{total} \leq m_{lp} - (m_{lp} - 1) \cdot U_{lp}^{max}$ *is satisfied, this task set* $\mathcal{T}$ *is schedulable by fpEDF method.*

The work of fpEDF-Virtual Deadline (fpEDF-VD) [10] is an extension of fpEDF to mixed-criticality systems. Virtual deadlines are enforced for high-criticality tasks. Each high-criticality task τ_j is mapped to $(\hat{T}_j, R_j^{lo})$ in low-criticality mode, where $\hat{T}_j = x \cdot T_j$ $(0 < x < 1)$ is the virtual deadline that is enforced in both offline schedulability test and online execution. Each low-criticality task τ_i is mapped to a regular implicit deadline task (T_i, R_i^{lo}) in low-criticality mode, and all low-criticality tasks are dropped in high-criticality mode. The schedulability conditions for fpEDF-VD are as follows.

- Task set $(\bigcup_{\tau_i \in \mathcal{T}_L}(T_i, R_i^{lo})) \bigcup (\bigcup_{\tau_j \in \mathcal{T}_H}(x \cdot T_j, R_j^{lo}))$ is schedulable on m processors in low-criticality mode according to Lemma 1.
- Task set $\bigcup_{\tau_j \in \mathcal{T}_H}((1 - x) \cdot T_j, R_j^{hi})$ is schedulable on m processors in high-criticality mode according to Lemma 1.

For a high-criticality task $\tau_j \in \mathcal{T}_H$ in high-criticality mode, its implicit deadline $(1 - x) \cdot T_j$ is used in the offline schedulability check. However, only its original deadline T_j needs to be enforced during online execution.

The schedulability condition in high-criticality mode leads to the following important conclusion.

Lemma 2 ([10]) *If a mixed-criticality task set* $\mathcal{T}$ *is schedulable by fpEDF-VD, each of its high-criticality job* J_j^k *in high-criticality mode can start from its virtual deadline* $\hat{d}_j^k = a_j^k + x \cdot T_j$ *and guarantee to finish by actual deadline* d_j^k *with execution time* R_j^{hi} *following fpEDF-VD scheduling.*

10.2.2.4 DP-Fair Scheduling on Multiprocessors

DP (Deadline Partition)-Fair [11] is a scheduling method based on proportional fairness for regular (non-MC) multiprocessor systems, and also allows a task to be allocated onto different processors. In DP-Fair, the density of each task τ_i is computed as $\delta_i = \frac{R_i}{T_i}$, where R_i and T_i are the worst-case execution time and period of task τ_i, respectively. DP-Fair divides time into slices, where boundaries between slices are formed by release time and deadlines of all jobs. Jobs are allocated in small pieces that have the same deadline in each slice. Assuming the length of a slice is l, $\delta_i \cdot l$ is executed for task δ_i in this slice. The schedulability condition for DP-Fair is as follows.

Theorem 2 (Lemma 14 in [12]) *A non-MC task set $\mathcal{T}$ is schedulable under DP-Fair iff $\sum_{\tau_i \in \mathcal{T}} \delta_i \leq m$, where m is the number of processors.*

MC-DP-Fair [12] is an extension of DP-Fair for mixed-criticality systems. A main change is that each task τ_i is assigned a virtual deadline $0 < \hat{T}_i \leq T_i$. Let Γ be the earliest job release time or deadline after a system is switched to high-criticality mode. The virtual deadlines and the original deadlines are enforced before and after Γ, respectively. MC-DP-Fair has schedulability equivalent to MC-Fluid [12], which is a theoretically optimal approach.

10.2.3 Approximations in Mixed-Criticality Systems

The treatment to low-criticality tasks in the classic MC model is controversial. All low-criticality tasks are dropped in high-criticality mode to facilitate the guarantee for high-criticality tasks. Despite their low-criticality, these tasks are still very much needed and completely abandoning them is a significant loss of Quality of Service (QoS).

Approximate computing [13], a.k.a. imprecise computing, can help reduce such complete loss of low-criticality tasks in high-criticality mode. Approximate computing does not provide precise results, but costs relatively short execution time, which allows low-criticality tasks to continue in high-criticality mode. In such an approach, each low-criticality job consists of two parts: a mandatory part and an optional part. Since more processing time is needed for high-criticality tasks in high-criticality mode, low-criticality tasks only need to complete their mandatory parts in high-criticality mode, while they need to complete both mandatory and optional parts in low-criticality mode. As such, we have $R_i^{lo} > R_i^{hi} \geq 0$ for each low-criticality task τ_i.

Approximate computing facilitates two additional system models.

1. Imprecise MC (IMC) system: a low-criticality task τ_i has a precise computing realization with execution time $\hat{R}_i$ and an approximate implementation with execution time $\tilde{R}_i < \hat{R}_i$. Let $R_i^{lo} = \hat{R}_i$ and $R_i^{hi} = \tilde{R}_i$, which is a non-zero

constant. When the system is switched to high-criticality mode, if a low-criticality job has executed longer than R_i^{hi}, it would be aborted for this period, otherwise it would continue to execute till R_i^{hi}.

2. Variable-Precision MC (VPMC) system: for a low-criticality task τ_i, $R_i^{lo} = \hat{R}_i$ and $R_i^{hi} \in \{\hat{R}_i, \tilde{R}_i\}$ corresponds to a decision variable. Since approximate computing leads to errors, the objective of VPMC is to minimize computing errors, or maximize precise computing, for low-criticality tasks in high-criticality mode.

The idea of exploiting approximate computing in real-time scheduling appeared more than two decades ago [14]. However, the adoption of approximate computing for MC systems is fairly recent and most works are built upon existing MC scheduling techniques. A uniprocessor scheduling considering approximate computing for low-criticality tasks is [15], which is an extension to EDF-VD (Virtual Deadline) scheduling [6]. Its schedulability condition in low-criticality mode is the same as EDF-VD. In high-criticality mode, the method of [15] continues to execute low-criticality tasks with approximate computing, and derives the following sufficient schedulability condition.

Theorem 3 (Theorem 2 in [15]) *If the following condition is satisfied, sporadic task set $\mathcal{T}$ is schedulable with EDF-VD method on uniprocessor in high-criticality mode.*

$$xU_L^{lo} + (1 - x)U_L^{hi} + U_H^{hi} \leq 1, \tag{10.2}$$

where x is the scaling factor for virtual deadlines.

Theorem 4 (Theorem 3 in [15]) *Given a task set, if $\frac{U_H^{lo}}{1-U_L^{lo}} \leq \frac{1-(U_H^{hi}+U_L^{hi})}{U_L^{lo}-U_L^{hi}}$, where $U_H^{hi} + U_L^{hi} < 1$ and $U_L^{lo} < 1$ and $U_L^{lo} > U_L^{hi}$, then this task set can be scheduled by EDF-VD with a deadline scaling factor x chosen in the following range*

$$x \in \left[\frac{U_H^{lo}}{1 - U_L^{lo}}, \frac{1 - (U_H^{hi} + U_L^{hi})}{U_L^{lo} - U_L^{hi}}\right]. \tag{10.3}$$

All these methods [15–17] execute low-criticality tasks with approximate computing in high-criticality mode. By contrast, a later work [18] allows some low-criticality tasks to be executed with full precision in high-criticality mode. It formulates an integer linear programming to maximize the number of low-criticality tasks that can continue with precise computing. Like [17], this work is also based on the fluid model and is applied on multiprocessors.

10.2.4 Multiprocessor Scheduling of Mixed-Criticality Systems with Approximate Computing

We introduce how approximate computing helps improve partitioned scheduling, global scheduling based on fpEDF-VD, and MC-DP-Fair scheduling in multiprocessor MC systems.

10.2.4.1 Partitioned Scheduling for IMC/VPMC Systems

For EDF-VD on uniprocessor IMC/VPMC systems, we introduce a sufficient schedulability condition that has a form similar to that in conventional MC systems.

Lemma 3 *If a task set in IMC/VPMC system satisfies the condition* $max(U_L^{lo} + U_H^{lo}, U_L^{hi} + U_H^{hi}) \leq \frac{3}{4}$*, it is schedulable by EDF-VD on uniprocessor.*

Proof According to Lemma 2 in [15], if $max(b + \alpha c, \lambda b + c) \leq S(\alpha, \lambda)$, then $\frac{\alpha c}{1-b} \leq \frac{1-(c+\lambda b)}{b-\lambda b}$, where $U_H^{hi} = c$, $U_H^{lo} = \alpha c$, $U_L^{lo} = b$, $U_L^{hi} = \lambda b$ and $S(\alpha, \lambda) = \frac{(1-\alpha\lambda)((2-\alpha\lambda-\alpha)+(\lambda-1)\sqrt{4\alpha-3\alpha^2})}{2(1-\alpha)(\alpha\lambda-\alpha\lambda^2-\alpha+1)}$. Based on Theorem 4 in [15], $S(\alpha, \lambda) \geq \frac{3}{4}$. As such, if $max(U_L^{lo} + U_H^{lo}, U_L^{hi} + U_H^{hi}) \leq \frac{3}{4}$, then $\frac{U_H^{lo}}{1-U_L^{lo}} \leq \frac{1-(U_H^{hi}+U_L^{hi})}{U_L^{lo}-U_L^{hi}}$, which is the sufficient schedulability condition for EDF-VD according to Theorem 4. □

The given tasks are first partitioned onto m unit-speed processors in the same order as that described in Sect. 10.2.2.2. When a task is assigned to a processor, the schedulability check is based on Lemma 3 instead of the conventional approach [8]. This change is to accommodate the IMC/VPMC model. This partitioning method is called *VPMC partitioning*. After the partitioning, the tasks on each processor are scheduled in the same way as EDF-VD under the IMC model [15]. Under the same schedulability constraints, VPMC further allows some low-criticality task to execute with full precision in high-criticality mode.

We introduce two techniques to enhance the above VPMC partitioning. The first improvement is to change the schedulability check in the partitioning from Lemma 3 to Theorem 4. From the proof of Lemma 3, we can tell the schedulability condition in Lemma 3 is sufficient for the schedulability condition in Theorem 4. On the other hand, the Lemma 3 condition is not necessary for the Theorem 4 condition. Therefore, the condition of Theorem 4 is less conservative according to [19]. The second enhancement is to balance the utilizations of each processor between the two different criticality modes. More specifically, we attempt to make the difference between $U_L^{lo} + U_H^{lo}$ and $U_L^{hi} + U_H^{hi}$ on each processor as small as possible. The intuition is that a small difference or balanced utilization can avoid one criticality mode being a bottleneck of the whole system. Each time a task τ_i is to be assigned to a processor, all the processors are sorted in non-decreasing order of $U_L^{hi} + U_H^{hi} - U_L^{lo} - U_H^{lo}$ and indexed from 1 to m. If $\chi_i = H$, the attempts of assigning τ_i to a processor are in the order from 1 to m. Otherwise, the attempts follow the order from m to 1.

10.2.4.2 Global Scheduling for IMC/VPMC Systems

We extend the popular global scheduling technique, fpEDF-VD (Sect. 10.2.2.3) [10], for IMC/VPMC systems, where low-criticality tasks continue to be executed by approximate computing, or even precise computing, in high-criticality mode. As such, R_i^{hi} for each low-criticality task $\tau_i \in \mathcal{T}_L$ is no longer 0 and $R_i^{lo} \geqslant R_i^{hi} > 0, \forall \tau_i \in \mathcal{T}_L$ is satisfied.

The low-criticality mode of an IMC/VPMC system is handled in the same way as fpEDF-VD. More specifically, a task set $(\bigcup_{\chi_i=L}(T_i, R_i^{lo})) \bigcup (\bigcup_{\chi_i=H}(x \cdot T_i, R_i^{lo}))$ is schedulable on m processors by fpEDF according to Lemma 1 in Sect. 10.2.2.3. Please note by scaling T_i by $x \in (0, 1)$, virtual deadline $x \cdot T_i$ is applied for all high-criticality tasks.

The transition from low-criticality to high-criticality mode is subtle and deserves a lot of attention [15]. We introduce three techniques for handling such transitions as well as high-criticality mode: (A) direct application of fpEDF-VD without any modifications; (B) fpEDF-DVD (fpEDF with Dual Virtual Deadlines); (C) service preserving.

(A) Direct Application of fpEDF-VD This technique may result in one-time loss of some low-criticality jobs during the transition. Except this loss, everything else is the same as fpEDF-VD in classic MC systems. Let t^* be the moment when the system enters high-criticality mode. We define d^{hi} as the earliest deadline (virtual deadline for high-criticality tasks) among all jobs that are active right after t^*. We further define a^{hi} to be the earliest release time among jobs released after t^*. We call $\Gamma = \min(d^{hi}, a^{hi})$ the *critical moment*. After Γ, the schedulability check of high-criticality mode is the same as that in classic MC systems. However, the transition time interval $[t^*, \Gamma]$ needs particular attention for IMC/VPMC systems. During $[t^*, \Gamma]$, there can exist carry-over jobs, which are jobs that are released before t^* and have not been completed at t^*. By the fpEDF-VD algorithm design, high-criticality carry-over jobs can be guaranteed to complete before their deadlines if the schedulability check is passed. If a low-criticality carry-over job J_i^j has already executed at least $\tilde{R}_i$ amount of time (time of approximate computing) at t^*, we take its approximate computing result [14] and quit this job. If J_i^j has executed less than $\tilde{R}_i$, we continue it till Γ and then quit. By disallowing low-criticality carry-over jobs after Γ, the schedulability of all high-criticality jobs can be maintained. In the worst case, a low-criticality task may lose its job once during $[t^*, \Gamma]$.

(B) fpEDF-DVD (Dual Virtual Deadline) Like the virtual deadlines for high-criticality tasks in fpEDF-VD, we apply virtual deadlines for low-criticality tasks in order to avoid their one-time loss during the transition. More specifically, each low-criticality task τ_i has deadlines $y \cdot T_i$ and $(1 - y) \cdot T_i$ for low-criticality mode and high-criticality mode, respectively, where y is a scaling factor between 0 and 1. The value of y is found by sweeping between 0 and 1 and selecting the one that satisfies schedulability conditions, i.e., both task systems $(\bigcup_{\chi_i=L}(y * T_i, R_i^{lo})) \bigcup (\bigcup_{\chi_i=H}(x * T_i, R_i^{lo}))$ and

$(\bigcup_{\chi_i=L}((1-y)*T_i, R_i^{hi}))\bigcup(\bigcup_{\chi_i=H}((1-x)*T_i, R_i^{hi}))$ are each (separately) schedulable on m processors by fpEDF.

Theorem 5 *If the virtual deadline based utilization of all tasks satisfy schedulability conditions in both low-criticality and high-criticality mode, the fpEDF-DVD scheduling guarantees all job completions before their deadlines and no job is abandoned.*

Proof If the schedulability conditions are satisfied, all tasks are evidently schedulable by fpEDF in low-criticality mode and high-criticality mode. Special attention needs to be paid to carry-over jobs, which are released before the moment t^* entering high-criticality mode and have not been completed at t^*. Then, the low-criticality mode virtual deadline for each carry-over job must be after t^*. The virtual deadlines partition a task period into low-criticality mode portion, which are $x \cdot T_i$ and $y \cdot T_i$, and high-criticality mode portion, which are $(1-x) \cdot T_i$ and $(1-y) \cdot T_i$, respectively. For the carry-over jobs, one can treat their low-criticality mode virtual deadlines as their high-criticality mode release times, which are after t^*. As the schedulability conditions are satisfied, even if the carry-over jobs start execution at their low-criticality virtual deadlines, they are all schedulable for completion by their actual deadlines. □

Virtual deadline is effective for providing guarantee on meeting deadlines. However, it is basically a conservative resource reservation approach that makes schedulability condition more strict and hence causes under-utilization of resources. Applying virtual deadlines for both low-criticality tasks and high-criticality tasks would exacerbate the inefficiency and is an expensive price paid for avoiding one-time loss of low-criticality jobs.

(C) Service Preserving Method This method can avoid the one-time loss of low-criticality jobs and is less conservative than the fpEDF-DVD technique. In this method, the treatment of high-criticality tasks is the same as fpEDF-VD [10]. Consider a high-criticality job J_j^k that is active at moment t^* of mode switching. Its virtual deadline satisfies $\hat{d}_j^k = a_j^k + x \cdot T_j \geq t^*$, otherwise this job would have finished. Right after time t^*, the system enters high-criticality mode and the actual deadline $d_j^k = a_j^k + T_j$ is enforced. According to Lemma 2, the extra time budget $(1-x) \cdot T_j$ is sufficient for J_j^k to finish with execution time R_j^{hi}. Therefore, high-criticality tasks are guaranteed to satisfy their deadlines.

The challenging part is how to handle low-criticality tasks during the transition interval at the beginning of high-criticality mode. The non-zero R_i^{hi} for low-criticality tasks makes the schedulability guarantee quite difficult. We suggest a *service preserving interval* $[t^*, t^* + P]$, when only the active (carry-over) low-criticality jobs are executed by DP-Fair scheduling [11], all active high-criticality jobs are suspended and no newly arrival jobs are started. This is to facilitate that all active low-criticality jobs can be finished with approximate computing during the transition. Meanwhile, the interval P is designed in a way that the schedulability of all the other jobs are still maintained. A critical basis for the service preserving interval is that execution time R_j^{hi} is accommodated after virtual deadline $\hat{d}_j^k$ for a high-criticality job J_j^k in high-criticality mode according to Lemma 2 by fpEDF-VD [10]. The service preserving interval length is defined as

$$P = \min_{\forall \tau_j \in \mathcal{T}_H} R_j^{lo}. \tag{10.4}$$

Next, we will discuss schedulability of active jobs and those involving the service preserving interval.

Lemma 4 *By following fpEDF-VD, all high-criticality jobs can guarantee to meet their deadlines in high-criticality mode even if they are not executed in* $[t^*, t^* + P]$.

Proof The high-criticality jobs involving the service preserving interval $[t^*, t^* + P]$ can be categorized into three cases, which will be discussed as follows.

Case 1 Overrun jobs. These are the high-criticality jobs that have executed R^{lo} time but have not finished (see Fig. 10.1). At the end of the R^{lo} time, the system enters high-criticality mode when the moment is t^*. According to the schedulability conditions of fpEDF-VD, the virtual deadline $\hat{d}_j^o$ of an overrun job J_j^o satisfies $\hat{d}_j^o \geq t^*$. The method of fpEDF-VD (Lemma 2) also indicates that all high-criticality jobs can execute R^{hi} after their virtual deadlines and finish before their actual deadlines in high-criticality mode. Since time R_j^{lo} has already been executed for job J_j^o at t^*, deferring the rest of its execution by R_j^{lo} maintains the schedulability. In other words, the rest of the overrun job can start from $\hat{d}_j^o + R_j^{lo} = \hat{d}_j^o + R_j^{lo} + t^* - t^* = t^* + P_j$, where $P_j = \hat{d}_j^o + R_j^{lo} - t^*$. Since $\hat{d}_j^o \geq t^*$, $P_j \geq R_j^{lo}$. Therefore, postponing the execution of the rest of J_j^o by R_j^{lo} will maintain the schedulability of overrun jobs.

Case 2 Active high-criticality jobs without overrun (see Fig. 10.2). A high-criticality job J_j^k has been executed $q_j^k < R_j^{lo}$ by t^*. Then, its rest portion can start from $\hat{d}_j^k + q_j^k$ with guarantee of meeting its deadline according to fpEDF-VD.

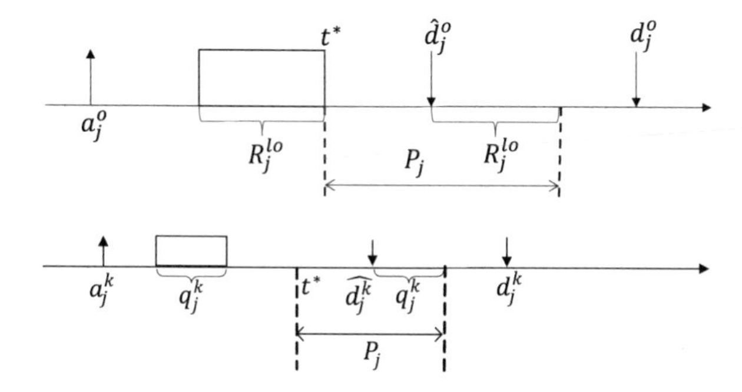

Fig. 10.1 Case 1: service preserving interval for an overrun job

Fig. 10.2 Case 2: service preserving interval for an active high-criticality job without overrun

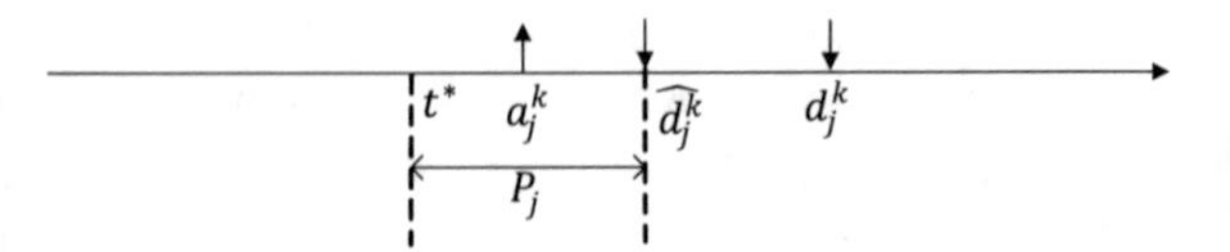

Fig. 10.3 Case 3: service preserving interval for an immediate newly coming high-criticality job

Like Case 1, $\hat{d}_j^k + q_j^k = t^* + P_j$, where $P_j = \hat{d}_j^k + q_j^k - t^*$. By schedulability condition in low-criticality mode, $q_j^k + \hat{d}_j^k - t^* \geq R_j^{lo}$, then $P_j \geq R_j^{lo}$. Hence, such a job can be suspended in $[t^*, t^* + R_j^{lo}]$ without affecting its schedulability.

Case 3 High-criticality jobs arriving during the service preserving interval (see Fig. 10.3). The arrival time a_j^k of such a job J_j^k satisfies

$$t^* \leq a_j^k \leq t^* + P. \tag{10.5}$$

The schedulability conditions in fpEDF-VD require that

$$a_j^k + R_j^{lo} \leq \hat{d}_j^k. \tag{10.6}$$

Combing inequality (10.5) and (10.6), we have

$$R_j^{lo} \leq \hat{d}_j^k - a_j^k \leq \hat{d}_j^k - t^* = P_j.$$

Since J_j^k can guarantee to finish before its deadline even if it starts from $\hat{d}_j^k$ according to fpEDF-VD, its start time can be deferred by P_j, which is lower bounded by R_j^{lo}.

Overall, all high-criticality jobs involving the service preserving interval can be deferred by R^{lo} without affecting their schedulability. Hence, deferring by $P = \min_{\forall \tau_j \in \mathcal{T}_H} R_j^{lo}$ for all these jobs can still guarantee to meet their deadlines. □

Next, we describe schedulability conditions for active low-criticality jobs during the service preserving interval $[t^*, t^* + P]$. At t^*, if a low-criticality job J_i^k has already been executed for at least R_i^{hi}, it is terminated with approximate computing result. An active (carry-over) low-criticality job J_i^k means that it has been executed for $q_i^k < R_i^{hi}$ by t^*. The active low-criticality jobs are scheduled by the fluid-based DP-Fair method (see Sect. 10.2.2.4) in the service preserving interval, while fpEDF-VD is employed for all the other time. Although fluid scheduling tends to entail frequent job preemptions, it is utilized only within the limited service preserving interval. The schedulability for DP-fair method is largely decided by the job density.

Lemma 5 *The density δ_i^k of an active low-criticality job J_i^k in $[t^*, t^* + P]$ is no greater than* $\max\left(\frac{R_i^{hi}}{P}, \frac{R_i^{hi}}{R_i^{lo}}\right)$.

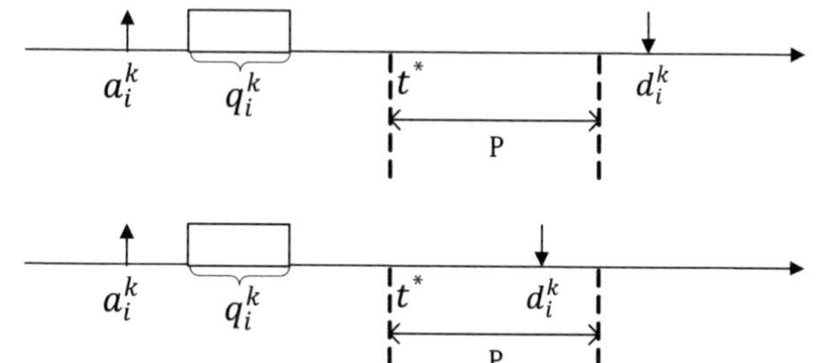

Fig. 10.4 Active low-criticality job with deadline after $t^* + P$

Fig. 10.5 Active low-criticality job with deadline before $t^* + P$

Proof This bound is derived from two cases. In one case, deadline $d_i^k \geq t^* + P$ as shown in Fig. 10.4. In the worst-case for the service preserving interval, entire job R_i^{hi} is executed by $t^* + P$, the density of this case is upper bounded as

$$\delta_i^k|_{d_i^k \geq t^*+P} \leq \frac{R_i^{hi}}{P}. \tag{10.7}$$

In the other case, $d_i^k < t^* + P$ as shown in Fig. 10.5. If q_i^k has been executed by t^*, the density is estimated by

$$\delta_i^k|_{d_i^k < t^*+P} = \frac{R_i^{hi} - q_i^k}{d_i^k - t^*}. \tag{10.8}$$

By the schedulability condition in low-criticality mode, $R_i^{lo} - q_i^k \leq d_i^k - t^*$. Therefore,

$$\delta_i^k|_{d_i^k < t^*+P} = \frac{R_i^{hi} - q_i^k}{d_i^k - t^*} \leq \frac{R_i^{hi} - q_i^k}{R_i^{lo} - q_i^k}. \tag{10.9}$$

Consider a function

$$f(x) = \frac{R_i^{hi} - x}{R_i^{lo} - x}, \quad 0 \leq x < R_i^{hi} < R_i^{lo}. \tag{10.10}$$

Since derivative $f'(x) = \frac{R_i^{hi} - R_i^{lo}}{(R_i^{lo} - x)^2} < 0$, $f(x)$ is a monotone decreasing function and its maximum is at $x = 0$. Hence,

$$\delta_i^k|_{d_i^k < t^*+P} \leq \frac{R_i^{hi} - q_i^k}{R_i^{lo} - q_i^k} \leq \frac{R_i^{hi}}{R_i^{lo}}. \tag{10.11}$$

By combining the two cases, we have

$$\delta_i^k \leq \max\left(\frac{R_i^{hi}}{P}, \frac{R_i^{hi}}{R_i^{lo}}\right). \tag{10.12}$$

□

In the worst case, every low-criticality task has an active job at t^*. According to Theorem 2 and Lemma 5, a sufficient condition for DP-Fair method to successfully schedule all the active jobs on m processors in $[t^*, t^* + P]$ is

$$\sum_{\forall \tau_i \in \mathcal{T}_L} \max\left(\frac{R_i^{hi}}{P}, \frac{R_i^{hi}}{R_i^{lo}}\right) \leq m. \tag{10.13}$$

Last, we discuss new low-criticality jobs that arrive in $[t^*, t^* + P]$. Our method does not allow such jobs to be executed until $t^* + P$. In other words, their execution is deferred by at most P. We specify that task set

$$\left(\bigcup_{\tau_i \in \mathcal{T}_L} (T_i - P, R_i^{hi})\right) \bigcup \left(\bigcup_{\tau_j \in \mathcal{T}_H} ((1 - x) \cdot T_j, R_j^{hi})\right)$$

must be schedulable according to Lemma 1 in high-criticality mode. More specifically, a low-criticality task τ_i is scheduled with period (implicit deadline) $T_i - P$. Thus, with deferral of P, a low-criticality job arriving in $[t^*, t^* + P]$ is still schedulable.

Putting everything together, the service preserving policy is stated as follows.

Service Preserving Policy *From the moment t^* switching to high-criticality mode to $t^* + P$, where $P = \min_{\forall \tau_j \in \mathcal{T}_H} R_j^{lo}$, only active low-criticality jobs are executed with DP-Fair scheduling and all the other jobs cannot be executed.*

From Lemmas 4 and 5, we can reach the following conclusion.

Theorem 6 *When applying the service preserving policy with fpEDF-VD scheduling, a task set $\mathcal{T}$ is schedulable on m identical processors if $\mathcal{T}$ satisfies the following schedulability conditions.*

- *task set*

$$\left(\bigcup_{\tau_i \in \mathcal{T}_L} (T_i, R_i^{lo})\right) \bigcup \left(\bigcup_{\tau_j \in \mathcal{T}_H} (x \cdot T_j, R_j^{lo})]\right)$$

must be schedulable on m processors according to Lemma 1.

- $\sum_{\forall \tau_i \in \mathcal{T}_L} \max(\frac{R_i^{hi}}{P}, \frac{R_i^{hi}}{R_i^{lo}}) \leq m$ *during* $[t^*, t^* + P]$, *where* $P = \min_{\forall \tau_j \in \mathcal{T}_H} R_j^{lo}$.

- *task set*

$$\left(\bigcup_{\tau_i \in \mathcal{T}_L} (T_i - P, R_i^{hi})\right) \bigcup \left(\bigcup_{\tau_j \in \mathcal{T}_H} ((1-x)\cdot T_j, R_j^{hi})\right)$$

must be schedulable on m processors according to Lemma 1.

10.2.4.3 Extension of MC-DP-Fair Scheduling for IMC/VPMC Systems

MC-DP-Fair is one realization of the fluid-based scheduling [12], which is not directly implementable by itself. Fluid-based scheduling improving QoS for low-critical tasks has been studied for VPMC in [18] and the method is called MCFQ, however, MC-DP-Fair scheduling for VPMC is barely discussed in [18]. Here, we show how to extend MC-DP-Fair scheduling to VPMC-DP-Fair scheduling. In DP-Fair scheduling, an important concept is task density δ_i for task τ_i, which is usually equal to $\frac{R_i}{T_i}$ with a few exceptions. Fluid-based scheduling uses another concept, execution rate θ_i for τ_i, which is the fraction of a unit-speed processor allocated for executing τ_i.

For a low-critical task τ_i in VPMC-DP-Fair, $\delta_i^{lo} = \theta_i^{lo} = U_i^{lo}$ and $\delta_i^{hi} = \theta_i^{hi} = U_i^{hi}$. Its virtual deadline $\hat{T}_i = T_i$. Please note $\delta_i^{hi} = 0$ in MC-DP-Fair. Let w_i be the length of time interval from job release time of τ_i to Γ, which is the earliest deadline or new job release time after the system enters high-criticality mode.

Lemma 6 *In VPMC-DP-Fair scheduling, a low-criticality carry-over job of τ_i can be executed for at least $\tilde{R}_i$ time, where $\tilde{R}_i$ is the execution time of approximate computing.*

Proof Let R_i^{TR} denote the actual execution time of a carry-over job of τ_i.

$$R_i^{TR} = w_i \cdot \delta_i^{lo} + (T_i - w_i)\delta_i^{hi} = w_i \cdot U_i^{lo} + (T_i - w_i)U_i^{hi}$$

$$\geq w_i \cdot U_i^{hi} + (T_i - w_i)U_i^{hi} = T_i \cdot U_i^{hi} = \tilde{R}_i.$$

□

For a high-criticality task τ_i, $\delta_i^{lo} = \theta_i^{lo}$, which is proved to be no greater than U_i^{hi} [18], virtual deadline $\hat{T}_i = R_i^{lo}/\theta_i^{lo}$. Its density in high-criticality mode is specified by Lee et al. [12]

$$\delta_i^{hi} = \frac{R_i^{hi} - \delta_i^{lo} \cdot w_i}{T_i - w_i}. \tag{10.14}$$

Lemma 7 *Given a task set that is deemed to be schedulable by MCFQ [18], if it is scheduled by VPMC-DP-Fair, then $\delta_i^{lo} \leq \theta_i^{lo}$ and $\delta_i^{hi} \leq \theta_i^{hi}$ for each task τ_i.*

Proof For each task τ_i, we have $\delta_i^{lo} = \theta_i^{lo}$. For each low-criticality task τ_i, we have $\delta_i^{hi} = \theta_i^{hi}$. For each high-criticality task τ_i, since δ_i^{hi} is a variable depending on w_i according to Eq. (10.14), we need to show that the maximum value of δ_i^{hi} is no greater than θ_i^{hi}. Consider the derivative of δ_i^{hi} with respect to w_i

$$\frac{d\delta_i^{hi}}{dw_i} = \frac{R_i^{hi} - \delta_i^{lo} \cdot T_i}{(T_i - w_i)^2} = \frac{U_i^{hi} - \delta_i^{lo}}{(T_i - w_i)^2 / T_i}. \tag{10.15}$$

Since $\delta_i^{lo} = \theta_i^{lo} \leq U_i^{hi}$ [18], the derivative is non-negative and function (10.14) is monotonically increasing. By definition, we know $w_i \leq \hat{T}_i$. Thus, δ_i^{hi} has the maximum value when $w_i = \hat{T}_i$,

$$\delta_{i,max}^{hi} = \frac{U_i^{hi} - U_i^{lo}}{1 - U_i^{lo}/\theta_i^{lo}}. \tag{10.16}$$

In MCFQ [18], $\theta_i^{hi} = \frac{U_i^{hi} - U_i^{lo}}{1 - U_i^{lo}/\theta_i^{lo}}$, which is equal to $\delta_{i,max}^{hi}$, then we have $\delta_i^{hi} \leq \theta_i^{hi}$ for high-criticality tasks. □

Lemma 8 *Given a task set that is deemed to be schedulable by MCFQ, it is schedulable by VPMC-DP-Fair.*

Proof Given a task set that is deemed to be schedulable by MCFQ, we have $\sum_{\tau_i \in \mathcal{T}} \theta_i^{lo} \leq m$ and $\sum_{\tau_i \in \mathcal{T}} \theta_i^{hi} \leq m$, then we have $\sum_{\tau_i \in \mathcal{T}} \delta_i^{lo} \leq \sum_{\tau_i \in \mathcal{T}} \theta_i^{lo} \leq m$ and $\sum_{\tau_i \in \mathcal{T}} \delta_i^{hi} \leq \sum_{\tau_i \in \mathcal{T}} \theta_i^{hi} \leq m$ from Lemma 7. Hence, low-criticality mode schedulability and high-criticality mode schedulability by Theorem 2 are satisfied and the task set is schedulable by VPMC-DP-Fair. □

10.2.5 Precision Optimization for Variable-Precision Mixed-Criticality Systems

10.2.5.1 Optimization Kernel

Under the VPMC model, there can be utilization slack for some processors when schedulability conditions are satisfied. The slack allows some low-criticality tasks to be executed with precise computing in high-criticality mode while the schedulability conditions are still satisfied. For a low-criticality task τ_i, the error of its approximate computing is denoted by e_i. The error of a low-criticality task τ_i execution in high-criticality mode is denoted by e_i^{hi}, which is equal to e_i if it is executed with approximate computing and otherwise 0. If each task τ_i has a weighting factor η_i indicating its importance, the precision optimization problem is stated as follows.

Problem 1 Given a set of independent sporadic tasks $\mathcal{T} = \{\tau_1, \tau_2, \ldots\}$ in VPMC model and a scheduling method $\mathcal{S}$, decide if each low-criticality task τ_i is executed with precise or approximate computing in high-criticality mode such that the total weighted error $\sum_{\chi_i = L} \eta_i \cdot e_i^{hi}$ is minimized while the schedulability conditions for $\mathcal{S}$ are maintained.

For each low-criticality task τ_i, let ΔU_i denote the additional processor utilization when its execution is changed from approximate to precise computing and thus

$$\Delta U_i = \frac{\hat{R}_i - \tilde{R}_i}{T_i}. \tag{10.17}$$

Let $\bar{U}_L^{hi}$ denote the maximal possible U_L^{hi} under the schedulability constraint for a scheduling method. The *utilization slack* Ψ for low-critical tasks in high-criticality mode is defined as

$$\Psi = \bar{U}_L^{hi} - U_L^{hi}. \tag{10.18}$$

Then, Problem 1 is essentially 0-1 knapsack problem. Let z_i be a binary decision variable for each low-criticality task τ_i. When $z_i = 1$, task τ_i is assigned to precise computing; otherwise it is executed with approximate computing in high-criticality mode. The knapsack problem formulation is as follows.

$$\begin{aligned} &\text{maximize} \sum_{\chi_i = lo} \eta_i \cdot e_i \cdot z_i \\ &\text{subject to} \sum_{\chi_i = lo} \Delta U_i \cdot z_i \le \Psi \\ &\qquad z_i \in \{0, 1\}, \quad \forall \tau_i \in \mathcal{T}_L. \end{aligned} \tag{10.19}$$

In this formulation, the objective is to maximize the error reduction obtained from using precise computing compared to IMC model. The 0-1 knapsack problem is a well-known NP-complete problem. It can be optimally solved by dynamic programming with pseudo-polynomial time complexity.

10.2.5.2 Utilization Slack Estimation and Customization for Different Scheduling Methods

For partitioned scheduling, if $U_L^{lo} + U_H^{hi} \le 1$, all tasks can be scheduled with EDF and all low-criticality tasks can be executed with precise computing. Hence, the slack estimation and precision optimization is necessary only when $U_L^{lo} + U_H^{hi} > 1$. For the partitioned scheduling methods introduced in Sect. 10.2.4.1, utilization slack is estimated for individual processors. On each processor, the maximal schedulable utilization $\bar{U}_L^{hi}$ can be derived according to Theorem 1 and Theorem 3.

Theorem 7 *The utilization slack of a processor after the VPMC partitioning is* $\frac{1-U_L^{lo}-U_H^{lo}U_L^{lo}-U_H^{hi}+U_L^{lo}U_H^{hi}}{1-U_L^{lo}-U_H^{lo}} - U_L^{hi}$.

Proof From inequality (10.1), we can find the range of the scaling factor as

$$x \geq \frac{U_H^{lo}}{1 - U_L^{lo}}. \tag{10.20}$$

Further, we know from inequality (10.2) that

$$U_L^{hi} \leq \frac{1 - xU_L^{lo} - U_H^{hi}}{1 - x}. \tag{10.21}$$

Taking derivative with respective to x on right-hand-side of inequality (10.21), we have

$$\frac{1 - U_L^{lo} - U_H^{hi}}{(1 - x)^2}. \tag{10.22}$$

Since $U_L^{lo} + U_H^{hi} > 1$, the right-hand-side of inequality (10.21) is a decreasing function with respect to x. Then, $\bar{U}_L^{hi}$ can be obtained by plugging RHS of inequality (10.20) into inequality (10.21):

$$\bar{U}_L^{hi} = \frac{1 - U_L^{lo} - U_H^{lo}U_L^{lo} - U_H^{hi} + U_L^{lo}U_H^{hi}}{1 - U_L^{lo} - U_H^{lo}}. \tag{10.23}$$

Therefore, the utilization slack is given by:

$$\Psi = \frac{1 - U_L^{lo} - U_H^{lo}U_L^{lo} - U_H^{hi} + U_L^{lo}U_H^{hi}}{1 - U_L^{lo} - U_H^{lo}} - U_L^{hi}. \tag{10.24}$$

□

Under fpEDF, a subset $\mathcal{T}_{hp} \subset \mathcal{T}$ of tasks are designated with the highest priority and $m_{hp} = |\mathcal{T}_{hp}|$ processors are allocated for them. Please note this allocation is not static, i.e., the m_{hp} processors at one time may be different from the m_{hp} processors at another time. The other tasks $\mathcal{T}_{lp} = \mathcal{T} - \mathcal{T}_{hp}$ follow EDF priority and are executed on $m_{lp} = m - m_{hp}$ processors. Each low-criticality task $\tau_i \in \mathcal{T}_{hp}$ can always execute with precise computing in high-criticality mode, since an entire processor is allocated to one task in $\mathcal{T}_{hp}$ and this allocation is sufficient for precise computing.

For the fpEDF-VD-VPMC method, the utilization slack of $\mathcal{T}_{lp}$ is estimated by the following statement according to Lemma 1.

Proposition 1 *The utilization slack for* $\mathcal{T}_{lp}$ *on the* m_{lp} *processors under fpEDF-VD-VPMC scheduling is* $m_{lp} - (m_{lp} - 1) \cdot U_{lp}^{max} - U_{lp}^{total}$, *where* U_{lp}^{max} *and* U_{lp}^{total} *are the maximal task utilization and total utilization for* $\mathcal{T}_{lp}$*, respectively.*

This estimation can be applied with fpEDF-DVD-VPMC method. However, the partition between $\mathcal{T}_{hp}$ and $\mathcal{T}_{lp}$ in fpEDF-DVD-VPMC is different from that in the direct application of fpEDF-VD in VPMC due to the virtual deadlines applied to low-criticality tasks.

The utilization slack for VPMC-DP-Fair Scheduling is estimated by $\Psi = m - \sum_{\tau_i \in \mathcal{T}} \theta_i^{hi}$, where θ_i^{hi} is the execution rate of task τ_i in high-criticality mode, which is computed according to [18].

10.2.6 Experiments and Result

In our experiments, we evaluate the schedulability and computing errors of the following methods through software simulations.

- **Partition-MC**: Partitioned scheduling with the conventional MC model [8]. Since this method does not incorporate any approximations, its results are used to provide a reference level for schedulability, but cannot be used for comparing computing errors.
- **Partition-VPMC**: The partitioned scheduling method with precision optimization.
- **Partition-VPMC-E**: Enhanced partitioned scheduling with precision optimization.
- **fpEDF-VD-MC**: fpEDF-VD scheduling with the conventional MC model [8]. Since this method drops all low-criticality tasks in high-criticality mode, it is not included for error analysis.
- **fpEDF-VD-VPMC**: fpEDF-VD scheduling with precision optimization.
- **fpEDF-DVD-VPMC**: fpEDF dual virtual deadline method, with precision optimization.
- **Service Preserving**: Our service preserving method based on fpEDF-VD scheduling. In this method, low-criticality tasks continue to execute with approximate computing in high-criticality mode.
- **Fluid-VPMC**: The MCFQ method [18] with precision optimization replaced by the dynamic programming-based knapsack solution described in Sect. 10.2.5.

The testcases are randomly generated as follows. For each task set, the probability of a task being low-criticality (high-criticality) is 0.5. For a low-criticality (high-criticality) task τ_i, its utilization in low-criticality (high-criticality) mode U_i^{lo} (U_i^{hi}) is randomly chosen within the interval [0.05, 0.9] under uniform distribution. The period T_i of each task is randomly chosen from a uniform distribution in [50, 500]. For a low-criticality task τ_i, we set $R_i^{lo} = \hat{R}_i = T_i \cdot U_i^{lo}$, $\tilde{R}_i = k_L \cdot \hat{R}_i$. The scaling factor k_L is randomly chosen from a uniform distribution in $[K_L, 0.9]$, where K_L is a parameter. For a high-criticality task τ_i, we set $R_i^{hi} = T_i \cdot U_i^{hi}$, $R_i^{hi} = k_H \cdot R_i^{lo}$ and $1.1 \leq k_H \leq K_H$, where K_H is a parameter. For each low-criticality task τ_i,

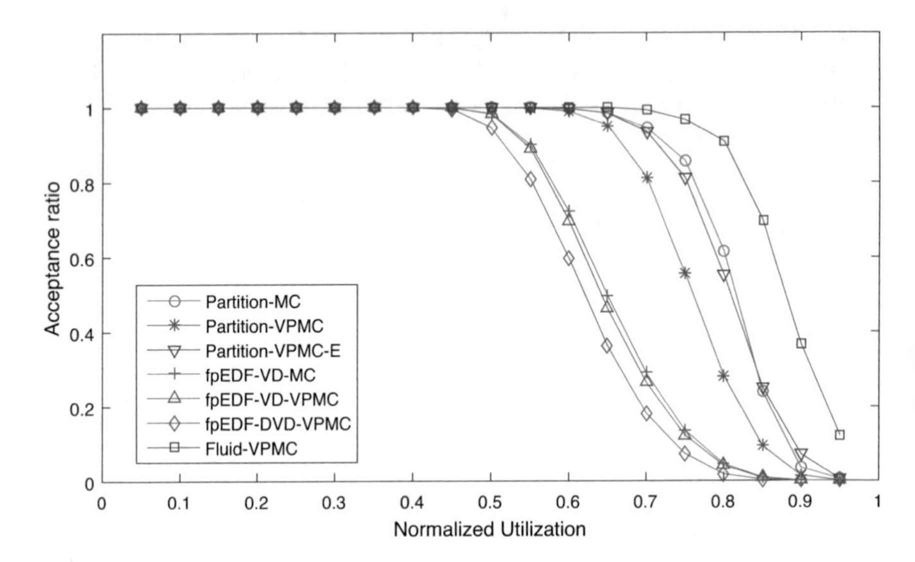

Fig. 10.6 Acceptance ratio vs. normalized utilization of 4 processors ($K_L = 0.1$, $K_H = 5$)

its approximate computing error is randomly chosen from a uniform distribution between 1 and 10. We set error weighting factors (defined in Sect. 10.2.5) $\eta_i = 1$.

Evaluation of the Acceptance Ratio Acceptance ratio is the ratio of schedulable tasks among all given tasks. We first evaluate the acceptance ratio at several values of the utilization, U_i. For each U_i, we generate 10,000 testcases, and for each testcase, we iteratively add new tasks till $\max(U_L^{lo} + U_H^{lo},\ U_L^{hi} + U_H^{hi})$ reaches U_i. The acceptance ratios on 4 processors are depicted in Fig. 10.6.

We see from the plot that Fluid-VPMC provides the best acceptance ratio (this is not surprising as the fluid-based scheduling is optimal in theory), while the three variants of fpEDF-VD have the lowest acceptance ratio due to their very conservative schedulability conditions. The acceptance ratio of fpEDF-VD-VPMC is very close to that of fpEDF-VD-MC. This implies that continuing low-criticality tasks at high-criticality mode hardly degrades schedulability. The dual virtual deadline technique reduces acceptance ratio, but it guarantees that no low-criticality job is dropped while the fpEDF-VD-VPMC cannot provide such guarantee. The result also shows that the enhancement techniques can indeed improve schedulability of partitioned scheduling.

The simulation for Fig. 10.6 does not consider overhead, which is important in practice. Overhead includes the time on context switching, job migration among processors, execution monitoring, scheduling job executions, etc. For each of the VPMC methods, we estimate its overhead according to data from Linux prototyping. Then, the overhead is added into the task execution time for the simulation. The acceptance ratio result with consideration of overhead is shown in Fig. 10.7. One can see that Fluid-VPMC is no longer the best due to its large overhead, and the best

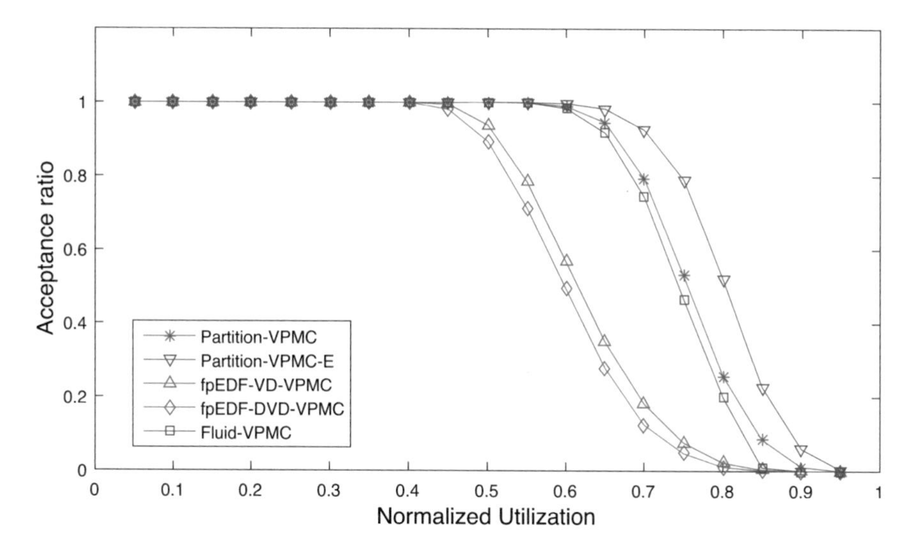

Fig. 10.7 Acceptance ratio versus normalized utilization of 4 processors with consideration of overhead

results are obtained from partitioned scheduling. The gap between Fluid-VPMC and fpEDF-VD-VPMC also becomes smaller.

We evaluate the acceptance ratio of our service preserving technique and compare with the dual-VD method. The testcase generation is similar to Fig. 10.6 with a few small changes: (1) each task utilization is obtained randomly from interval $[0.1, 0.9]$ under uniform distribution; (2) period T_i of each task τ_i is randomly chosen in $[100, 500]$ according to uniform distribution; (3) scaling factor k_L is randomly chosen in $[0.1, 0.9]$ following uniform distribution. The result for 8 processors is shown in Fig. 10.8. The difference between the two methods mainly exhibit around utilization 0.6, where service preserving can improve as much as 50%.

Evaluation of Errors Next, we evaluate computing errors of low-criticality tasks in high-criticality mode for different methods. Following the same testcase generation for evaluating the acceptance ratio, 1000 *schedulable* testcases are obtained at each utilization value. Figure 10.9 shows the mean error with standard deviation among tasks as function of the normalized utilization. For a single testcase, minimizing mean error is equivalent to minimizing the total error as the number of tasks is a constant for the precision optimization. When evaluating multiple testcases, mean error is more like a normalized result that can avoid the result being dominated by a few cases. In both of the figures, errors from IMC is plotted besides those from other methods. IMC is the model where all low-criticality tasks continue with approximate computing in high-criticality mode. Hence, its error is the same for different scheduling methods. One can see that the VPMC model can provide large error reductions. Again, Fluid-VPMC provides the lowest error levels as its optimality allows more utilization slack for error reduction.

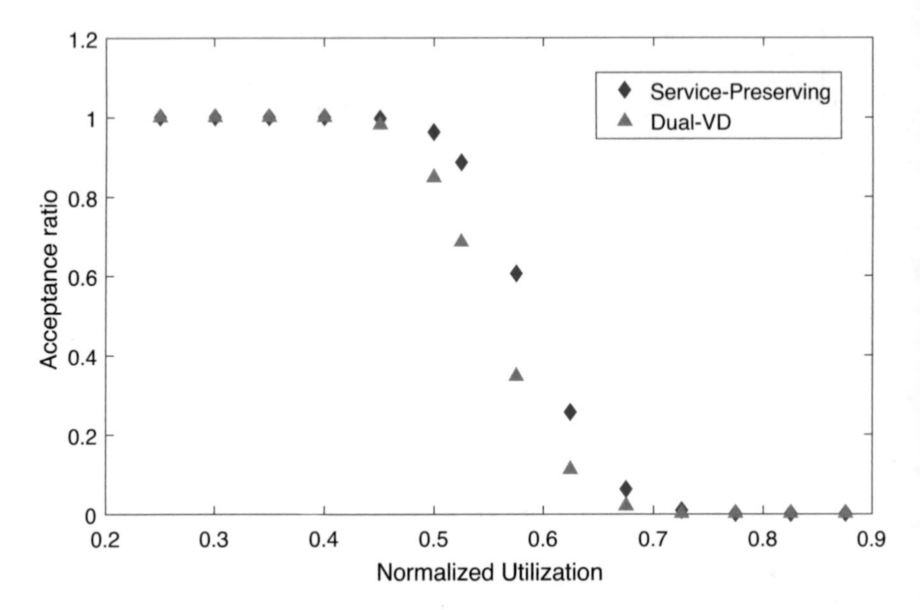

Fig. 10.8 Acceptance ratio vs normalized utilization of 8 processors

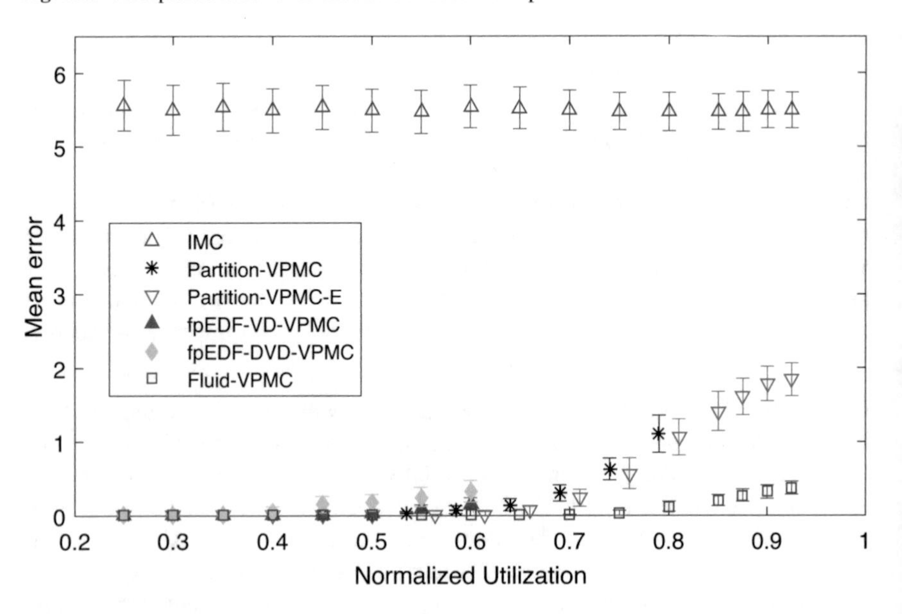

Fig. 10.9 Mean error (with standard deviation) vs. normalized utilization of 8 processors ($K_L = 0.1$, $K_H = 5$)

10.3 Scheduling Dependent Tasks

In this section, we consider the real-time scheduling of tasks with dependencies and communication. Such tasks are often modeled as directed acyclic graphs (DAGs), named task graphs, where each vertex corresponds to a task and edges represent the dependencies. Scheduling these graphs, in addition to traditional mapping and scheduling, requires budgeting of computation and communication times. Traditionally, such budgeting is done to account for worst-case execution times and communication delays. This can be difficult or impossible especially in case of distributed settings with communication over open and wireless networks that can have losses and unpredictable or potentially unbounded delays. In an approximate context, the goal is to tighten the budgets in exchange for accepting losses in communicated data as well as imprecision or dropping of computations [20].

10.3.1 Task Graph System Model

Formally, given a task graph $G(\mathcal{T}, C)$, with tasks $\mathcal{T}$ and communication dependencies C, scheduling such a graph can be expressed as a mapping of tasks $\tau_i \in \mathcal{T}$ to a set of processing hosts H and assignment of start times t_i to each τ_i such that dependencies $(\tau_i, \tau_j) \in C$ are not violated. The goal of real-time execution of task graphs is to set a constraint or deadline on the latency between the start times of the first source and last sink actor that execute in the graph.

Traditionally, when exact bounds for computation and communication times are known, start times are chosen according to those bounds. However, in approximate settings, an assignment of start times can be made that might occasionally lead to violations of associated computation and/or communication deadlines. This in turn requires subsequent and dependent tasks to execute without or with incomplete data, which results in errors and thus a degradation in quality of computed results at the outputs of the task graph. Existing task graph models do not inherently account for such scenarios.

In [21], we proposed Reactive and Adaptive Data Flow (RADF) as an extension of synchronous data flow (SDF) models to incorporate the notion of losses. RADF, in addition to traditional *lossless channels*, provides *lossy channels* that do not require communication to be reliable. Losses in these channels are represented by replacing lost token(s) with *empty token(s)*. This simple extension allows preserving analyzability and determinism of the underlying data flow model even in the presence of unreliable communication. Following SDF semantics, every actor has a firing rule that specifies firing conditions in terms of the number of tokens consumed from input channels and the number of tokens produced in output channels. Upon firing, an RADF actor can consume empty tokens as well as non-empty tokens but is required to produce non-empty tokens regardless.

Although RADF is based on an SDF model, task graphs are a special case of acyclic and homogeneous SDF graphs. As such, the concept of empty tokens can be similarly used to model communication losses as well computation losses in task graphs. In real-time scheduling of distributed task graphs, we therefore assume that deadline violations will be realized by empty computation and communication tokens and lossy executions of dependent tasks.

10.3.2 Approximations in Task Graph Scheduling

Figure 10.10a shows a simple task graph with a linear chain of tasks executed in a distributed fashion on three network nodes or hosts (h_0 through h_2). To provide a real-time latency guarantee, we assign a fixed start time to the sink task τ_C and allow it to potentially fire without data. Given that task graphs often execute in steady state periodically, the period of task τ_C between firings translates into an offset between its firing time and that of task τ_A. This offset can be statically calculated by subtracting τ_C's worst-case execution time from the overall latency constraint. Assigning fixed offsets and start times is equivalent to determining timeouts based on the deadline for the sink task. The more relaxed this relative constraint and timeout is, the more likely it is that input data will arrive at τ_C before it fires.

In addition to the sink task, instead of just waiting for data to arrive, intermediate tasks might have or require their own deadlines and timeouts, e.g., in cases where data order is important, such as tasks with multiple inputs or with state. Figure 10.10b shows an example of such a graph, where task τ_A fuses data from

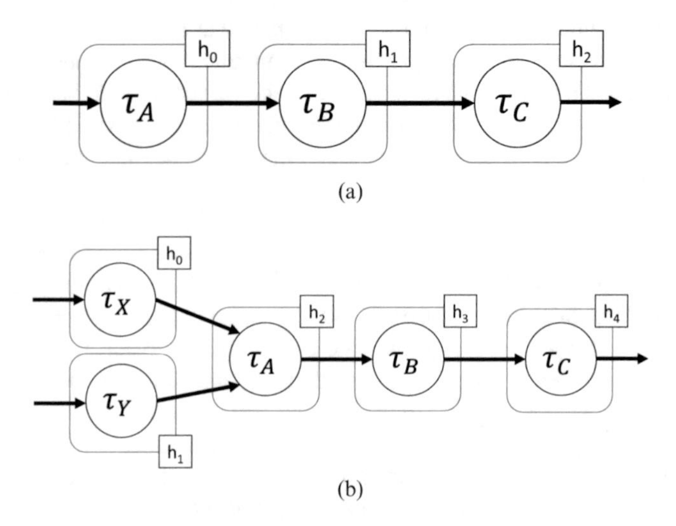

Fig. 10.10 Examples of distributed task graphs. (**a**) A graph with linear task chain. (b) A graph with multiple source tasks

τ_X and τ_Y. If τ_A receives data on one input, it would need to buffer and wait until matching data is received on the other input. The RADF model is based on allowing not only the sink but any task in a graph to fire without (or only with partial) data and in turn produce outputs with reduced quality, e.g., by interpolating results from previous data. If delays exceed overall latency constraints, at least one of the tasks in the chain needs to fire without data. Since it is better to allow partial computation with a single input at τ_A rather than letting τ_C fire without any data, τ_A can be assigned its own deadline and timeout. This introduces further tradeoffs: the smaller the offset of τ_A relative to τ_X/τ_Y is, the higher the data loss at τ_A and the lower the loss at τ_C, and vice versa. Further extending the deadline concept, even task τ_B can time out to react to network delay and loss earlier in the chain.

Figure 10.11 shows three different schedules for the graph of Fig. 10.10b under period and latency constraints of 1 and 13 ms, respectively. Figure 10.11a shows a pure data-driven schedule, where tasks execute only once dependencies are satisfied and data is available. Due to increased network delay in the second iteration, the overall constraint and deadline of task τ_C is violated and hence no data is produced in the second, third and, subsequently, all following iterations. Note that if tasks τ_A, τ_B and τ_C are allowed to fire out of order, violations in the third and subsequent iterations can be avoided. However, in general task graphs, e.g., in case of tasks with state, deterministic token and task firing orders need to be maintained, where large network delays or network losses can lead to long-lasting or permanent latency violations in a data-driven schedule.

In Fig. 10.11a, we can allow sink task τ_C to time out and thus avoid deadline violations. However, this would require τ_C to fire without data in the second and all subsequent iterations. By contrast, Fig. 10.11b shows a schedule where the latency budget is equally distributed between input channels of tasks τ_A, τ_B and τ_C, giving each a timeout of 3 ms. As seen in the figure, the constraint is met in all four iterations, but in the second and third iteration, τ_A fires without any data. Crucially, however, rather than violating the deadline in all subsequent iterations, this schedule avoids any losses in the fourth and following iterations. Noticing that increasing the timeout of task τ_A can eliminate its loss in the third iteration, Fig. 10.11c shows an alternative schedule where delay budgets are shifted by 1ms from τ_C to τ_A. This schedule prevents τ_A from firing without data in the third iteration and improves overall quality. This example shows that, depending on the network and application characteristics, there is a non-trivial schedule that optimizes the quality/latency tradeoff.

For a simple graph such as the one in Fig. 10.10a, the output quality can be defined as the number of iterations that execute without a loss. We can assign a probability of a task firing without a loss as the probability of the random network delay on its input channel being smaller than the timeout. The probability of an iteration executing without a loss then becomes the product of loss-free probabilities over all tasks and links in the chain. To maximize this product, all probabilities should be made equal. This requires timeouts to be selected in accordance with differences in network delays across channels. In general task graphs, however, not all losses are equal and the impact of losses on overall application quality depends

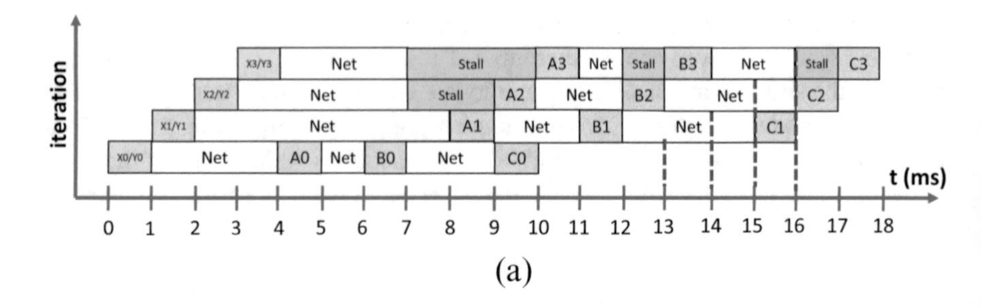

(a)

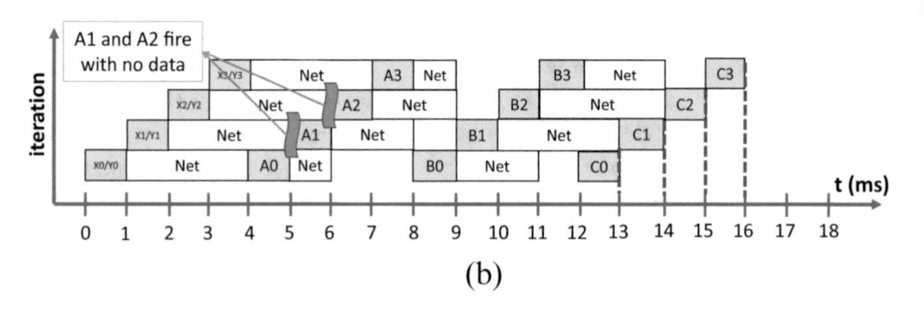

(b)

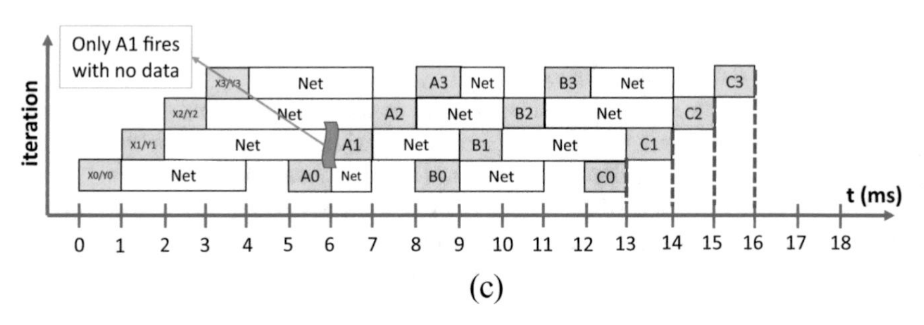

(c)

Fig. 10.11 Comparison of different schedules for graph of Fig. 10.10b. (**a**) A pure data-driven schedule. (**b**) A schedule with uniform latency budget distribution. (**c**) A schedule with optimized latency budget distribution

on their location in and how they propagate through the graph, the execution history, and input data in general. As such, more complex, application-specific quality models will be needed to derive optimal timeouts.

In the aforementioned schedules, we also assumed that the communication delay between τ_X running on host h_0 and τ_A running on host h_2 is similar to that of τ_Y running on h_1 and τ_A on h_2. However, in a scenario, where either h_0 or h_1 has a better connectivity and thus lower delay in communicating with h_2, it is easy to see that task mapped to the node with worse connectivity will experience more losses. As such, by optimizing the mapping of tasks to host, one can ensure that tasks that produce data with higher impact on overall application quality will have less losses and the overall quality will be maximized. This requires corresponding extensions of traditional task graph mapping and scheduling approaches.

10.3.3 Quality/Latency-Aware Task Graph Scheduling

For distributed execution of task graphs under given latency constraints, we need to map tasks and derive start times. In addition to the graph, this requires task execution times on each host, latency constraints and a network specification to be known. The network specification lists delay and loss characteristic of network paths between hosts. Network delays are assumed to be continuous random variables that are specified in terms of a Network Delay Distribution (NDD), i.e., a probabilistic distribution model that specifies the likelihood of a given one-way network delay in absence of any retransmission [22]. A similar model of task execution times as random variables can then be assumed and incorporated.

10.3.3.1 Scheduling Formulation

Analyzing a graph to derive a schedule that provides static guarantees requires instantaneous timeouts of all intermediate tasks to be statically derived. In this work, we perform a conservative analysis assuming a fixed schedule in which all intermediate tasks execute with a constant period as given by the overall period of the task graphs. This reduces the timeout problem to determining offsets between periodic task executions while allowing for a static analysis that provides upper bounds on latency and data losses. In practice, a schedule can be dynamically adjusted to further optimize latency or quality at runtime, e.g., by firing tasks and sending outputs early if input data arrives before the start of the next period.

Figure 10.12 shows a task graph with a chain of N tasks, each mapped to a separate host. Given the execution time r_i of task τ_i and communication delay $d_{i,j}$ between tasks τ_i and τ_j, the total latency of this task graph is:

$$l = \sum_{i=0}^{N-1} r_i + \sum_{i=0}^{N-2} d_{i,i+1} \leq l', \tag{10.25}$$

where l' is the end-to-end latency constraint for the graph. Satisfying this constraint requires bounds on execution times $r_i \leq r'_i$ and communication delays $d_i \leq d'_i$ to be known or assigned. The choice of these bounds determines the probability of deadlines being missed and results or data being dropped.

Fig. 10.12 A linear task graph with N tasks mapped to N hosts

Assuming that the execution time of task τ_i mapped to host h_i is modeled as a continuous random variable R_{τ_i,h_i}, the probability p_i^r that it will not violate its computation deadline r_i' can be calculated as:

$$p_i^r = F_{R_{\tau_i,h_i}}(r_i'), \tag{10.26}$$

where $F_{R_{\tau_i,h_i}}$ is the cumulative distribution function (CDF) of random variable R_{τ_i,h_i}.

Similarly, the probability $p_{i,j}^d$ that data sent from task τ_i on host h_i to task τ_j mapped to host h_j will not violate the deadline $d_{i,j}'$ can be calculated as:

$$p_{i,j}^d = F_{D_{h_i,h_j}}(d_{i,j}'), \tag{10.27}$$

where D_{h_i,h_j} is the random variable modeling the communication delay between hosts h_i and h_j, and $F_{D_{h_i,h_j}}$ is the cumulative distribution function (CDF) of random variable D_{h_i,h_j}.

To minimize the probability of violations, we need to maximize r_i' and $d_{i,j}'$. However, given a total end-to-end latency constraint, the total latency budget needs to be partitioned among computation and communication budgets r' and d' such that the total latency constraint is satisfied while overall quality is maximized. Towards this end, in the remainder of this section, we first develop a quality model that, given a mapping and scheduling, describes the relationship between overall application quality and computation and communication budgets. Using this model, we then propose heuristics for general mapping and scheduling of tasks to hosts. Finally, we formulate real-time budgeting of task graphs as an optimization problem that determines assignments of computation and communication budgets to maximize overall quality while meeting a total latency constraint.

10.3.3.2 Quality Model

The output quality of a distributed application is inversely related to the impact of data losses on output quality and the frequency of such computation and communication losses. Given a mapping, scheduling, and budgeting in real-time execution of a task graph, a deadline violation and data loss in any intermediate task in the graph can be handled at runtime by using replacement functions to estimate values to compute with instead of unavailable data. Such replacement functions can be based on using the last value previously computed or received, or some more complex history-based prediction. In case of computation deadline violations, replacement functions can also take the form of the output computed by a shorter imprecise or reduced-precision part of a task as described in Sect. 10.2. Given that replacement functions will not be accurate in case of all violations, they introduce errors and degrade the overall application quality. The magnitude of errors in computation or communication outputs depends on the relative accuracy of the replacement functions, while their frequency is determined based on the probability of computation and communication budget violations p^r and p^d described in the previous section.

The output quality Q of a distributed application is then a function of (1) the magnitude and frequency of errors $e_{i,j}$ in the data values $v_{i,j}$ generated by task τ_i and communicated to task τ_j, (2) the amplification, reduction, combination or cancelling out of errors as they propagate through the graph to primary outputs, and (3) the metric used to translate errors in primary outputs into an overall application-specific quality Q. Developing quality models that capture the relationship between computation and communication budgets and overall application quality Q thus require developing appropriate models for error generation and propagation all the way to final quality metrics. Errors $e_{i,j}$ can be modeled as random variables that account for their dependencies on input data and loss probabilities. From error distributions at each location in the graph, a variety of metrics such as min/max errors can then be extracted and propagated through the graph. In special cases, such as linear systems using well-defined replacement functions and statistical quality metrics such as SNR, this allows closed-form quality models to be derived [20], e.g., by applying well-known variance-based noise generation and propagation methods from fixed-point optimization of digital signal processing systems [23, 24].

In general non-linear applications with arbitrary replacement functions and quality metrics, however, developing closed-form models of application quality Q will be infeasible. Instead, we can develop models that estimate the expected quality degradation ΔQ due to reduced computation and communication budgets. The quality impact $\Delta q_{i,j}$ of an individual error $e_{i,j}$ in data exchanged between task τ_i and its consumer τ_j can by calculated as the product of error $e_{i,j}$ and the derivative $\partial Q/\partial v_{i,j}$ of Q with respect to values $v_{i,j}$ communicated between τ_i and τ_j:

$$\Delta q_{i,j} = e_{i,j} \cdot \partial Q/\partial v_{i,j}. \tag{10.28}$$

Gradients $\partial Q/\partial v_{i,j}$ can be computed using methods similar to back-propagation in neural network training. Due to the random nature of computation and communication losses, the quality impact is in practice a random variable that depends on loss probabilities p_i^r and $p_{i,j}^d$ and hence, in turn, computation and communication budgets r_i' and $d_{i,j}'$. Assuming that errors $e_{i,j}$ will be incurred at the input of consuming task τ_j in case of computation τ_i or communication between τ_i and τ_j violating deadlines, their quality impact is a discrete random variable that is zero with probability $p_i^r \cdot p_{i,j}^d$, and $\Delta q_{i,j}$ otherwise. Hence, the expected quality impact $\Delta Q_{i,j}$ can be quantified as:

$$\begin{aligned}\Delta Q_{i,j}(r_i', d_{i,j}') &= \Delta q_{i,j} \cdot \left(1 - p_i^r \cdot p_{i,j}^d\right) \\ &= \Delta q_{i,j} \cdot \left(1 - \left(F_{R_{\tau_i,h_i}}(r_i') \cdot F_{D_{h_i,h_j}}(d_{i,j}')\right)\right).\end{aligned} \tag{10.29}$$

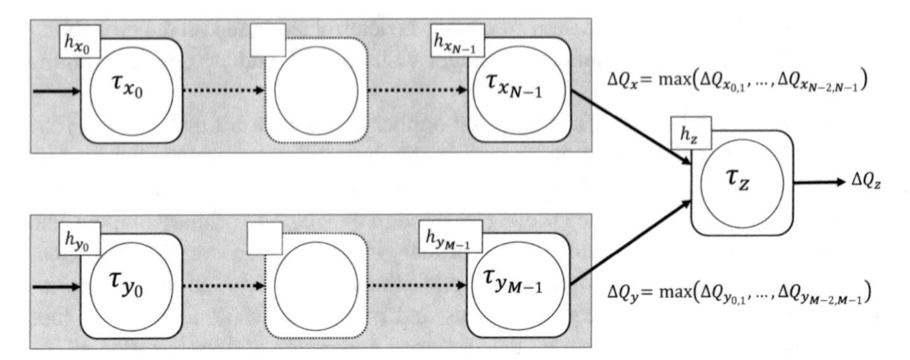

Fig. 10.13 A task graph with two parallel task chains

Equation (10.29) gives the quality impact of errors at a single location in the graph. The combined quality impact across multiple locations depends on the topology of the task graph. For dependent tasks with sequential producer-consumer relationship, such as those depicted in Fig. 10.12, errors, gradients and hence loss impacts interact and are dependent on each other. The total combined impact will be bounded by the maximum individual impacts on the lower end and the sum of individual impacts at the upper end. Optimistically, we can use the maximum to derive total impact:

$$\Delta Q = max\left(\Delta Q_{0,1}, \Delta Q_{1,2}, \ldots, \Delta Q_{N-2,N-1}\right). \tag{10.30}$$

In case of parallel sub-graphs as shown in Fig. 10.13, the gradients and losses are independent and hence, the total combined impact will be the sum of individual impacts:

$$\Delta Q = \Delta Q_x + \Delta Q_y. \tag{10.31}$$

By combining Eqs. (10.30) and (10.31), we can estimate the quality impact for arbitrary task graph topologies. Note that if task τ_i and its consumer τ_j are mapped to the same host, $F_{C_{h_i,h_j}}(c'_{i,j})$ will be equal to one and there will be no communication losses nor quality impact. Likewise, as in traditional data flow models, if upper bounds on task execution times are known, $F_{R_{\tau_i,h_i}}(r'_i)$ will be equal to one and there will be no computation losses.

10.3.3.3 Mapping Heuristics

Mapping and scheduling of DAGs and task graphs across a target platform consisting of multiple, potentially heterogeneous processors or hosts to minimize total schedule makespan is a classic and widely studied problem. Multiprocessor DAG scheduling to minimize makespan is in general NP-complete. List schedulers and their variants are commonly used as effective heuristics. In list schedulers, tasks are scheduled on a processing element that provides maximum performance where choices among tasks that are ready to execute are resolved based on a notion of priority that reflects their inherent criticality.

Minimizing makespan under a total latency constraint will increase the total latency budget available, i.e., will inherently increase quality. In addition, quality can be further optimized by taking the quality impact of different mapping choices into account. Specifically, network delay distributions and hence communication delays $d_{i,j}$ will generally depend on the connectivity of each host. As such, mapping tasks with higher quality impact onto hosts that are better connected and hence less likely to experience losses will significantly improve overall quality. This can be achieved by folding estimates of quality impacts of different host mappings into the list scheduler's priority function, e.g., to resolve choices among tasks that otherwise would have the same priority.

Quality impact estimates of different mappings can be calculated following Eqs. (10.30) and (10.31). Computation of quality impacts, however, requires an initial estimate of computation and communication budgets r_i' and $d_{i,j}'$ for each task τ_i and each dependency (τ_i, τ_j). Such initial estimates for mapping purposes can be derived by assuming a uniform budget assignment at each τ_j or by incorporating additional application-specific information, such as partitioning the total end-to-end latency constraint l' according to the maximum loss impact $\Delta q_{i,j}$ at each consuming task τ_j. Following Eq. (10.29), local budgets at each τ_j can be further partitioned into optimal computation and communication budgets r_i' and $d_{i,j}'$ by ensuring that computation and communication losses impact quality equally, i.e., such that $F_{R_{\tau_i,h_i}}(r_i') = F_{D_{h_i,h_j}}(d_{i,j}')$. Using budget estimates $d_{i,j}'$, r_j' and $d_{j,k}'$ when mapping the ready task τ_j to the set of available hosts $h_0, \ldots, h_{N-1}$, we can then rank target hosts based on the least quality impact. Note that since any successor task τ_k will not yet have been mapped, the loss probability and ultimately quality impact of the (τ_j, τ_k) communication will be unknown. In practice, one can optimize for the worst case by assuming a mapping h_k for τ_k with the worst connectivity, i.e., minimum $F_{D_{h_j,h_k}]}(d_{j,k}')$.

10.3.3.4 Budgeting Formulation

Given a mapping $\mathbf{h}$ of tasks τ_i to hosts h_i, assignment of computation and communication budgets $\mathbf{r}'$ and $\mathbf{d}'$ can finally be formulated as an optimization problem that minimizes total quality degradation under a total latency constraint l':

$$\begin{aligned} \underset{\mathbf{r}',\mathbf{d}'}{\text{minimize}} \quad & \Delta Q(\mathbf{h}, \mathbf{r}', \mathbf{d}') \\ \text{subject to} \quad & \sum_{(\tau_i,\tau_j)\in w} r_i' + d_{i,j}' \leq l', \ \forall w \in \mathbf{w}, \end{aligned} \tag{10.32}$$

where $\mathbf{w}$ is the set of all paths from source to sink of the graph. Equations (10.30) and (10.31) enable calculating quality impact for a given mapping $\mathbf{h}$ and budgets

$\mathbf{r}'$ and $\mathbf{d}'$. Total quality degradation $\Delta Q(\mathbf{h}, \mathbf{r}', \mathbf{d}')$ can be expressed in (max, +)-algebra following a traversal of the task graph, but individual sub-expressions will contain non-linear and non-convex terms. However, an iterative gradient-based approach based on tasks with maximum impact in each chain and derivatives of probabilities p^r and p^c in Eqs. (10.26) and (10.27), respectively, can be used to solve this optimization problem. Note that when quality Q can be directly expressed in terms of computation and computation budgets, budget assignment can be expressed as a similar optimization problem that aims to directly maximize $Q(\mathbf{h}, \mathbf{r}', \mathbf{d}')$ [20].

10.3.4 Experiments and Results

We evaluated the quality/latency-aware task graph mapping, scheduling, and budgeting approaches described in this chapter for distributed real-time execution of neural network inference applications in edge computing settings. We applied our approach to an image segmentation and two object detection neural networks from [25–27] implemented on top of the Darknet deep learning framework [26]. We distribute the convolutional neural networks by tiling, fusing, and executing layers in a map-reduce scheme similar to [28]. In our evaluation, we focus on communication budgeting by assuming fixed worst-case bounds for compute tasks. Budgeting such a distributed neural network task graph requires partitioning the total latency budget among each layer groups' scatter and gather operations. Mapping of this graph entails assigning fused tile stacks of each layer group to one of the edge devices in the local network, where a master node coordinates all task distribution.

We use the Zurich Urban Micro Aerial Vehicle Dataset [29] for all three applications. Given edge computation and communication limits, we downsample the dataset to 500 frames at 0.2 fps. For all three applications, we generate ground truth by processing the input dataset via baseline versions of each neural network. We use mean average precision (mAP) and intersection over union (IoU) as quality metrics for objection detection and image segmentation, respectively. We compute gradients for quality estimation using a custom loss function and back-propagation pass.

We evaluated distributed real-time execution both in simulation as well as in a real-world deployment. Our custom network simulator emulates timeouts by injecting random losses assuming normally distributed packet delays with an MTU of 1500 bytes, an average bandwidth of 600 Mbps corresponding to the IEEE 802.11n Wi-Fi standard, and a coefficient of variation of 0.2 ($\sigma = 0.2 * \mu$). The simulator assumes six nodes with client hosts placed at increasing distances from the master node with a 20% increase in average delays and variances from node to node.

For real-world experiments, we deployed our approach on a cluster of six Raspberry Pi 3 devices with a quad-core ARM Cortex-A53 CPU and 1 GB of RAM using a custom middleware framework for task distribution and budgeting. Given the limited amount of memory available for back-propagation, we only evaluated the

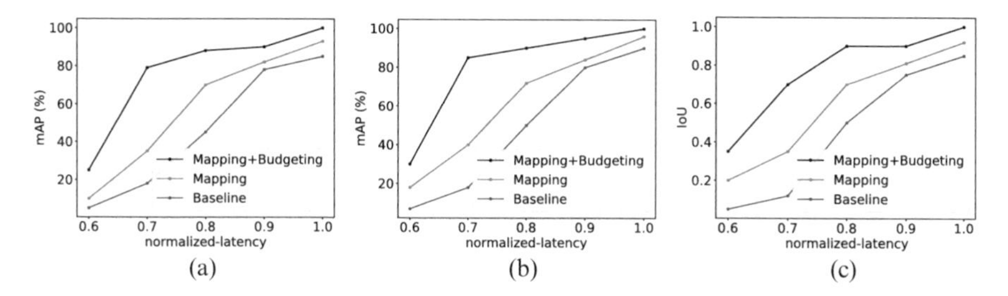

Fig. 10.14 Quality/latency tradeoff in simulated real-time execution of distributed neural network applications under optimized mapping and budgeting. (**a**) YOLOv2. (**b**) YOLOv3. (**c**) Image segmentation

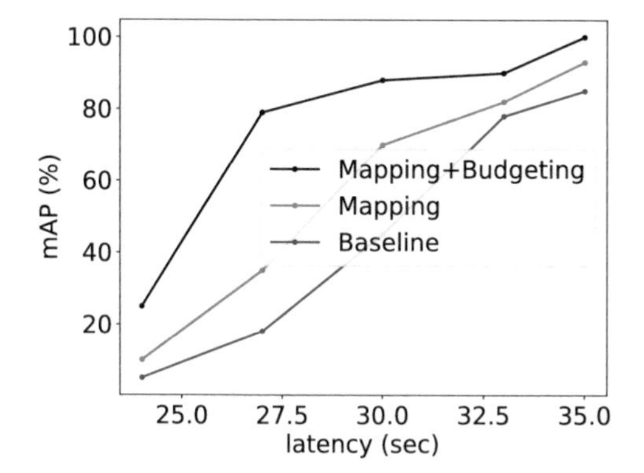

Fig. 10.15 Quality/latency tradeoff in distributed real-time execution of YOLOv2 object detection network on a Raspberry Pi3 cluster under different optimizations

32-layer YOLOv2 object detection network. To provide consistent measurements and avoid interference, a 54 Mbps 801.11g wireless network with Nakagami fading model was emulated using CORE/EMANE [30] running over Ethernet interfaces. Emulated nodes were placed at uniformly spaced distances ranging between 3m and 17m from the master node.

Simulation Results Figure 10.14 shows the tradeoff in the classification accuracy of simulated applications as a function of latency constraints normalized against 99-percentile network delays. We compare a baseline configuration that uses uniform budgeting and random mapping against optimizations applied every 25 frames. As seen in this Figure, mapping greatly improves accuracy (up to 20%) across different applications and latency constraints. Optimized budgeting further improves accuracy by up to 10%. The impact of optimizations is more significant under tight latency constraints and becomes smaller as constraints are relaxed.

Deployment Results Figure 10.15 shows the quality/latency tradeoff of the YOLOv2 object detection network when executed in a distributed real-time fashion on a Raspberry Pi3 edge cluster. Results are similar to simulations and confirm that optimizations can significantly improve classification accuracy under tight latency constraints in real-world distributed real-time edge machine learning deployments.

10.4 Summary and Conclusions

In this chapter, we discussed how approximate computing ideas can be applied to real-time scheduling by trading off tightness of bounds on computation and communication deadlines for quality degradations. By establishing tighter than worst-case bounds and allowing for occasional deadline violations, schedule admissibility or reaction times can be improved in exchange for data losses or a reduction in data precision. We showed how scheduling of such systems for both independent and dependent task sets can be formulated as an optimization problem to maximize quality while satisfying schedule admissibility or real-time guarantees. For independent tasks, we extend traditional mixed-criticality (MC) system models to allow low-criticality tasks to execute with lowered precision such that combined quality is maximized while satisfying schedule admissibility. For dependent tasks, we find optimized mapping, scheduling, and budgeting of task graph models that minimizes overall quality degradation while meeting end-to-end latency constraints.

Overall, exploiting approximations opens new avenues for real-time system design. Given the challenges in determining tight worst-case bounds, e.g., on task execution times, hard real-time systems are often over-designed. Incorporating approximations will allow dynamically recovering available deadline slack to perform more useful work. This opens new opportunities in connecting real-time system design to approximation techniques at software and hardware implementation levels.

References

1. Wilhelm, R., Engblom, J., Ermedahl, A., Holsti, N., Thesing, S., Whalley, D., Bernat, G., Ferdinand, C., Heckmann, R., Mitra, T., Mueller, F., Puaut, I., Puschner, P., Staschulat, J., & Stenström, P. (2008). The worst-case execution-time problem—overview of methods and survey of tools. *ACM Transactions on Embedded Computing Systems (TECS), 7*(3), 1–53.
2. Burns, A., & Davis, R. (2013). *Mixed criticality systems - a review*. Tech. Rep., Department of Computer Science, University of York.
3. Derler, P., Feng, T. H., Lee, E. A., Matic, S., Patel, H. D., Zhao, Y., & Zou, J. (2008). *PTIDES: A programming model for distributed real-time embedded systems*. Tech. Rep. UCB/EECS-2008-72, EECS Department, University of California, Berkeley.
4. Schmidt, D., & Kuhns, F. (2000). An overview of the real-time CORBA specification. *Computer, 33*(6), 56–63.
5. Schulzrinne, H., Casner, S., Frederick, R., & Jacobson, V. (2003). *RTP: A transport protocol for real-time applications*. RFC 3550. RFC Editor.

6. Baruah, S., Bonifaci, V., DAngelo, G., Li, H., Marchetti-Spaccamela, A., Van Der Ster, S., & Stougie, L. (2012). The preemptive uniprocessor scheduling of mixed-criticality implicit-deadline sporadic task systems. In *Euromicro Conference on Real-Time Systems (ECRTS)*.
7. Liu, C. L., & Layland, J. W. (1973). Scheduling algorithms for multiprogramming in a hard-real-time environment. *Journal of the ACM, 20*, 46–61.
8. Baruah, S., Chattopadhyay, B., Li, H., & Shin, I. (2014). Mixed-criticality scheduling on multiprocessors. *Real-Time Systems, 50*(1), 142–177.
9. Baruah, S. (2004). Optimal utilization bounds for the fixed-priority scheduling of periodic task systems on identical multiprocessors. *IEEE Transactions on Computers (TC), 53*(6), 781–784.
10. Li, H., & Baruah, S. (2012). Global mixed-criticality scheduling on multiprocessors. In *Euromicro Conference on Real-Time Systems (ECRTS)*.
11. Funk, S., Levin, G., Sadowski, C., Pye, I., & Brandt, S. (2011). DP-Fair: A unifying theory for optimal hard real-time multiprocessor scheduling. *Real-Time Systems, 47*(5), 389–429.
12. Lee, J., Phan, K.-M., Gu, X., Lee, J., Easwaran, A., Shin, I., & Lee, I. (2014). MC-Fluid: fluid model-based mixed-criticality scheduling on multiprocessors. In *Real-Time Systems Symposium (RTSS)*.
13. Han, J., & Orshansky, M. (2013). Approximate computing: an emerging paradigm for energy-efficient design. In *IEEE European Test Symposium (ETS)*.
14. Liu, J. W.-S., Lin, K.-J., Shih, W. K., Yu, A. C.-S., Chung, J.-Y., & Zhao, W. (1991). *Algorithms for scheduling imprecise computations* (pp. 203–249). New York: Springer.
15. Liu, D., Spasic, J., Guan, N., Chen, G., Liu, S., Stefanov, T., & Yi, W. (2016). EDF-VD scheduling of mixed-criticality systems with degraded quality guarantees. In *Real-Time Systems Symposium (RTSS)*.
16. Burns, A., & Baruah, S. (2013). Towards a more practical model for mixed criticality systems. In *Workshop on Mixed-Criticality Systems*.
17. Baruah, S., Burns, A., & Guo, Z. (2016). Scheduling mixed-criticality systems to guarantee some service under all non-erroneous behaviors. In *Euromicro Conference on Real-Time Systems (ECRTS)*.
18. Pathan, R. M. (2017). Improving the quality-of-service for scheduling mixed-criticality systems on multiprocessors. In *LIPIcs-Leibniz International Proceedings in Informatics* (vol. 76).
19. Huang, L., Hou, I., Sapatnekar, S. S., & Hu, J. (2018). Graceful degradation of low-criticality tasks in multiprocessor dual-criticality systems. In *International Conference on Real-Time Networks and Systems (RTNS)*.
20. Barijough, K. M., Zhao, Z., & Gerstlauer, A. (2019). Quality/latency-aware real-time scheduling of distributed streaming IoT applications. *ACM Transactions on Embedded Computer Systems (TECS), 18*, 83:1–83:23.
21. Francis, S., & Gerstlauer, A. (2017). A reactive and adaptive data flow model for network-of-system specification. *IEEE Embedded System Letters (ESL), 9*, 121–124.
22. Bovy, C., Mertodimedjo, H., Hooghiemstra, G., Uijterwaal, H., & Van Mieghem, P. (2002). Analysis of end-to-end delay measurements in internet. In *The Passive and Active Measurement Workshop (PAM)*.
23. Lee, S., Lee, D., Han, K., Kim, T., Shriver, E., John, L. K., & Gerstlauer, A. (2016). Statistical quality modeling of approximate hardware. In *IEEE International Symposium on Quality Electronic Design (ISQED)*.
24. Lee, S., & Gerstlauer, A. (2013). Fine grain word length optimization for dynamic precision scaling in DSP systems. In *IFIP/IEEE International Conference on Very Large Scale Integration (VLSI-SoC)*.
25. ArtyZe. (2018). Image segmentation in darknet. Retrieved 15 Nov, 2020, from https://github.com/ArtyZe/yolo_segmentation
26. Redmon, J., Divvala, S., Girshick, R., & Farhadi, A. (2016). You only look once: unified, real-time object detection. In *IEEE Conference on Computer Vision and Pattern Recognition (CVPR)*.
27. Redmon, J., & Farhadi, A. (2018). Yolov3: An incremental improvement. arXiv:1804.02767.

28. Zhao, Z., Barijough, K. M., & Gerstlauer, A. (2018). DeepThings: distributed adaptive deep learning inference on resource-constrained IoT edge clusters. *IEEE Transactions on Computer-Aided Design of Integrated Circuits and Systems, 37*(11), 2348–2359.
29. Majdik, A. L., Till, C., & Scaramuzza, D. (2017). The Zurich urban micro aerial vehicle dataset. *The International Journal of Robotics Research, 36*(3), 269–273.
30. Ahrenholz, J. (2010). Comparison of core network emulation platforms. In *Military Communications Conference (Milcom)* (pp. 166–171). Piscataway: IEEE.

Chapter 11
Security in an Approximated World: New Threats and Opportunities in the Approximate Computing Paradigm

Paolo Palmieri, Ilia Polian, and Francesco Regazzoni

11.1 Introduction

Approximate Computing (AxC) was originally intended for low-cost systems, such as low-end IoT devices, where security did not play a significant role. However, recent progress in AxC makes this technology attractive for applications which are resource-constrained and security-critical at the same time. For example, wearables are processing sensitive health information that needs to be protected, and emerging autonomous systems (automotive and beyond) incorporate deep-learning engines that can benefit from AxC [1]. Therefore, security of AxC implementations can be no longer neglected. This chapter provides an overview of security threats that apply to approximate circuits and systems, and on potential of AxC to improve security. It goes beyond existing overviews and position papers [2–4] by discussing the entire universe of security aspects of AxC technology that are foreseeable today.

As with many new topics, the exact scope of AxC is not universally agreed upon among researchers. For example, some consider *stochastic computing* [5] a subclass of AxC while others regard it as a separate concept. Furthermore, some researchers count non-deterministic techniques such as voltage overscaling as AxC

P. Palmieri
School of Computer Science & IT, University College Cork, Cork, Ireland
e-mail: p.palmieri@cs.ucc.ie

I. Polian
Institute of Computer Architecture and Computer Engineering, University of Stuttgart, Stuttgart, Germany
e-mail: ilia.polian@informatik.uni-stuttgart.de

F. Regazzoni (✉)
University of Amsterdam, Amsterdam, The Netherlands

Università della Svizzera italiana, Lugano, Switzerland
e-mail: f.regazzoni@uva.nl; francesco.regazzoni@usi.ch

A. Bosio et al. (eds.), *Approximate Computing Techniques*,
https://doi.org/10.1007/978-3-030-94705-7_11

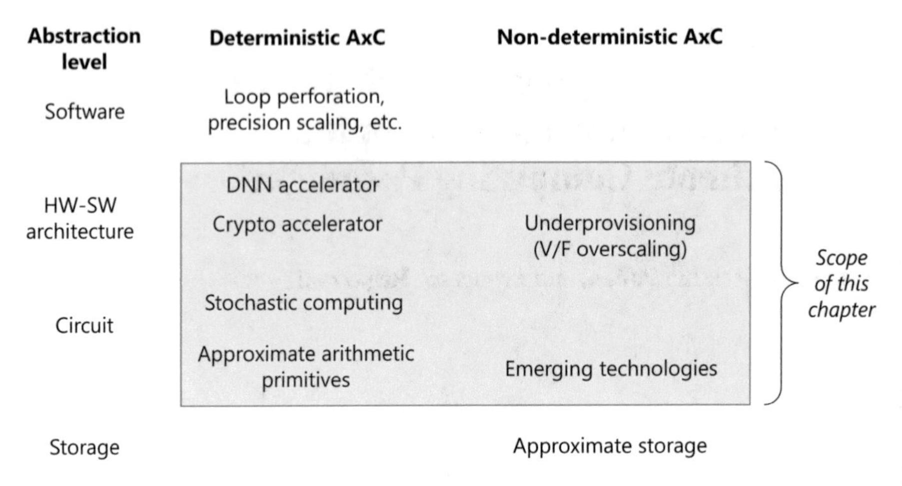

Fig. 11.1 Scope of AxC techniques discussed in this chapter

and some do not. Figure 11.1 shows a simple taxonomy of various AxC techniques with an indication which of them are relevant in the context of this chapter. We briefly describe the different techniques next, motivating why some of them are not further discussed with respect to their security. We include both stochastic computing and non-deterministic AxC, yet we do not include traditional computing procedures associated with some accuracy loss, such as lossy compression or processing digitized analog data represented using a floating-point or fixed-point numbers.

A central aspect in assessing security implication of an AxC class is whether it introduces *non-determinism* on the hardware level. In general, a non-deterministic system can develop a multitude of behaviors, some of which may compromise security, thus enlarging the potential attack surface and making it more difficult to rule out security loopholes. A deterministic system is easier to predict and analyze for security vulnerabilities, yet even deterministic AxC schemes can open up new possibilities for attackers, as discussed in the next section.

We start discussing Fig. 11.1 with its left part, namely the deterministic AxC techniques, moving from upper to lower level of abstraction. Several of them are defined on software level [6]. Examples of software-only techniques include *loop performation* (executing only a some of an iterative algorithm's loops) [7], *precision scaling* (replacing floating-point by fixed-point operations or reducing the bitwidth of operands) [8], speculative *approximation of load values* to avoid cache misses [9], or *relaxed synchronization* [10]. While these approaches unfold the accuracy-cost trade-off, they do not seem to introduce new security vulnerabilities *per se*. If an approximate algorithm (and its implementation) need to be protected against a security threat, the same techniques as for an exact algorithm are applicable.

Specific architectures, often containing both hardware and software components, can give rise to AxC. One prominent class of such architectures are *deep neural network* (DNN) accelerators for machine learning applications. Their relationship to AxC is twofold. First, DNN-based classification as such can be seen as an approximate algorithm, as it exhibits non-trivial false-positive and false-negative rates for most practical applications. Second, DNN accelerators are a prime instance of systems that can clearly benefit from approximate circuits. They consist of a large number of arithmetic primitives, first and foremost adders and multipliers, so even small savings in area or power consumption for an individual primitive are effective with respect to the whole network. Moreover, DNNs tend to tolerate inaccuracies that occur during processing. Another relevant class of architectures are *cryptographic accelerators*. While cryptography has historically been found "not amenable for approximation" [6], Sect. 11.3.1 will discuss how AxC can be helpful in (complex) cryptographic schemes.

An interesting architecture that is intrinsically approximate is *stochastic computing* (SC) [5], offering extremely low-area (and therefore low-power) realizations of important arithmetic operations. While stochastic circuits as such are rarely used for security-relevant calculations, they turn out to be helpful in protecting neural networks against adversarial attacks, as discussed in Sect. 11.3.2.

The perhaps most prominent class of AxC on hardware level are *approximate arithmetic primitives*, such as approximate adders or multipliers [11]. These circuits offer lower implementation cost at the expense of not working correctly for certain inputs. The inaccuracy is typically bounded by a certain error rate (proportion of the inputs for which the operation is incorrect) and/or error magnitude (maximal extent of the deviation).

Non-deterministic AxC techniques form the right hand side of Fig. 11.1. They are not well-defined on the pure software level, since the non-determinism of the discussed AxC approaches originates from hardware. The main technique here is *underprovisioning*, where a system is operated at either too-fast frequency or too-low voltage for all circuit paths to finish switching before the cycle time. One can distinguish between *voltage overscaling* [12], *frequency overscaling* [13] and *adaptive techniques* where the operating point is continuously updated until error are detection, and yet the errors that happen are not corrected [14, 15].

A further source of approximation is the use of *emerging technologies* that are not yet sufficiently well controlled and result in non-trivial error rates on device level. Following the early ideas by von Neumann [16], architectures for executing software on unreliable substrates are being designed in, e.g., [17]. A similar idea is the usage of analog properties of (inherently noisy) emerging devices in *neuromorphic computing* [18].

Finally, *approximate storage* has recently received some attention [3]. This term subsumes memories (static RAMs, dynamic RAMs and emerging nonvolatile storage solutions such as phase-change memories) that are operated under conditions that do not guarantee their reliable operation, not unsimilar to underprovisioned approximate circuits. For instance, they can use an aggressively low voltage or slow refresh rate. The failure patterns of an approximate memory are unpredictable, and therefore this class of AxC falls under non-deterministic schemes. While approximate storage can introduce errors and inaccuracies and thus disrupt the

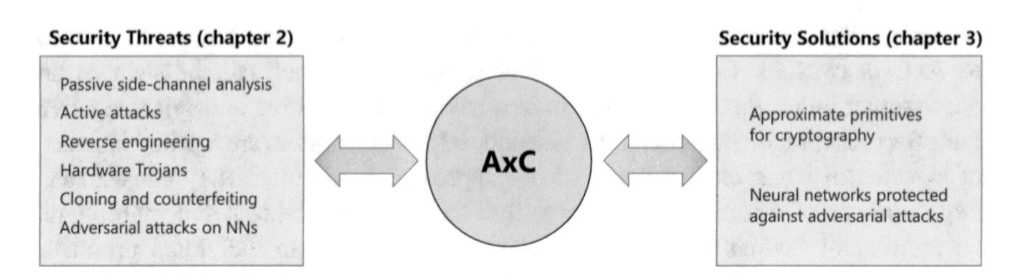

Fig. 11.2 Overview of this chapter

operation of the system using the memory, the uncontrollable and unpredictable nature of such disturbances makes them less security-critical. Therefore, we decided to omit them from this chapter.

After the AxC techniques in scope of this chapter have been outlined above, the subsequent chapters will describe both the (potentially) negative and positive impacts of AxC with respect to security. Section 11.2 will focus on security threats and vulnerabilities related to approximate computing. Section 11.3 shows how AxC can help in providing solutions to make systems more secure. Figure 11.2 visualizes this chapter's structure.

11.2 Approximate Computing and Security Threats

Approximate computing is a relatively new computing paradigm and its implications on security are not well known. In this section we summarize the main security threats for classical computing paradigm and we discuss possible effects that approximate computing can have on them.

11.2.1 Passive Side Channel Attacks

The first class of security threat that we explore is the one of passive side channel attacks. Passive side channel attacks are attacks in which the adversary observes a physical quantity, for instance, the power consumed [19] by the device during an encryption operation (or the time needed to complete it [20]) and uses this physical quantity to extract secret information. The attacks usually take their name from the physical quantity used to mount the attack. The most popular ones are timing attacks, which exploit the different times needed by computations, power analysis attacks, which exploit the power consumed by the computation, and electromagnetic attacks, which exploit the electromagnetic emissions. Passive side channels have been presented to the scientific community in 1996, and from that moment they have been subject of extensive research.

One of the most active topic of research in the field of passive side channel is the evaluation of the robustness of certain implementation against specific attacks. The topic has been widely explored in the context of power analysis, with few metrics being proposed and a number of common criteria and "de facto" standards for the security assessment of devices. Among the widely accepted methods for security evaluation are mutual information, which is a metric to quantify the leakage of information, and the t-test methodology, which is a statistical test to evaluate the dependency between the power trace and the secret key.

Statistical tests and metrics are typically carried out on traces collected at a specific operation voltage. However, this might not hold in the context of approximate computing. Aggressive power saving techniques used in the approximate paradigm can involve a self-adaptive voltage scaling. In that case, one should ensure that the evaluation carried out at a specific operational point is still valid at all the other ones. Alternatively, test and security assessment procedures should be revised to include evaluation at multiple operational points. Indeed, this would increase the time required and the complexity of the security evaluation when compared to the classical computing paradigm.

Self-adaptation circuits can also be the source of leakage. Certainly, they can be paired with an hardware Trojan to build covert channels, communicating secrets from the inner core to the external world by modulating the information as high or low voltage. Furthermore, when approximate circuits using voltage scaling are used to compute sensitive data, the self-adaptation circuit can, as all the other electronic components, unintentionally cause the leakage of information. To avoid this, protections against side channel attacks should be applied also to self-adaptation and power scaling circuits, exactly as it is done for the main circuits carrying out security computations.

Approximate computing can also have positive effects on robustness against passive side channel attacks. Firstly, voltage scaling could add another dimension to the protection space, since it gives designers the possibility to alter the power traces during the computation, making the adversary's task of measuring and verifying her attack hypothesis more complex. Secondly, approximate computing paradigm could help in realizing circuits that are intrinsically more robust. For instance, to defeat timing attacks it is necessary to realize circuits that compute in constant time. This could be hard (or expensive) to achieve in classical computing paradigm. Approximate circuits could instead simplify the guaranteed of constant computation time. In cryptography, these circuits could be used, for instance, in lattice-based schemes, where classical multiplies could be replaced by approximate multipliers achieving protection against timing attacks while having only a minimal effect on the performance. Constant-time multiplier realized using the classical computing paradigm incurs, in fact, in a quite high overhead.

11.2.2 Active Attacks

The adversary can also directly tamper with the device with the goal of inducing it into a wrong state or with the goal of exploiting the faulty behavior. We call the physical attacks mounted in this way "active attacks", since the adversary actively intervenes on the device instead of just passively observing its behavior. The most common active attack is fault attack [21], where the adversary injects a fault into a device computing a cryptographic function to reduce the security of the algorithm (for instance, by reducing the number of rounds that are computed) or to extract the secret key comparing the correct and the faulty outputs. These attacks often require a precise faulty pattern to be successful. The precision and the granularity of the faults largely depends on the equipment available to the attacker. It is, however, important to mention that fault attacks can be successfully carried out also with an extremely cheap equipment, for instance, with a supply voltage used to underfeed the cryptographic device.

These types of faults are very likely to be quite simple to be caused in approximate circuits. In fact, approximate circuits often operate at the edge of the operation conditions, making the injection of fault a much easier task compared to the one of injecting faults in regular circuits. This simplification, however, does not necessary brings advantages to the attacker. This is because the sweet spot for injecting the right fault to mount the attack is generally reduced, such that an exploitable fault has to be injected using much more expensive and much more precise injection devices. Ultimately this would increase the cost of the attack [22]. Also, the very nature of approximate circuits could make certain types of fault attacks, the ones that require well-defined faulty patterns, much harder to be carried out.

Another important point is related to the deployment of countermeasures. The typical way to counteract fault attacks is to limit the access to faulty outputs. This is obtained by using error detecting and, when possible, error correcting codes. These codes, however, have been studied and developed so far only for exact circuits. Their application to approximate computing paradigm, even if some codes can tolerate randomness [23], is largely unexplored and their effectiveness in this context has not been proved yet.

11.2.3 Reverse Engineering

Reverse engineering [24], in the context of hardware security, is a threat that attacks confidentiality, meaning that a certain information should be accessible only to legitimate parties. Here, the information to be kept secret is the hardware circuit itself, since it is the outcome of a costly and time consuming work. During reverse engineering, an adversary analyzes the circuit itself or its behavior (or a combination of both) to infer the original design. This operation is often done maliciously, with the goal of reconstruct the original circuit for stealing the intellectual property associated with it. However, reverse engineering can also be carried out with the legitimate goal of discover contract violations or patent infringements.

Circuit analysis usually start with the depackaging of the chip. High-resolution images of the chip are then collected. If the adversary has advanced capabilities and equipment that makes it possible, he can take high-resolution images of each layer of the circuit. The attempt of reconstructing the original circuit is then carried out using a pattern-matching tool, to match the collected images with the expected layout, and with machine learning techniques that are sufficiently powerful to allow the precise reconstruction of the initial netlist.

It is not yet known how reverse engineering could behave on approximate circuits. As discussed, one of the phases of reverse engineering involves pattern matching, which is typically done starting from known structures and basic blocks. Examples of such blocks are finite state machines, multipliers, and control flow blocks. Since the computation in approximate computing is not exact, it is possible that approximated structures do not precisely match the reference structures used in exact computation (this is true, for instance, in control flow). As a result, the task of the adversary could be more complicated, since it could not be able to detect the approximation or, at least, the selected approximation.

Certainly, it would be simpler for an attacker to identify blocks that do not use approximated circuits. These could be immediately identified as critical blocks, implementing, for instance, cryptographic functions (since they cannot, at least generally, be implemented in an approximated way—see Sect. 11.3.1 for an interesting class of cryptographic functions that *can* benefit from AxC). The identification of cryptographic blocks could point the adversary to relevant components of the cryptographic module, for instance, the memory cells where the key is stored.

It is also not known whether and to which extent approximation, by its very nature, could help the adversary. Approximation, in fact, is designed to "tolerate" errors. These errors, in approximate circuits, can come from aggressive power scaling or similar optimizations. However, approximation could also stem from a not completely correct reverse engineering process. If approximation is tolerant to errors, it can be that the same paradigm offers one additional weapon to the adversary, since he knows that, even if imperfectly reverse engineered, the circuit can be still working in a satisfactory way.

11.2.4 Hardware Trojans

Hardware Trojans are malicious and deliberate changes to an hardware design or circuit inserted with the goal of modifying its original behavior [25]. Often, the goal of an attacker is to cause a leak of the information handled by the circuit or to cause a denial of service. Hardware Trojans can be inserted at any point of the design flow, from the early stages [26] to the alteration of the polarity of the doping in transistors [27]. Insertion of Trojans at early stages of the hardware design can be performed by rogue designers (in principle, even by as single one) by simply modifying the circuit specifications or the circuit architecture. Trojans can also be inserted by malicious foundry (that has anyway to reverse engineer the circuit to understand where the alteration should be inserted and to cause a "meaningful" effect) or by malicious design tools that can potentially automatically insert circuitry during the various steps of the design flow.

Approximate computing usually requires the presence of dedicated circuits to control the approximation. These additional circuits represent an extension of the surface available to the attacker. It can be even argued that circuits controlling the approximation are an appealing target for injection of hardware Trojans. For example, it is quite simple to modify the circuit controlling the voltage scaling to cause a denial of service attack. In this case, since the controlling circuit is already present, the Trojan is very likely to be simple and small, ultimately difficult to be detected. Circuits realized using the approximate computing paradigm often require, in addition to the already mentioned controls for approximation, also error correction circuits. These circuits can be used to create covert channels, modulating the information that needs to be leaked through the error rate or the number of errors

The intrinsic tolerance of approximate circuits against errors could, on the positive side, make the realization of effective hardware Trojans more difficult. As a result, it is possible that, to be meaningful, hardware Trojans should be quite complex. As a direct consequence, such Trojans would also occupy a larger amount of area and consume an higher amount of power. This fact would simplify the task of Trojan detection [28].

11.2.5 Cloning and Counterfeiting

The final class of threats for hardware security analyzed in this chapter are cloning and counterfeiting. These two threats, which are direct consequences of an highly fragmented design and fabrication chain, cause a huge economic loss every year. Cloning is the attempt to illegitimately copy an hardware design, with the purpose of reselling or using it without paying the required amount. Fabricating more chips than the ordered amount (usually called overproduction) is an example of cloning. Counterfeiting is an attempt of selling an illegitimate product as a regular one. A possible example of this is the fabrication of a reverse engineered circuit, or the selling of an used circuit as a regular one.

Locking and metering are possible countermeasure against these attacks. The key idea behind these approach is that a circuit operates correctly only of the correct unlocking sequence is provided. The sequence is often paired with a unique ID stored in (or extracted from) the circuit itself. The sequence is known only to the circuit designer and to the legitimate user, and it is unknown from all the other parties involved in the fabrication process. Illegitimate users will not have the unlocking sequence and thus should not be able to operate the circuit (or, at least, to operate it properly).

Applying locking and metering scheme to approximate circuit could not be trivial. In fact, approximation naturally means that multiple execution paths can be followed to reach an acceptable result. On the contrary, locked circuits are attempting exactly the opposite, namely limiting the number of acceptable paths. The situation is even more complex in circuits designed to operate also in presence of errors (such as non-deterministic circuits). Since these errors can be caused by the wrong unlocking, it could be that the circuit itself is capable of operating, at least to some extent, even when the correct locking sequence is not provided.

11.3 Approximate Computing and Security Solutions

In this section, we discuss two distinctive security techniques which benefit from approximate computing. The first, in the domain of cryptography, the second one in the area of designing secure neural network implementations. While diverse in nature, these two techniques highlight the—sometimes unexpected—potentials that AxC offers for making systems more secure at a reasonable cost.

11.3.1 Cryptography Based on Approximate Primitives

In Mittal's survey of approximate computing [6], he discusses some of the limitations of the paradigm, in particular with regards to AxC's applicability. Mittal believed that "due to their nature, some applications are not amenable to approximation, for example, cryptography" ([6], page 62:5). However, some specific novel techniques in cryptography can potentially work, and indeed benefit, from approximate computing, being already based on approximation themselves. The most important example is homomorphic encryption, which incidentally, is particularly suited as a security mechanism to protect some of the main applications running on AxC, namely those based on artificial intelligence technology (e.g., machine learning, neural networks).

11.3.1.1 Homomorphic Encryption

Homomorphic encryption is a form of encryption that allows calculations to be performed on the encrypted data (the *ciphertext*) without decrypting it first. For example, if we have two integers a and b, and their encryptions $Enc(a)$ and $Enc(b)$, the encryption scheme (or *cipher*) is homomorphic with respect to the addition operation if there is an operation $\boxplus$, for which

$$Enc(a) \boxplus Enc(b) = Enc\,(a + b) \tag{11.1}$$

is true for any (a, b). The operation $\boxplus$ is the equivalent on ciphertexts of addition on *plaintexts* (the unencrypted data). We say that $\boxplus$ operates in the ciphertext space: in fact, both the inputs and the result of the computation are in encrypted form. When the result is decrypted, it is the same as if the equivalent operation in the plaintext space $(+)$ had been performed on the unencrypted data.

The potential of homomorphic encryption is self-evident: if an encryption scheme is found that is homomorphic for a wide range, or potentially all operations, then computation could be securely outsourced. The outsourcing party could keep their inputs in the computation private, only providing them in encrypted form to the party performing the computation. As long as the encryption scheme is secure, the party performing the computation cannot learn the inputs or indeed the outputs of the computation. More broadly, homomorphic encryption can provide security in any secure multi-party computation scenario, that is, when two or more parties are interested in computing a function cooperatively, but want to maintain their inputs private. For these reasons, the design of homomorphic encryption schemes has become one of the main research challenges in modern cryptology.

The problem of constructing an homomorphic encryption scheme was first proposed in 1978, within a year of publishing of the popular and still widely used RSA public-key encryption scheme [29]. The RSA scheme, in fact, is homomorphic with respect to multiplication. Other schemes also exhibit homomorphic properties, such as ElGamal (multiplication), or the Benaloh and Paillier cryptosystems (addition). For over 30 years, however, it was unclear whether a cipher allowing any operation to be performed homomorphically could be created. For such a scheme to exist, it would have to support both addition and multiplication operations on ciphertexts, from which it would then be possible to construct circuits for performing arbitrary computations.

A scheme that supports only addition or multiplication, but not both, is called *partially* homomorphic. If we consider computations as circuits composed of arithmetic operations (addition, multiplication), partially homomorphic schemes support the evaluation of circuits consisting of only one type of operation, that is, either addition or multiplication. Certain schemes may support both addition and multiplication, but be limited in their capabilities: *somewhat* homomorphic encryption schemes can evaluate the two types of operations, but only for a subset of circuits; while *leveled* fully homomorphic encryption supports the evaluation of arbitrary circuits of bounded (pre-determined) depth. Finally, *fully* homomorphic encryption (FHE) allows the evaluation of arbitrary circuits of unbounded depth, and is the strongest form of homomorphic encryption.

For the majority of homomorphic encryption schemes that would theoretically support both additions and multiplications, the limitation in the kind of circuits to be evaluated (somewhat) or their depth (leveled) derives from the fact that each ciphertext is noisy in some sense, and the noise grows as ciphertexts are added and multiplied, ultimately making the final output ciphertext indecipherable. In particular, the main limitation in performing computations over encrypted data is the multiplicative depth of circuits. The specific definition of noise varies slightly depending on the specific public-key encryption scheme that is at the base of the homomorphic construction, but in general a ciphertext c is considered a function $c = \gamma(v, x)$ where vector v fulfils some (algebraic) property and vector x represent the noise. It is possible to correctly decrypt $\gamma(v, x)$ as long as x is not too large. Applying operations (additions and/or multiplications) to encrypted data will yield new ciphertexts $c' = \gamma(v', x')$. For a good construction, the increase of noise during one operation should be bounded, but successive operations will ultimately result in an undecipherable c. The presence of this noise is what makes computations performed over homomorphic encryption somewhat similar to those performed in approximate computing; but it is also what limited, until recently, the practical applicability of the technology.

The first major breakthrough came in 2009, when Craig Gentry proposed the first construction for a fully homomorphic encryption scheme, based on lattice-based cryptography [30]. The construction is based on a somewhat homomorphic scheme, which is limited to evaluating low-degree polynomials due to the noise in the ciphertext. In the scheme, v is in the ideal lattice, and the vector x grows with each addition and (especially) multiplication operation: similarly to other schemes, x eventually becomes so long that it causes a decryption error. However, Gentry's intuition was that the noise could be reduced before it made the ciphertext undecipherable, by using an appropriate technique at some point in the computation. If the noise reduction technique can be applied multiple times, then theoretically a circuit of unlimited depth or complexity could be evaluated. Gentry's strategy is to make the cipher *bootstrappable*, i.e., capable of evaluating its own decryption circuit (homomorphically) and then at least one more operation. Any bootstrappable somewhat homomorphic encryption scheme can be then converted into a fully homomorphic encryption scheme by recursive self-embedding.

The function of the bootstrapping procedure is to "refresh" the ciphertext by applying to it the decryption procedure homomorphically: this results in a new ciphertext that encrypts the same value as the previous one, but has lower noise. Bootstrapping is a *noise reduction* function, i.e., it replaces $c = \gamma(v, x)$ by a value $c_{\text{boot}} = \gamma(v, x_{\text{boot}})$ with $x_{\text{boot}} \ll x$. By applying the bootstrapping procedure sufficiently often, whenever the noise exceeds a certain threshold, it is possible to compute an arbitrary number of additions and multiplications while keeping the processed ciphertext decryptable. Intuitively, it is easy to see how an appropriately designed bootstrapping scheme could potentially also be used to allow homomorphic encryption to be performed over approximate computing, which similarly introduces "noise" in the computation in the form of approximation. Since the scheme inherently tolerates non-determinism associated with noise x, it is suitable, in principle, to tolerate effects of approximate arithmetic operations as well.

11.3.1.2 Approximate Homomorphic Encryption

Following Gentry's first proposal for fully homomorphic encryption, a number of improvements and new schemes have been proposed in the literature. The new schemes focus on two main areas: a different security assumption (e.g., the hardness of a different problem), and increased efficiency. The latter is normally achieved through improvements in the efficiency of the bootstrapping procedure (e.g., in [31, 32]), or in schemes that achieve a slower growth of the noise during the homomorphic computations (e.g., [33, 34]), which in turn means bootstrapping has to be performed less frequently, if at all.

The Brakerski-Gentry-Vaikuntanathan scheme [33], first proposed in 2011, and a number of following constructions base their security on the hardness of the (Ring) Learning With Errors (RLWE) problem. Learning with errors is the computational problem of inferring a linear n-ary function over a finite ring from a set of given samples, some of which may be erroneous [35]. RLWE is a specialization of the learning with errors problem to polynomial rings over finite fields [36]. Because of the presumed difficulty of solving the RLWE problem on a quantum computer, cryptographic schemes based on this security assumption are assumed to be post-quantum secure. (R)LWE-based cryptosystems are inherently approximate, and the decryption function is probabilistic by design. Therefore, an appropriate selection of parameters may permit approximate hardware or approximate computing techniques to be deployed in the implementation of homomorphic encryption schemes based on this paradigm, although this is limited to the decryption function.

In 2017, the CKKS (Cheon-Kim-Kim-Song) scheme was proposed, marking yet one more step towards the realization of a practical fully homomorphic scheme [37]. The CKKS scheme is also derived from the RLWE problem, and supports efficient rounding operations. Rounding controls the noise increase over encrypted multiplications, thus reducing the number of bootstrapping operations in a circuit. Interestingly, CKKS performs additions and multiplications on encrypted real numbers but yields approximate results (the evaluation of an approximate circuit over a ciphertext returns an approximation rather than an exact result), and is constructed to deal efficiently with the errors arising from these approximations. The CKKS approach is suited for fast polynomial approximation and floating-point computations. In particular, fast polynomial approximation in CKKS can evaluate the multiplicative inverse, exponential, and logistic functions, as well as the discrete Fourier transform. These properties make the schemes particularly suited for machine learning applications [38], which have inherent noises in their structure, and also open the way to the implementation of fully homomorphic encryption over approximate computing.

Preliminary results of the application of approximate computing to approximate homomorphic encryption demonstrate the promise of this research direction. In [39] Bian et al. apply hardware-based approximate computing techniques to an appropriately designed approximate decryption technique for (R)LWE-based cryptosystems. Their results indicate that speed increase, area reduction, power reduction, and ciphertext size reduction can be achieved simultaneously, with efficiencies of 20% or higher on the decryption function. Khanna et al. [40] apply a variation of the approximate computing techniques of task skipping and depth reduction (derived from loop perforation) to functions evaluated using the CKKS scheme, thus further approximating the function, with the goal of increasing the performance of the homomorphic evaluation. Their experiment targets logistic (commonly used in statistics, neural networks and machine learning) and exponential functions, which are implemented with and without approximate computing optimizations over CKKS. The results indicate a speed up in running time for homomorphic encryption evaluation with task skipping is between 12.1% and 45.5%.

The significance of these results is linked to the fact that current homomorphic encryption techniques imply large computational overheads, making widespread deployment of the schemes still far from being realistic. Approximate computing can substantially increase the efficiency of fully homomorphic schemes, and therefore contribute to the ultimate goal of widespread deployment of FHE in real-world systems.

11.3.1.3 Implementations and Standardization

As evidence of the increasing viability and practicality of fully homomorphic encryption, a number of related libraries have been developed and released.

Some libraries are the implementation of specific schemes, such as HEAAN (Homomorphic Encryption for Arithmetic of Approximate Numbers) [41], which implements the approximate homomorphic encryption scheme proposed by Cheon, Kim, Kim and Song (CKKS) [37]. FHEW (Fully Homomorphic Encryption library) [42] implements the Fully Homomorphic Encryption scheme by Ducas and Micciancio [43]. Their work was later improved by Chillotti et al. in [44], which resulted in the TFHE library [45].

Recently, a number of large industry players have also displayed increasing interest in the development of homomorphic encryption solutions, including Microsoft and IBM. Instead of proposing yet another cryptographic scheme, the industry has focused on developing multi-scheme frameworks that implement a number of existing FHE primitives. The most notable frameworks are presented in Table 11.1. These include HElib [46], developed by Shai Halevi and Victor Shoup at IBM; SEAL [47], by Microsoft Research; and PALISADE [49], developed by a consortium of industry partners and universities lead by Duality Technologies and New Jersey Institute of Technology, and funded by DARPA, the US Defense Advanced Research Projects Agency. The same consortium is also working on a DARPA project to develop an application-specific integrated circuit (ASIC) chip for hardware-accelerated homomorphic encryption, currently codenamed TRE-

Table 11.1 A list of current libraries supporting multiple FHE schemes

Library	*Developer*	*Supported FHE schemes*
HElib [46]	IBM	BGV [33], CKKS [37]
SEAL [47]	Microsoft	CKKS [37], BFV [48]
PALISADE [49]	Consortium of DARPA-funded defense contractors and academics	BGV [33], CKKS [37], BFV [48], FHEW [43], TFHE [44]

BUCHET. Worthy of note is also SHEEP [50], by the Alan Turing Institute, which is a homomorphic encryption evaluation platform aimed at providing a tool for practitioners to evaluate the state-of-the-art of (fully) homomorphic encryption technology in the context of their concrete application.

Another indication of the readiness of homomorphic encryption for real-world computation scenarios is the joint standardization effort by a government, industry, and academia consortium promoted through two Homomorphic Encryption Standardization Workshops in 2017 and 2018. The first workshop produced three white papers addressing the security [51] and applications [52] of homomorphic encryption as well as the development of an API [53]. After a public comment period and peer-review, the security white paper was publicly endorsed at the second standardization workshop, resulting in the first draft of the Homomorphic Encryption Standard [54].

11.3.2 Defending Against Adversarial Attacks on Neural Networks

The explosive proliferation of artificial intelligence (AI) technology has given rise to new security threats. *Adversarial attacks against neural networks* (NNs) is one such threat. In contrast to attacks considered in Sect. 11.2, these adversarial attacks are not hardware-oriented; they apply to hardware and software implementations of NNs alike. We will discuss, however, that approximate hardware can provide protections against such attacks.

11.3.2.1 Adversarial Attacks

An NN takes an object, e.g., an image, a video or an audio sample, as input and produces a *classification* for this object. A classification can be understood as the index of one out of a finite list of predefined classes, such as letters, road signs, or control instructions for a personal assistant device. During an adversarial attack, the attacker (adversary) makes minimal perturbations to the input with the malicious purpose to obtain a mis-classification [55, 56]. These perturbations are often called *adversarial noise*, but it is important to distinguish them from natural noise due to, e.g., radiation or electromagnetic disturbances. Perturbations due to natural noise are random and happen with no particular intent; perturbations due to adversarial noise are carefully calculated by the adversary and guided by the intent to achieve mis-classification. Figure 11.3 shows an example of an adversarial attack on an image. Such attacks can also affect AI in the context of text or audio processing. For example, Chen et al. [57] reported that an image correctly captioned "*A red stop sign sitting on the side of a road*" was captioned "*A brown teddy bear laying on top*

Fig. 11.3 Adversarial attack example: Image classified as `trombone` with confidence 24.67% (**a**); perturbed image classified as `mousetrap` with confidence 55.07% (**b**); perturbations (differences) between the two images (**c**). Original image from https://commons.wikimedia.org/wiki/Trombone

of a bed" due to an adversarial attack. A collection of playable adversarial audio examples can be found on the website [58].

The potential consequences of a successful adversarial attack can be grave. Imagine a person saying to the personal assistant device "Open the window", whereas the adversary adds to the recorded sound a perturbation making the device understand the sentence as "Transfer $100 to bank account number such-and-such". Even more dangerous, consider a vehicular network of autonomous self-driving cars that are exchanging images. One car arriving at a crossing makes a photo of the traffic situation and forwards it to the following cars which have no direct line of sight to the crossing, in order to provide them with situational awareness and improve their autonomous decision-making, ultimately improving safety. If the image contains a stop sign, but an adversary manages to hijack the communication and to replace this image with one that looks identically but where the stop sign is interpreted as, e.g., turn-right sign, the following car will likely cause an incident even though the available images look unsuspicious.

One can distinguish between different types of attacks. The most important distinctions are [56]:

Targeted vs. untargeted attack: In a *targeted attack*, the adversary's intent is to change the classification of an input from correct class t_{corr} to one particular class t_{target}. In an *untargeted attack*, the adversary simply wants to achieve misclassification; an attack is considered successful if the classification changes from t_{corr} to *any* class $t \neq t_{\text{corr}}$.

White-box vs. black-box attack: In a *white-box attack*, the adversary knows the structure of the NN and its parameters (weights, biases, etc.). In a *black-box attack*, the adversary can observe the outcomes of the classification but does not know the NN's internals.

From the adversary's perspective, a targeted attack is more difficult than a untargeted attack, and a black-box attack is more difficult than a white-box attack. Consequently, designing an effective countermeasure against untargeted and white-box attacks is harder compared with targeted and black-box attacks.

Mathematically, the adversary is trying to solve the following optimization problem in the untargeted, white-box case [59]:

$$\min_{x\in[0,1]^d} f(x) = (x - x_0)^2 + \lambda \cdot \mathrm{Loss}(x, Net(x), t_{\mathrm{corr}}). \tag{11.2}$$

Here, $(x - x_0)^2$ is the difference between the sought input x and the original, unperturbed input x_0; this part of (11.2) guides the search towards values that are as close to x_0 as possible. $Net(x)$ is a function that represents the neural network; it assigns every class t a score $Net(x)_t$. The expression $\mathrm{Loss}(x, Net(x), t_{\mathrm{corr}})$ stands for

$$\mathrm{Loss}(x, Net(x), t_{\mathrm{corr}}) = \max\left\{Net(x)_{t_{\mathrm{corr}}} - \max_{t\neq t_{\mathrm{corr}}}\{Net(x)_t\} + \kappa, 0\right\}. \tag{11.3}$$

This part of (11.2) incurs a penalty when x is still classified in the correct class t_{corr}, with κ serving as a safety margin and enforcing that x has "crossed the boundary" to the neighboring class by at least this amount. λ balances between the two objectives.

In the targeted case, the loss term in (11.2) would be defined with respect to the target class t_{target} rather than $t \neq t_{\mathrm{corr}}$. For the black-box attack, the function $Net(x)$ is unavailable, but the network can still be used for arbitrary values of x. One workaround is to approximate f's gradients by perturbing each of the input's d components. For example, Ting and Hayes [59] define a small constant ε and approximate gradients by

$$\frac{\partial f(x)}{\partial x_i} \approx \frac{f(x + \varepsilon e_i) - f(x - \varepsilon e_i)}{2\varepsilon}, \tag{11.4}$$

where e_i is the i-th standard basis vector.

Figures 11.4 and 11.5 show the example application of the white-box untargeted attack against MobileNetV2 trained with the ImageNet dataset. One can see how adding perturbations calculated the method discussed in [60] weighted by parameter ε first leads to a rather high-confidence mis-classification (Fig. 11.4). For larger values of ε (Fig. 11.5), the visual quality of perturbed images is reduced, the classification is still wrong and the confidence low. Note that the case $\varepsilon = 0.01$ in Fig. 11.4 corresponds to Fig. 11.3b and c.

The key idea of the protection from [59] is to use NNs implemented based on *stochastic computing*, a variant of AxC associated with a certain degree of randomness, and actually *increase* this randomness by adding a small dedicated circuit to the NN hardware. This will corrupt, to some extent, the gradient calculation according to (11.4) and interfere with its usage in (11.2), thus thwarting the attack in most cases. Employing too much randomization will deteriorate the actual performance of the implemented NN. However, a careful control of the

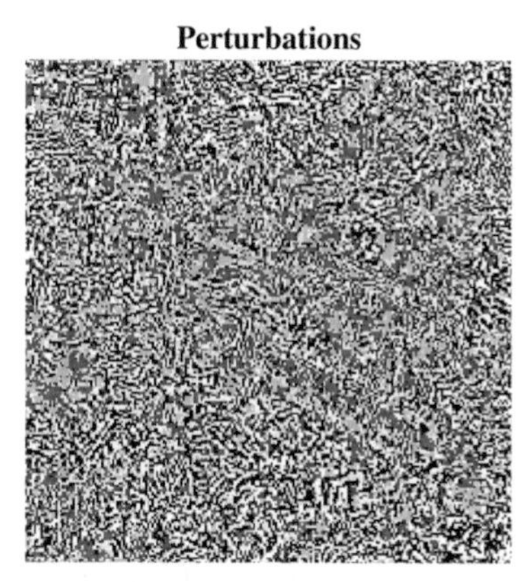

Fig. 11.4 Adversarial attack example: Perturbations calculated by the method from [60] for image from Fig. 11.3a, and the results of adding these perturbations weighted by $\varepsilon = 0.01$ and 0.02

$\varepsilon = 0.05$: **Classified `stethoscope` with confidence 7.02%**

$\varepsilon = 0.10$: **Classified `dragonfly` with confidence 10.64%**

Fig. 11.5 Adversarial attack example cont'd: Results of adding perturbations from Fig. 11.4 to image from Fig. 11.3a, weighted by $\varepsilon = 0.05$ and 0.10

additional injected randomness can deliver a good compromise between greatly improved attack protection and insignificantly reduced classification performance.

11.3.2.2 Neural Networks Based on Stochastic Computing

Stochastic computing (SC) is a special type of approximate computing that leads to very area- and power-efficient arithmetic operations and is therefore very useful for implementing NNs. The focus of this section is how SC is useful in protecting NN implementations against adversarial attacks. To this end, instead of giving a full introduction into different flavors of SC, the following paragraphs will concentrate on SC concepts and primitives that are useful for implementing NNs. More information about SC in general can be found in, e.g., [5]; for a more in-depth coverage of SC-based NNs, see the overview paper [61] and the upcoming (at the time of writing this chapter) Special Issue of the IEEE Design & Test magazine on Stochastic Computing for Neuromorphic Architectures.

SC is based on the notion of a *stochastic number* (SN). While different definitions of an SN exist, the focus here is on *bipolar SNs* and refer to them as SNs in the following. An N-bit SN is a string of N bits (0s or 1s). The value $\mathrm{val}(a)$ of an SN a is defined to be $(N_1 - N_0)/N$, where N_1 and N_0 are the number of 1 and 0 values in a, respectively. For example, the value of a 4-bit SN 1110 is $\mathrm{val}(1110) = (3-2)/4 = 0.5$, the value of another 4-bit SN 0100 is $(1-4)/4 = -0.5$, and the value of a 10-bit number 0100110000 is $(3-7)/10 = -0.4$. It is apparent that SNs can assume values between -1 and 1; all calculations with number exceeding this range must be *scaled* into this range. It is also clear that SN values are not unique and depend only on the numbers of 0s and 1s and not on their precise positions.

The striking feature of SN representations is that the basic arithmetic operations, addition and multiplication, have an extremely compact representation. Given two SNs $a = a_1 \dots a_N$ and $b = b_1 \dots b_N$, applying these SNs, bit by bit, to the inputs of an XNOR gate will produce an N-bit SN c at the outputs ($c = c_1 \dots c_N$ where $c_i = \overline{a_i \oplus b_i}$ for $1 \le i \le N$). c's value is the product of values of a and b: $\mathrm{val}(c) = \mathrm{val}(a) \cdot \mathrm{val}(b)$. Moreover, applying a and b to the data inputs of a multiplexer with the select input randomly switching between 0 and 1 (this is equivalent to applying an SN with value 0 to the select input) will result in the multiplexer's output having value $(\mathrm{val}(a)+\mathrm{val}(b))/2$. This function is called *scaled addition*; note that regular addition cannot be implemented in SC because adding two numbers in range $[-1, 1]$ will, in general, lead to leaving this range. With these constructions, SC can represent arbitrary polynomials; non-polynomial functions such as division can either be approximated or designed using sequential elements. One prominent example is the *stochastic hyperbolic tangent* function stanh [62], which is used as activation function in neural network.

Figure 11.6a shows the SC design of one 4-input neuron, consisting of a multiply-accumulate part and the activation function. The first row of XNOR gates adds weights w_j to inputs i_j. The 4-input multiplexer performs the scaled addition (note that the multiplexer has two select inputs). The activation function outlined in Fig. 11.6a is stanh from [62], implemented by an up-down counter. The gate-level netlist of the activation function is not shown; note that some adjustment within the stanh block is required to compensate for the scaling factor of 4 from the multiplexer-based addition.

Figure 11.6b illustrates the multiply-accumulate part of the neuron for specific values using the SN length $N = 10$. It starts with *stochastic number generators* (SNGs) which take a binary number and convert it into the SC domain. In the simplest case, an SNG consists of a comparator and a pseudo-random number generator, even though more sophisticated designs exist [63]. It outputs N bits such that their value equals (in the general case, approximates) the binary value on the

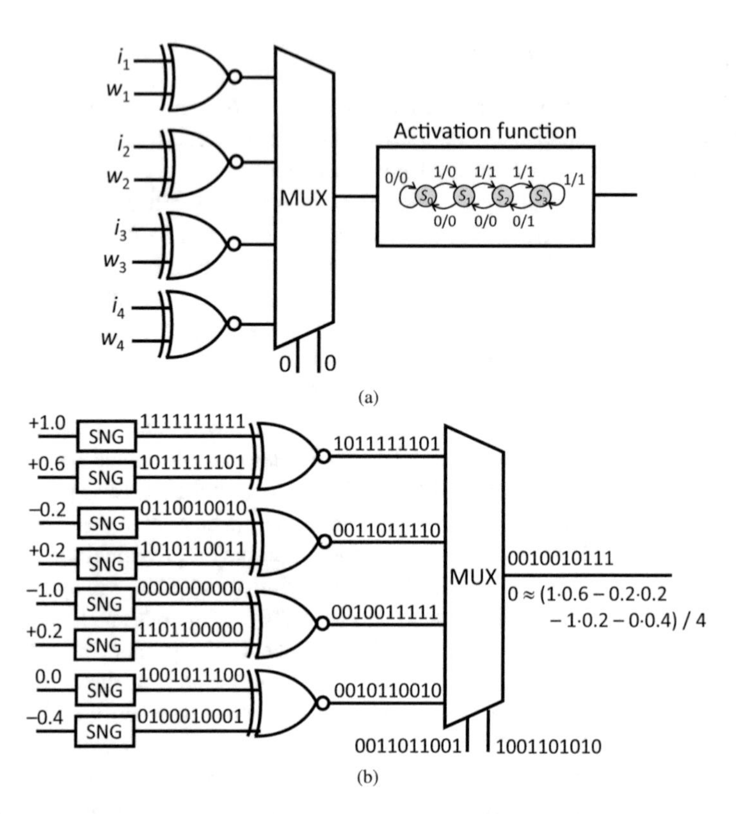

Fig. 11.6 4-input neuron based on stochastic computing with stanh FSM as activation function (**a**) and example calculation using its multiply-accumulate part (**b**)

SNG's input. The circuit from Fig. 11.6b takes the values from the SNG outputs, adds and multiples them.

One notices immediately that stochastic computing is associated with some precision loss. For example, the second product ($-0.2 \cdot 0.2 = -0.04$) in the example of Fig. 11.6b has no exact representation as a 10-bit SN. Moreover, the XNOR gate does not calculate the closest representable approximation of -0.04, which would have been 0; instead it calculates 0011011110, which has the value +0.2. One can define several sources of imprecision in SC, including approximation, quantization, random fluctuations [64] and correlations [65]. In general, imprecisions are considered undesired and limiting the performance of stochastic circuits in general and stochastic NN in particular [66]. However, here these imprecisions are utilized to defend NNs against adversarial attacks, and the extent of such imprecisions is even artificially increased.

11.3.2.3 Adversarial Attack Defense Based on Stochastic Computing

The NN considered in [59] is the 19-layer version of the VGG network introduced in [67], called VGG19-SC. This network's structure is shown in Fig. 11.7. The last fully connected layer of this network is implemented using SC whereas the first layers are using the conventional binary representation. Note that this hybrid approach necessitates binary-to-stochastic conversion by means of SNGs briefly mentioned in the last section *within* the NN, rather than at its beginning, as would be the case in a fully-SC implementation. The intrinsic imprecision of SC complicates adversarial attacks against the network, yet [59] adds a further protection mechanism: the randomness-injection circuit (RIC).

The RIC, shown in Fig. 11.8, takes a stochastic number x of arbitrary length N and a parameter c, also represented as an SN. If c has the minimum value (i.e., it consists of only 0 bits), the RIC does not modify x at all and simply outputs it one clock cycle later at the output y. If, however, c has bits equal to 1, then y does not track x but rather retains its value for every such bit. It was proven in [59] that this construction guarantees, under certain assumptions, that the expected value of y will be equal to that of the input x for all values of c. However, the "imprecision", or variance, of y increases with an increasing c.

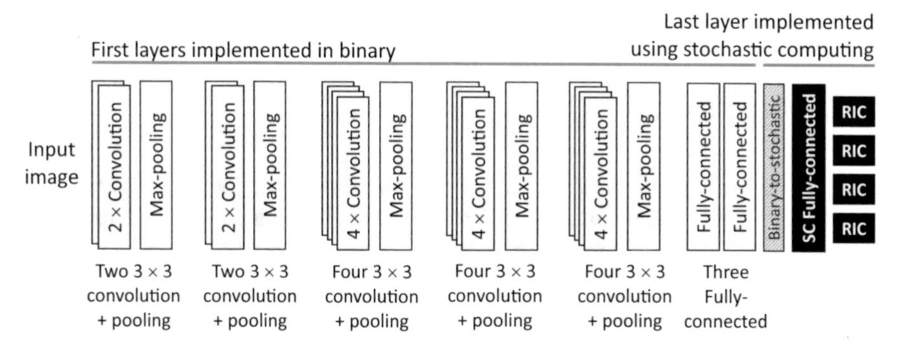

Fig. 11.7 Hybrid binary-stochastic VGG19-SC network used for analysis of adversarial attacks. The network structure corresponds to VGG-19 [67]. The stochastic layer and the randomness-injection circuits (RICs) are shown in black

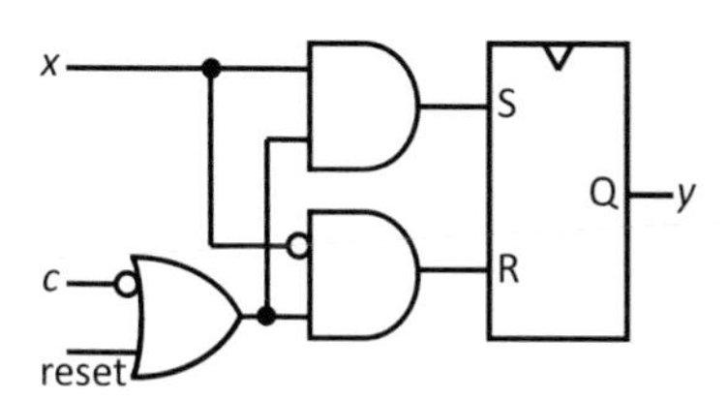

Fig. 11.8 Randomness-injection circuit (RIC) [59]

The experiment in [59] was based on the VGG19-SC network from Fig. 11.7 with the last fully connected layer implemented in SC (using SNs of length $N = 32$) and RICs added to its outputs. Since making parts of the network stochastic (i.e., approximate) and adding RIC may degrade its classification accuracy, the authors compared both: the resistance against adversarial attacks and the classification accuracy, when SC was used or not and for different values of parameter c. To counter the accuracy degradation, they trained the network while adding Gaussian noise to the inputs of fully connected layer. With the NN thus trained, the classification accuracy was around 86% for the completely non-SC architecture with the last fully connected layer implemented in binary and therefore no RICs. For the SC version with different levels of c, the accuracy declined minimally, by up to about 1%.

To quantify the protective capabilities of VGG19-SC, the *attack success rate* ASR was calculated for all considered architectures. An attack was considered successful if after performing 5,000 steps the mis-classification happened with confidence parameter κ from (11.3) exceeding 0.2 and the distortion $(x - x_0)^2$ from (11.2) was below 10. The ASR reported is simply the proportion of successful attacks according to the above definition among 100 attack attempts. It turned out that the non-SC NN lacking any protections had an ASR of 76%. For the largest value of c, the ASR went down to 59%, demonstrating the effectiveness of SC as a protective mechanism.

11.4 Conclusions

This chapter discussed the implications of approximate computing on security. It provided a systematic overview of threats potentially caused or exacerbated by adopting the AxC paradigm and the impact of AxC on the applicability of known countermeasures. Moreover, it provided two examples how AxC based solutions can provide benefits for emerging cryptographic approaches and for making neural network implementations more secure. We believe that more research is required in order to fully understand AxC challenges and potentials related to security and to make this technology applicable in next-generation security-critical systems.

Acknowledgments This work was partially supported by the German Research Foundation (DFG) under grant 1220/12-1 and by the European Union Horizon 2020 research and innovation program under CPSoSAware project (grant no. 871738). The authors are thankful to John P. Hayes of University of Michigan, Ann Arbor, and Florian Neugebauer, University of Stuttgart, for provided materials and helpful discussions.

References

1. Vogel, S., Guntoro, A., & Ascheid, G. (2017). Efficient hardware acceleration for approximate inference of bitwise deep neural networks. In *DASIP* (pp. 1–6). New York: IEEE.
2. Regazzoni, F., Alippi, C., & Polian, I. (2018). Security: The dark side of approximate computing? In *ICCAD* (p. 44). New York: ACM.
3. Yellu, P., Boskov, N., Kinsy, M. A., & Yu, Q. (2019). Security threats in approximate computing systems. In *ACM Great Lakes Symposium on VLSI* (pp. 387–392). New York: ACM.
4. Liu, W., Gu, C., O'Neill, M., Qu, G., Montuschi, P., & Lombardi, F. (2020). Security in approximate computing and approximate computing for security: Challenges and opportunities. *Proceedings of the IEEE, 108*(12), 2214–2231.
5. Alaghi, A., Qian, W., & Hayes, J. P. (2018). The promise and challenge of stochastic computing. *IEEE Transactions on CAD of Integrated Circuits and Systems, 37*(8), 1515–1531.
6. Mittal, S. (2016). A survey of techniques for approximate computing. *ACM Computing Surveys, 48*(4), 62:1–62:33.
7. Baek, W., & Chilimbi, T. M. (2010). Green: A framework for supporting energy-conscious programming using controlled approximation. In *PLDI* (pp. 198–209). New York: ACM.
8. Shim, B., Sridhara, S. R., & Shanbhag, N. R. (2004). Reliable low-power digital signal processing via reduced precision redundancy. *IEEE Transactions on Very Large Scale Integration (VLSI) Systems, 12*(5), 497–510.
9. Yazdanbakhsh, A., Pekhimenko, G., Thwaites, B., Esmaeilzadeh, H., Mutlu, O., & Mowry, T. C. (2016). RFVP: Rollback-free value prediction with safe-to-approximate loads. *ACM Transactions on Architecture and Code Optimization, 12*(4), 62:1–62:26.
10. Renganarayana, L., Srinivasan, V., Nair, R., & Prener, D. (2012). Programming with relaxed synchronization. In *Proceedings of the 2012 ACM Workshop on Relaxing Synchronization for Multicore and Manycore Scalability, RACES '12*, New York, NY (pp. 41–50). New York: ACM.
11. Jiang, H., Liu, C., Liu, L., Lombardi, F., & Han, J. (2017). A review, classification, and comparative evaluation of approximate arithmetic circuits. *Journal on Emerging Technologies in Computing Systems, 13*(4), 60:1–60:34.
12. Hegde, R., & Shanbhag, N. (2004). A voltage overscaled low-power digital filter IC. *IEEE Journal of Solid-State Circuits*, 39(2), 388–391 (2004)
13. Uppu, R. T., Uppu, R. K., Singh, A. D., & Chatterjee, A. (2013). A high throughput multiplier design exploiting input based statistical distribution in completion delays. In *VLSI Design* (pp. 109–114). Washington: IEEE Computer Society.
14. Krause, P. K., & Polian, I. (2011). Adaptive voltage over-scaling for resilient applications. In *DATE* (pp. 944–949). New York: IEEE.
15. Uppu, R. K., Uppu, R. T., Singh, A. D., & Polian, I. (2014). Better-than-worst-case timing design with latch buffers on short paths. In *VLSI Design* (pp. 133–138). New York: IEEE Computer Society.
16. von Neumann, J. (1956). Probabilistic logics and the synthesis of reliable organisms from unreliable components. In *Automata studies* (pp. 43–98).
17. Cho, H., Leem, L., & Mitra, S. (2012). ERSA: Error resilient system architecture for probabilistic applications. *IEEE Transactions on CAD of Integrated Circuits and Systems, 31*(4), 546–558.
18. Li, B., Gu, P., Shan, Y., Wang, Y., Chen, Y., & Yang, H. (2015). RRAM-based analog approximate computing. *IEEE Transactions on CAD of Integrated Circuits and Systems, 34*(12), pp. 1905–1917.
19. Kocher, P., Jaffe, J., & Jun, B. (1999). *Differential power analysis* (Vol. 1666, pp. 398–412). Berlin: Springer.
20. Kocher, P. C. (1996). *Timing attacks on implementations of Diffie-Hellman, RSA, DSS, and other systems* (Vol. 1109, pp. 104–13). Berlin: Springer.

21. Barenghi, A., Breveglieri, L., Koren, I., & Naccache, D. (2012). Fault injection attacks on cryptographic devices: Theory, practice and countermeasures. *Proceedings of the IEEE, 100*(11), 3056–3076 (2012).
22. Barenghi, A., Hocquet, C., Bol, D., Standaert, F., Regazzoni, F., & Koren, I. (2014). A combined design-time/test-time study of the vulnerability of sub-threshold devices to low voltage fault attacks. *IEEE Transactions on Emerging Topics Computing, 2*(2), 107–118 (2014)
23. Wang, Z., & Karpovsky, M. (2011). Algebraic manipulation detection codes and their applications for design of secure cryptographic devices. In *IEEE 17th Int'l On-Line Testing Symposium (IOLTS)* (pp. 234–239). New York: IEEE.
24. Torrance, R., & James, D. (2009). The state-of-the-art in IC reverse engineering. In *CHES. Lecture Notes in Computer Science* (Vol. 5747, pp. 363–381). New York: Springer.
25. Bhunia, S., Hsiao, M. S., Banga, M., & Narasimhan, S. (2014). Hardware Trojan attacks: Threat analysis and countermeasures. *Proceedings of the IEEE, 102*(8), 1229–1247.
26. Polian, I., Becker, G., & Regazzoni, F. (2016). Trojans in early design steps—An emerging threat. In *TRUDEVICE - 6th Conference on Trustworthy Manufacturing and Utilization of Secure Devices*. http://hdl.handle.net/2117/99414.
27. Becker, G. T., Regazzoni, F., Paar, C., & Burleson, W. P. (2013). Stealthy dopant-level hardware trojans. In *International Conference on Cryptographic Hardware and Embedded Systems* (pp. 197–214). New York: Springer.
28. Bhasin, S., & Regazzoni, F. (2015). A survey on hardware trojan detection techniques. In *2015 IEEE International Symposium on Circuits and Systems (ISCAS)* (pp. 2021–2024). New York: IEEE.
29. Rivest, R. L., Adleman, L., & Dertouzos, M. L. (1978). On data banks and privacy homomorphisms. *Foundations of secure computation* (pp. 169–179). New York: Academia Press.
30. Gentry, C. (2009). Fully homomorphic encryption using ideal lattices. In M. Mitzenmacher (Ed.), *Proceedings of the 41st Annual ACM Symposium on Theory of Computing, STOC 2009, Bethesda, MD, May 31–June 2, 2009* (pp. 169–178). New York: ACM.
31. Han, K., & Ki, D. (2020). Better bootstrapping for approximate homomorphic encryption. In S. Jarecki (Ed.), *Topics in Cryptology - CT-RSA 2020 - The Cryptographers' Track at the RSA Conference 2020, San Francisco, CA, February 24–28, 2020, Proceedings. Lecture Notes in Computer Science* (Vol. 12006, pp. 364–390). New York: Springer.
32. Alperin-Sheriff, J., & Peikert, C. (2014). Faster bootstrapping with polynomial error. In J. A. Garay & R. Gennaro (Eds.), *Advances in Cryptology - CRYPTO 2014 – 34th Annual Cryptology Conference, Santa Barbara, CA, August 17–21, 2014, Proceedings, Part I. Lecture Notes in Computer Science* (Vol. 8616, pp. 297–314). New York: Springer.
33. Brakerski, Z., Gentry, C., & Vaikuntanathan, V. (2014). (Leveled) fully homomorphic encryption without bootstrapping. *ACM Transactions on Computation Theory, 6*(3), 13:1–13:36.
34. Gentry, C., Sahai, A., & Waters, B. (2013). Homomorphic encryption from learning with errors: Conceptually-simpler, asymptotically-faster, attribute-based. In R. Canetti & J. A. Garay (Eds.), *Advances in Cryptology - CRYPTO 2013 – 33rd Annual Cryptology Conference, Santa Barbara, CA, August 18–22, 2013. Proceedings, Part I. Lecture Notes in Computer Science* (Vol. 8042, pp. 75–92). New York: Springer.
35. Regev, O. (2005). On lattices, learning with errors, random linear codes, and cryptography. In H. N. Gabow & R. Fagin (Eds.) *Proceedings of the 37th Annual ACM Symposium on Theory of Computing, Baltimore, MD, May 22–24, 2005* (pp. 84–93). New York: ACM.
36. Lyubashevsky, V., Peikert, C., & Regev, O. (2013). On ideal lattices and learning with errors over rings. *Journal of the ACM, 60*(6), 43:1–43:35.
37. Cheon, J. H., Kim, A., Kim, M., & Song, Y. S. (2017). Homomorphic encryption for arithmetic of approximate numbers. In T. Takagi & T. Peyrin (Eds.), *Advances in Cryptology - ASIACRYPT 2017 - 23rd International Conference on the Theory and Applications of Cryptology and Information Security, Hong Kong, December 3–7, 2017, Proceedings, Part I. Lecture Notes in Computer Science* (Vol. 10624, pp. 409–437). New york: Springer.
38. Wood, A., Najarian, K., & Kahrobaei, D. (2020). Homomorphic encryption for machine learning in medicine and bioinformatics. *ACM Computing Surveys, 53*(4), 70:1–70:35.

39. Bian, S., Hiromoto, M., & Sato, T. (2018). DWE: Decrypting learning with errors with errors. In *Proceedings of the 55th Annual Design Automation Conference, DAC 2018, San Francisco, CA, June 24–29, 2018* (pp. 3:1–3:6). New york: ACM.
40. Khanna, S., & Rafferty, C. (2020). Accelerating homomorphic encryption using approximate computing techniques. In P. Samarati, S. D. C. di Vimercati, M. S. Obaidat, & J. Ben-Othman (Eds.), *Proceedings of the 17th International Joint Conference on e-Business and Telecommunications, ICETE 2020 - Volume 2: SECRYPT, Lieusaint, Paris, July 8–10, 2020* (pp. 380–387). Setubal: ScitePress.
41. K. Crypto Lab Inc (2021). HElib - An implementation of homomorphic encryption. https://github.com/snucrypto/HEAAN. Last accessed on 15 March 2021).
42. Ducas, L., & Micciancio, D. (2021). FHEW - A fully homomorphic encryption library. https://github.com/lducas/FHEW. Last accessed on 15 March 2021).
43. Ducas, L., & Micciancio, D. (2015). FHEW: Bootstrapping homomorphic encryption in less than a second. In E. Oswald & M. Fischlin (Eds.), *Advances in Cryptology - EUROCRYPT 2015 - 34th Annual International Conference on the Theory and Applications of Cryptographic Techniques, Sofia, April 26–30, 2015, Proceedings, Part I. Lecture Notes in Computer Science* (Vol. 9056, pp. 617–640). New York: Springer.
44. Chillotti, I., Gama, N., Georgieva, M., & Izabachène, M. (2020). TFHE: Fast fully homomorphic encryption over the torus. *Journal of Cryptology, 33*(1), 34–91.
45. Chillotti, I., Gama, N., Georgieva, M., & Izabachène, M. (2021). TFHE - Fast fully homomorphic encryption over the torus. https://tfhe.github.io/tfhe/. Last accessed on 15 March 2021.
46. Halevi, S., & Shoup, V. (2021). HElib - An implementation of homomorphic encryption. https://doi.org/10.1145/2535925. (Last accessed on 15 March 2021).
47. W. Microsoft Research, Redmond (2021). Microsoft SEAL. https://www.microsoft.com/en-us/research/project/microsoft-seal/. Last accessed on 15 March 2021
48. Fan, J., & Vercauteren, F. (2012). Somewhat practical fully homomorphic encryption. *IACR Cryptology ePrint Archive, 2012*, 144.
49. D. Technologies and N. J. I. of Technology (2021). PALISADE lattice homomorphic encryption software library. https://palisade-crypto.org/. Last accessed on 15 March 2021.
50. U. Alan Turing Institute (2021). SHEEP - a homomorphic encryption evaluation platform. https://github.com/alan-turing-institute/SHEEP. (Last accessed on 15 March 2021).
51. Chase, M., Chen, H., Ding, J., Goldwasser, S., Gorbunov, S., Hoffstein, J., Lauter, K., Lokam, S., Moody, D., Morrison, T., Sahai, A., & Vaikuntanathan, V. (2017). *Security of Homomorphic Encryption*, Tech. Rep., HomomorphicEncryption.org, Redmond WA, July 2017.
52. Archer, D., Chen, L., Cheon, J. H., Gilad-Bachrach, R., Hallman, R. A., Huang, Z., Jiang, X., Kumaresan, R., Malin, B. A., Sofia, H., Song, Y., & Wang, S. (2017). *Applications of Homomorphic Encryption*, Tech. Rep., HomomorphicEncryption.org, Redmond WA, July 2017.
53. Brenner, M., Dai, W., Halevi, S., Han, K., Jalali, A., Kim, M., Laine, K., Malozemoff, A., Paillier, P., Polyakov, Y., Rohloff, K., Savaş, E., & Sunar, B. (2017). A standard API for RLWE-based homomorphic encryption, Tech. Rep., HomomorphicEncryption.org, Redmond WA, July 2017.
54. Albrecht, M., Chase, M., Chen, H., Ding, J., Goldwasser, S., Gorbunov, S., Halevi, S., Hoffstein, J., Laine, K., Lauter, K., Lokam, S., Micciancio, D., Moody, D., Morrison, T., Sahai, A., & Vaikuntanathan, V. (2018). *Homomorphic Encryption Security Standard*, Tech. Rep., HomomorphicEncryption.org, Toronto, November 2018.
55. Chen, P., Zhang, H., Sharma, Y., Yi, J., & Hsieh, C. (2017). ZOO: Zeroth order optimization based black-box attacks to deep neural networks without training substitute models. In B. M. Thuraisingham, B. Biggio, D. M. Freeman, B. Miller, & A. Sinha (Eds.), *Proceedings of the 10th ACM Workshop on Artificial Intelligence and Security, AISec@CCS 2017, Dallas, TX, November 3, 2017* (pp. 15–26). New York: ACM.

56. Papernot, N., McDaniel, P. D., Goodfellow, I. J., Jha, S., Celik, Z. B., & Swami, A. (2017). Practical black-box attacks against machine learning. In R. Karri, O. Sinanoglu, A. Sadeghi, & X. Yi (Eds.), *Proceedings of the 2017 ACM on Asia Conference on Computer and Communications Security, AsiaCCS 2017, Abu Dhabi, April 2-6, 2017* (pp. 506–519). New York: ACM.
57. Chen, H., Zhang, H., Chen, P., Yi, J., & Hsieh, C. (2018). Attacking visual language grounding with adversarial examples: A case study on neural image captioning. In I. Gurevych & Y. Miyao (Eds.), *Proceedings of the 56th Annual Meeting of the Association for Computational Linguistics, ACL 2018, Melbourne, July 15-20, 2018, Volume 1: Long Papers* (pp. 2587–2597). Stroudsburg: Association for Computational Linguistics.
58. Carlini, N. (2018). *Audio adversarial examples.*
59. Ting, P., & Hayes, J. P. (2019). Exploiting randomness in stochastic computing. In D. Z. Pan (Ed.), *Proceedings of the International Conference on Computer-Aided Design, ICCAD 2019, Westminster, CO, November 4–7, 2019* (pp. 1–6). New York: ACM.
60. TensorFlow Tutorial (2019). *Adversarial example using FGSM.* https://www.tensorflow.org/tutorials/generative/adversarial_fgsm
61. Liu, Y., Liu, S., Wang, Y., Lombardi, F., & Han, J. (2020). A survey of stochastic computing neural networks for machine learning applications. *IEEE Transactions on Neural Networks and Learning Systems, 32*, pp. 1–16.
62. Brown, B. D., & Card, H. C. (2001). Stochastic neural computation I: Computational elements. *IEEE Transactions on Computers, 50*(9), 891–905.
63. Neugebauer, F., Polian, I., & Hayes, J. P. (2017). Building a better random number generator for stochastic computing. In H. Kubátová, M. Novotný, & A. Skavhaug (Eds.), *Euromicro Conference on Digital System Design, DSD 2017, Vienna, August 30–September 1, 2017* (pp. 1–8). Washington: IEEE Computer Society.
64. Qian, W., Li, X., Riedel, M. D., Bazargan, K., & Lilja, D. J. (2011). An architecture for fault-tolerant computation with stochastic logic. *IEEE Transactions on Computers, 60*(1), 93–105.
65. Neugebauer, F., Polian, I., & Hayes, J. P. (2018). Framework for quantifying and managing accuracy in stochastic circuit design. *ACM Journal on Emerging Technologies in Computing Systems, 14*(2). https://doi.org/10.1145/3183345
66. Neugebauer, F., Polian, I., & Hayes, J. P. (2019). On the limits of stochastic computing. In *2019 IEEE International Conference on Rebooting Computing, ICRC 2019, San Mateo, CA, November 6–8, 2019* (pp. 98–105). New York: IEEE.
67. Simonyan, K., & Zisserman, A. (2015). Very deep convolutional networks for large-scale image recognition. In Y. Bengio & Y. LeCun (Eds.) *3rd International Conference on Learning Representations, ICLR 2015, San Diego, CA, May 7–9, 2015, Conference Track Proceedings.*

Chapter 12
Design, Verification, Test, and In-Field Implications of Approximate Digital Integrated Circuits

Alberto Bosio, Stefano Di Carlo, Patrick Girard, Annachiara Ruospo, Ernesto Sanchez, Alessandro Savino, Lukas Sekanina, Marcello Traiola, Zdenek Vasicek, and Arnaud Virazel

12.1 Introduction

Despite significant energy efficiency improvements in the semiconductor industry, computer systems keep consuming more and more energy [1]. Interestingly, many widely used applications—such as Recognition, Mining, and Synthesis (RMS) applications—now target a deployment toward mobile devices and on Internet of Things (IoT) structures. Therefore, it is necessary to improve the next-generation silicon devices and architectures on which these applications will run. The *inherent resiliency property* of RMS applications has been thoroughly investigated over the last few years [1–4]. This interesting property leads applications to be already (partially) tolerant to errors—as long as their results remain close enough to the

A. Bosio (✉)
University of Lyon, ECL, INSA Lyon, CNRS, UCBL, CPE Lyon, INL, UMR5270, Écully, France
e-mail: alberto.bosio@ec-lyon.fr

M. Traiola
University of Rennes, Inria, CNRS, IRISA, UMR 6074, Rennes, France
e-mail: marcello.traiola@inria.fr

S. Di Carlo · A. Ruospo · E. Sanchez · A. Savino
Control and Computer Eng. Dep., Politecnico di Torino, Torino, Italy
e-mail: stefano.dicarlo@polito.it; annachiara.ruospo@polito.it; ernesto.sanchez@polito.it; alessandro.savino@polito.it

P. Girard · A. Virazel
Université de Montpellier, CNRS, FR, Montpellier, France
e-mail: patrick.girard@lirmm.fr; arnaud.virazel@lirmm.fr

L. Sekanina · Z. Vasicek
Faculty of Information Technology, Brno University of Technology, Brno, Czech Republic
e-mail: sekanina@fit.vutbr.cz; vasicek@fit.vutbr.cz

A. Bosio et al. (eds.), *Approximate Computing Techniques*,
https://doi.org/10.1007/978-3-030-94705-7_12

expected ones. As reported in [4], the main sources of error tolerance for these applications are:

- noisy real-world inputs,
- redundant data,
- perceptual limitations of individuals who will use the computation output,
- non-deterministic algorithms which lead to non-unique outcomes, and
- self-healing capable systems.

As already pointed out in previous chapters, *Approximate Computing* (AxC) [1, 2] is an emerging computing paradigm that takes advantage of the inherent resiliency property. AxC has garnered increasing interest in the scientific community in the last years. It leverages the intuitive observation that selectively relaxing non-critical specifications may lead to improvements in power consumption, execution time, and/or chip area. AxC has been applied to the whole digital system stack, from hardware to applications.

This chapter focuses on *Approximate digital Integrated Circuits (AxICs)* design and manufacturing flow. Figure 12.1 depicts the main phases of the design flow. The starting points are the requirements, i.e., which functionalities have to be designed coupled with the energy, performances and area requirements, and the AxC metrics, i.e., how to estimate the quality of the outcomes due to approximation. **AxC design** stems from the application of AxC at hardware level. A widely used method to design those circuits is *functional approximation* of conventional integrated circuits (ICs) as described in Chaps. 3 and 4.

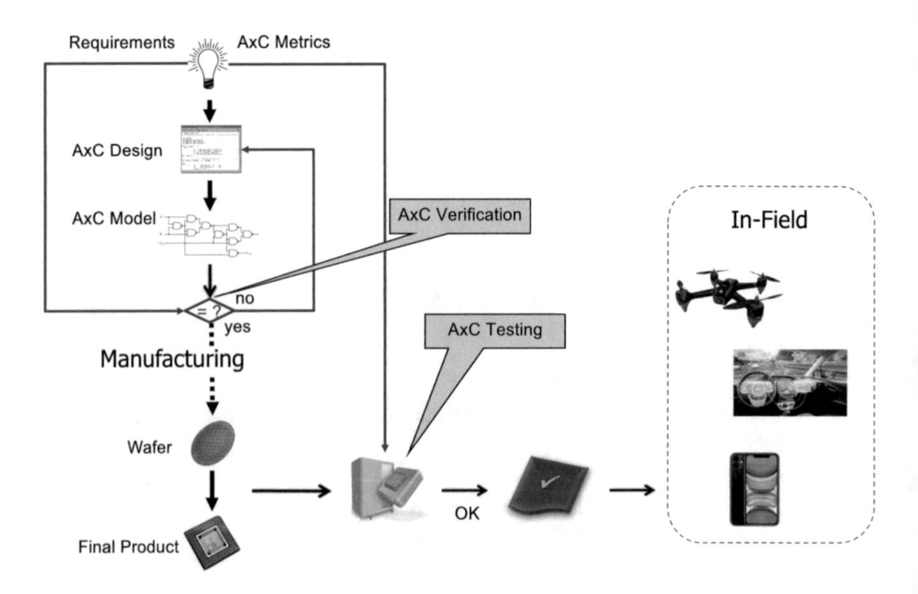

Fig. 12.1 AxIC design and manufacturing flow

AxC verification aims at verifying that the approximate design satisfy both the requirements and AxC metrics. If so, the AxC design goes through the manufacturing flow, and the fabricated AxIC will be eventually tested. While **AxC Testing** aims at screening defective circuits, it is interesting to note that AxC metrics have to be considered during the testing phase. Indeed, manufacturing defects may not significantly impact the functionality of the AxIC. AxC testing thus will focus on detecting only defects causing unacceptable degradation of the circuit, usually referred to as **critical defects**. All critical-defect-free circuits will be ready to be employed in the **in-field** application.

This chapter overviews the AxIC design and manufacturing phases by presenting the main challenges and state-of-the-art solutions. The chapter is structured as follows: Sect. 12.3 considers the design phase, Sect. 12.4 the verification phase, Sect. 12.5 the testing phase, and Sect. 12.6 the implication of Approximate Computing on in-field operation. Finally, Sect. 12.7 summarizes the main contributions of the chapter.

12.2 Background

This section introduces the basic concepts about testing, fault modeling, test generation, and fault simulation. These concepts will serve as background in the rest of the chapter.

12.2.1 Conventional IC Testing

To understand the impact of approximate computing on the design and manufacturing, it is necessary to recall some basic principles of conventional IC testing. The reported concepts are not intended to be exhaustive; for an extensive introduction to them, readers may refer to [5]. As sketched in Fig. 12.2, in digital testing, the test is carried out in the form of binary patterns (or *test patterns*) applied to circuit's inputs. The circuit's outputs are compared with the expected ones (*golden* responses): if they do not match, the circuit is marked as faulty. The idea came in 1959, when Eldred proposed tests capable of observing the internal state of signals in large digital system, by propagating their effect to primary outputs [6].

Very large-scale integration (VLSI) testing can be classified depending on the goal it is intended to serve:

- **Production testing:** After chip manufacturing, the production testing determines whether the actual manufacturing process produced correct devices or not. This process is performed by the device manufacturer that owns full details about the internal structure of the manufactured system and usually exploits Automated Test Equipment (ATEs) for performing the tests. Different test types are usually

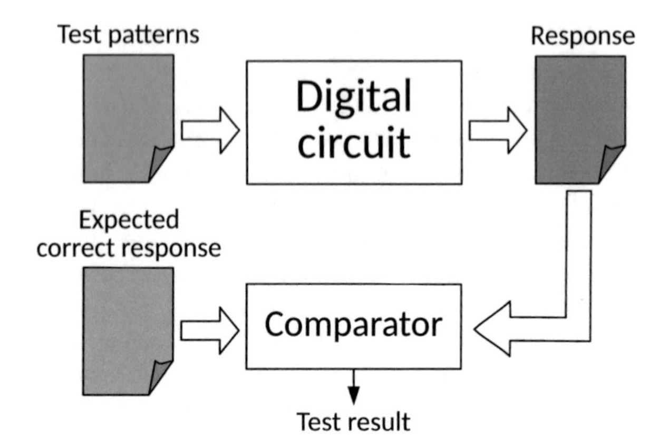

Fig. 12.2 Digital testing [5]

performed at the end of manufacturing and their implementation or not may depend on the targeted market of the device. Some of the most common testing processes are: Wafer testing before the wafer is sliced and any device is packaged as a standalone device. Manufacturing testing that checks for the key parameters of the device and its main functionalities. Burn-in testing, where the produced chips are stressed by placing them in high temperature environments, while applying functional and post-production tests; by doing this over a certain period of time, it is possible to guarantee the reliability of the tested devices since the weak devices are eliminated during the burn-in process. During these testing procedures, the devices are stressed by using structural and functional approaches targeting different fault models and the test procedures are performed either at nominal speed or at the speed required by the testing process.

- **In-field testing:** On the other side, when the device is already integrated in the final application and under certain conditions, it is necessary to test the device during its normal operational life, it is required to implement a periodic testing strategy named In-field testing. In this case, the test cannot be performed supported by an ATE, but it should be done through the available mechanisms included in the device itself. In-field testing may require to test the device at the turn-on and turn-off, or periodically while running concurrently with the actual application. Today, some industrial standards such as ISO26262 for automotive, and DO254 for avionics provide the guidelines for implementing these kind of test strategies for different safety-critical applications.

Usually, two are the types of tests performed on VLSI chips:

- **Functional test:** it is possible to define functional test considering the way in which the test procedure is applied and the information used to develop the whole testing procedure. In the first case, the test procedure is performed acting on the functional inputs of the device under test and observing the functional outputs, only, without resorting to any kind of special mechanisms as the ones called Design-for-Testability (DfT). On the other side, the test procedure is developed on the basis of the functional information regarding the module under test, only:

therefore, it aims at testing the functions rather than the faults, this kind of test can be considered as a kind of black box testing strategy.
- **Structural test:** a structural test exploits the structural information of the device under test to generate the resulting stimuli set. In most of the cases, the strategies based on structural test use the circuit netlist, that represents the topological distribution of all the logic gates composing the circuit, and the circuit fault list to create specific test patterns able to cover the complete list of faults. Very sophisticated algorithms and strategies have been proposer to efficiently exploit the structural information of the circuit to generate the stimuli set, this is the case, for example, of modern ATPG tools.

Structural testing is usually considered opposed to functional test since in general these strategies do not use the functional inputs of the device to apply the test patterns, but exploit, for example, the circuit scan chains and more sophisticated strategies such as logic Built-In Self-Test (BIST), being in this way contrary to the first definition of functional test. On the other side, since the stimuli set is generated resorting to the structural circuit information and not use the circuit functionalities, this makes this test strategy contrary to the second definition given before.

12.2.2 Defect Modeling

To correctly describe a *faulty* electronic circuit, different terms have to be defined. Below, we report the common definitions of *Defect*, *Error*, and *Fault*.

- **Defect**: Unintended difference between the implemented hardware and its design. Defects can occur during manufacture, as well as during the device lifetime.
- **Error**: A wrong output signal produced by a defective system. An error is caused by some defect in the hardware.
- **Fault**: An abstraction model of a defect.

It is important to highlight that even if a defect is present within an IC, its effects might not affect the IC behavior. In general, given the list of all possible defects (modeled as faults) that can occur within an IC, a fault is defined as *detectable* if it exists an input pattern sensitizing and propagating the fault effect to outputs. All faults being detectable are referred to as *detectable faults*. From now on in the text, we will refer to defect and to its model – the fault – interchangeably.

Fault modeling can be described at different levels of abstraction:

- **Behavioral level**: Sometimes referred to as *high level faults*, behavioral level fault models may not have correspondence in manufacturing defects because the model the general behavior. Mostly, they are used in design verification rather than testing.
- **Logic level or Register-transfer level (RTL)**: At this level, we find fault models usually built by considering the *netlist*, i.e., the circuit component list and their

interconnections. **Stuck-at** fault model is the most popular and used one in digital testing. Among others, we find *delay* fault model and *bridging* fault model.
- **Component level**: this is the lower abstraction level, such as the transistor level. Stuck-open fault model, which is a technology-dependent model, is mainly used at this level. Mostly, analog circuit testing resorts to component level fault models.

In this chapter, we focus on logic level fault models, since we address digital integrated circuit testing. In the following, we report some definition's concerning faults, in order to provide readers with some useful terms for the rest of the chapter.

- **Stuck-at fault model (SaF)**—In this abstraction, a circuit net is considered to be permanently set at a constant value. By assigning a fixed (0 or 1) value to an input or an output of a logic gate or to a flip-flop in the circuit, the SaF model represents this condition. The SaF model is the most popular fault model used in practice for digital IC testing. The most popular forms are the *single stuck-at faults*. In this abstraction, a single faulty line is assumed to be present in the IC, either stuck-at-1 (Sa1) or stuck-at-0 (Sa0).
- **Delay fault model**—Defects modeled by delay fault model prevent the correct data from reaching outputs at the right time. Among different types of delay faults models we find transition faults, gate-delay faults, path-delay faults.
- **Redundant fault**—In a combinational circuit, a redundant fault does not modify the circuit's output for any input combination. Thus, a test detecting a redundant fault cannot exist. Redundant faults are a subset of the more general *untestable faults*. In sequential circuits, faults for which no test pattern can be found fall into the untestable fault category.
- **Multiple fault**—The condition that simultaneous single faults affect the same circuit is referred to as multiple fault. Multiple Stuck-at faults model is usually not considered, due to the tremendous complexity. Moreover, a very high percentage of these faults are covered by single stuck-at faults tests.
- **Equivalent faults**—If two faults f_1 and f_2 lead a circuit to exhibit the exact same behavior, they are defined as *equivalent*. A test detecting f_1 detects also f_2 and vice versa. This leads to *fault collapsing*: partitioning all the faults of a circuit into disjoint equivalence sets and selecting one fault from each equivalence set to test. For a circuit having n lines (thus $2n$ single stuck-at faults) the equivalence between $2(n^2 - n)$ pairs of faults should be determined, which is complex. Therefore, for stuck-at fault model, the fault equivalence is usually determined between faults affecting each Boolean gate.

12.2.3 Fault Simulation

In the design of VLSI circuits, the concept of *simulation* is of great importance. Firstly, it serves the purpose of verifying the circuit correctness. Secondly, it verifies whether and how efficiently a test set fulfills its purpose.

The circuit correctness verification is a fundamental step of the design activity. After the synthesis process, the produced netlist is verified by a *true-value simulator*, i.e., it produces the responses of the defect-free circuit. Since the goal is to verify the circuit functionality according to the specification, the input stimuli (or vectors) applied by the simulator to the circuit are based on the specification. The main assumption is that circuit errors lead to change the design to make responses to all stimuli different than the ones expected by the specification.

Simulation is also used for the development of manufacturing tests. A *fault simulator* acts like a true-value simulator with the capability to simulate a faulty circuit. Once the verified circuit netlist is available, the fault simulator can measure the percentage of faults that are detected when a given set of input stimuli (usually, the verification ones) is applied to the circuit. Faults are organized in *fault lists* and input stimuli in *test sets*. Faults covered by the given test set are marked as *detected* and the *Fault Coverage* is measured.

Definition 12.1 Fault Coverage (FC): the ratio of the number of faults detected by a set of test patterns to the total number of faults in the fault list.

An adequate FC (98%–100%) is usually required in order to ship high quality devices to the customers. A good-quality test is a test that can minimize the number of faulty circuits sold, while keeping the test cost acceptable.

Definition 12.2 Test quality: the fraction of chips that, despite having passed the test, are actually faulty. It is usually referred to as *defect level* (or *field reject rate*). Defect level is expressed as *parts per million (ppm)*. High quality tests are considered as providing chips with a defect level of 100 ppm or lower.

Process variations, such as impurities in materials, dust particles, etc., can produce defects during the manufacture. In turn, defects can cause circuits to fail. Process variation effects reflect on the *process yield* defined as follows:

Definition 12.3 Process yield: the fraction (or percentage) of acceptable parts (thus, sold) among all fabricated parts. It is also commonly referred to simply as *yield*.

In a typical case, a newly designed chip has a low yield at its early manufacturing period. Thanks to process diagnosis and correction, a higher process maturity is achieved and, thus, a significantly higher yield.

12.2.4 Test Generation

In late-fifties, Eldred highlighted the necessity for the structural testing of logic circuits to prevail over the classic functional test [6]. He argued that formulating test conditions at the level of the components is *"the only way in which all conditions of operation of each logical function can be uniquely [...] defined and all logical components within each logical function can be made to perform the task to which they are assigned [...] thereby producing a minimum program which tests and detects failure"*. The goal of structural test is to verify the presence of the minimal set of faults in the circuit. Therefore, the application of fault equivalence is important to reduce the final set of faults to test. The *Automatic Test Pattern Generation (ATPG)* serves the purpose of producing patterns to test the internal structure of a digital circuit, starting from its netlist description. The commonly used method in ATPG, namely *path sensitization*, is based on three steps:

1. **fault *injection*** in the circuit netlist;
2. **fault *activation***;
3. **fault effect *propagation*** toward circuit outputs.

To briefly describe path sensitization, let us resort to the stuck-at fault model (see Sect. 12.2.2). Let us assume that we want to test if a line l is stuck to a constant value (say 1). The test vector v detecting that fault is composed of input values such that:

- the line l is set to the opposite value of the fault (say 0). This is commonly referred to as fault *sensitization* or *activation* or *excitation*;
- the effect of the previous action is propagated to circuit outputs. This is commonly referred to as fault *propagation* or *path sensitization*.

By simulating the pattern with the fault-free circuit, we obtain the fault-free output value (expected output). Now, let us assume that an actual stuck-at fault (say Sa1) occurs at line l. In presence of the fault, the circuit outputs will be different from expected. Therefore, by applying the test vector v to the circuit and knowing the expected output, we are able to detect the fault by observing a difference between actual and expected outputs.

In the context of conventional IC test, even a little difference between the nominal behavior and the manufactured IC behavior leads to reject the circuit. Later in this chapter, we will discuss this aspect when approximate computing is considered. In this particular context, the magnitude of the difference between the nominal behavior and the manufactured IC behavior is important. In fact—under specific conditions—the manufactured circuit may be still accepted even if some defects occur.

The above described ATPG method works correctly only for combinational circuits, i.e., without cycles. In fact, any circuit with cycles will lead the aforementioned method to fall into an infinite loop. ATPG methods for sequential circuits exist but are usually very resource-consuming and sometimes inefficient. The main difficulty for *sequential ATPG* is to control and to observe the internal state of the circuit.

Therefore, *design-for-testability (DfT)* comes into play. As stated by Agrawal and Seth [7],"*testability is the property of a circuit that makes it easy (and sometimes possible!) to test*". DfT refers to the set of design techniques for ICs aiming at improving the testability of the target design. The most popular DfT technique is the scan design. However, DfT techniques are out the scope of this chapter. More details can be found in [7].

12.3 Approximate Integrated Circuits Design Phase

The AxC paradigm has been successfully applied to digital ICs. The first technique was referred to as *over-scaling based approximation*. Basically, the IC is forced to work outside its specified operating conditions [2]. The classical example is the reduction of the supply voltage under the minimum value. This will result in energy saving but it will introduce timing errors. The second technique is the *functional approximation* [2]. Functional approximation aims at modifying the circuit so that its original behavior F is replaced by a similar one F', whose implementation leads to area/energy reduction at the cost of a reduced accuracy. In other words, being some responses of F' different compared to F, the circuit output is sometimes erroneous but it is computed more efficiently. The accuracy loss is always measured by means of quality metric(s). The most adopted ones are the Error Rate (i.e., how many times an error is observed at circuit outputs) and the Error Magnitude (i.e., the difference between the golden and erroneous outputs), formally defined in [2].

In the literature, several approaches for functional approximation have been proposed so far, and they can be classified as manual or automated [2]. Manual techniques target specific (small) circuit designs like adders and multipliers [8, 9]. On the other hand, to manage more complex circuit designs, automated approaches are mandatory. State-of-the-art techniques for Approximated Logic Synthesis (ALS) can be summarized as follows.

The approaches in [10] and [11] target two-level circuits and considers both the error rate and error magnitude as quality metrics. Concerning multi-level circuits, SALSA [12] encodes the quality metric as a function and it further simplifies the circuit exploiting the resulting *do not cares*. This approach can only be applied by taking into account error magnitude metrics. In [13], the authors propose to consider both error rate and error magnitude during the ALS. SASIMI [14] aims at identifying internal circuit net pairs with high probability to have the same logical value. Then, it replaces one net with the other in order to simplify the circuit structure. It considers Error Rate as the quality metric. In [15], the authors propose an approach based on local node simplification in the circuit structure.

Shin et al. introduce in [16] a different approach based on the idea to inject stuck-at-faults into the circuit to simplify it under a composite constraint on error rate and error magnitude. All the above approaches target combinational circuits only. To the best of our knowledge, ASLAN [17] is the only one targeting ASL for sequential circuits. The basic idea is to “unroll” the sequential circuit in time frames also called an iterative logic array of the circuit. For each time frame, the flip-flop inputs from the previous time frame are often referred to as pseudo primary inputs with respect to that time frame, and the output signals to feed the flip-flops to the next time frame are referred to as pseudo primary outputs [18]. In this way, the sequential circuit can be represented as a combinational one on which it is possible to apply the SASIMI approach (on each time frame). Clearly, the complexity of ASLAN is very high and generally depends on the sequential depth (i.e., how many time frames have to be considered). Therefore, a faster and simpler ASL approach needs to be introduced.

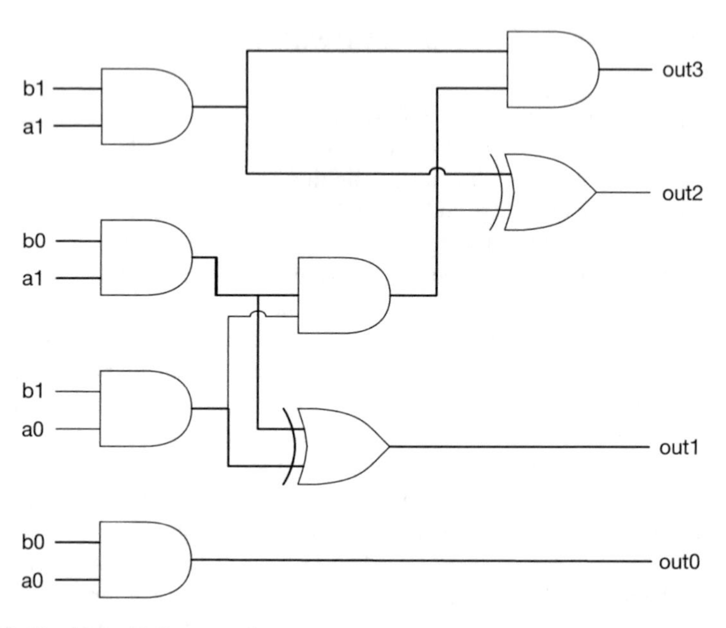

Fig. 12.3 Two-bit multiplier example

A different approach for functional approximation of multi-level circuits leverages on the concept of **fault simulation** [16]. To better show the main principle, let us resort to the example depicted in Fig. 12.3. The circuit example is a two-bit multiplier described in [9]. On this simple circuit, we first define as the quality metric the error magnitude and, more in detail, we set to 4 the maximum acceptable error magnitude. In order to approximate the two-bit multiplier structure using the approach of [16], we consider the output *out3* as affected by a stuck-at-0 fault. In other words, we force this output to be always at logic '0'. The example will clarify why considering the stuck-at-0 fault affecting *out3* ensures that the error constraint will be satisfied. Now, the approach of [16] performs a back-track propagation of the forced fault from the affected output back to a "barrier". Each time that a logic gate is traversed, faulty values are forced to its input to justify the faulty value at its output, and the traversed gate is removed. The barrier is either a primary input or a branch node. In Fig. 12.4, we report in bold the back-track propagation. When a branch is found, a forward propagation is performed. The faulty value is now propagated to reach other outputs and each time that a logic gate is traversed, it is "simplified" depending on the traversed gate and the faulty value. The forward propagation is reported in bold red in Fig. 12.4. Here, the logic value '0' is set as input of the XOR gate, that can be simply removed since $x \oplus 0 = x$. Finally, Fig. 12.5 reports the final result.

Looking at the obtained circuit, the error margin constraint can be satisfied by removing the most significant output bit (*out3* in figures). Table 12.1 gives all the possible results for all the possible inputs. The only error (highlighted in **red**)

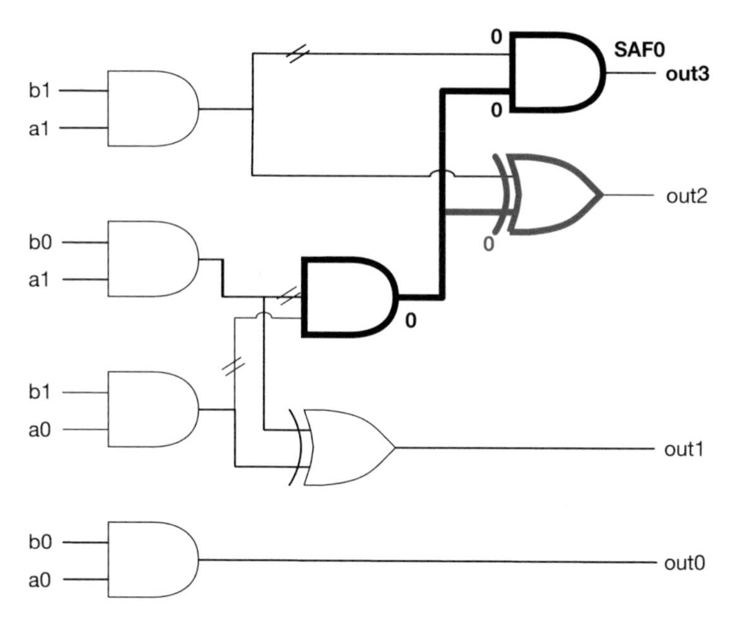

Fig. 12.4 Fault Injection based functional approximation

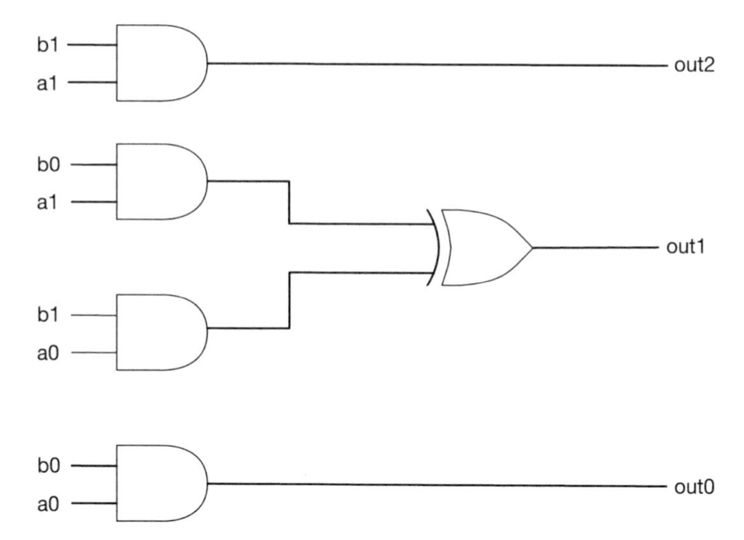

Fig. 12.5 Approximate two-bit multiplier

appears during the computation of 3 x 3. The result should be 9, but rather we get 5 due to the approximation. However, the erroneous result is still acceptable since the error magnitude is 4 (9 - 5).

Please note that in this simple example only one fault is considered. However, if more than one output has to be considered, multiple faults have to be used (one

Table 12.1 Error magnitude example

A x B	0	1	2	3
0	0	0	0	0
1	0	1	2	3
2	0	2	4	6
3	0	3	6	5

fault per output), thus leading to increase the complexity of the fault analysis (back-track as well as forward propagation). Moreover, the fault affecting output *out3* has been selected because it allows to achieve the best approximate solution. Actually, we performed the same analysis for the other faults and we found out that the area reduction was lower compared to the one shown in the reported example. In general, there are no guidelines for the selection of the fault. In [19], an extension has been published including the strategy of considering a single fault instead of multiple ones. Therefore, both processes of fault selection and fault analysis are significant simplified compared to [16]. Additionally, the whole methodological flow does not involve any iteration, but rather it requires to run a fault simulation once. Moreover, it can be taken into consideration for approximating both sequential and combinatorial circuits and it can be used with an arbitrary quality metrics, including the Error Rate and Error Magnitude.

12.4 Approximate Integrated Circuits Verification Phase

The verification phase of a digital circuit design typically employs a method capable of determining whether the circuit exhibits the same behavior as the so-called golden model. The verification can be conducted by means of simulation, but reaching all possible states of a complex circuit is usually intractable for any simulation algorithm, i.e., the simulation does not guarantee that the circuit perfectly meets all requirements. Hence, formal equivalence checking methods have been developed that try to formally prove that two representations (the golden one and the proposed one) of a circuit design exhibit exactly the same behavior. In the context of approximate computing, the verification problem is reformulated in such a way that we try to prove that the golden model and an approximate circuit are equal up to some bound with respect to a chosen distance (error) metric [12]. A particular equivalence checking method's success depends on several factors, primarily including the circuit type, the circuit complexity, and the error metric. Current methods are capable of an exact error analysis only for some circuits and error metrics. On the other hand, complex approximate circuits (such as 128-bit adders, 32-bit multipliers, 32-bit Multiply and Accumulate circuits, and 31-bit dividers) have already been reported together with determining their exact worst-case errors [20]. Most research deals with formal error analysis of arithmetic circuits as they frequently appear in the most popular error-resilient applications such as deep learning and video processing. Formal methods can also be applied to analyze the errors of other combinational circuits effectively (e.g., complex median networks [21]) as well as sequential systems [22].

Two main approaches have been developed for equivalence checking techniques based on Reduced Ordered Binary Decision Diagrams (ROBDD) and satisfiability (SAT) solvers [23]. In both cases, an auxiliary circuit, the so-called *miter*, is constructed and then analyzed. The miter connects corresponding outputs of the candidate circuit (to be checked), the golden circuit, and an error-specific circuit to determine the approximation error. As ROBDDs are inefficient in representing classes of circuits for which the number of nodes in the BDD is growing exponentially with the number of input variables (e.g., multipliers and dividers), their use in equivalence checking of approximate circuits is typically possible for adders and other less structurally complex functions [23].

If the error analysis is based on SAT solving, the miter is represented as a logic formula in Conjunctive Normal Form (CNF) for which SAT solver decides whether it is satisfiable or unsatisfiable. The interpretation of this outcome depends on the construction of the miter, see Chap. 4. Common SAT solvers are, in principle, applicable to the worst-case analysis only. However, this approach is more scalable than ROBDDs for the error analysis of multipliers [24]. Specialized SAT solvers (#SAT) are capable of counting the number of satisfiable assignments. Still, their scalability is very limited, and thus they are currently less practical for the exact error analysis [23].

The performance of verification algorithms is critical if the circuit approximation process is based on a fully automated search in the space of approximate implementations and every candidate implementation has to be verified. A detailed overview of formal verification techniques for approximate arithmetic circuits is provided in Chap. 4.

12.5 Approximate Integrated Circuits Testing Phase

This section focuses specifically on the testing aspects of functionally approximate circuits. These circuits are referred to as *Approximate Integrated Circuits (AxICs)*. Since approximating circuits alters their functional behavior, techniques to test them must be revisited [25–35]. As a matter of fact, extending the basic testing concepts to AxICs is not straightforward. In particular, as mentioned in Sect. 12.2.4, during the test of a conventional circuit, any change in its functional output signals with respect to the expected values leads to labeling the circuit as faulty and discarding it. When dealing with AxICs, the presence of a fault may lead the circuit to behave differently than expected, yet still in an acceptable manner. In this case the circuit should not be discarded. Acceptable behaviors are defined according to one or more error metrics and corresponding bounds, fixed in the design phase and usually expressed as thresholds. Mastering these mechanisms may lead to increase the production

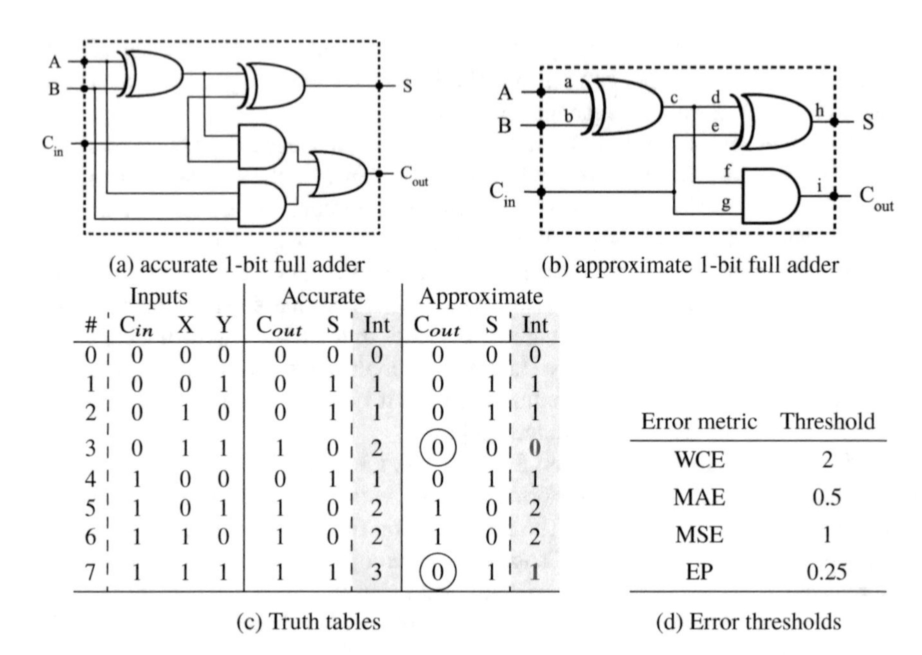

#	Inputs C_{in}	X	Y	Accurate C_{out}	S	Int	Approximate C_{out}	S	Int
0	0	0	0	0	0	0	0	0	0
1	0	0	1	0	1	1	0	1	1
2	0	1	0	0	1	1	0	1	1
3	0	1	1	1	0	2	(0)	0	**0**
4	1	0	0	0	1	1	0	1	1
5	1	0	1	1	0	2	1	0	2
6	1	1	0	1	0	2	1	0	2
7	1	1	1	1	1	3	(0)	1	**1**

(c) Truth tables

Error metric	Threshold
WCE	2
MAE	0.5
MSE	1
EP	0.25

(d) Error thresholds

Fig. 12.6 (**b**) Example of an approximate 1-bit full adder, obtained from the accurate 1-bit full adder in (**a**). The subfigure (**c**) shows the truth tables of the two circuits: for each input, the output bit values are reported (S and C_{out}), as well as their unsigned integer representation, calculated as $S * 2^0 + C_{out} * 2^1$. Subfigure (**d**) reports the error thresholds for the approximate circuit, for different error metrics

process yield, i.e., increase the percentage of sold AxICs among all fabricated AxICs.

To illustrate the issue related to the AxIC test, throughout the whole section we refer to a simple arithmetic circuit, shown in Fig. 12.6. The figure depicts a 1-bit Full Adder (FA) (Fig. 12.6a) and an approximate version of it (12.6b), which is more efficient, i.e., with reduced area (3 logic gates instead of 5) and lower delay (2 logic levels instead of 3), but shows some errors at outputs. Figure 12.6c reports the truth tables of both the circuits. We also report the integer representation of both the circuit outputs ("Int" column), calculated as $S * 2^0 + C_{out} * 2^1$. As reported in Fig. 12.6d, by considering all the possible circuit inputs, we can calculate the error values according to the following metrics, *Worst-Case Error* (WCE), *Mean Absolute Error* (MAE), *Mean Squared Error* (MSE), and *Error Probability* (EP) [36], defined as follows:

$$\text{WCE} = \underset{\forall i \in I}{max} \left| O_i^{\text{approx}} - O_i^{\text{precise}} \right| \tag{12.1}$$

$$\text{MAE} = \frac{\sum_{\forall i \in I} \left| O_i^{\text{approx}} - O_i^{\text{precise}} \right|}{2^n} \tag{12.2}$$

$$\text{MSE} = \frac{\sum_{\forall i \in \mathcal{I}} \left| O_i^{\text{approx}} - O_i^{\text{precise}} \right|^2}{2^n} \tag{12.3}$$

$$\text{EP} = \sum_{\forall i \in \mathcal{I}:\, O_i^{\text{approx}} \neq O_i^{\text{precise}}} \frac{1}{2^n} \tag{12.4}$$

where:

$i \in \mathcal{I}$ is the input value within the set of all possible inputs $\mathcal{I}$,
$O_i^{precise}$ is the precise output integer representation, for input i,
O_i^{approx} is the approximate output integer representation, for input i, and
n is the number of input signals to the circuit.

Values reported in Fig. 12.6d are a direct consequence of the approximation. They constitute the error threshold values of the AxIC, fixed by specification and known at design time. Thus, after manufacturing, the produced AxIC must produce outputs respecting the boundaries set by the error thresholds. The issues shown in this section and the approaches to face them are generic and applicable to all kind of combinational circuits, provided that a measure of their approximation error is available and reproducible.

The following section illustrates a test flow—called *Approximation-Aware (AxA) test flow*—to properly deal with the test of AxICs. The flow is composed of three main steps: (i) AxA fault classification, (ii) AxA test pattern generation and (iii) AxA test set application. Briefly, the *fault classification* divides faults producing critical effects on the circuit behavior from those producing acceptable effects. The *test pattern generation* produces test stimuli able to cover all the critical faults and, at the same time, to leave acceptable faults undetected, as much as possible. Finally, the *test set application* labels AxICs under test as critically faulty, acceptably faulty, or fault-free. Only AxICs falling into the first group will be discarded, thus minimizing over-testing (i.e., minimizing AxICs discarded due to acceptable faults). Next subsections describe each AxA test step.

12.5.1 AxA Fault Classification

The first step of the AxA testing is the *fault classification*. It aims at separating acceptable faults from critical ones. The outputs of this phase are two fault lists (critical and acceptable). The part of detectable faults classified as acceptable constitutes the *expected Yield Increase* (eYI) with respect to the conventional test. The eYI is expressed as follows:

$$\text{eYI} = \frac{\text{acceptable faults}}{\text{total faults}}, \tag{12.5}$$

The measure of the eYI is another outcome of the AxA fault classification. The purpose of such a metric is to establish an upper bound to the achievable yield gain. To turn eYI in an actual gain, we have to go through the other AxA testing phases, discussed throughout this chapter.

The key aspect to consider in the fault classification is the AxICs' output deviation measure. As mentioned in the previous subsection, different error metrics exist to measure AxIC output deviations [36]. In Table 12.2, in the left part we report the error threshold value alterations caused by all possible Stuck-at faults in the approximate FA (Fig. 12.6b). The fault list was generated with a commercial tool [37] with the fault collapsing option active. We highlight in red solid-bordered boxes the non-acceptable error values, i.e., higher than the respective thresholds, shown in Fig. 12.6d. We use the notation *SaX@N* to indicate a "stuck-at-X affecting the net N", where X can be either the value 1 or 0 and N is the label of the net. Please, refer to Fig. 12.6b for the net labels. By observing the table, we can firstly

Table 12.2 Error metric values in presence of different faults affecting the approximate circuit in Fig. 12.6b, and fault coverage report for the exhaustive test set

Fault	Error metrics											
	WCE	MAE	MSE	EP	Test vectors[a]							
Fault-free	2	0.5	1	0.25	0	1	2	3	4	5	6	7
Sa1@a	2	1	1.5	0.75	X	X			X	X		
Sa0@a	1	0.5	0.5	0.5			X	X			X	X
Sa1@b	2	1	1.5	0.75	X		X		X		X	
Sa0@b	1	0.5	0.5	0.5		X		X		X		X
Sa1@c	1	0.5	0.5	0.5	X			X	X			X
Sa0@c	2	1	1.5	0.75		X	X			X	X	
Sa1@d	3	0.75	1.5	0.5	X			X	X			X
Sa0@d	2	1	1.5	0.75		X	X			X	X	
Sa1@e	2	0.75	1	0.625	X	X	X	X				
Sa0@e	3	1	2	0.625					X	X	X	X
Sa1@h	2	0.75	1	0.625	X			X		X	X	
Sa0@h	3	1	2	0.625		X	X		X			X
Sa1@f	2	0.5	1	0.25					X			X
Sa1@g	2	1	2	0.5		X	X					
Sa1@i	2	1	2	0.5	X	X	X	X	X			X
Sa0@i	2	1	2	0.5						X	X	
eYI(%)	81.25%	25%	37.5%	6.25%								

[a]0="000", 1="001",. . ., 7="111"
(solid box) = critical effect, (dashed box) = beneficial effect

remark that not all the metrics are impacted by the same faults. Furthermore, in some particular cases, faults even reduce the observed error (green dash-bordered boxes in Table 12.2).

The complexity of the classification task depends on the complexity of the metric computation. For instance, in the example in Fig. 12.6, let us suppose that the approximate circuit had a defect whose effect set the net h to 0 (Sa0@h). To measure the impact on the different metrics described by Eqs. (12.1), (12.2), (12.3), and (12.4) different procedures are required. To find out that the WCE threshold is not respected, it is only necessary to find a single input stimulus generating an erroneous output having $WCE > 2$, e.g., input 7 (111) that should give 3 rather than 0 ($3 - 0 = 3 > 2$). Conversely, to find out that the thresholds for MAE MSE and EP are not respected either, we should test all the possible input stimuli to observe the outcome and then calculate a mean to obtain the final results. More in general, for the WCE we have to demonstrate the existence or nonexistence of a single input vector whose application leads the circuit to produce an out-of-bounds error. This task is well achieved by Satisfiability solvers (SAT) or by using ATPG for integrated circuits. By using the previously mentioned auxiliary *miter* module, SAT solvers or ATPGs can fairly easily manage the problem. Conversely, to measure the impact of a fault on metrics based on average calculation, simulation approaches are preferred. When the complexity of the circuit does not allow using an exhaustive analysis, a random or workload-dependent subset of input is used to estimate the measure. SAT solvers counting the number of satisfiable assignments exist but they suffer of scalability issues.

Figure 12.7 sketches the necessary modules to obtain a classification similar to the one in Table 12.2: the original (precise) circuit, the AxIC under test, and the *miter* module that performs the evaluation on the circuit outputs with respect to the chosen error metric(s). The final output reports an erroneous condition when the AxIC produce output values outside the error metric bounds. Thus, when the AxIC

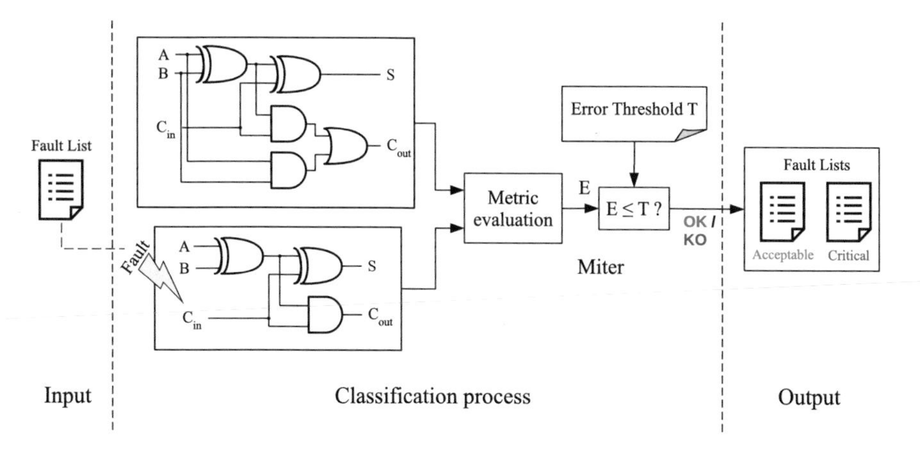

Fig. 12.7 Approximation-Aware (AxA) fault classification concept

is fault-free, no erroneous conditions are reported. The underlying idea is masking acceptable fault effects by using a filter (implemented by the miter). In this way, only critical faults generate an error condition at the output of the miter. Hence, the conventional test approaches mentioned at the beginning of the chapter can be used to classify the AxIC possible faults. In particular, for the WCE metric—and for all the metrics for which only a punctual condition must be demonstrated—a conventional ATPG approach (or a SAT-based one) can be used: given a fault, the procedure proves whether an input stimuli generating an error condition at the output of the miter exists or not and classifies the fault accordingly. For MAE, MSE, and EP—and for all the metrics requiring the calculation of a mean—fault injection and input simulation can be used: given a fault, it is injected in the AxIC, a set of input stimuli (exhaustive, random or workload-dependent) is simulated and the metric is calculated in the miter considering all the results. More details on the approaches are reported in [28, 31, 34]. Finally, it is worth mentioning that the above-discussed miter is never manufactured. It is only used in simulation to classify the AxIC faults.

12.5.2 AxA Test Pattern Generation

The second step of the AxA testing is the *test pattern generation*. Historically, a lot of effort has been spent in providing test generation methodologies achieving higher *Fault Coverage (FC)* for conventional integrated circuits. In the context of AxICs, test patterns must cover all critical faults and as few as possible acceptable ones. Respecting both these conditions is crucial to discard AxICs affected by critical defects and, at the same time, to avoid discarding those affected by acceptable defects. To achieve this goal, it is necessary to find, among the input vectors, the smallest subset covering all the critical faults and minimizing the acceptable fault coverage. Therefore, the concept of *Fault Coverage (FC)*, defined in Sect. 12.2.3 Definition 12.1, needs to be divided into *acceptable FC* and *critical FC*, as defined below:

$$\text{acceptable FC} = \frac{\text{acceptable faults detected}}{\text{total acceptable faults}} \tag{12.6}$$

$$\text{critical FC} = \frac{\text{critical faults detected}}{\text{total critical faults}} \tag{12.7}$$

Naturally, for conventional approaches a good test set aims at detecting as much faults as possible without considering their classification. Conversely, in the AxIC case, an ideal test set should achieve 100% critical FC and 0% acceptable FC. If no effort is spent toward achieving the second condition, a still-good AxIC affected by an acceptable fault will be rejected during the test phase. The phenomenon due to which a good product is considered as faulty by the test process is commonly referred to as *over-testing*. If not properly managed. If not properly managed, the over-testing will eventually cause some yield reduction.

Let us refer again to our example. The right part of Table 12.2 reports the input stimuli detecting (i.e., sensitizing and propagating to outputs) each possible Stuck-at fault for the AxIC in Fig. 12.6b. Firstly, let us assume that the fault classification is performed by using the EP metric (threshold = 0.25). Table 12.2 shows that all the faults lead the error to be critical, except for Sa1@f, that leaves it to 0.25. Therefore, vectors 4 and 7 must be avoided, since they are the only ones detecting that fault. An example of test set achieving 100% critical FC and 0% acceptable FC is {2,3,5}. However, it is not always possible to find a suitable test set satisfying these conditions. For example, let us consider that the fault classification is performed by using the WCE metric (threshold = 2). In Table 12.2 we can observe that three faults lead the error to be critical, i.e., Sa1@d, Sa0@e, and Sa0@h increase the WCE to 3. Both vectors 4 and 7 independently detect the three faults. However, both vectors detect also five acceptable faults (38% acceptable FC). Moreover, there is no input vector combinations achieving 100% critical FC and 0% acceptable FC all at once. A further analysis highlights that vector 4 covers four acceptable faults leaving the WCE to 2 and one lowering it to 1, while vector 7 covers three acceptable faults lowering the WCE to 1. Therefore, this consideration would lead to the selection of vector 4 over vector 7 to avoid detecting too much faulty conditions that actually improve (i.e., lower) the WCE. In conclusion, it is not always possible to achieve 0% acceptable FC and 100% critical FC at all at once. As a consequence, an ideal *AxA test pattern generation* approach should produce a test set achieving 100% critical FC and an acceptable FC as close as possible to 0%. Unfortunately, conventional ATPG algorithms are not designed to produce test set with such particular properties.

One way to achieve an improved test set (i.e., with 100% critical FC and low acceptable FC) is to develop an exploration methodology to find, among the input vectors, the smallest subset covering all the critical faults and minimizing the acceptable FC. Such a methodology, sketched in Fig. 12.8, measures both critical and acceptable FCs of the AxIC input vector set and formulates and resolves an *Integer Linear Programming (ILP)* optimization problem to find the smallest subset achieving 100% critical FC and minimizing the acceptable FC. The ILP solution is the final *ax-aware test set*. More in details, firstly a (sub)set $\mathcal{S}$ of the AxIC input vector set is generated. Then, fault simulation is used, taking as input $\mathcal{S}$, the AxIC and the two fault lists (critical and acceptable) generated in the AxA fault classification phase. The output of the fault simulation is a *fault coverage (FC) report* which records, for each fault, all the input vectors in $\mathcal{S}$ covering it, as shown in Table 12.2. Finally, an optimization problem is formulated, by using the fault coverage report and the fault lists. This leads to a system of linear inequalities whose solution will be the final *ax-aware test set*, i.e., the smallest subset $\mathcal{V} \subset \mathcal{S}$ which minimizes the acceptable FC and achieves 100% critical FC. If $\mathcal{S}$ corresponds to the exhaustive input set of the AxIC, the output solution will be the *globally optimal* one (i.e., the best possible vector combination). When $\mathcal{S}$ is a subset of the exhaustive vector set, the ILP solution will be *locally optimal* (i.e., the best

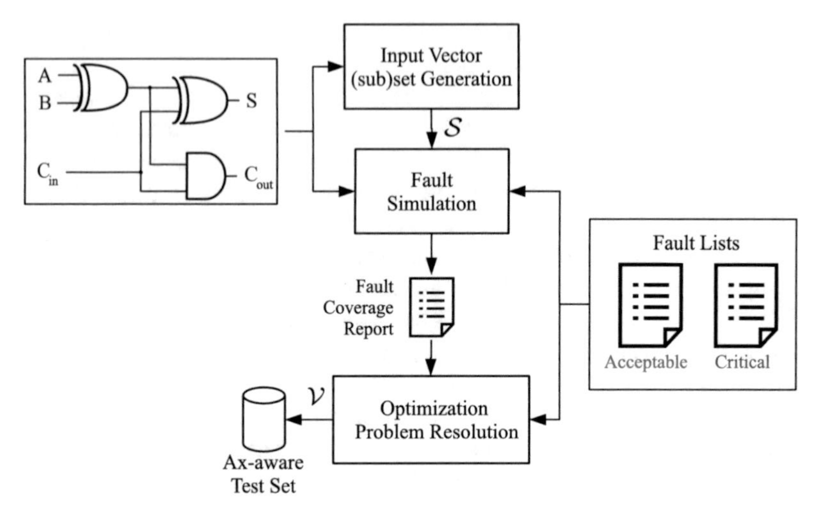

Fig. 12.8 Approximation-Aware (AxA) test generation methodology

combination, among vectors in $\mathcal{S}$). This approach is independent of the specific fault classification technique and of the error metrics and thresholds. Indeed, as long as a fault classification is correctly produced, the methodology is applicable. Further details on the approach and its mathematical basis are available in [32]. Although the methodology guarantees finding an optimal test set, the ideal outcomes (i.e., 100% covered critical faults and 0% covered acceptable faults) cannot be guaranteed, due to the structure of the AxIC. Therefore, further efforts must be spent in the test application phase, as shown in the next subsection.

12.5.3 AxA Test Set Application

To push further the test outcomes, the third step of AxA testing, the *test pattern application*, comes into play. In the conventional test set application phase, observing a circuit response different from the expected one always leads to reject the circuit. On the contrary, in AxA testing, whether the erroneous response is due to an acceptable fault or to a critical fault must be taken into account. The test must reject the AxIC only if a critical fault caused the error. As shown in the previous subsection, often it is not possible to avoid detecting acceptable faults. Thus, the main solution is to verify, after the test application, whether the detected fault was acceptable or not. Another metric is used to evaluate the effect of the AxA testing procedures on the yield, the *Yield Increase Loss* (YIL), defined below:

$$\text{YIL} = \frac{\text{acceptable faults detected}}{\text{total faults}} \tag{12.8}$$

It describes the value of the yield increase **not achieved** due to the detection of acceptable faults. The YIL is in the range $[0, \text{eYI}]$. By considering Eqs. (12.5) and (12.8), we can observe that the YIL can be expressed also as follows:

$$\text{YIL} = \text{acceptable FC} \cdot \text{eYI} \tag{12.9}$$

This means that the acceptable FC metric represents the part of the possible yield increase (eYI) that **is not** actually achieved, due to the detection of acceptable faults. Therefore, if acceptable FC = 0 then YIL = 0 (i.e., maximum yield increase). On the contrary, if acceptable FC = 1 then YIL = eYI, thus the achieved yield increase is null.

We need a methodology able to observe the circuit's responses and distinguish between the detection of an acceptable fault (i.e., the test passes) and a critical one (i.e., the AxIC is rejected). Let us observe Table 12.3, which reports the output

Table 12.3 Values of the AxIC in Fig. 12.6b when the input vectors are applied in presence of the faults

Input vector		0	1	2	3	4	5	6	7	Classification
Output	Precise	0	1	1	2	1	2	2	3	WCE (2)
	Fault-free	0	1	1	0	1	2	2	1	
	Sa1@a	**1**	**0**	1	0	2	**1**	2	1	acceptable
	Sa0@a	0	1	**0**	**1**	1	2	**1**	**2**	acceptable
	Sa1@b	**1**	1	**0**	0	2	2	**1**	1	acceptable
	Sa0@b	0	**0**	1	**1**	1	**1**	2	**2**	acceptable
	Sa1@c	**1**	1	1	**1**	2	2	2	**2**	acceptable
	Sa0@c	0	**0**	**0**	0	1	**1**	**1**	1	acceptable
	Sa1@d	**1**	1	1	**1**	0	2	2	**0**	critical
	Sa0@d	0	**0**	**0**	0	1	**3**	**3**	1	acceptable
	Sa1@e	**1**	**0**	**0**	**1**	1	2	2	1	acceptable
	Sa0@e	0	1	1	0	0	**3**	**3**	**0**	critical
	Sa1@h	**1**	1	1	**1**	1	**3**	**3**	1	acceptable
	Sa0@h	0	**0**	**0**	0	0	2	2	**0**	critical
	Sa1@f	0	1	1	0	3	2	2	**3**	acceptable
	Sa1@g	0	**3**	**3**	0	1	2	2	1	acceptable
	Sa1@i	**2**	**3**	**3**	**2**	3	2	2	**3**	acceptable
	Sa0@i	0	1	1	0	1	**0**	**0**	1	acceptable

bold = output value obtained when an input vector (column) detects a fault (row)

□ = value produced by the test set (input vector 4) when detecting a fault

integer values of the AxIC in Fig. 12.6b, obtained in presence of the different faults. For each fault, we report also its classification according to the WCE metric. In bold are reported the value observed when a particular vector (column) detects the presence of a particular fault (row), i.e., there is a difference between the expected output (fault-free) and the obtained output. In particular, we highlight with blue solid-bordered boxes the faulty values obtained by applying the vector 4. In Sect. 12.5.2, we observed that vector 4 is a good solution to achieve 100% critical FC for the AxIC in Fig. 12.6b. Unfortunately, it also achieves 38% acceptable FC (5 acceptable faults over 13). This means losing a part of the possible yield gain, meaning $YIL = \frac{5}{16} = 31.25\%$.

However, by looking at the critical values produced by applying vector 4 (i.e., in presence of Sa1@d, Sa0e, Sa0@h), we can notice that they are different from the values produced when an acceptable fault is present (Sa1@a, Sa1@b, Sa1@c, Sa1@f, Sa1@i). Therefore, if we observe an unexpected value when applying vector 4, depending on its value we can understand whether it is due to an acceptable or to a critical fault. This, in turn, avoids rejecting circuits due to acceptable faults, i.e., it reduces the YIL to 0%.

Starting from this observation, we can build a test application methodology to further improve the test results. The well-known *signature analysis* concept—successfully applied to built-in self-test (BIST) architectures in the seventies [38] and still used in modern BIST architectures—can be applied in this context. The conventional signature analysis approach compacts test responses of a fault-free circuit into a *golden signature* (i.e., the reference behavior). In the test phase, the test responses of the circuit under test are compacted together into a signature (i.e., the actual behavior). Hence, the latter is compared with the golden one. If the two signatures are identical, the circuit under test is considered fault-free; otherwise, a malfunction is detected.

This concept can be applied to AxIC test, as depicted in Fig. 12.9. It is divided in two phases, as follows:

1. At design time (left branch in Fig. 12.9), we perform a fault simulation by using test patterns and the AxIC's faults. For each fault, we compact simulation responses into a signature. We obtain acceptable and critical signatures. We remove from acceptable signatures those overlapping with critical ones (if any). We add the golden signature (i.e., fault-free) and end up having an *ax-aware signature set*.
2. At test time (right branch in Fig. 12.9), manufactured AxIC test responses are compacted into a signature. The latter is compared with the ones in the ax-aware signature set. If there is at least one match, then the AxIC is considered acceptable. Otherwise, the circuit is rejected.

The discussed approach can be used for external test, i.e., test are applied by using an Automatic Test Equipment (ATE) and it can be also adapted to a BIST context. With this approach, results close to the ideal ones are achieved. Further details are available in [33].

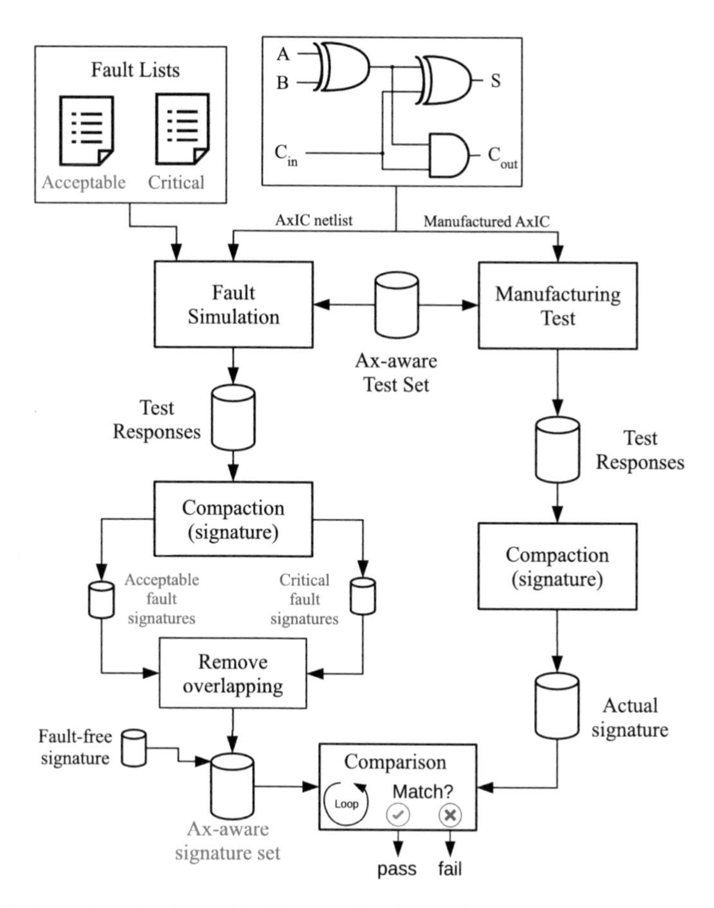

Fig. 12.9 Approximation-Aware (AxA) test application methodology

12.6 In-Field Applications: DNN as Case Study

Approximate Computing techniques have been positively introduced thanks to the intrinsic resilience of many applications [4]; as a collateral resiliency effect, it has been stated that a resilient application is able to provide good enough outputs (i.e., acceptable) despite of the presence of hardware faults. The previous section described in detail how to characterize the impact of hardware faults on a given approximate circuit resorting to well established metrics, i.e., WCE, MAE, MSE, and EP. In this section, we intend to discuss the usage of the same fault impact characterization, but at application level instead of component/circuit level. Indeed, presented metrics may not be valid at application level and thus new application-dependent metrics and even characterization methods have to be defined. To support the investigation, we resort to a Deep Neural Network (DNN) as case study. The target DNN is the LeNet-5 [39] used as classifier for handwritten digit recognition task. We resort to the MNSIT database of training and validation. The end-goal is to characterize the impact of permanent faults affecting the LeNet through fault injection campaigns. The characterization is done on an original version of the DNN, i.e., without any approximation, and on approximate versions of the same DNN.

12.6.1 DNN Data Type Approximation

Chapter 15 presented several approximation techniques for the development of approximate neural networks. In this section, we exploit the reduction of the precision and data type for weight and activation's values. More in details, we intend to use custom floating-point and fixed-point representations with different precision (i.e., bit-width) at inference time. To explore DNNs data type approximation we leverage the *darknet* open source DNN framework [40]. Implemented in C language, the library allows end-to-end deployment of neural network architectures in a very simple way. It further supplies a very simple environment where several configurations of DNNs, including CNNs, can be executed either to perform training or inference tasks. We modified *darknet* framework to (i) approximate the DNN and (ii) inject Stuck-at Faults at inference time. The targets of injections are the DNN weights. The description of how injection is implemented is out the scope of this chapter, the reader can found all details in [41].

Originally, the *darknet* framework leverages on 32 floating-point data types only. We thus modified the *darknet* source code to allow data type conversions. All the conversions between the standard 32 floating-point and **custom data types** have been carried out by integrating two open source libraries into the *darknet* framework: the *libfixmath* library [42] for managing fixed-point and the *FloatX* library.

Figure 12.10 represents our custom data type defined as following:

- N defines the data bit-width;
- i sets the dynamic and the precision of the data type depending on data representation:
  - Floating-Point: i is the mantissa width, $N - 1 - i$ is the exponent width;
  - Fixed-Point: i is the fractional width, $N - 1 - i$ is the integer width.

Figure 12.11 sketches the scenario in which the fault injection campaign is executed. The first step allows to determine the **Accuracy Loss** due to approxi-

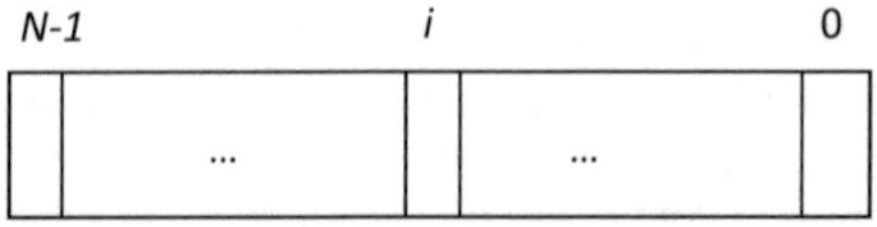

Fig. 12.10 Custom data type

N-1 i 0

... ...

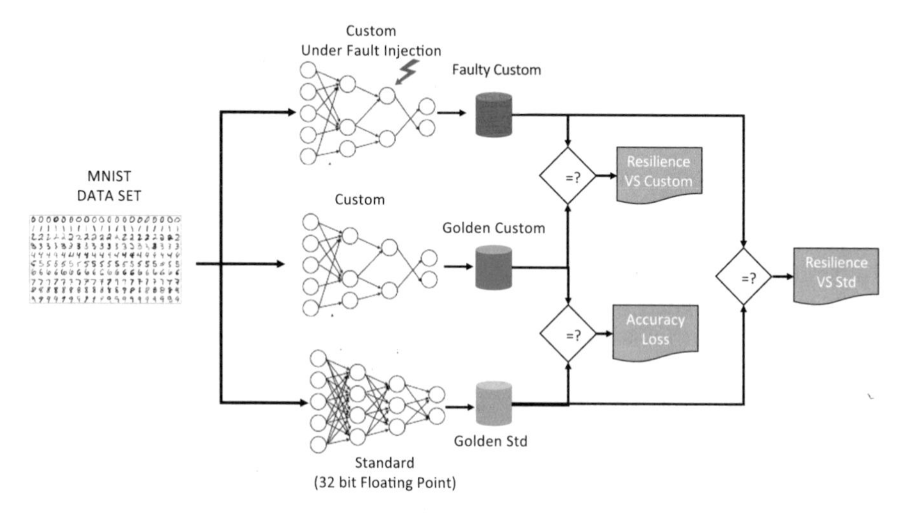

Fig. 12.11 Fault injection scenario

mation. Starting from the trained DNN with the 32 bit floating-point data types, called **Golden Standard**, the network is approximated using custom data type representation, called **Golden Custom**. The inference outputs are then compared to determine the **Accuracy Loss** induced by the approximation. Once the Golden Custom network has been built, the fault injection campaign is carried out on this DNN, and the inference outputs are compared with the faulty free "Golden Custom" inference to first assess the accuracy loss due to fault injection. Eventually, the same outputs are compared with "Golden Standard" ones to assess the inner resiliency due to the approximation.

The final purpose of customizing a DNN, i.e., by approximating it, aims at replacing the Standard DNN in edge/resource limited devices. Thus, it is crucial to assess the faults impact on the Custom DNN with respect to the standard DNN. On the other hand, it is also possible to apply the custom-data type to DNN before training. In this case, the reference DNN to compare the faults impact to be assessed is the Custom DNN itself.

To classify the faults impact, we will define the following application-dependent outcome:

- **Masked:** no difference is observed from the faulty DNN and the golden one.
- **Observed:** a difference is observed from the faulty DNN and the golden one. Depending on how much the results diverge, we further classify these as:
  - **Good:** the confidence score of the top ranked is higher with respect to the golden DNN. In other words, the faulty DNN provides a better inference than the golden one;
  - **Accept:** the confidence score of the top ranked element is reduced by less than 5% with respect to the golden DNN;

– **Warning:** the confidence score of the top ranked element is reduced by more than 5% with respect to the golden DNN;
– **Data Corruption:** the top-1 prediction is different. In other words, the faulty DNN makes a wrong inference.

It can be easy seen that the above classification is slightly different than the one of Sect. 12.5.1. This is mainly because of the need of considering the final application outputs, and the fact that in some cases the faults impact lead to an improvement of the output quality. The latter point is specific to the case of DNNs. Indeed, a DNN can be considered as an approximation of a given function (in our case a classifier). The approximation is implemented by determining the parameters (i.e., weights) values through the training. Since the training cannot be exhaustive, it is still possible that a different weights values' distribution provides better results. As mentioned before, in the provided experiments, we target the DNN weights, so, the faults impact can be seen as a modification of weights values that may improve the output accuracy. For the sake of clarity, we can also classify faults as proposed in Sect. 12.5.1 for further comparison:

- Acceptable Faults: Masked $\bigcup$ Observed $\bigcup$ Accept
- Critical Faults: Warning $\bigcup$ Data Corruption

It is important to specify that "Warning" can or cannot be considered as Critical Faults depending on a user threshold.

12.6.1.1 Experiments

For running the experiments, the MNIST database [43], a well-known dataset used to evaluate the accuracy of new emerging models, has been selected. The dataset includes 60,000 images for training and 10,000 for test/validation of the model, encoded in 28×28 grayscale pixels. Since we are focusing on the inference phase and on the response of the network in a faulty scenario, a set of pre-trained weights has been adopted. The set comes from the *darknet* website, and defines all the weights as 32-bit floating-point.

To carefully select the custom data type representation, we first analyze the LeNet-5 weight values distribution. All the values are in the range -0.6 to 0.6 with the most of them around zero. From this analysis, we simply deduce that the data type does not need a high dynamic range while a high precision is preferred. We thus select the custom data types reported in Table 12.4.

Two data types are used: the fixed and floating-point with different bit-width. Moreover, we computed the accuracy loss of the network resulting from the adoption of custom data types weights. As highlighted in Table 12.4, five different scenarios have been analyzed. The second column of the table reports the data type used in each campaign, while the third column reports the bit-width of the weights. The fourth column shows the amount of bits allocated to encode the different part of the number, i.e., sign, exponent, and fractional part in the case of floating-point

Table 12.4 LeNet-5 data type accuracy loss [%]

Scenario	Data type	Bit-width	Bit encoding	[%] Accuracy loss
FP32	Floating-point	32	1 sign, 8 exponent, 23 fractional	Ref.
FP16	Floating-point	16	1 sign, 5 exponent, 10 fractional	0%
FP8	Floating-point	8	1 sign, 4 exponent, 3 fractional	0.02%
FxP32	Fixed-point	32	1 integer, 31 fractional	0%
FxP16	Fixed-point	16	1 integer, 15 fractional	0%
FxP8	Fixed-point	8	1 integer, 7 fractional	0.04%

representations, and integer and fractional part in the case of fixed-point. To compute the accuracy of the network in the different scenarios, the inference of all the images belonging to the validation set of the MNIST database (10,000 images) have been run on LeNet-5, without injecting any faults, i.e., in a golden scenario. The results show that only when reducing the bit-width to 8 bits the network exhibits an appreciable level of accuracy loss. In details, for the network with weights encoded by using 8-bit floating-point variables (FP8) the accuracy loss was 0.02%, while it was 0.04% when the weights were encoded by using 8-bit fixed-point variables (FxP8).

We evaluated the faults impact by using as reference the Standard 32 bit floating-point DNN (see Fig. 12.11), and considering the whole set of workloads (10,000 images). This is useful in the case where a designer wants to approximate the DNN (i.e., change its data type and/or bit-width) after that it has been trained.

To discuss the results, we refer to the classification presented in Sect. 12.6.1. In particular, we want to evaluate the safety of the different DNN versions, when subject to faults. Therefore, we consider faults in the *Accept*, *Warning*, and *Data Corruption* classes as events reducing the DNN safety; these classes include the faults defined before as *Critical Faults*. The sum of these contributions is referred to as ***Safety Decrease***. Conversely, we consider the faults in the *Masked* and *Good* classes as events either leaving the safety of the DNN unaltered or improving it. The sum of these contributions is referred to as ***Safety Increase***.

Results are shown in Table 12.5 where each row corresponds to one of the DNN variants (FP32-FP16-FP8-FxP32-FxP16-FxP8 defined in Table 12.4). Each column corresponds to a faulty behavior class as described in Sect. 12.6.1.

We can firstly note a different resilience to faults depending on the data type. More in detail, **the safety decreasing effect is lower for fixed-point than for floating-point**, comparing the same bit-width. As representative of this fact, let us discuss scenarios with FP32 and FxP32 (32-bit DNNs): the *safety increasing (decreasing)* effect varies from 69% (31%) of the floating-point version (scenario FP32) to 74% (26%) of the fixed-point version (scenario FxP8). This corresponds to a difference of 5%. The average difference between floating- and fixed-point versions with respect to safety increasing/decreasing effect over the three variants (32- 16- 8- bits) is 10.64%. This can be seen by comparing the scenarios FP32 with FxP32, FP16 with FxP16, and FP8 with FxP8, in terms of the average safety

Table 12.5 LeNet-5 fault injection outcomes w.r.t. Golden Std.

CNN	Observed				Masked	Safety		Gain w.r.t. FP32	
	Data corruption	Warning	Accept	Good		Decrease	Increase	Safety[a]	Memory
FP32	1.32%	0.06%	29.96%	28.98%	39.68%	31.34%	68.66%	–	–
FP16	2.61%	0.12%	53.28%	41.27%	2.71%	56.01%	43.98%	−24.67%	2X
FP8	3.41%	0.91%	64.13%	31.53%	0.02%	68.45%	31.55%	−37.11%	4X
FxP32	0.03%	0.04%	25.93%	25.54%	48.47%	26.00%	74.01%	+5.34%	0
FxP16	0.05%	0.08%	49.98%	46.94%	2.96%	50.11%	49.90%	−18.77%	2X
FxP8	0.45%	0.60%	46.74%	52.18%	0.03%	47.79%	52.21%	−16.45%	4X

[a] Safety increasing effect difference between a given scenario and FP32

increase/decrease effect variation (columns 7 and 8). In general terms, the safety decreasing effect is critical only in few cases. In fact, the percentage of Data Corruption is always lower than 3.42% for all variants. In particular, fixed-point variants have a very small percentage of critical faults, always lower than 0.46%. Moreover, the increasing contribution of *Good* faults to the safety turns out to be significant, especially for 16- and 8-bit versions. As an example, in the scenario FxP8, we observed a safety increasing effect for 52.21% of the cases, with a 52.18% of *Good* faults.

Furthermore, also the bit-width plays an important role for the reliability: **the lower the bit-width the lower the resilience**. Therefore, a designer who wants to use a more efficient version of the DNN (reduced memory footprint) has to be aware that it would also be less resilient than the original DNN (FP32). However, it is worth also remarking that using fixed-point data representation, instead of the floating-point counterpart, provides the better results in terms of trade-off between resilience and efficiency. This is reported in the last two columns of Table 12.5 where the difference in *Safety* and *Memory* footprint is reported considering the FP32 DNN as the reference. For instance, we may compare scenarios FP8 and FxP8 (8-bit DNNs, 1 bit for integer and 7 bit for fractional): we observed a safety loss compared with the FP32 DNN of 37% in the floating-point version (FP8) and only of 16% in the fixed-point version (FxP8). Therefore, choosing the DNN in the scenario FxP8 allows the designer to compact the memory footprint by a 4x factor while reducing the safety only by 16%. By looking more closely, the occurrence of critical faults in scenario FxP8 even decreases from 1.32% of FP32 to 0.45%, while in the case of the floating-point scenario (FP8) it increases to 3.41%. Additionally, for scenario FxP32 (32-bit fixed-point DNN), it has been observed that the DNN achieves 5% improved safety with respect to the FP32 scenario for the same memory footprint. Thus, simply changing the DNN data type to fixed-point improves its resiliency.

12.6.1.2 Discussion

The above results have been obtained resorting to fault injection campaigns that are highly time consuming because of the huge number of faults that have to be considered. To reduce the number of faults, we consider the layers on the LeNet-5 that perform arithmetic computations involving the trained weights, i.e., the two Convolutional and the two Fully Connected layers. Indeed, we consider the resilience of the DNN against faults affecting the memory, where the weights are stored.

Table 12.6 provides details about the configuration as well as the fault list of each layer. The first two rows (labeled *"Layer"* and *"Detail"*) of the table present the target layers; the third one (*"Connections"*) specifies the amount of their connection weights. The number of possible faults is computed as the multiplication between the connections number (*"Connections"*) and the weight size (*"Bit-width"*).

As the rows *"#Faults"* point out, the overall number of possible faults is very high and this reflects in a non-manageable fault injection campaign execution time.

Table 12.6 LeNet-5 fault list for injection campaigns

	Layer	L0	L2	L4	L6
	Detail	Convolutional	Convolutional	Fully connected	Fully connected
	Connections	2400	51,200	3,211,264	10,240
Scenarios FP32, FxP32	Bit-width	32	32	32	32
	#Faults	76,800	1,638,400	102,760,448	327,680
	#Injections	13,678	16,474	16,638	15,837
Scenarios FP16, FxP16	Bit-width	16	16	16	16
	#Faults	38,400	819,200	51,380,224	163,840
	#Injections	11,610	16,310	16,636	15,107
Scenarios FP8, FxP8	Bit-width	8	8	8	8
	#Faults	19,200	409,600	25,690,112	8,1920
	#Injections	8,915	15,991	16,630	13,831

Thus, in order to reduce the fault injection execution time, we can randomly select a subset of faults. To obtain statistically significant results with an error margin of 1% and a confidence level of 99%, an average of 15.6k fault injections have to be considered for 32-bit scenarios (FP32 and FxP32), 15k for 16-bit scenarios (FP16 and FxP16), and 13.8k for 8-bit scenarios (FP8 and FxP8). The precise numbers are given in the rows of Table 12.6 labeled *"#Injections"* and they have been computed by using the approach presented in [44]. In details, we resorted to the following formula:

$$fault\_injections = \frac{N}{1 + e^2 \times \frac{N-1}{t^2 \times 0.25}} \tag{12.10}$$

where N is the total number of fault locations (i.e., row *#Faults* of Table 12.6), e is the desired error margin (1%), and t depends on the desired confidence level (t=2.58 corresponds to 99% confidence level [44]). Equation (12.10) has an horizontal asymptotic value ($N \rightarrow \infty$) equal to 16,641, thus limiting the number of fault injections necessary to achieve an evaluation with an error margin of 1% and a confidence level of 99%. Moreover, it is worth underlining that the injections are performed by randomly selecting the bit to inject among all the bits of all the connection weights.

Despite the fact that we used Eq. (12.10), it is quite clear that for bigger DNNs the faults number can literally explode. In the next subsection we thus propose a novel method to avoid the need of carrying out fault injection campaign on the whole DNN.

12.6.2 Probabilistic Approach

In the previous section, we showed how we could investigate the fault effect within the system under analysis using fault injection techniques and quantify the deviation with respect to the expected results. In the literature, several other fault injection approaches exist [45]. The common drawback of existing techniques is the high cost of fault injection campaigns in terms of time. We conclude this chapter by presenting a stochastic approach to predict the impact of faults on a DNN's accuracy. The proposed approach builds a Bayesian model of the neural network and, by analyzing the network using the Bayesian inference theory, estimates the neural network's error distribution.

The idea is to first model the DNN topology as a Bayesian Network (BN). In BN, the nodes represent random variables, each defined over a set of states, while edges model the conditional relationships. Figure 12.12b shows a simple NN neuron having two inputs and processing the results using a sigmoid activation function. The neuron can be modeled based on the data flow, having inputs, weights, and biases as data sources, which feed a set of multiplications and sums to produce a final result that is the input of the activation function. Figure 12.12 reports the transformation into a BN model, having the data and the operators play as nodes and the edges modeling the flow between nodes of the network.

Figure 12.12 includes three color classes of nodes: the black ones represent the input of the node, the green ones represent weights and biases, and the yellow ones are the intermediate nodes representing the mathematical operation required within the neuron. It is easy to notice how the graph in Fig. 12.12 has all yellow nodes depicting the multiplications, sums, and sigmoid function response, with the edges showing the data flow. The color distinction is necessary to handle the neuron

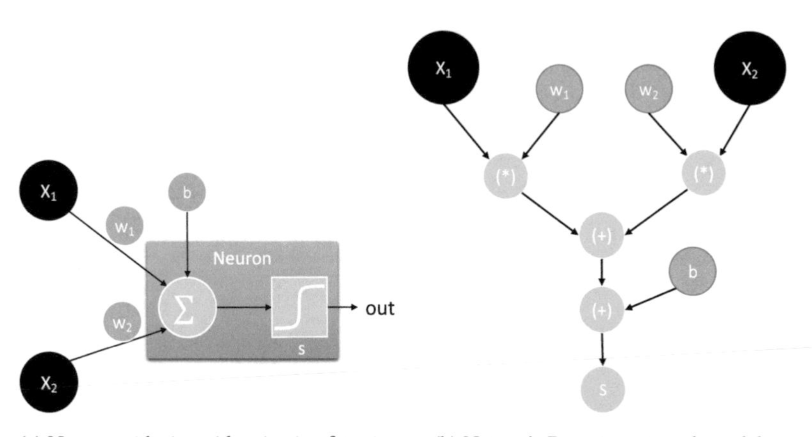

(a) Neuron with sigmoid activation function (b) Neuron's Bayesian network model

Fig. 12.12 Two input neuron Bayesian Network example. (**a**) Neuron with sigmoid activation function. (**b**) Neuron's Bayesian network model

behavior in Bayesian Theory terms properly. In fact, every single node must be a random variable to be compliant with the Bayesian Theory. To define the states representing the behavior of every type of node in the neuron model, we propose a classification approach to the tricky point of DNN faults impact assessment because not all deviations of the output lead to an inference error. The classification flows what already published in [46–48]. We classify the data associated with the node, i.e., the value of an input or weight or the result of multiplication, with the possibility of representing three different error's states:

- *Masked* (M), when the value is error-free regardless of the HW fault;
- *Acceptable* (A), when the introduced error remains under a user-defined threshold (α);
- *Critical* (C), when the introduced error rises above the threshold, making the value not acceptable.

The use of a user-define threshold allows to support a flexible evaluation, which can be adapted to the specific problem. The data error ($\tilde{y}$), easily evaluable by difference with the faulty free one, reflect the classification using Eq. (12.11):

$$Class(\tilde{y}) = \begin{cases} M \text{ (Masked)} & if\ \epsilon_y = 0 \\ A \text{ (Acceptable)} & if\ \epsilon_y \leq \alpha \\ C \text{ (Critical)} & if \epsilon_y > \alpha \end{cases} \tag{12.11}$$

Resorting to the classification in Eq. (12.11), the model can include the necessary set of Conditional Probability Tables (CPTs) that should describe each node's probabilistic behavior, i.e., express every node's probability belonging to each of these classes, eventually conditioned to the inputs states. Figure 12.13 shows how the classification translates in terms of information. The reported two tables demonstrate the two types of CPT associated with a node of the BN. The first one (the CPT associated with X_2) displays how black and green nodes of the network are probabilistically characterized: three probabilities define the distribution of all possible faulty values among the three classes. This table is easily computed following an enumeration approach.

Since ϵ_y depends on the actual value, we already know that in a DNN this value is fixed for weights and biases or is distribution-based for inputs. Therefore, for every single value, we can compute the probability of having $\tilde{y}$ in M, A, or C, from all possible faults ($f \in F$), as in Eq. (12.12).

$$\begin{aligned} P(\tilde{y}\ is\ M) &= P(\epsilon_y|_f = 0) = \frac{\sum_{\forall F \rightarrow \epsilon_y|_f)=0} 1}{\#F} \\ P(\tilde{y}\ is\ A) &= P(\epsilon_y|_f) \leq \alpha) = \frac{\sum_{\forall k \rightarrow \epsilon_y|_f \leq \alpha} 1}{\#F} \\ P(\tilde{y}\ is\ C) &= P(\epsilon_y|_f) > \alpha) = \frac{\sum_{\forall k \rightarrow \epsilon_y|_f > \alpha} 1}{\#F} \end{aligned} \tag{12.12}$$

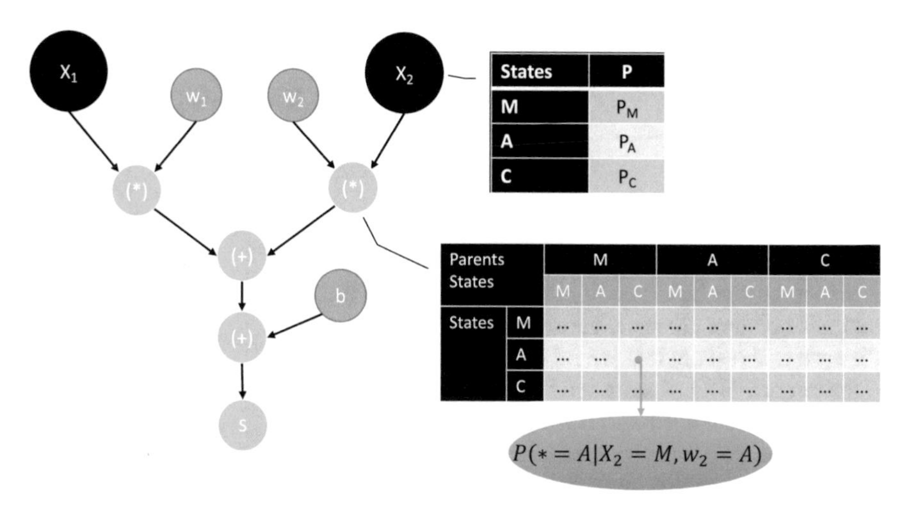

Fig. 12.13 Bayesian conditional probability tables meaning example

This evaluation perfectly stands for all fixed values, while for distributed values, the evaluation has to spawn over all possible values in the distribution (or a statistically significant subset). The second CPT in Fig. 12.13 refers to the operators, i.e., operations and functions, case (the CPT associated with the multiplication $(*)$). When the random variable depends on other random variables, i.e., its parents, those probabilities are defined for each possible combination of states of its parents [46]. In the BN model, all operations and functions (yellow nodes) are always nodes that depend on the states of their input to produce the output. In our example, we have two operations, i.e., sum and multiplication, and one function, i.e., the sigmoid ($S(x)$ in Eq. (12.13)).

$$S(x) = \frac{1}{1 + e^{-x}} \tag{12.13}$$

Since the input error can be either masked or amplified, depending on how the operation handles the inputs, the operations and function characterization determine how this propagation occurs. In general, the manipulation done by an operator may change the way the results of the application distribute among the three accuracy classes (M, A, or C). It is also crucial to understand that those probabilities are independent of the probability expressed by the input states. In the sense that they reflect the probability of the node being in a particular state knowing that its parents express a specific combination of states (as in Fig. 12.13 the $P(* = A)$ expresses the probability of the output of the multiplication to be an acceptable value knowing that X_2 value belongs to Masked and w_2 value to Acceptable).

All the yellow nodes' CPTs have been evaluated through simulations, one per type. In detail, we characterize the operator's output when we feed as input values corresponding to the three error classes. In order to produce those CPTs, we devise

Table 12.7 Prediction average error

	Prediction Classes (avg. error)		
α	M	A	C
1	0.0477	0.0477	0.0000
0.1	0.0675	0.0030	0.0705
0.0001	0.0845	0.0049	0.0893

a computational task that characterizes a library of **faulty** operators quantifying the error introduced by the occurrence of a fault. The most important message here is that this operator characterization has to be done **only one time for each operator** for each α threshold that we might need to evaluate. The consequence is that even if the target DNN has thousands of neurons, we need to work on a single operator.

After the model is built, it is time to use the Bayesian theory to evaluate it. The Bayesian inference allows the analysis of the model to predict the approximation at the output of the DNN [49]. We compute the posterior probability of the leaves of the network to be in one of the three error classes defined in Eq. (12.11) to have the necessary prediction. We used a publicly available BN library and engine [50] to implement the Bayesian inference.

To demonstrate the modeling's prediction capability when applied to DNNs, we compared the iterative exploration against the BN model. The comparison covers different α values. This process requires generating the CPT tables of all operators for each α, and then use them on the BN modeled only once (by replacing the CPTs). The experimental setup works under the hypothesis of having the faults appearing into the weights and biases memory; thus, the weights and biases distribution have been computed and assigned to the proper nodes of the CPT. Table 12.7 shows the three classifications for all three α used.

Reported results confirm the reliability of the BN prediction, showing a maximum percentage point error of 5.42pp. The variability in the absolute error reflects the data dependency of the multiplication. Nonetheless, since the outcome comes from a non-linear function such as the sigmoid, it is interesting to notice that we could adequately model it using the BN. Moreover, having the three alpha expressing a considerable variation in the quality tolerance, the absolute errors confirm the BN prediction's precision with respect to a whole fault injection campaign.

12.7 Conclusions

This chapter presented an overview of different approaches to handle the design, verification, testing, and in-field operation of approximate computing systems. The presented solutions are not exhaustive and new publications and approaches will appear while the field becomes mature. The chapter leverages on the experience of the authors to overview the major challenges that still represent a barrier to transform this interesting research field into real solutions ready to the market.

Acknowledgments This work was partly supported by the Czech Science Foundation project 21-13001S.

References

1. Xu, Q., Mytkowicz, T., & Kim, N. S. (2016). Approximate computing: A survey. *IEEE Design Test, 33*, 8–22.
2. Mittal, S. (2016). A survey of techniques for approximate computing. *ACM Computing Surveys, 48*, 62:1–62:33.
3. Han, J., & Orshansky, M. (2013). Approximate computing: An emerging paradigm for energy-efficient design. In *2013 18th IEEE European Test Symposium (ETS)*, May 2013 (pp. 1–6).
4. Chippa, V. K., Chakradhar, S. T., Roy, K., & Raghunathan, A. (2013). Analysis and characterization of inherent application resilience for approximate computing. In *2013 50th ACM/EDAC/IEEE Design Automation Conference (DAC)*, May 2013 (pp. 1–9).
5. Bushnell, M. L., & Agrawal, V. D. (2000). *Essentials of Electronic Testing for Digital, Memory, and Mixed-Signal VLSI Circuits.*
6. Eldred, R. D. (1959). Test routines based on symbolic logical statements. *Journal of ACM, 6*, 33–37.
7. Agrawal, V., Seth, S., & Society, I. C. (1988). *Tutorial test generation for VLSI chips.* Washington: Computer Society Press.
8. Momeni, A., Han, J., Montuschi, P., & Lombardi, F. (2015). Design and analysis of approximate compressors for multiplication. *IEEE Transactions on Computers, 64*, 984–994.
9. Kulkarni, P., Gupta, P., & Ercegovac, M. (2011). Trading accuracy for power with an underdesigned multiplier architecture. In *2011 24th International Conference on VLSI Design* (pp. 346–351).
10. Shin, D., & Gupta, S. K. (2010). Approximate logic synthesis for error tolerant applications. In *Design, Automation Test in Europe Conference Exhibition (DATE)* (pp. 957–960).
11. Miao, J., Gerstlauer, A., & Orshansky, M. (2013). Approximate logic synthesis under general error magnitude and frequency constraints. In *2013 IEEE/ACM International Conference on Computer-Aided Design (ICCAD)* (pp. 779–786).
12. Venkataramani, S., Sabne, A., Kozhikkottu, V., Roy, K., & Raghunathan, A. (2012). Salsa: Systematic logic synthesis of approximate circuits. In *DAC Design Automation Conference 2012*, June 2012 (pp. 796–801)
13. Miao, J., Gerstlauer, A., & Orshansky, M. (2014). Multi-level approximate logic synthesis under general error constraints. In *2014 IEEE/ACM International Conference on Computer-Aided Design (ICCAD)* (pp. 504–510).
14. Venkataramani, S., Roy, K., & Raghunathan, A. (2013). Substitute-and-simplify: A unified design paradigm for approximate and quality configurable circuits. In *Design, Automation Test in Europe Conference Exhibition (DATE)* (pp. 1367–1372).
15. Wu, Y., & Qian, W. (2016). An efficient method for multi-level approximate logic synthesis under error rate constraint. In *2016 53rd ACM/EDAC/IEEE Design Automation Conference (DAC)*, June 2016 (pp. 1–6)
16. Shin, D., & Gupta, S. K. (2011). A new circuit simplification method for error tolerant applications. In *Design, Automation Test in Europe (DATE)* (pp. 1–6).
17. Ranjan, A., Raha, A., Venkataramani, S., Roy, K., & Raghunathan, A. (2014). ASLAN: Synthesis of approximate sequential circuits. In *Design, Automation Test in Europe Conference Exhibition (DATE)*.
18. Wang, L.-T., Wu, C.-W., & Wen, X. (2006). *VLSI test principles and architectures* (1st ed.). Amsterdam: Elsevier.
19. Traiola, M., VirazelA., Girard, P., Barbareschi, M., & Bosio, A. (2017) Towards digital circuit approximation by exploiting fault simulation. In *2017 IEEE East-West Design Test Symposium (EWDTS)* (pp. 1–7).

20. Ceska, M., Matyas, J., Mrazek, V., Sekanina, L., Vasicek, Z., & Vojnar, T. (2020). Adaptive verifiability-driven strategy for evolutionary approximation of arithmetic circuits. *Applied Soft Computing, 95*, 1–17.
21. Vasicek, Z., & Mrazek, V. (2017). Trading between quality and non-functional properties of median filter in embedded systems. *Genetic Programming and Evolvable Machines, 18*(1), 45–82.
22. Chandrasekharan, A., Soeken, M., Große, D., & Drechsler, R. (2016). Precise error determination of approximated components in sequential circuits with model checking. In *Proceedings of the DAC'16* (pp. 1–6). New York: ACM.
23. Vasicek, Z. (2019). Formal methods for exact analysis of approximate circuits. *IEEE Access, 7*(1), 177309–177331.
24. Sekanina, L., Vasicek,Z., & Mrazek, V. (2019). *Automated search-based functional approximation for digital circuits* (pp. 175–203). Cham: Springer International Publishing.
25. Wali, I., Traiola, M., Virazel, A., Girard, P., Barbareschi, M., & Bosio, A. (2017). Can we approximate the test of integrated circuits? In *3rd Workshop On Approximate Computing (WAPCO)*, January 2017 (pp. 1–7).
26. Wali, I., Traiola, M., Virazel, A., Girard, P., Barbareschi, M., & Bosio, A. (2017). Towards approximation during test of integrated circuits. In *2017 IEEE 20th International Symposium on Design and Diagnostics of Electronic Circuits Systems (DDECS)*, April 2017 (pp. 28–33).
27. Traiola, M., Virazel, A., Girard, P., Barbareschi, M., & Bosio, A. (2018). Testing integrated circuits for approximate computing applications. In *4rd Workshop On Approximate Computing (WAPCO)*, January 2018 (pp. 1–7).
28. Traiola, M., Virazel, A., Girard, P., Barbareschi, M., & Bosio, A. (2018). Testing approximate digital circuits: Challenges and opportunities. In *2018 IEEE 19th Latin-American Test Symposium (LATS)*, March 2018 (pp. 1–6).
29. Traiola, M., Virazel, A., Girard, P., Barbareschi, M., & Bosio, A. (2018). On the comparison of different atpg approaches for approximate integrated circuits. In *2018 IEEE 21st International Symposium on Design and Diagnostics of Electronic Circuits Systems (DDECS)*, April 2018 (pp. 85–90).
30. Anghel, L., Benabdenbi, M., Bosio, A., Traiola, M., & Vatajelu, E. I. (2018). Test and reliability in approximate computing. *Journal of Electronic Testing, 34*, 375–387.
31. Traiola, M., Virazel, A., Girard, P., Barbareschi, M., & Bosio, A. (2018). Investigation of mean-error metrics for testing approximate integrated circuits. In *2018 IEEE International Symposium on Defect and Fault Tolerance in VLSI and Nanotechnology Systems (DFT)*, October 2018 (pp. 1–6).
32. Traiola, M., Virazel, A., Girard, P., Barbareschi, M., & Bosio, A. (2019). A test pattern generation technique for approximate circuits based on an ILP-formulated pattern selection procedure. In *IEEE Transactions on Nanotechnology* (pp. 1–1).
33. Traiola, M., Virazel, A., Girard, P., Barbareschi, M., & Bosio, A. (2020). Maximizing yield for approximate integrated circuits. In *2020 Design, Automation Test in Europe Conference Exhibition (DATE)*.
34. Traiola, M., Virazel, A., Girard, P., Barbareschi, M., & Bosio, A. (2020). A survey of testing techniques for approximate integrated circuits. *Proceedings of the IEEE, 108*(12), 2178–2194.
35. Bosio, A., Carlo, S. D. , Girard, P., Sanchez, E., Savino, A., Sekanina, L., Traiola, M., Vasicek, Z., & Virazel, A. (2020). Design, verification, test and in-field implications of approximate computing systems. In *2020 IEEE European Test Symposium (ETS)* (pp. 1–10).
36. Liang, J., Han, J., & Lombardi, F. (2013). New metrics for the reliability of approximate and probabilistic adders. *IEEE Transactions on Computers, 62*, 1760–1771.
37. [Online]. *Tetramax*. https://www.synopsys.com/
38. Frohwerk, R. A. (1977). *Signature analysis: A new digital field service method.*
39. Lecun, Y., Bottou, L., Bengio, Y., & Haffner, P. (1998). Gradient-based learning applied to document recognition. *Proceedings of the IEEE, 86*, 2278–2324.
40. Joseph, R. (2016). Darknet: Open source neural networks in C. Available online: https://pjreddie.com/darknet/

41. Ruospo, A., Bosio, A., Ianne, A., & Sanchez, E. (2020). Evaluating convolutional neural networks reliability depending on their data representation. In *2020 23rd Euromicro Conference on Digital System Design (DSD)* (pp. 672–679).
42. [Online], *Libfixmath library*. https://github.com/Petteri-Aimonen/libfixmath, 2020.
43. LeCun, Y., et al. (2020). The MNIST database.
44. Leveugle, R., Calvez, A., Maistri, P., & Vanhauwaert, P. (2009). Statistical fault injection: Quantified error and confidence. In *2009 Design, Automation Test in Europe Conference Exhibition*, April 2009 (pp. 502–506).
45. G. D. Natale, D. Gizopoulos, S. D. Carlo, A. Bosio, & R. Canal (Eds.), *Cross-layer reliability of computing systems*. London: Institution of Engineering and Technology.
46. Savino, A., Traiola, M., Carlo, S. D., & Bosio, A. (2021). Efficient neural network approximation via Bayesian reasoning. In *2021 24th International Symposium on Design and Diagnostics of Electronic Circuits Systems (DDECS)* (pp. 45–50).
47. Traiola, M., Savino, A., Barbareschi, M., Carlo, S. D., & Bosio, A. (2018). Predicting the impact of functional approximation: From component- to application-level. In *24th International Symposium on On-Line Testing And Robust System Design*, July 2018 (pp. 61–64)
48. Traiola, M., Savino, A., & Di Carlo, S. (2019). Probabilistic estimation of the application-level impact of precision scaling in approximate computing applications. *Microelectronics Reliability, 102* (p. 113309).
49. Box, G. E., & Tiao, G. C. (2011). *Bayesian inference in statistical analysis* (Vol. 40). New York: Wiley.
50. BayesFusion, LLC (2015). *SMILE Engine*.

Chapter 13
Approximate Computing for Fault Tolerance Mechanisms for Safety-Critical Applications

Gennaro S. Rodrigues, Fernanda L. Kastensmidt, and Alberto Bosio

13.1 Introduction

Today's computing is a true continuum that ranges from consumer-level IoT objects or smartphones to Safety- and Mission-critical systems such as space, civil avionics, or autonomous vehicles running crucial tasks, often with human lives at stake. In this context, **reliability** and **safety** of computing systems are thus key challenges for the whole information and communication technology and must be guaranteed. The concepts are deeply intertwined. Reliability describes a characteristic of the system itself by defining the probability of a system not being subjected to failures. Safety is more focused on the interaction with the environment, and therefore on its usage, in order to assure that even in the presence of such failures, the system will not generate any dangerous outcomes. Both are negatively affected by faults mainly caused by physical manufacturing imperfections, environmental perturbations, and aging-related phenomena. Faults affecting hardware components (e.g., microprocessor, memory, ...) are then propagated through the software executed by the computing system and can induce failures such as information loss, wrong behavior, up to complete system unavailability, with a direct effect both in terms of reliability (i.e., assurance of the system to execute correctly its tasks) and safety (i.e., the absence of dangerous behavior in presence of faults).

Safety-critical systems such as aerospace and avionics applications are often exposed to space radiation. Indeed, even systems that operate at ground level can

G. S. Rodrigues · F. L. Kastensmidt
Universidade Federal do Rio Grande do Sul (UFRGS), Porto Alegre, Brazil
e-mail: gsrodrigues@inf.ufrgs.br; fglima@inf.ufrgs.br

A. Bosio (✉)
University of Lyon, ECL, INSA Lyon, CNRS, UCBL, CPE Lyon, INL, UMR5270, Écully, France
e-mail: alberto.bosio@ec-lyon.fr

A. Bosio et al. (eds.), *Approximate Computing Techniques*,
https://doi.org/10.1007/978-3-030-94705-7_13

be subject to space radiation [1], and some of those are also categorized as safety-critical systems (e.g., self-driven cars and their collision avoidance algorithms).

Radiation effects in semiconductor devices vary from data disruptions to permanent damage. The state of a memory cell, register, latch, or any part of the circuit that holds data can be changed by a radiation event. Radiation events might cause *soft* or *hard* errors. Soft errors are the primary concern for commercial applications [2] and occur when this radiation event is strong enough to change the data state without permanently damaging the system [3], manifesting as many types of errors. In software applications, those errors can be categorized into two major groups: SDCs (silent data corruption) and FIs (functional interruption) [4]. An SDC occurs when the application finishes properly, but its output differs from the expected gold output. FIs are considered when the application hangs or terminates unexpectedly. Hard errors are permanent damages to the system and are often related to dose-rate radiation effects (i.e., associated with the accumulation of radiation and its impacts on the behavior of the transistor).

The new transistor technologies' reduction of the dimensions and operation thresholds have improved their energy consumption and performance. Their sensitivity to radiation, however, is often not a concern for the industry that focuses their efforts on higher transistor density and functionality at low cost. Indeed, the reduction of the transistor sizes on new technologies can now lead to radiation-induced faults, that would otherwise occur on space environments, to occur at ground level [5]. Although those fault-induction-related issues are not a significant concern for the traditional consumer, which can accept sparse little errors, they are indeed a severe concern for safety-critical systems.

The traditional hardware manufacturers are not motivated to develop new radiation hardened technologies because of their high development cost and, consequently, low-profit margin due to limited production volume [1]. On the other hand, the safety-critical industry is also often not interested in radiation hardened hardware, which is expensive and does not provide the same performance as the state-of-the-art hardware devices. This is the reason why the industry has turned to COTS (commercial off-the-shelf) embedded processors and systems-on-a-chip (SoC) combined with fault tolerance techniques [6]. COTS is typically low cost, very flexible, and low power consumption. For example, we can cite the Zynq-7000 All Programmable SoC [7] as a COTS system composed of two ARM processor cores and a field programmable gate array (FPGA) that is capable of serving a wide range of safety-critical applications, such as avionics communications, missile control, and smart ammunition control systems.

They do not provide, however, inherent fault tolerance (apart from traditionally methods such as memory error correction codes, which alone does not provide all the required reliability level for safety-critical systems), and therefore specific hardware- and software-based fault tolerance must be integrated. Fault tolerance can be applied at the hardware level by duplicating or triplicating an entire component and adding voters and checkers that verify the consistency of the processed data. Those techniques, however, introduce a prohibitive area and power overhead. Software-based fault tolerance does not need extra hardware and is widely presented and discussed in the literature [8, 9]. In that case, redundancy is applied at the task level and executed in single or multiple processing cores. Although software-based techniques may present no hardware area overhead, they pay the cost on execution time and memory footprint, as well as power consumption (as a consequence of those). One example of a fault tolerance mechanism that can be applied to both hardware and software is duplication with comparison (DWC), which duplicates the application and implements a checker to compare any discrepancy between the data generated by the two independent executions. DWC is capable of finding errors, but not to correct them. A third execution would be needed to tolerate the error, making a vote for the correct data.

Approximate computing has emerged as a computing paradigm capable of achieving good performances on execution time and energy consumption, as well as inherent reliability. However, it pays the price in terms of precision loss and has to consider "good enough" results as acceptable (i.e., near the expected traditional computation output). Using approximate computing on safety-critical systems could improve their performances while also making them inherently more reliable. However, it can be conflicting with some of the safety-critical systems requirements, such as accuracy. Concerning fault tolerance, approximate computing can mask a higher number of errors by relaxing data precision requirements. On systems that do not need high accuracy or quality, the approximation can be used because the small errors it introduces are not big enough to be considered a problem. Besides, the execution time reduction attained by approximate computing can improve an application reliability by reducing its exposure time: it is evident that an application that executes faster will be subject to less radiation, and therefore less to radiation-induced faults. SoCs arise as perfect implementation platforms for approximate computing.

Industry-leading companies offer SoC presenting both an FPGA logic layer (PL) and an embedded processor as a processing system (PS). Approximate computing projects can profit from the hardware-software co-design made available from COTS systems to implement any level of approximation, or as means of co-processing.

This chapter investigates the use of approximate computing on safety-critical systems. The approximate computing paradigm can be used to achieve several fundamental requirements of embedded safety-critical systems, such as low power consumption and high performance. Those, however, are achieved at the cost of precision and accuracy, which are serious concerns regarding critical applications. Another significant point of interest is reliability: approximation methods shall be able to tolerate errors or at least support traditional fault tolerance techniques. Therefore, it is essential to study not only the improvements approximate computing can bring to a project, but also its costs, and how it affects the dependability of those projects.

Approximation is presented in this work applied both at the hardware and software level. On hardware projects, the techniques are implemented in hardware description language (HDL) with and without the aid of high-level synthesis. In a first analysis, the implementation cost and precision loss of approximation methods are assessed. Then, they are subject to fault tolerance analysis by fault injection experiments. Those experiments are intended to evaluate both the approximation fault tolerance by itself and the efficacy of traditional fault tolerance mechanisms when applied to an approximate application.

The fault injections are performed in four different methodologies: fault injection emulation, fault injection simulation, and laser and heavy-ion radiation experiments. Each one of those methodologies serves better for a specific evaluation purpose. The fault injection emulation on programmable hardware, for example, can be used to evaluate the behavior of the design under a situation of cumulative faults, in an effort to find out on which point (given the number of accumulated injected faults) the design begins to present errors. It can also be used to perform exhaustive studies on programmable hardware, to find out which bits of the bitstream used to program the FPGA are critical (i.e., a bit-flip on this bit will provoke errors). On the other hand, fault injection simulation can be used to inject faults on the register file of the processor to analyze which are the most critical registers and how faults affecting the register file propagate to become errors in a given context.

The chapter is organized as follows. The Sect. 13.2 presents the source of external perturbation leading to errors in the hardware, the way of model it and how to evaluate their impact on the system. Section 13.3 presents state of the art techniques about fault-tolerant techniques. Section 13.4 presents the use of approximation to implement low-cost fault tolerance techniques. Conclusions are drawn in Sect. 13.5.

13.2 External Perturbation

Radiation can affect electronics devices in multiple ways. Single event effects (SEE) are non-cumulative and caused by single events that trigger transient upsets also called Soft Errors. Total ionizing dose (TID) and displacement damage (DD) are cumulative, which means their effects get worse over time as the system is exposed to radiation. Notice that not all radiation effects are ionizing. DD is caused by the kinetic energy of particles. TID and DD lead to permanently damage electronic devices, their effect is also called Hard Error.

The rate at which soft errors occur in a system is called Soft Error Rate (SER). SER is caused in semiconductor devices mainly because of three sources of radiation: alpha particles, high-energy cosmic rays, and low-energy cosmic rays [1]. An ion traveling through a silicon substrate loses energy, generating one electron-hole pair for each 3.6 eV lost. The linear energy transfer (LET) of an ion defines how much it can interfere with the proper device operation. It depends not only on the mass and energy of the particle but also on the material it is traveling in (represented in units of $MeVcm^2/mg$).

When radiation particles transfer enough energy into the silicon of circuits, they generate transient upsets. Upsets are manifested as bit-flips fault in any part of the circuit that holds data, causing errors [1, 10]. In microprocessors, bit-flips can occur in all registers and memories of the processor. Bit-flips induced by SEE can lead to following error classification:

- **Single Event Upset (SEU):** as soft errors are commonly referred to. Those are non-permanent errors affecting one single bit of one word of data.
- **Multiple Bit Upset (MBU):** occurs when the radiation event has energy high enough to flip multiple bits on a single word. This can be especially problematic for memory circuits that make use of error correction codes, compromising those that cover and mask only one bit [11].
- **Multiple Cell Upset (MCU):** occurs when the radiation event has energy high enough to affect multiple bits on different localities. The difference between and MBU and an MCU is that the latter consists of bit-flips affecting various parts of a system (e.g., different memory words), while the former consists of multiple bit-flips in a single word [12].
- **Single Event Transient (SET):** considered when transient upsets occur in the combinational logic part of the circuit. If propagated and latched into memory elements, those can lead to soft errors [13].

Errors induced by SEE can pass by unperceived, or be corrected (i.e., masked) by a fault tolerance mechanism. When the system misbehaves, and this is noticed by the user or propagated to another part of the system that, in its turn, shows a problematic external behavior, we say that a failure happened. Taking the example of a fault affecting the memory circuit, the definition of the events would be the following: the bit-flip on the memory data is a fault, the error is the impact it has in the data being stored in the word where the fault was raised, and the failure would be the malfunction of the software that could, for example, use this data as a control variable of a loop, causing the application to never finish its execution. Notice that the fault could have happened in an unused word of memory, causing no errors. Similarly, the error could have been overwritten by a store instruction shortly before happening, and never turn into a failure.

13.2.1 Error Analysis

The reliability of a system to radiation-induced soft errors can be measured in many different ways, depending on the available data and experiments performed. Reliability is today mostly quantified through fault injection campaigns. During each campaign, the system is forced to behave as faulty (i.e., a fault is injected in the system itself) in order to observe the impact on the application outputs. The most widely adopted fault injection methods can be mainly classified as:

- **Physical Fault Injection** is based on the realization of controlled experiments to evaluate the system behavior in the presence of artificial faults. Hardware-based fault injection technique allows injecting physical faults (e.g., bit-flip fault, stuck at fault) in the target hardware system. It uses either a manufactured system prototype or an implementation of the system on an FPGA board. Commonly used physical fault injection methods are: pin-level injection [14], heavy-ion radiation, power or electromagnetic disturbances [15], and non-destructive laser fault injection [16]. Physical fault injection techniques closely imitate real fault situations and they can access locations that are not easy to access by other techniques, by providing accurate evaluation of the system reliability. However, they introduce a high risk to damage the system under test and they are extremely expensive and applicable only after the physical chip is available.

- **Simulation-based fault injection** resorts to a hardware model of the computing system. The fault injection in the hardware models can be performed either at run-time or at compile-time, and the hardware model of the system can be described either at circuit level [17] or at micro-architectural level [18]. In the first case, the results of the fault simulation will be very accurate but they require much higher execution time.

Some of the most used metrics for reliability and fault tolerance of safety-critical systems under radiation are the mean work to failure (MWTF) [19], the cross-section, and failure in time (FIT) [1], alongside with the already discussed SER (soft error rate). The cross-section (σ) is defined as the area of the device that is sensitive to radiation, with (13.1). A larger cross-section means that a particle that hits the device is more likely to produce a failure. Thus, a design of smaller area (such as an approximate one) will typically present a smaller cross-section. The FIT is commonly as a means to express SER and is equivalent to one failure in 10^9 hours of device operation. MWTF is particularly interesting for this works discussing because it presents a correlation between performance and the fault tolerance of a technique, and is presented in (13.2).

$$\sigma = \frac{\text{number of errors/failures}}{\text{fluence of particles}} \tag{13.1}$$

$$MWTF = \frac{\text{amount of work completed}}{\text{number of errors encountered}} \tag{13.2}$$

When analyzing data from simulation experiments, the error occurrence is often presented as a simple percentage. In this type of analysis, faults are injected into the system, and it is often possible to trace the types of errors and their origin. Thus, it is easy to calculate the percentage of faults that caused errors (and failures) and their types. When analyzing fault tolerance techniques, especially those implemented on embedded software, metrics like cross-section might not be the most appropriate (in fact, using this type of metric would need an adaptation, because there is no particle fluence in this type of experiment). In those cases, data might be better presented merely as the reduction of the percentage of faults capable of inducing failures.

13.3 Fault Tolerance Techniques

Faults caused by radiation on electronic devices can become errors that need to be treated before evolving into failures. The more usual way of doing it on complex systems is with fault tolerance techniques implemented in programmable hardware or embedded software [20, 21]. Fault-tolerant devices are usually expansive, therefore the industry tends to turn to in-house developed fault tolerance techniques. Those techniques shall be able to detect errors and masking (or correcting) them when possible.

Figure 13.1 classifies fault tolerance techniques in three major groups concerning their capability. A fault tolerance technique shall be able to detect errors. What it does with this information, however, may vary. For some systems, fault detection is enough. Nevertheless, safety-critical systems often call for error masking or correction. The difference between an error masking and correction is that masking an error consists of keeping the system safe and hiding the error from the end-user (or the rest of a more complex system). An excellent example of this type of fault tolerance technique is triple modular redundancy (TMR) [22], which avoids the use of an erroneous data value, outputting a correct one. Correcting an error is a much harder and complex task, and from a system point of view, the impact would be the same as masking it. As an example, the lockstep technique [23] finds an error and rolls all the system execution back to a safe-state before the error happened, and then resumes the system execution with the hopes that the error has forever vanished.

The literature presents an enormous set of techniques implemented in software to protect applications against hardware errors. Those are called software-implemented hardware fault tolerance (SIHFT) techniques [24], and achieve pro-

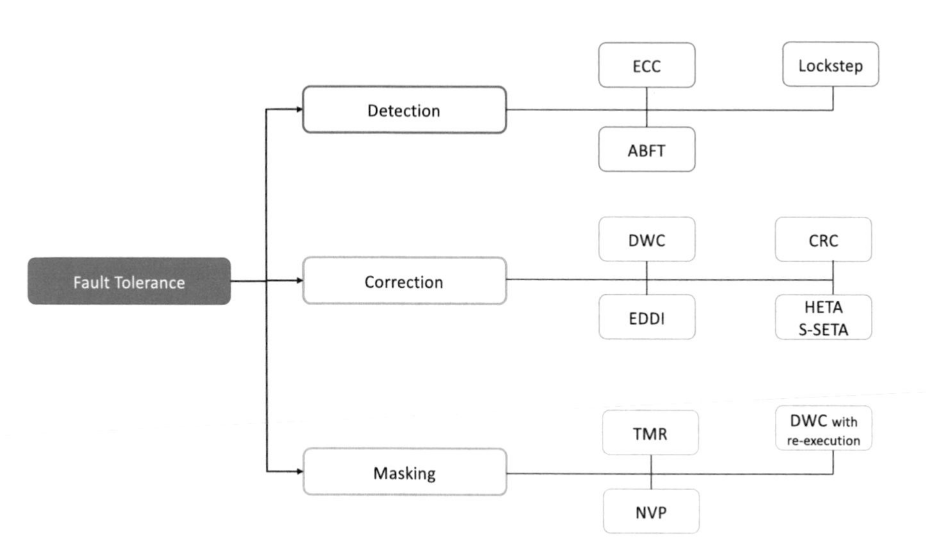

Fig. 13.1 Fault tolerance techniques classification

tection with function redundancy and variables replication. An example of this type of technique is EDDI (error detection by duplicated instructions) [25]. EDDI detects faults by comparing two different executions of the program, mapping all numbers on the original computation to new values, and applying transformations to the program so that it can be backward comparable with the original calculation. Techniques like HETA [26] and S-SETA [27] detect control-flow faults and put the system in a fail-safe state. The CFT-tool [28] is capable of combining these techniques to detect both SDCs and functional interruption (FI) errors. CFT-tool inserts the fault tolerance methods directly on the Assembly level code of the program to be protected (after the compilation). Nevertheless, and it can present some limitations for complex applications and those which are supposed to run on top of operating system. Techniques called application-based fault tolerance (ABFT) encode the used data, profiting from unique application characteristics [29]. ABFT shall be specifically designed for the application under protection. Therefore, it is not scalable to a high range of applications and tends to be costly in design. Both SIHFT and ABFT come with the cost of execution time overhead.

Redundancy methods such as triple modular redundancy (TMR) and duplication with comparison (DWC) are employed in a multitude of systems, both to provide error detection and masking. TMR can be implemented bot to protect a hardware module [30] or software code [31]. It consists of triplicating the hardware (or software code) and voting the output of the redundancies: if at least two results match, it is considered the correct output. When TMR is used on hardware, it mainly implies area and power overhead. When it is applied to software, it mostly provokes execution time overhead. DWC techniques are only capable of detecting errors, but can implement re-execution methods to provide error masking. This way, an error can be detected and mitigated before becoming a failure. DWC techniques have an overhead of two times the execution time of the original application for pure redundancy and three times when applying re-execution for error masking. N-version programming (NVP) [32] is a programming strategy that consists of developing a number of different (but equivalent) algorithms from the same specification. With this method, designers hope to achieve fault tolerance via code redundancy (voting the results from each one), expecting that two different programmers independently generating code would not produce software that is susceptible to the same errors.

Cyclic redundancy check (CRC) [33] is commonly used on network and storage systems to detect errors affecting the stored data. This error checking method is broadly used on network systems because it is easy to implement on hardware and perfect to detect burst errors, as well as those caused by noises in the transmission. A multitude of CRC designs is proposed in the literature, but it consists of check values based on the calculations of polynomials, that shall be re-calculated to verify if the check value remains the same. If it is not, there is an error in the data. CRC can be used as a first step for error correction. Error correction codes (ECC) [34] is also presented in the literature in various forms. Hamming ECC, for instance, is extensively used to protect NAND flash memories against errors. This method provides the correction of one error and the detection of up to two errors (with no correction possible in this case).

13.3.1 Approximate Fault Tolerance

In most of the cited works, fault tolerance approaches are proposed to deal with the errors. Nevertheless, a plethora of safety-critical applications may not need it. As stated in [35], real-time systems have to deal with *data freshness* requirements, which defines the time interval on which data is considered valid. For instance, an automatic navigation system may have an error during its execution, but because its data freshness has a minuscule time interval, the error will soon disappear as the algorithm keeps its execution generating a new value. Because of that, an error correction procedure is not always necessary. However, the system shall be aware of the error to put itself in a fail-safe mode. Indeed, in some cases, it is better to warn the user about the error and let him decide how to handle it. Such is the case of some errors that might affect an airplane system, for example. Trying to correct an error in an airplane can cause the system to overwrite the pilot's demands and cause a catastrophe. In those cases, it is often better to warn the pilot that certain data is not to be trusted or alert for a malfunction and let him deal with the situation in the most suitable manner. This type of situation calls for an error detection system (without the masking capability). In those cases, the values of the redundant re-computations are only used for comparison and error checking. That is where a designer may profit from approximate computing.

Approximation itself implies the idea of inherent error tolerance. On approximate systems, a specified error tolerance has to be considered, but that is not the same error definition used when discussing radiation effects and safety-critical systems. Approximation errors are caused by the system itself and manifested as quality or accuracy degradation. Also, when dealing with approximation, the decision of whether an error caused a failure or not is a matter of definition related to what would be considered a "correct" application output, which is often hard to be defined. Taking, for instance, the example of image outputs, the correctness of the output is tied to an image quality definition, which is different from one human being to another because of biological reasons. This accuracy relaxation from the approximate system can, however, be used in favor of fault tolerance on safety-critical systems: a system that accepts some accuracy degradation can ignore errors in memory that have a low impact on the data value, for example. Also, the reduction of the complexity, achieved by approximation, can help to reduce the system's susceptibility to faults (e.g., by reducing the critical area of a hardware circuit).

On safety-critical systems, however, the definition of error is related to the occurrence of a fault. In this scope, the approximation can be used in two manners. First, it can be used to improve the application execution time, energy consumption, and even reliability. Secondly, approximate computing can also be used to reduce the costs of fault tolerance techniques. The impact of using approximation on those two levels, however, is different. As already discussed, the approximation of the application directly impacts its accuracy, and therefore reliability. Approximating fault tolerance techniques may, however, be developed in such a way to avoid affecting the accuracy of the application, or affecting it only up to the acceptable level that is defined by its quality (or accuracy) requirements.

TMR is one of the most traditional fault tolerance techniques presented by the literature. Approximate TMR (ATMR) [36] is based on implementing each redundancy task with a different architecture or algorithm to provide the capability of masking multiple errors. When applied to hardware designs, ATMR has been presented as a way to achieve fault coverage almost as good as traditional TMR but avoiding the huge area overhead that it costs [37]. Designers might accept a lower fault coverage if the area overhead of the project is to drop significantly. Also, a smaller hardware area implies higher fault tolerance due to the reduction of the critical area. Therefore ATMR might be, in some cases, not only less costly but also more reliable than traditional TMR. In traditional TMR, at least two redundancies need to have the same correct value at a given time so that the correct output can be voted. Using approximations on TMR is not trivial, because of the errors caused by the accuracy loss: even in the absence of a fault, two TMR redundancies of different accuracies will present different outputs. At [38], the authors present an ATMR approach that guarantees that the result of at least two redundancy circuits will always be the same (at the absence of a fault). The idea is using different forms of approximation on each redundancy so that two of them will not be affected by approximation errors at the same time, and the ATMR will be able to mask that error. The authors present their approximation method and prove mathematically that the errors introduced by the approximation will not harm the normal behavior of the ATMR. They also propose a full ATMR (FATMR) approach where all the three circuits are approximations (instead of having one non-approximate circuit and two approximations). This ATMR technique can also be used alongside tools that generate the best possible approximate functions with genetic algorithms [39]. The evolutionary algorithm is capable of generating the best combination of approximate functions possible for a given system. However, the ATMR and FATMR methodologies are still limited by their mathematical and theoretical constraints.

Most of the approximation techniques presented in the literature are application-specific. Therefore, it is very hard or impossible to apply the same approximation technique to any possible design or code. Knowing all the possible approximation methods and which type of design is a better fit for each of them is barely impossible work. Also, some approximation methods are applicable to multiple types of applications and hardware designs. Therefore the designer should test all of them before deciding for the one with better performance. All that would demand design time that most developers cannot afford. This work tries to solve those issues by presenting easy-to-implement approximation methods that can be applied both to programmable hardware and embedded software. Approximate fault tolerance techniques are also proposed by applied those methods to traditional TMR.

13.4 Approximate Triple Modular Redundancy (ATMR)

Given the proposed approximate computing methods, we believe that some of them can be used to improve traditional fault tolerance methods. The most classical fault tolerance method presented in the literature is TMR, as already discussed. Therefore, to evaluate how approximate computing can improve fault tolerance methods in different parts of the computing stack, two approximate TMR (ATMR) techniques are proposed: one based on hardware implementation and another based on software.

Fault tolerance techniques often introduce a high execution time or hardware area overhead. Such is the case of TMR, which costs an overhead of at least 200%. This section proposes an ATMR method to deal with that issue without highly compromising fault tolerance. Differently from [38, 39] and [37], the ATMRs presented in this section deal with the concept of *approximation intensity*, where a function can be more (or less) accurate, having a direct impact on the method fault coverage, the final answer accuracy and the application execution time.

13.4.1 Hardware ATMR Based on Data Precision Approximation

The ATMR benefits from the data precision approximation to generate redundancies that are less accurate than the classical ones, but smaller in area [40]. This ATMR is expected to achieve fault tolerance near to the traditional ones, but with less area overhead. The ATMR is applied to simple codes (two matrix multiplication algorithms). This is intended to evaluate how the studied type of approximation affects data operations its effects on hardware. Using a sophisticated code could mask that information. The fault tolerance of the proposed technique is assessed with fault injection on the FPGA configuration memory. Details about the fault injection are out the scope of this chapter the reader can found all details in [41].

In previous chapters has been proved that the proposed data precision reduction approximation saves resources. This indicates that the proposed approximation can be used to provide an ATMR design with a lower area overhead. If that turns to be true, it may even be possible to improve general use designs, achieving better performance and resources usage, as well as fault tolerance (given the lower hardware area).

Listing 13.4.1 presents a pseudo-C code that summarizes the ATMR implementation. Vivado HLS is used to synthetize hardware design as explained in [40, 41]. Some less important parts of the code are left out for simplification purposes. The ATMR is implemented as three operations, in different functions at the C code, so that Vivado HLS is forced to implement specific hardware for each one of them. Otherwise, it could re-use hardware, which is not desired for the TMR implementation. The voter is implemented as a single independent function and consists of Boolean operations that perform a bitwise check between three values.

ATMR Pseudo-C Code Using 24-bit Variables for Vivado HLS

```
#include <ap_fixed.h>
typedef ap_fixed<32,9> tsize_32;
typedef ap_fixed<24,7> tsize_24;

void main(tsize_32 input_A[2][2], tsize_32 input_B[2][2],
tsize_32 output[2][2])
{
   tsize_24 result1[2][2], result2[2][2], result3[2][2];
   result1 = matrixMult1(matrixA, matrixB);
   result2 = matrixMult2(matrixA, matrixB);
   result3 = matrixMult3(matrixA, matrixB);
   output = bitwiseVoter(result1, result2, result3);
}
```

Between the matrix multipliers and the voter, converters may or may not be needed: depending on the sizes of the data in use. That is because the voter cannot vote values of different bit-sizes. Converters may also be needed inside the matrix multipliers functions implementation, in case that the input matrices are of different sizes from the ones used in the ATMR redundancies. At the Listing 13.4.1 code, for example, the ATMR uses 24-bit variables. Therefore, additional hardware will be implemented by Vivado HLS to handle the conversion from 32-bit (size of the inputs) to 24-bit variables. Each of the ATMR redundancies can be implemented with different data sizes. The data bit-sizes will affect the final result accuracy and hardware usage. Typically, if a specific data bit-size is applied to two redundancies, it will define the overall accuracy (because of the bitwise voter). However, a designer may choose different approaches to profit from the hardware cost improvement without losing precision (e.g., comparing the values considering an acceptable difference threshold and taking the output from the best accuracy redundancy as the final result). The conversions between different data sizes and types are handled by Vivado HLS. A simple cast from a different data size in the C code is enough. A more complex and probably less costly conversion could be designed, but this type of improvement is not studied. This is also the case of the ATMR voter implementation: it is left for Vivado HLS to transform the code in hardware implementation, and possible improvements are not in the scope of this work.

Six ATMR designs were implemented, varying the data precision of the operations. A non-approximate TMR version is also presented (with the three modules work with 32-bit data). The designs are named following the data precision of each redundancy module to simplify the results analysis. For example, the ATMR design called “32-24-16” is composed of a module with 32-bit, one with a 24-bit and another with 16-bit precision data and operations. Those ATMR designs were applied to two matrix multiplication algorithms, one with matrices of size 3×3 and other of size 2×2.

13.4.1.1 Accuracy Assessment

Figure 13.2 presents the inaccuracy generated by the use of approximation for each ATMR applied to the matrix multiplication operation. The data is shown in percentages and log scale. The inaccuracy value is obtained by comparing the output values of the ATMR with the one that gives better accuracy (which is the 32-bit data size multiplication due to its higher bit-size). From Fig. 13.2, it is clear that the use of fewer representation bits impacts the accuracy. As expected, if a data bit-size is applied to two different ATMR modules, it determines the inaccuracy. This is due to the ATMR voter applied to the output, which ends up considering the results from this data precision as the final one because of that behavior, using a 24-24-24 ATMR design results in the same output accuracy as a 32-24-24 one, but with lower area usage. Another interesting outcome is the inaccuracy data for the 32-24-16 ATMR design. In this case, the inaccuracy seems to hover between the ones from the three modules.

The inaccuracy, however, is usually not high. Even in the worst case, the inaccuracy is of less than 0.04%, which means the result is more than 99.96% correct. However, the increase in inaccuracy from one design to another may be a warning for more complex systems. If the inaccuracy for a complex system applying the proposed method would be significant for a 32-24-24 ATMR case, it could be considered unacceptable for the 32-16-16 one (or any situation with two modules employing 16-bit data). The ATMR variants presenting two 16-bit size modules are two orders of magnitude more inaccurate. The inaccuracy of the 3×3

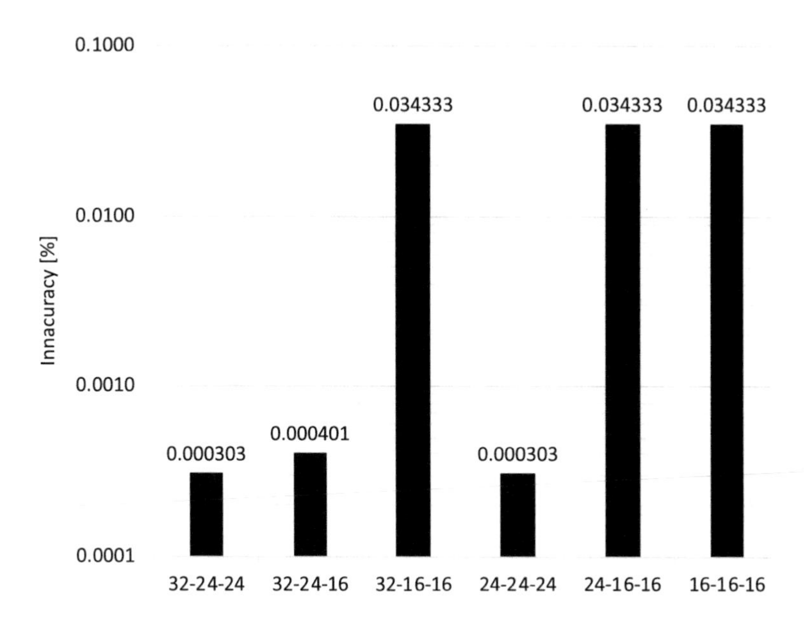

Fig. 13.2 Inaccuracy for each ATMR by data precision design applied to a 2×2 matrix multiplication. Source: [40]

matrix multiplication design follows the same trend observed for the 2 × 2 matrix multiplication and therefore is not presented.

13.4.1.2 Area Usage Assessment

Table 13.1 presents the FPGA area consumption of each ATMR design for the 2 × 2 and 3 × 3 matrix multiplication operation. FPGA resources are presented as DSP48E, FF, and LUT. DSP48E is a digital signal processing logic element included FPGA device families (e.g., Xilinx). It can be used to implement different kinds of arithmetic operations, including a multiply-accumulator and multiply-adder. The FF is Flip-Flop element, while LUT stands for look up table used to implement a given Boolean function. Data shows that approximation saves DSP usage. The FF usage can be explained by the needed converters between the matrix multiplication operations and the voter function. The LUT area follows almost the same trend of the DSPs, decreasing with the precision reduction. The variation at the LUT usage can also be explained by the needed converters. The DSP usage for the 32-32-32 TMR design was considerably high, taking into consideration that the FPGA used in this work contains 220 DSPs. This fact highlights the importance of the approximation method presented in this work.

The first 3 × 3 matrix multiplication TMR design is bold to highlight the number of DSPs used. The FPGA used in this work contains 220 DSPs, while the 32-32-32 TMR design for the 3 × 3 matrix multiplication would require 324 DSPs (bold text

Table 13.1 Area usage and performance latency of the ATMR by data reduction designs for 2 × 2 and 3 × 3 matrix multiplications

Benchmarks		Area			Max latency
TMR design	Matrices size	DSP48E	FF	LUT	Target clock: 10 ns
32-32-32	2×2	96	1985	888	9
	3×3	**324(*)**	**7560**	**3541**	**15**
32-24-24	2×2	64	1859	761	9
	3×3	216	6543	2964	14
32-24-16	2×2	56	1763	595	9
	3×3	189	5735	2138	14
32-16-16	2×2	48	1759	945	9
	3×3	162	4576	1368	14
24-24-24	2×2	48	1815	1609	8
	3×3	162	5649	2673	12
24-16-16	2×2	32	1841	1305	6
	3×3	108	3653	1165	11
16-16-16	2×2	24	1032	689	6
	3×3	81	2257	346	9

Source: [40]

in Table 13.1). Therefore, this design could not be implemented on this hardware, needing a more expensive one. With the proposed approximation, however, the implementation of an ATMR-protected 3×3 matrix multiplication is now possible. All the 3×3 matrix multiplication ATMR designs fit in the FPGA.

Comparing the data from Fig. 13.2 and Table 13.1, it is clear how the data precision reduction method is capable of reducing the area usage of the design with low effect on accuracy. The 2-24-16 ATMR design is capable of reducing the DSP usage to almost half of the 32-32-32 TMR design while introducing an inaccuracy of only 0.0004%. Another excellent example of the proposed approximation method efficiency is the results for the 16-16-16 ATMR design. It was able to reduce the DSP usage to a fourth and the FF usage in half while maintaining an accuracy of more than 99.96% comparing with the 32-32-32 design. From Sect. 13.4.1.1 it is known that the 32-16-16, 24-16-16, and 16-16-16 ATMR designs have all the same accuracy. However, it is clear from Table 13.1 that the 16-16-16 ATMR design is a better choice not only because of the area usage but also due to its lower latency.

13.4.1.3 Random Accumulated Fault Injection

This section presents the results for the randomly injected accumulated faults. In this experiment, bit-flips caused by ionizing radiation are emulated by injecting faults randomly in the FPGA resources. These bit-flips are accumulated over time, as would happen if the system were under ionizing radiation. As described in previous section, approximation leads to always have a difference between the hardware modules. This acceptable difference between values is henceforth called acceptance threshold (ε). This means that if the difference between the ATMR and the golden output value is equal or higher than ε the output will be considered as an SDC, in other words the ATMR was not able to mask the error(s).

Figure 13.3 depicts results of fault injection on all the ATMR configurations for a threshold $\varepsilon = 0.01$. The graph presents in the $y-$axis the reliability of the system and, in the $x-$axis, the number of faults accumulated on that point. The reliability is defined as the inverse of the occurrence of errors at a given number of accumulated injected faults (e.g., if the reliability at the point is of 0.9 it means that 10% of the observed errors occurred with that number of accumulated injected faults or less). As expected, the ATMR configuration with three redundancies with 16-bit data is the one more reliable. It is clear that its curve is well detached from the other ones. Another expected result is the lower reliability of the full precision ATMR configuration (32-32-32) due to its larger area. However, the 32-32-32 curve is very similar to the 32-24-16 curve.

Figure 13.4 presents the results for the randomly injected accumulated faults on all the ATMR configurations for an $\varepsilon = 1$. That is a very high acceptance threshold, that would only be acceptable on real case scenarios where accuracy is not a strong concern. The ATMR configuration with the highest reliability is again the one with three redundancies with 16-bit data. It is evident the difference between the two extremes of data precision. Nevertheless, the middle-term configurations seem to

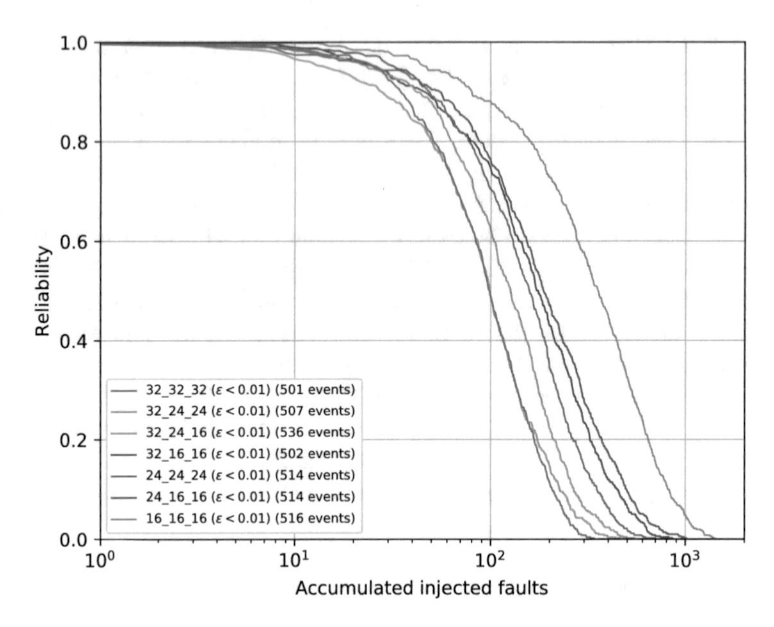

Fig. 13.3 Reliability for each ATMR configuration for an acceptance threshold of 0.01. Source: [40]

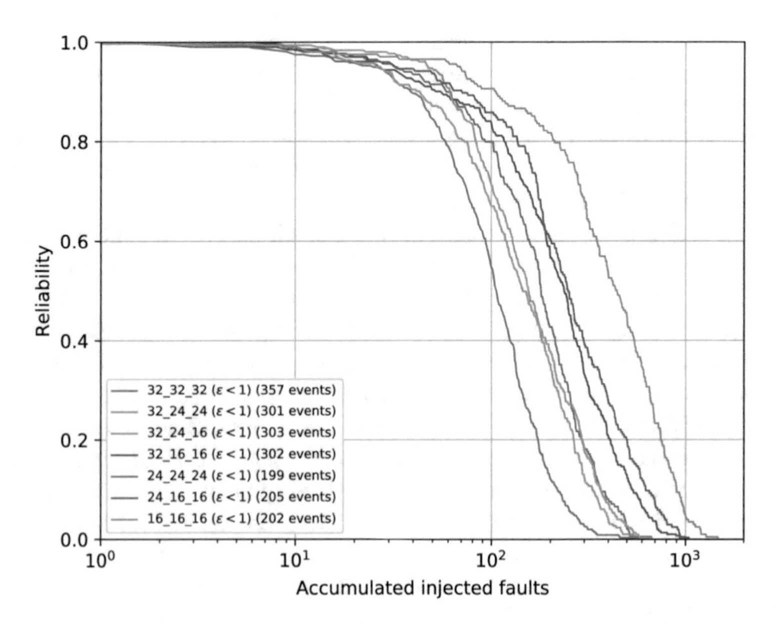

Fig. 13.4 Reliability for each ATMR configuration for an acceptance threshold of 1. Source: [40]

have similar reliabilities. It is also evident by comparing Figs. 13.3 and 13.4 that the behavior of the reliability curve is the same. However, the number of errors (number of events, on each figure legend) has dropped considerably.

The 32-32-32 ATMR configuration arises as to the worst one in terms of reliability for $\varepsilon = 1$, distancing itself from the other curves. The fact that the 32-24-16 configuration is no more as bad as the 32-32-32 one indicates that this ATMR implementation is terrible when dealing with low-ε errors (Fig. 13.3), but is able to handle higher ones (Fig. 13.4). This is because this variable of the benchmark has to deal with the low precision of the 16-bit variables and the higher area of the 32- and 24-bit ones. Because the 32-24-16 configuration does not have two redundancies with the same precision, the 16-bit redundancy has a negative effect on accuracy without significant improvement on the fault tolerance with the area reduction.

13.4.1.4 Exhaustive Fault Injection

Table 13.2 presents the results from the exhaustive fault injections. Because of how the random fault injection works, not all the injected faults affect the FGPA area occupied by the ATMR design. It is thus interesting to focus the analysis only on the area really used by the design. Therefore, the table presents the number of essential bits (which are the ones used by the ATMR design) and critical bits (the ones that caused errors when flipped) of the ATMR. The last column of the table presents the variation of the number of critical bits in relation to the 32-32-32 TMR configuration.

As expected due to the previous observations, the 16-16-16 ATMR design is the one with the lowest number of critical bits. That is reflected in its high reliability concerning the other configurations. It is interesting to notice, however, that this ATMR configuration has a high percentage of critical bits in relation to essential bits. It indicates that a design of a smaller area tends to be more reliable, even if a higher percentage of this design is critical. This idea is also backed by the fact

Table 13.2 Exhaustive onboard fault injection emulation results for a 2×2 matrix multiplication

TMR design	Essential bits	Critical bits	Critical bits variation[a]
32-32-32	540,454	7126	0%
32-24-24	355,164	3296	−53.47%
32-24-16	299,456	4016	−43.64%
32-16-16	228,122	4178	−41.36%
24-24-24	305,093	6343	−10.98%
24-16-16	165,172	3724	−47.74%
16-16-16	88,253	1764	−75.24%

[a] In relation to the 32-32-32 TMR design
Source: [40]

that the 32-32-32 and 32-24-16 ATMR configurations are the ones with the worst reliability (as presented at Sect. 13.4.1.3) and also a high number of critical bits.

The 24-24-24 ATMR configuration is the one with the second highest number of critical bits (being the 32-32-32 the one with the highest). Given this fact, it could be expected that it would also be the one with the second worst reliability. That, however, is not the case. Both Figs. 13.3 and 13.4 show that the 24-24-24 configuration is actually between the worse and the best ones, which proves that the precision and accuracy of the design also play a significant role in the system reliability.

13.4.2 Software ATMR Based on Successive Approximation

Successive approximation algorithms are numerical calculations for which an exact, straightforward solution is not computationally achievable. Those algorithms are iteration-based and get closer to an acceptable result on every iteration. In this section we use Newton-Raphson method as case study of successive approximation. The Newton-Raphson method is an algorithm used to find the roots of a function. It calculates the intersection of the tangent line of the function in an initial guess point x_0 with the x-axis. It is calculated iteratively, as stated in (13.3), until it reaches a sufficient approximation or maximum number of iterations is reached.

$$x_{n+1} = x_n - \frac{f(x_n)}{f'(x_n)} \tag{13.3}$$

The unique behavior of successive approximation algorithms arises as an opportunity to improve traditional redundancy fault tolerance methods. The number of iterations of a successive approximation algorithm impacts not only the accuracy of the output but also its execution time. When applying a TMR method to a successive approximation algorithm, there is no need to have three tasks with high accuracy. Because only one of the outputs will be taken as the final "correct" one, the others can have a lower accuracy (i.e., fewer iterations). Tasks with lower accuracy and execution time cause less overhead [42].

Figure 13.5 presents the proposed ATMR method. In the figure, R1' and R2' are redundant tasks of R0 with fewer iterations, while R1 and R2 are hard copies of R0. The overhead of a TMR consists of the extra execution time it costs. Unfortunately, the overhead of the checker (represented at the figure by the CKR box) is constant. However, reducing the execution time of the tasks, the overhead of the TMR can be lowered. Because R1' and R2' execute faster than R1 and R2, the ATMR presents a speedup in relation to the TMR.

Table 13.3 presents five different ATMR configurations applied to the Newton-Raphson algorithm running in a single ARM Cortex-A9 processor with data cache enabled. This algorithm is an excellent example of successive approximation used to calculate the roots of a function. The example reported in Table 13.3 is not related

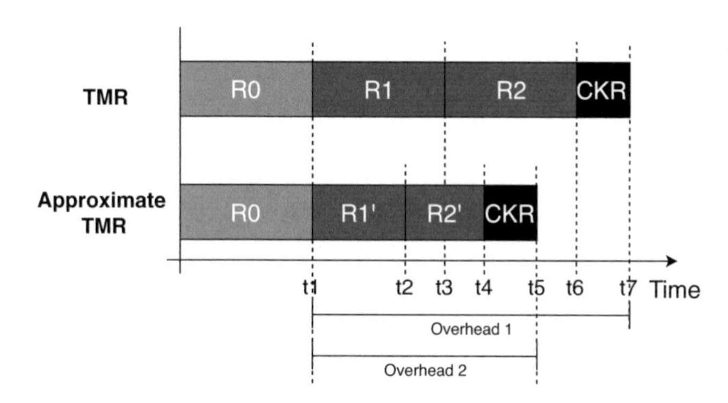

Fig. 13.5 Diagram of the proposed ATMR method. Source: [42]

Table 13.3 Execution time overheads of ATMR configurations applied to the Newton-Raphson algorithm

ATMR configuration	Execution time overhead (factor)	Execution time [ms]
71-71-71	3.09	963.268
71-71-37	2.48	771.479
71-71-14	2.22	690.381
71-37-37	1.86	579.986
71-37-14	1.60	496.237
71-14-14	1.33	414.201

Source: [41]

to a specific application. The reader has therefore to be aware of the fact that, depending on the application, the root computation can be more or less complex thus requiring more or less iterations.

The execution time overhead is presented at the table as a factor and is calculated in relation to a single execution of the Newton-Raphson algorithm with 71 iterations. The execution time on the last column is the total execution time of that ATMR configuration. The benchmarks are named following the number of iterations of each ATMR task (N_0-N_1-N_2, being N_n the number of iterations of the n-th ATMR task). For example, the ATMR configuration called "71-37-14" is composed of one task with 71 iterations, one with 37 iterations and another with 14 iterations. Each ATMR task may have a different number of iterations, but the algorithm remains the same. The number of iterations of each task differs because they start at different starting points and have different stop conditions. As the table shows, the configurations with tasks that contain fewer iterations presented a lower execution time overhead.

The checker plays a critical role in the ATMR method. In a traditional TMR, the checker would make a bitwise comparison between the three outputs, changing every bit that is different from the other two to the same value. However, with approximate computing, the checker needs to be more complex. The value of the three outputs may be different even in the absence of errors, because of the varying accuracy of each ATMR task. To deal with this issue, the ATMR checker is programmed to generate as system output a midterm between the three output values of the redundant tasks. However, this imply that the ATMR output is approximated as well. We have to consider a threshold of acceptable difference between the ATMR output value and the expected golden value (i.e., without any approximation). The acceptable difference is computed at design time (i.e., eventually it depends on the user requirements). If the ATMR checker output value differs from the golden value inside this threshold limit, the ATMR is considered to have masked the error. In the other cases, we consider that the fault has not been masked. This acceptance threshold might be different for each application or system and impacts the ATMR error masking performance.

Another way of providing approximate computing in software is through variable data size reduction. When working with embedded software, data precision reduction is imposed by software variables that are subject to predefined types. Programming new data types in software is possible, but implies on a large execution overhead, given that all the operations that would otherwise be native to the hardware in use now have to be software-processed. For this reason, we will use *float* (32-bit, single-precision) and *double* (64-bit, double-precision) variables. Because those two types of variables are capable of achieving different accuracies, they are expected to influence the behavior of the successive approximation method. Changing the variable type for a more precise one can, for example, reduce the accuracy difference between more and less precise ATMR tasks, or make the successive approximation algorithm converge faster.

13.4.2.1 Evaluation

Figures 13.6, 13.7 and 13.8 present the "Error Distribution" of the ATMR tasks (i.e., the number of ATMR tasks with errors) applied to the single-precision version of the Newton-Raphson algorithm. They, respectively, present data for ≈0%, 2% and 5% difference thresholds between the outputs of the tasks and the golden value. The ≈0% data presented actually stands for a difference of 0.000013%, which is the difference between the values from the 71- and the 14-iterations executions (without errors). It is written as ≈0% for simplification, and because it is the maximum difference that will always be present due to the usage of approximation in this application. Data is presented in percentage and calculated concerning the number of the executions that had any difference between the ATMR output and the expected golden value. For example, at Fig. 13.6 the white bar on the graphs (called "1 of 3") presents the percentage of the ATMR executions showing an error in one of the three tasks (i.e., R0, R1', and R2'), considering a ≈0% difference threshold between the outputs of the tasks and the golden value. To gather this data, the output of each task is compared to the golden value and checked for errors.

Figure 13.6 shows that a considerable amount of errors are not masked by the ATMR (because most cases presented two or more tasks with errors represented by the gray and black bars). This result is expected because of the natural variation of

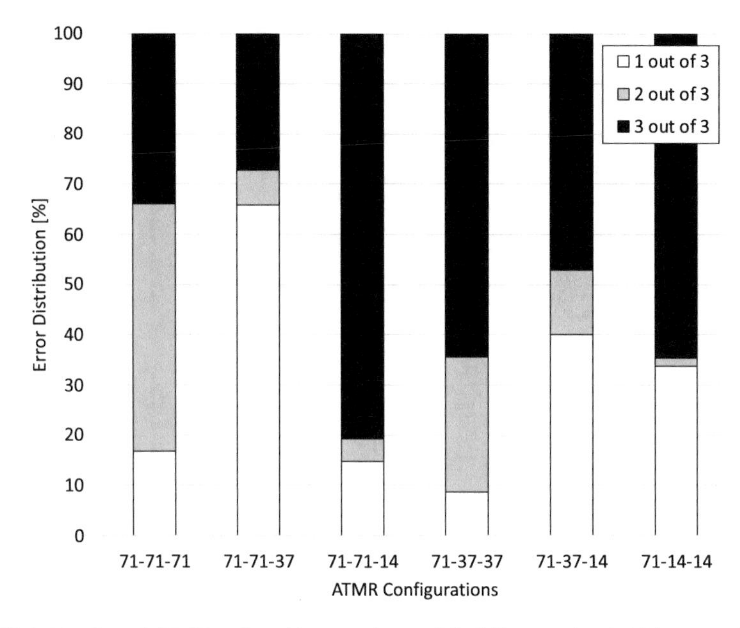

Fig. 13.6 Number of ATMR tasks with errors for a ≈0% difference threshold between the tasks outputs and golden value, on the single-precision version of the Newton-Raphson algorithm. Source: [41]

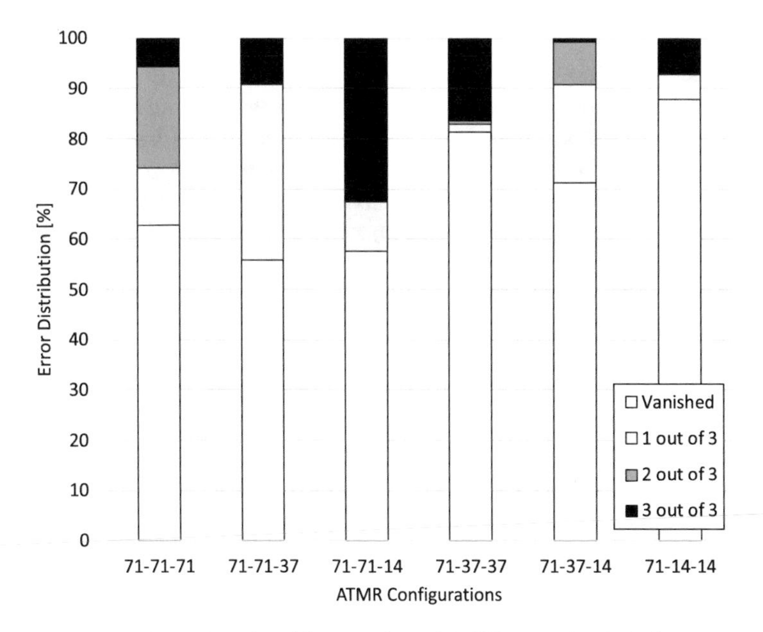

Fig. 13.7 Number of ATMR tasks with errors for a 2% difference threshold between the tasks outputs and golden value, on the single-precision version of the Newton-Raphson algorithm. Source: [41]

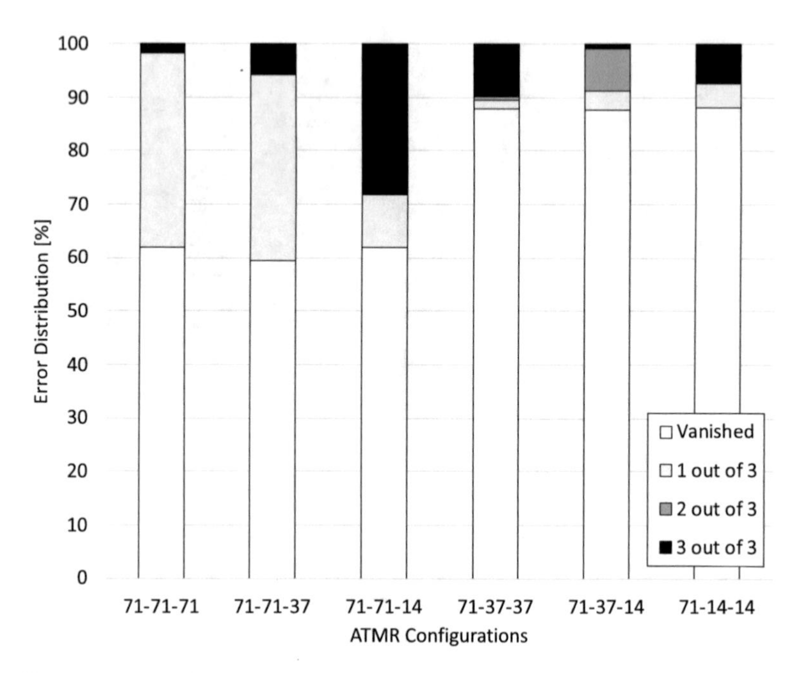

Fig. 13.8 Number of ATMR tasks with errors for a 5% difference threshold between the tasks outputs and golden value, on the single-precision version of the Newton-Raphson algorithm. Source: [41]

approximate computing algorithms outputs. When using single-precision, the 71-71-14 ATMR is the one with the highest percentage of errors affecting three out of three tasks. Two factors can explain it. First, the 14-iterations task is the one most susceptible to faults. Secondly, the 71-iterations task is the one with higher execution time. A higher execution time means more exposition to faults (because the laser pulse frequency is constant for all benchmarks). Those two factors contribute to a very inefficient ATMR configuration.

At Figs. 13.7 and 13.8, the "Vanished" bars represent the amount of errors that are no more present when the difference threshold increased (respectively, from ≈0% to 2% and from ≈0% to 5%). Figure 13.7 shows that increasing the acceptable difference threshold between the outputs and the golden value not only masks some errors but also decreases the number of erroneous tasks. This same behavior is also observed in Fig. 13.8, where the difference threshold increased to 5%. Comparing the data from Figs. 13.7 and 13.8 it becomes evident that the amount of vanished errors cease to increase at a certain point. It indicates that there may be an optimal difference threshold point, capable of providing good fault tolerance while not compromising too much the output accuracy.

Figures 13.9, 13.10 and 13.11 present the error distribution of the ATMR tasks applied to the double-precision version of the Newton-Raphson algorithm, respectively, presenting data for ≈0%, 2% and 5% difference thresholds between

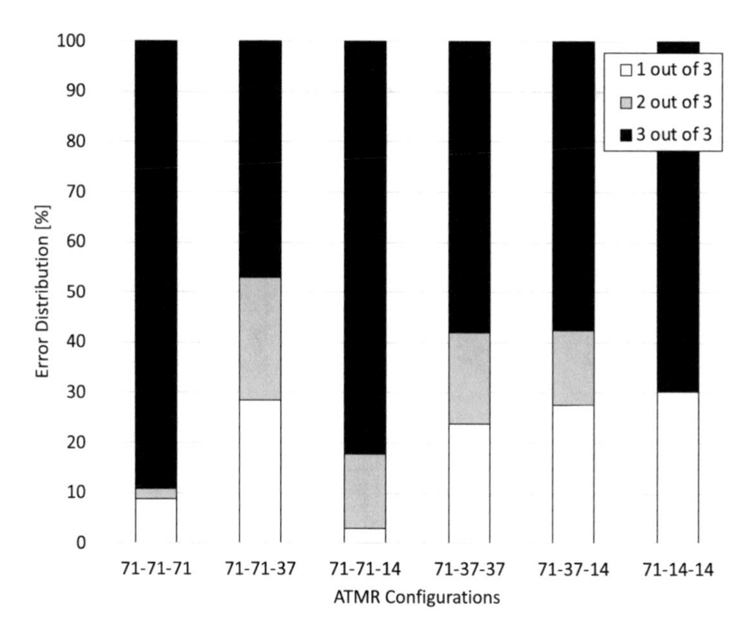

Fig. 13.9 Number of ATMR tasks with errors for a ≈0% difference threshold between the tasks outputs and golden value, on the double-precision version of the Newton-Raphson algorithm. Source: [41]

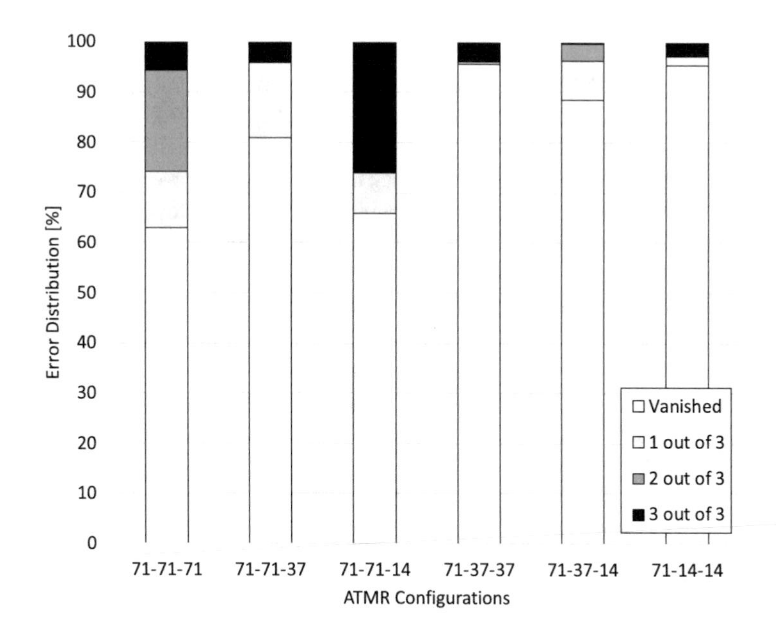

Fig. 13.10 Number of ATMR tasks with errors for a 2% difference threshold between the tasks outputs and golden value, on the double-precision version of the Newton-Raphson algorithm. Source: [41]

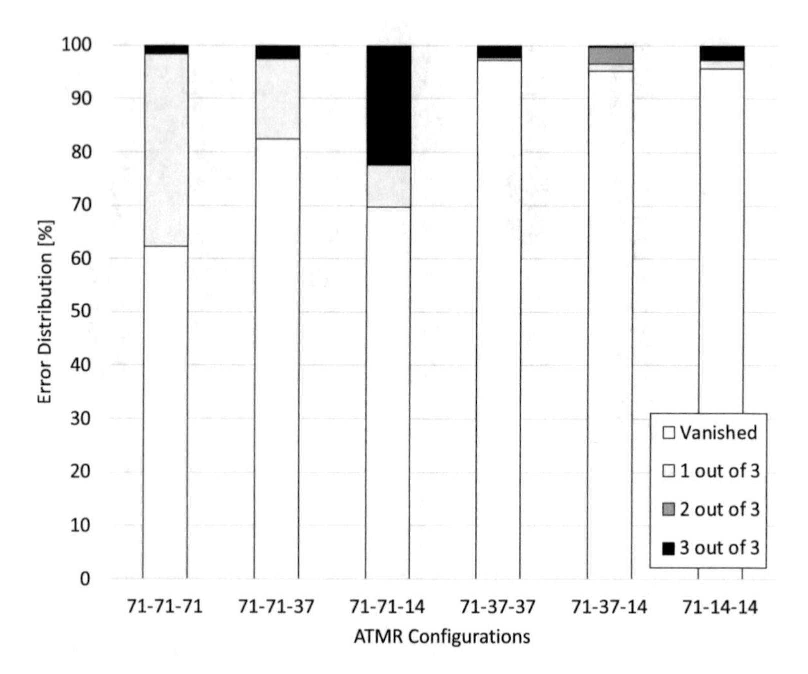

Fig. 13.11 Number of ATMR tasks with errors for a 5% difference threshold between the tasks outputs and golden value, on the double-precision version of the Newton-Raphson algorithm. Source: [41]

the outputs of the tasks and the golden value. Once again, the "Vanished" bars at Figs. 13.10 and 13.11 present the amount of errors that are no more present when the difference threshold increased. Comparing Figs. 13.9 and 13.6, using double-precision variables makes a ≈0% difference threshold ATMR even less appropriate. The number of executions with errors affecting two and three tasks is more relevant in that case. However, increasing the difference threshold between the outputs and the golden value highly increases the fault masking capability of the technique. Figure 13.10 shows that a 2% threshold is enough to provide a good fault masking. Figure 13.11 shows that increasing the threshold to 5% does not improve the fault masking performance very much in comparison with a 2% threshold.

Table 13.4 presents the percentage of masked errors for three thresholds of difference between the ATMR voted values and the golden value. Differently from the data shown at Figs. 13.6, 13.7, 13.8, 13.9, 13.10 and 13.11, this now concerns the value voted by the ATMR, not the outputs from the tasks. As discussed before, the more iterations successive approximation algorithms have, the more fault-tolerant we expect it to be. However, some unexpected results are present. Such is the case of the 71-14-14 ATMR configuration, due to its high performance both for the single and double-precision implementations. Because more iterations usually mean more fault tolerance, this is non-intuitive. Nevertheless, it can be explained by the execution time of this benchmark. It is the one with the lowest overhead

Table 13.4 Error masking for each ATMR configuration variating thresholds

	Single-precision			Double-precision		
ATMR config.	≈0% Thres.	2% Thres.	5% Thres.	≈0% Thres.	2% Thres.	5% Thres.
71-71-71	17.37	76.02	97.90	9.30	87.30	87.35
71-71-37	66.80	91.06	94.72	34.68	88.36	88.36
71-71-14	14.84	66.06	66.06	4.29	98.73	98.73
71-37-37	8.82	82.56	89.19	30.09	92.19	94.62
71-37-14	40.03	90.83	91.31	27.67	80.45	80.45
71-14-14	33.82	94.97	94.97	31.88	99.96	99.98

Source: [41]

(Table 13.3), being subject to fewer fault injections than the others. Literature shows that a high execution time implies in low fault tolerance, once the system is exposed to more faults, particularly on radioactive environments [19, 43].

Table 13.4 shows that by increasing the threshold, the ATMR was capable of masking many more faults. Even a small difference threshold of 2% is enough to make some configurations mask more than 90% of the errors. The ATMR configuration capable of masking most errors with single-precision with a high threshold is the 71-71-71. However, this configuration performs very poorly for a small threshold. This is probably due to the fact that this is the configuration with the highest execution time and therefore is subject to more faults per execution. In this case, increasing the number of iterations would, instead of improving the fault tolerance (by making the output converge), make it worse (because of the high execution time). The 71-14-14 configuration is the best one at double-precision, and it reaches a good error masking even for a 2% difference threshold. The double-precision implementations have worse performance than the single-precision ones for the ≈ 0% acceptable difference threshold. Nevertheless, increasing the threshold increases the error masking faster than it did on the single-precision cases.

13.5 Conclusion

This chapter presented the use of approximate computing for safety- and mission-critical systems. We first discussed the main source of external perturbation impacting the hardware and leading to errors. Such errors can be easily modelled and analyzed through fault injection campaign. Several fault tolerance techniques exist to increase the reliability of systems and among them we focused on those leveraging approximate computing. The presented ATMR technique shown a lower cost while maintaining reliability requirements.

One of the problems of approximate computing is that it is often not of easy implementation. Finding the best approximation method for a given algorithm is very consuming work. Future works in this topic is the development of a framework that can help software engineers to approximate their codes with minimal efforts. It is evident by the results presented that combining two or more approximation methods imply a multitude of different effects on system reliability. A designer might then ask himself, which is the optimal configuration between all possible approximation strategies that would achieve the best relation between cost and performance. Evolutionary algorithms could be used to test possible combinations of approximation configurations to find this optimal point between cost, performance, and reliability.

References

1. Baumann, R. C. (2005). Radiation-induced soft errors in advanced semiconductor technologies. *IEEE Transactions on Device and Materials Reliability, 5*(3), 305–316.
2. Poivey, C., Barth, J. L., LaBel, K. A., Gee, G., & Safren, H. (2003). In-flight observations of long-term single-event effect (SEE) performance on orbview-2 solid state recorders (SSR). In *2003 IEEE Radiation Effects Data Workshop* (pp. 102–107).
3. Tylka, A. J., Dietrich, W. F., Boberg, P. R., Smith, E. C., & Adams, J. H. (1996). Single event upsets caused by solar energetic heavy ions. *IEEE Transactions on Nuclear Science, 43*(6), 2758–2766.
4. Hsueh, M.-C., Tsai, T. K., & Iyer, R. K. (1997). Fault injection techniques and tools. *Computer, 30*(4), 75–82.
5. Tausch, J., Sleeter, D., Radaelli, D., & Puchner, H. (2007). Neutron induced micro sel events in cots sram devices. In *2007 IEEE Radiation Effects Data Workshop* (pp. 185–188).
6. Pignol, M. (2010). Cots-based applications in space avionics. In *2010 Design, Automation Test in Europe Conference Exhibition (DATE 2010)* (pp. 1213–1219).
7. Al Kadi, M., Rudolph, P., Gohringer, D., & Hubner, M. (2013). Dynamic and partial reconfiguration of ZYNQ 7000 under linux. In *2013 International Conference on Reconfigurable Computing and FPGAs (ReConFig)* (pp. 1–5). IEEE.
8. Saha, G. K. (2006). Software based fault tolerance: A survey. *Ubiquity, 2006*(July), 1:1–1:1.
9. Osinski, L., Langer, T., & Mottok, J. (2017). A survey of fault tolerance approaches on different architecture levels. In *ARCS 2017; 30th International Conference on Architecture of Computing Systems* (pp. 1–9).
10. Poivey, C., Barth, J. A., Reed, R., Stassinopoulos, E. G., LaBel, K. A., & Xapsos, M. (2001). Implications of advanced microelectronics technologies for heavy ion single event effect (SEE) testing. In *RADECS 2001. 2001 6th European Conference on Radiation and Its Effects on Components and Systems (Cat. No.01TH8605)* (pp. 328–331).
11. Maiz, J., Hareland, S., Zhang, K., & Armstrong, P. (2003). Characterization of multi-bit soft error events in advanced SRAMs. In *IEEE International Electron Devices Meeting 2003* (pp. 21.4.1–21.4.4).
12. Ibe, E., Chung, S. S., Wen, S., Yamaguchi, H., Yahagi, Y., Kameyama, H., Yamamoto, S., & Akioka, T. (2006). Spreading diversity in multi-cell neutron-induced upsets with device scaling. In *IEEE Custom Integrated Circuits Conference 2006* (pp. 437–444).
13. Benedetto, J., Eaton, P., Avery, K., Mavis, D., Gadlage, M., Turflinger, T., Dodd, P. E., & Vizkelethyd, G. (2004). Heavy ion-induced digital single-event transients in deep submicron processes. *IEEE Transactions on Nuclear Science, 51*(6), 3480–3485.
14. Constantinescu, C. (2000). Teraflops supercomputer: architecture and validation of the fault tolerance mechanisms. *IEEE Transactions on Computers, 49*(9), 886–894.
15. Miremadi, G., & Torin, J. (1995). Evaluating processor-behavior and three error-detection mechanisms using physical fault-injection. *IEEE Transactions on Reliability, 44*(3), 441–454.

16. Samson, J., Moreno, W., & Falquez, F. (1997). Validating fault tolerant designs using laser fault injection (LFI). In *1997 IEEE International Symposium on Defect and Fault Tolerance in VLSI Systems* (pp. 175–183).
17. Bosio, A., & Natale, G. D. (2008). Lifting: A flexible open-source fault simulator. In *2008 17th Asian Test Symposium* (pp. 35–40).
18. Chatzidimitriou, A., Kaliorakis, M., Gizopoulos, D., Iacaruso, M., Pipponzi, M., Mariani, R., & Di Carlo, S. (2017). Rt level vs. microarchitecture-level reliability assessment: Case study on arm(r) cortex(r)-a9 cpu. In *2017 47th Annual IEEE/IFIP International Conference on Dependable Systems and Networks Workshops (DSN-W)* (pp. 117–120).
19. Reis, G. A., Chang, J., Vachharajani, N., Mukherjee, S. S., Rangan, R., & August, D. I. (2005). Design and evaluation of hybrid fault-detection systems. In *32nd International Symposium on Computer Architecture (ISCA'05)* (pp. 148–159).
20. Avizienis, A., Laprie, J., Randell, B., & Landwehr, C. (2004). Basic concepts and taxonomy of dependable and secure computing. *IEEE Transactions on Dependable and Secure Computing, 1*(1), 11–33.
21. Kritikakou, A., Psiakis, R., Catthoor, F., & Sentieys, O. (2020). Binary tree classification of rigid error detection and correction techniques. *ACM Computing Surveys, 53*(4), 1–38.
22. Sanchez-Clemente, A. J., Entrena, L., & Garcia-Valderas, M. (2016). Partial TMR in FPGAs using approximate logic circuits. *IEEE Transactions on Nuclear Science, 63*(4), 2233–2240.
23. de Oliveira, A. B., Rodrigues, G. S., Kastensmidt, F. L., Added, N., Macchione, E. L. A., Aguiar, V. A. P., Medina, N. H., & Silveira, M. A. G. (2018). Lockstep dual-core ARM A9: Implementation and resilience analysis under heavy ion-induced soft errors. *IEEE Transactions on Nuclear Science, 65*(8), 1783–1790.
24. Goloubeva, O., Rebaudengo, M., Reorda, M. S., & Violante, M. (2003). Soft-error detection using control flow assertions. In *Proceedings 18th IEEE Symposium on Defect and Fault Tolerance in VLSI Systems* (pp. 581–588).
25. Oh, N., Mitra, S., & McCluskey, E. J. (2002). ED4I: Error detection by diverse data and duplicated instructions. *IEEE Transactions on Computers, 51*(2), 180–199.
26. Azambuja, J. R., Altieri, M., Becker, J., & Kastensmidt, F. L. (2013). Heta: Hybrid error-detection technique using assertions. *IEEE Transactions on Nuclear Science, 60*(4), 2805–2812.
27. Chielle, E., Rodrigues, G. S., Kastensmidt, F. L., Cuenca-Asensi, S., Tambara, L. A., Rech, P., & Quinn, H. (2015). S-seta: Selective software-only error-detection technique using assertions. *IEEE Transactions on Nuclear Science, 62*(6), 3088–3095.
28. Chielle, E., Barth, R. S., Lapolli, A. C., & Kastensmidt, F. L. (2012). Configurable tool to protect processors against see by software-based detection techniques. In *2012 13th Latin American Test Workshop (LATW)* (pp. 1–6).
29. Huang, K.-H., & Abraham, J. A. (1984). Algorithm-based fault tolerance for matrix operations. *IEEE Transactions on Computers, C-33*(6), 518–528.
30. Quinn, H., Baker, Z., Fairbanks, T., Tripp, J. L., & Duran, G. (2017). Robust duplication with comparison methods in microcontrollers. *IEEE Transactions on Nuclear Science, 64*(1), 338–345.
31. Quinn, H., Baker, Z., Fairbanks, T., & Tripp, J. L. (2015). Software resilience and the effectiveness of software mitigation in microcontrollers. *IEEE Transactions on Nuclear Science, 62*(6), 2532–2538.
32. Chen, L., & Avizienis, A. (1995). N-version programminc: A fault-tolerance approach to rellablllty of software operatlon. In *Twenty-Fifth International Symposium on Fault-Tolerant Computing, 1995, ' Highlights from Twenty-Five Years'* (p. 113).
33. Koopman, P., & Chakravarty, T. (2004). Cyclic redundancy code (CRC) polynomial selection for embedded networks. In *International Conference on Dependable Systems and Networks, 2004* (pp. 145–154).
34. Dong, G., Xie, N., & Zhang, T. (2011). On the use of soft-decision error-correction codes in nand flash memory. *IEEE Transactions on Circuits and Systems I: Regular Papers, 58*(2), 429–439.

35. de Freitas, E. P., Wehrmeister, M. A., Silva, E. T., Carvalho, F. C., Pereira, C. E., & Wagner, F. R. (2007). *DERAF: A High-Level Aspects Framework for Distributed Embedded Real-Time Systems Design* (pp. 55–74). Berlin:Springer.
36. Gomes, I. A. C., Martins, M., Kastensmidt, F. L., Reis, A., Ribas, R., & Novalès, S. P. (2014). Methodology for achieving best trade-off of area and fault masking coverage in atmr. In *2014 15th Latin American Test Workshop - LATW* (pp. 1–6).
37. Arifeen, T., Hassan, A. S., Moradian, H., & Lee, J. A. (2016). Probing approximate TMR in error resilient applications for better design tradeoffs. In *2016 Euromicro Conference on Digital System Design (DSD)* (pp. 637–640).
38. Gomes, I. A., Martins, M. G., Reis, A. I., & Kastensmidt, F. L. (2015). Exploring the use of approximate tmr to mask transient faults in logic with low area overhead. *Microelectronics Reliability, 55*(9), 2072 – 2076. *Proceedings of the 26th European Symposium on Reliability of Electron Devices, Failure Physics and Analysis.*
39. Albandes, I., Serrano-Cases, A., Martins, M., Martínez-Álvarez, A., Cuenca-Asensi, S., & Kastensmidt, F. (2018). Design of approximate-TMR using approximate library and heuristic approaches. *Microelectronics Reliability, 88–90*, 898–902. *29th European Symposium on Reliability of Electron Devices, Failure Physics and Analysis (ESREF 2018).*
40. Rodrigues, G. S., Fonseca, J., Benevenuti, F., Kastensmidt, F., & Bosio, A. (2019). Exploiting approximate computing for low-cost fault tolerant architectures. In *2019 32nd Symposium on Integrated Circuits and Systems Design (SBCCI)* (pp. 1–6).
41. Rodrigues, G., Fonseca, J., Kastensmidt, F., Pouget, V., Bosio, A., & Hamdioui, S. (2019). Approximate TMR based on successive approximation and loop perforation in microprocessors. *Microelectronics Reliability, 100–101*, 113385. *30th European Symposium on Reliability of Electron Devices, Failure Physics and Analysis.*
42. Rodrigues, G. S., Kastensmidt, F. L., Pouget, V., & Bosio, A. (2018). Approximate TMR based on successive approximation to protect against multiple bit upset in microprocessors. In *2018 18th European Conference on Radiation and Its Effects on Components and Systems (RADECS)* (pp. 1–5).
43. Quinn, H. (2014). Challenges in testing complex systems. *IEEE Transactions on Nuclear Science, 61*(2), 766–786.

Chapter 14
Approximate Computing for Scientific Applications

Hartwig Anzt, Marc Casas, A. Cristiano I. Malossi, Enrique S. Quintana-Ortí, Florian Scheidegger, and Sicong Zhuang

14.1 Introduction

This chapter illustrates the performance advantages that can be obtained from exploiting approximate computing techniques in the solution of scientific applications connected to linear algebra and deep learning. For the linear algebra case, we address the solution of sparse linear systems via iterative methods, giving experimental evidence of how approximate computing can help to reduce the dominant cost factor, in general due to memory access, for Jacobi solvers, Krylov subspace methods as well as simple block-Jacobi preconditioners.

In addition, this chapter proposes a technique to reduce the training cost of Deep Neural Networks by decreasing data movement across heterogeneous architectures composed of several GPUs and multicore CPU devices. In particular, this chapter proposes an algorithm to dynamically adapt the data representation format of network weights during training. This algorithm drives a compression procedure that reduces network parameters size before sending them over the parallel system. We run an exhaustive evaluation campaign considering several up-to-date deep

H. Anzt (✉)
Karlsruhe Institute of Technology, Karlsruhe, Germany
e-mail: hartwig.anzt@kit.edu

M. C. Guix · S. Zhuang
Barcelona Supercomputing Centre, Barcelona, Spain
e-mail: marc.casas@bsc.es; sicong.zhuang@bsc.es

A. C. I. Malossi · F. Scheidegger
IBM Research - Europe, Zurich, Switzerland
e-mail: acm@zurich.ibm.com; eid@zurich.ibm.com

E. S. Quintana-Ortí
Universitat Politècnica de València, Valencia, Spain
e-mail: quintana@disca.upv.es

A. Bosio et al. (eds.), *Approximate Computing Techniques*,
https://doi.org/10.1007/978-3-030-94705-7_14

neural network models and two high-end architectures composed of multiple GPUs and CPU multicore chips. Our solution reaches average performance improvements from 6.18% up to 11.91%.

14.2 Linear Algebra

Fundamental linear algebra problems –such as linear systems, eigenvalue problems, numerical rank-related computations, and least-squares problems, among others– lie at the heart of a myriad scientific applications, and sophisticated methods for the solution of these problems are a fundamental pillar of the Computational Sciences [1, 2].

Large-scale linear algebra problems represent a large fraction of the total cost-to-solution in many of these scientific applications. In order to tackle these expensive computations, during the past decades high-performance computing (HPC) techniques, including parallel algorithms, have been compiled into highly optimized math libraries that aim to squeeze up to the last drop of performance from the “latest” computer architecture of the moment [3].

LAPACK (Linear Algebra PACKage) [4] and BLAS (Basic Linear Algebra Subprograms) [5] are two well-known examples of standard interfaces for numerical *dense* linear algebra operations, with a considerable number of HPC realizations being available from companies as well as independent developer teams for a large variety of computer architectures. Although *sparse* linear algebra operations are even more commonly encountered in scientific applications than their dense counterparts, their standardization, unfortunately, lags far behind. As a result, the number of HPC libraries for the solution of sparse linear algebra problems is much scarcer.

In this section, we review a collection of Approximate Computing (AC) techniques that have been successfully applied in the iterative solution of sparse linear systems. These AC techniques propose some sort of “manipulation” of the floating point values that contain the problem data in order to compress them (with or without information loss). The ultimate goal is to reduce the data traffic between the memory and the processor registers/floating point units, in order to improve the performance (and reduce energy consumption) of the sparse linear algebra kernels, while maintaining the convergence rate of the iterative solver, resulting in the acceleration on current computer technologies. This is achieved under the reasonable assumption that, for sparse linear algebra kernels, the runtime (and energy cost) of accessing data in memory correlates with the number of bits employed by the precision format [6].

At this point, we note the following:

- This section does not aim to provide a complete survey on AC applied to (the iterative solution of sparse) linear systems. Instead, our goal is to provide a short review of existing techniques as well as their underlying ideas and concepts, and illustrate their practical benefits.

- Some of the AC techniques described in the following subsections can be easily assembled using fundamental building blocks for sparse linear algebra, with a few already being integrated into modern HPC libraries such as Ginkgo [7].
- Also, many of the following AC techniques are not exclusive of the specific sparse linear algebra methods described next. Indeed, we expect they carry over to other dense and sparse linear algebra operations and even to other problems beyond the scope of linear algebra.
- Finally, there exist some orthogonal compression techniques that, for example, reduce the storage needed to maintain the indexing information for a sparse matrix; see, e.g., [8]. While these schemes also aim to reduce the data movement, strictly speaking, they cannot be considered to be AC techniques.

The rest of this section on Linear Algebra for Scientific Computing problems is structured as follows. In Sect. 14.2.1, we provide a brief review of computer data representation formats in connection with the iterative solution of sparse linear systems via Krylov subspace methods enhanced with preconditioners. Then, in the next three subsections, we discuss AC in the context of iterative solvers, initially via the discussion of several relevant principles, in Sect. 14.2.2; and then through the application of these AC principles in order to accelerate the execution of iterative solvers on parallel computers, in Sects. 14.2.4 and 14.2.5. The section is closed with a short summary in Sect. 14.2.6.

14.2.1 Overview of Sparse Linear Systems and Iterative Solvers

14.2.1.1 Representation of Data

Consider the linear system

$$Ax = b, \tag{14.1}$$

where the $n \times n$ system coefficient matrix A is sparse; b is the right-hand side vector, consisting of n elements; and x is the sought-after solution vector, also with n entries. The main property of a sparse linear system matrix is that most of the entries of the coefficient matrix are zero. Even though the distinction between "dense" and "sparse" is blurry, a good reference threshold is usually that the amount of memory needed to store the matrix data can be reduced by moving from the convention of storing all matrix entries to a (more compact) sparse storage where some zero entries are not kept explicitly. The general motivation behind the adoption of sparse storage structures is to reduce the memory footprint by explicitly maintaining only a subset of the matrix entries (including all nonzero elements) along with the location information for these entries. The most intuitive example is the coordinate

(COO) format, which stores only the nonzero elements along with their respective coordinates. This results in a data structure that consists of a floating point array for the nonzero values and two integer arrays containing the row/column indices. Obviously, this COO format does not require the nonzero values to be organized in any specific order and allows for an easy modification of the matrix contents (for example, adding or removing nonzero entries). However, the performance of sparse matrix kernels operating on COO data structures often benefits from the nonzero values being ordered row-wise with increasing column indices. If a COO structure is so re-organized, the memory footprint can be further reduced by replacing the explicit row-indexing with a row pointer array indicating the start of each row. The resulting "compressed sparse row" (CSR) format is widely adopted as a data exchange layout. However, it lacks the flexibility of the COO format in terms of changing the nonzero matrix pattern and presents a more complex identification of the coordinates of a specific matrix entry. In general, both COO and CSR keep explicitly only the nonzero elements, but they can also maintain zero elements. The latter option obviously introduces some storage overhead, but may bring some advantages for algorithms acting on the sparse data structures. For example, the ELLPACK (ELL) format explicitly stores some zero elements so that the same number of elements reside in each matrix row. For balanced matrices where each matrix row contains about the same number of nonzero elements, the overhead of this padding is small, but if a sparse matrix contains a row with many nonzero elements, the overhead grows dramatically. While being counter-intuitive, padding can reduce the memory footprint for balanced problems, as it removes the need for maintaining any row index information. The goal of the ELL format, though, is not reducing the memory footprint but accelerating the execution of fundamental linear algebra kernels, such as the matrix-vector product, operating on ELL data structures in data-parallel (SIMD) fashion [9].

The other major components involved in the solution of sparse linear systems are dense vectors, which are usually maintained as a collection of floating point numbers consecutively stored in memory.

For most applications, the numeric values (including the sparse matrix entries) are represented in either IEEE (64-bit) double precision or IEEE (32-bit) single precision. In this standard representation of real (or complex) values, each floating point number is stored as a three-tuple of bits: sign (1 bit), exponent, and significand. The number of bits in the exponent determines the range of the representable space while the amount of bits in the significand dictates the accuracy (granularity) of the representation. The IEEE formats [10] represent well-accepted compromises between data range and accuracy, with high-performance support in the hardware, though some applications may benefit from different trade-offs. The indexing information is usually maintained in (32-bit) signed/unsigned integers.

The previous discussion of storage formats for sparse matrices and representation of floating point data is relevant because it exposes that the amount of indexing information which needs to be maintained in order to represent a sparse matrix is non-negligible. For example, the COO format maintains both a row index and a column index for each nonzero value of the matrix. If the floating point numeric values are stored in double precision (64 bits) and the indexing integers using the standard datatype (32 bits), the indices represent an overhead of 100%. In the popular CSR format, for each nonzero value there is an associated index representing the column position of that value, yielding an approximate overhead of 50%. (Indeed, the CSR format requires an additional integer array to mark the beginning/end of each row, increasing the overhead above that figure, especially for matrices with a small number of nonzeros per row.) Knowing the overhead that the indexing data represents provides a more informed perspective of the benefits that can be obtained by compressing only the floating point values.

14.2.1.2 Iterative Solvers and Preconditioners

Direct solvers for linear systems of equations apply a fixed number of matrix transformations after which the solution is readily available. Conversely, iterative solvers generate a sequence of solution approximations which successively approach the actual solution. In the latter case, the rate at which the accuracy of the approximation is improved depends on the type of iterative solver and the numerical condition of the linear system [1]. Krylov Subspace Methods (e.g., GMRES, CG, BiCG, etc.), or KSMs, are some of the most efficient iterative solvers and approximate the solution of a linear system in a subspace of increasing dimension [11].

The condition number of the linear system correlates to the ratio between the largest and the smallest eigenvalues. By modifying the matrix eigenvalues, for example, by multiplying both sides of the linear system of equations with another matrix M, the convergence of the iterative method can be accelerated. Particularly, for an identity coefficient matrix, the iterative method converges instantaneously, and the solution is readily available. Unfortunately, turning an arbitrary matrix A into the identity matrix requires multiplying with the inverse matrix $M = A^{-1}$, and computing the inverse of a matrix is prohibitively expensive (except for trivial cases). An alternative is given by computing an approximation of the matrix inverse ($M \approx A^{-1}$). In this context, the "preconditioner" is an operator that approximates the inverse of the system matrix. An efficient preconditioner for iterative solvers aims to balance an effective reduction of the condition number of the linear system (i.e., providing a fair approximation for the matrix inverse) with a low computational cost of generating the preconditioner operator.

A very basic, yet efficient preconditioner is the Jacobi preconditioner which scales the (entries of the) rows of the coefficient matrix by the inverse of the corresponding diagonal entry. The Jacobi preconditioner is appealing as it is inexpensive and can be generated and applied in row-parallel fashion, thus introducing negligible overhead to the iterative solver. The block-Jacobi generalizes the idea of diagonal scaling to extract the diagonal blocks from the coefficient matrix, and composes the preconditioner from the collection of inverted diagonal blocks.

14.2.2 Approximate Computing for Iterative Linear System Solvers

The AC techniques that we review in the following sections leverage one or more of the following five relevant properties:

P1. *Arithmetic is cheap, but memory accesses are expensive.*
In current computers, the memory access time is a few orders of magnitude more expensive than the processing (i.e., arithmetic) cost [12, 13].

P2. *Iterative solvers for sparse linear systems are mostly composed of memory-bound operations.*
This type of methods perform a small number of floating point operations (flops) per memory access. A clear example is the sparse matrix-vector product (SpMV), a key kernel present in both stationary solvers and KSMs, which concentrates a significant part of the total cost. The SpMV roughly requires three memory accesses in order to perform two flops. Thus, together with P1, this property dictates the memory-bound nature of these methods.
In practice, the irregular sparsity structure of the coefficient matrix dictates an irregular access pattern to the memory, further constraining the throughput of this type of solvers.

P3. *Data storage can be decoupled from the arithmetic (processor) format.*
Current processors support arithmetic in IEEE single, double and, in some cases, (16-bit) half-precision (SP, DP, and HP, respectively). The current convention "couples" the storage format with the arithmetic so that the values are stored in memory using the same exact format as they are operated in the processor. The main advantage is that no transformation is necessary when a value is retrieved from memory into a processor register (or sent back from the processor register to memory), avoiding the conversion overhead. However, given P1, the cost of these transformations can easily become negligible for memory-bound operations.

P4. *Iterative solvers usually produce a sequence of vectors which gradually approximate the solution of the system.*
In other words, the initial solution can be far (very distinct) from the real one, but as the iteration converges, the successive approximations get closer to (i.e., they have more digits in common with) the real solution.

P5. *Preconditioners only provide a rough approximation of the inverse of the coefficient matrix.*
The best preconditioner for the linear system $Ax = b$ is given by the explicit inverse $M = A^{-1}$. However, computing the inverse is even more expensive than directly solving the problem. In practice, a rough approximation of A^{-1} often yields a sufficiently good preconditioner.

14.2.3 *Mixed Precision Iterative Refinement*

14.2.3.1 Overview

Iterative refinement is a well-known technique that follows a “residual-correction” approach. In the case of linear systems, starting from an approximate solution to the problem, the idea is to compute the error of this initial solution (in the form of a residual), solve a linear system for the error, and then update the initial approximate solution (residual correction). Provided *(i)* the solution for the error equation is cheap to compute, and *(ii)* this solve is performed to some accuracy, iterative refinement yields an efficient approach that progressively approximates the correct solution [14].

One way to satisfy the previous two conditions, *(i)* and *(ii)*, consists in combining iterative refinement with reduced precision arithmetic. In particular, starting from an initial solution guess $x^{\{k\}}$, with $k = 0$, mixed precision iterative refinement (MPIR) can be formulated as the following 4-step procedure:

1. Compute the residual $r^{\{k\}} := b - Ax^{\{k\}}$ (full precision).
2. Solve the linear system $Ay^{\{k\}} = r^{\{k\}}$ for $y^{\{k\}}$ (reduced precision).
3. Update the solution $x^{\{k+1\}} := x^{\{k\}} + y^{\{k\}}$ (full precision).
4. $k = k + 1$; go back to step 1 until necessary.

MPIR can be applied independently of the type of linear solver employed in step 2. For example, one can leverage there either a direct (factorization-based) method or an iterative solver. The potential advantage of MPIR comes from exploiting the low cost of solving the system in step 2 in reduced precision. For direct methods, the costly part lies in the factorization of A, which can be performed in reduced precision. In this type of method, in general, the arithmetic cost can be as relevant as the memory accesses. In contrast, in the case of iterative solvers, the cost comes from reading the sparse matrix from memory; see properties P1–P2.

14.2.3.2 Discussion

MPIR has been exploited in the context of direct (dense) solvers in [15–17] and for iterative (sparse) solvers in [18, 19], among others.

More recently, the introduction of hardware support for HP in many recent architectures, motivated by the urge to accelerate deep learning models, has also conditioned the specialization of MPIR for the solution of linear systems. For example, in [20] the authors propose to leverage the tensor cores in NVIDIA GPUs to accelerate the direct solution of linear systems using three distinct precision formats.

MIPR can be viewed as an AC technique because it computes the solution to a system with low precision (inner solver), which provides an “approximate” solution to that system. This is then used to improve the approximation of the solution to the target system in full precision (iterative outer solver).

14.2.4 Adaptive Precision in Stationary Solvers

MPIR combines two (or more) distinct types of precision in different parts of an algorithm. However, this strategy of mixing precision formats does not really exploit the principle stated in P4 (iterative solvers gradually approximate the solution). This particular property poses the interesting question of whether we could start the iteration with a low precision format, to store an initial approximation that is still far from the actual solution, and gradually increase the accuracy of the format as the iteration converges. One problem of this idea is how to detect that the accuracy of the floating point arithmetic needs to be increased. Note that this has to be done during the iteration and, therefore, the detection overhead has to be low (compared with the cost of the iteration). A related problem is at which granularity the precision variations should operate.

Fortunately, stationary solvers, such as the Jacobi relaxation method reviewed next, exhibit a key property that solves the first problem stated in the previous paragraph. In addition, they are simple algorithms, which helps to address the second problem.

14.2.4.1 Overview of the Jacobi Relaxation Method

Given an initial solution guess $x^{\{0\}}$, the Jacobi relaxation method [11] applies a component-wise relaxation to update, at each iteration k, the individual components of the approximate solution $x^{\{k\}}$ as follows:

$$x^{\{k\}} := D^{-1}\left(b - (A - D)x^{\{k-1\}}\right) = D^{-1}b + Gx^{\{k-1\}}, \qquad k = 1, 2, \ldots \tag{14.2}$$

Here D is an $n \times n$ diagonal matrix that contains only the diagonal entries of A [11]. From the numerical perspective, the Jacobi relaxation iteration converges if the spectral radius of the iteration matrix G is smaller than one [11].

The *contraction property* of the Jacobi relaxation method states that, for any component $i \in [1, n]$ of the approximate solution vectors, at any two consecutive iterations, there exists a scalar θ_i, with $0 < \theta_i < 1$, such that

$$\left|x_i^{\{k\}} - x_i^{\{k-1\}}\right| \leq \theta_i \left|x_i^{\{k-1\}} - x_i^{\{k-2\}}\right| \leq \theta_i^2 \left|x_i^{\{k-2\}} - x_i^{\{k-3\}}\right| \ldots . \tag{14.3}$$

Furthermore, due to the linear convergence rate of the Jacobi relaxation iteration, the ratios

$$c_i^{\{k\}} := \frac{z_i^{\{k-1\}}}{z_i^{\{k\}}} = \frac{\left|x_i^{\{k-1\}} - x_i^{\{k-2\}}\right|}{\left|x_i^{\{k\}} - x_i^{\{k-1\}}\right|}, \quad k \geq 2, \tag{14.4}$$

remain constant up to convergence, say $c_i^{\{2\}} = c_i^{\{3\}} = c_i^{\{4\}} = \ldots = c_i$.

14.2.4.2 AC in the Jacobi Relaxation Method

The key to an adaptive-precision formulation of the Jacobi relaxation method is that a deviation in the component-wise convergence rate from c_i is an indicator the approximation has converged for this component in the current precision. Therefore, the length of the significand has to be extended.

In practice, rounding effects can produce differences between the observed and theoretical linear convergence rates. To account for these, the test to detect a deviation should integrate a tolerance-based criterion based on a threshold value, say $\tilde{\delta}$. The condition to decide whether an extension of the significand is necessary can then be based on a test such as

$$\left| \frac{z_i^{\{k-1\}}}{z_i^{\{k\}}} - c_i^1 \right| > \tilde{\delta}, \tag{14.5}$$

where, to avoid stagnation, we set

$$\tilde{\delta} := \delta \cdot (c_i - 1) \tag{14.6}$$

for some user-defined $0 < \delta < 1$. For a detailed discussion of the adaptive-precision formulation of the Jacobi relaxation method, see [21].

14.2.4.3 Practical Implementation: Flexible Format and CPMS

A practical important aspect when implementing the adaptive-precision Jacobi relaxation method is how to combine the theory underlying the method with the actual formats for storing and operating with floating point data that are supported in current hardware. A straight-forward possibility is to limit the number formats used in the Jacobi relaxation method to those defined by IEEE, for example, for 32-bit and 64-bit numbers. The clear advantage of this option is that there exists efficient hardware support for these formats in virtually all types of computer architectures, from multicore processors to graphics processing units (GPUs). The problem though is twofold: (1) this option constrains the number of different precisions that can be exploited in the algorithm; and (2) we need to maintain several copies of the data, one per number format, or perform an on-the-fly format conversion every time the data is retrieved from memory. In [22, 23], the authors propose a "modular precision ecosystem" that *stores the floating point numbers in flexible accuracy formats without replicating the information.*

In particular, the "customized precision format based on mantissa (significand) segmentation (CPMS)", originally introduced in [22], decomposes the significand of the standard IEEE floating point formats into two or more segments enabling a partial access to the information. For that purpose, the format to maintain the numbers in memory is completely decoupled from the arithmetic format used in the floating point units. Thus, instead of using the IEEE floating point formats in the memory operations, the significand is split into several segments which are stored separately. This allows reading the part of the significand (together with the sign and the full exponent) that the Jacobi relaxation method requires during the iteration. One additional feature of this approach is that the number of exponent bits remains unchanged with respect to the length of this information in DP, which virtually eliminates the danger of overflow and underflow, as segmentation cannot turn a valid number into "NaN" or "infinity."

In practice, the CPMS technique decouples the data storage from the arithmetic format, as stated in P3, in order to materialize the potential benefits from exploiting P1–P2 in the context of an iterative solver (and, therefore, a method that satisfies P4).

14.2.4.4 Experimental Evaluation

In this subsection, we provide some practical evidence of the benefits that can be attained when gradually incrementing the precision format in the iterative solution of linear systems via stationary solvers. For that purpose, we consider a simple problem obtained from a finite difference discretization of a Laplace equation using a mesh of 100×100 elements. The following experiment was performed on a single server equipped with an AMD Radeon 7 GPU with 16 GB of HBM2 memory. This particular accelerator can deliver a memory bandwidth of 1024 GB/s, 3.36 TFLOPS (10^{12} floating point operations per second) for FP64, and 13.44 TFLOPS for FP32.

The heat maps in Fig. 14.1 illustrate the performance of the Jacobi AC-variant with incremental precision, using a 2-segment realization of CPMS, relative to that of the conventional FP64-based implementation of the same method. Note that the test is carried out for a variety of nonzero entries per row (x-axis) and tolerance threshold for the stopping criterion (y-axis). The values in the tables highlighted in white correspond to speed ups and, therefore, show performance gains for the AC-variant while those in black indicate a slowdown. In general, the AC-variant of the stationary solver benefits from lower stopping thresholds and larger numbers of elements per row. Furthermore, the advantages are more visible when the sparse matrix is stored in the ELLPACK format instead of CSR. Concretely, for the former layout, we can observe a speed up of up to 27% while the highest speed up in the last

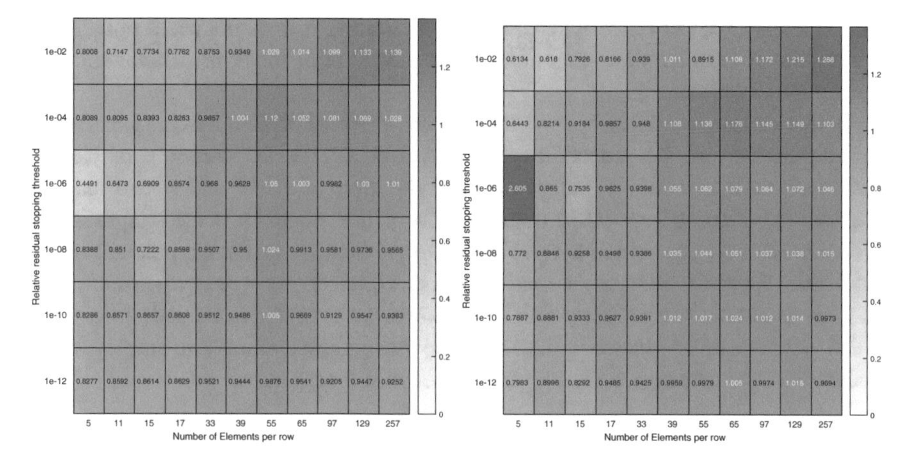

Fig. 14.1 Acceleration of the adaptive precision Jacobi in a 2-segment realization for a sparse matrix stored in CSR format (left) and ELLPACK format (right), respectively

case is only 14%. Higher performance advantages for the Jacobi AC-variant were observed on NVIDIA GPUs in [24].

14.2.4.5 Discussion

In this section, we have reviewed a practical approach to exploit P3–P4 in the context of a Jacobi relaxation method for the iterative solution of sparse linear systems. We close the section by noting that the same idea can be applied to other stationary methods (Gauss-Seidel, Successive Overrelaxation Method, etc.) [11] as well as to stencil computations. The same principles can also yield highly parallel, approximate sparse triangular solvers in order to realize incomplete factorization preconditioners [21]. Finally, a close approach is applicable to the acceleration of the PageRank algorithm for web information retrieval [25].

14.2.5 Adaptive Precision in the Preconditioner

Although the Jacobi relaxation method is a simple and highly parallel algorithm, the class of linear systems where it can be applied is strongly constrained, and its convergence to the solution is, in general, slow. In comparison, on the one hand KSMs are also based on SpMV and they present a degree of parallelism similar to that of stationary methods (except for the presence in KSMs of synchronization points in the form of "dot" products); on the other hand, they are applicable to a much wider spectrum of problems, and their convergence can be considerably accelerated via the integration of some type of preconditioner. Unfortunately, KSMs do not satisfy the contraction property and, therefore, it does not seem possible to formulate an AC technique that adjusts the precision of the approximate solution during the iteration.

In this section, we describe how to introduce adaptive precision in the calculation and application of the preconditioner instead of the "basic" iteration of KSMs. This technique is actually independent of the type of KSM and it can also be assembled with other non-KSM iterative solvers that benefit from the integration of a preconditioner. The principle this AC technique exploits is that the preconditioner itself is an approximation of the matrix inverse, which does not need to be computed to high accuracy; see property P5. This "lower accuracy" can be used to discard small elements during the computation of the preconditioner and/or to employ lower accuracy in the calculation/storage of the preconditioner entries.

14.2.5.1 Overview of Block-Jacobi Preconditioning

A block-Jacobi preconditioners can be formulated from a splitting of the coefficient matrix into

$$A = L + M + U, \tag{14.7}$$

where the preconditioner is given by the block-diagonal $n \times n$ matrix

$$M = \text{diag}(D_1, D_2, \ldots, D_m). \tag{14.8}$$

Here, D_i is an $m_i \times m_i$ block that contains the corresponding entries on the diagonal blocks of A, with $\sum_{i=1}^{m} m_i = n$, while L and U (of the same dimensions as A) comprise the elements of the coefficient matrix below and above those of M, respectively. The block-Jacobi preconditioner is well-defined if all the diagonal blocks D_i are nonsingular, and this preconditioning scheme is particularly effective if the block structure of the Jacobi preconditioner matches a block structure that is inherently present in the system matrix A.

In case the block-inverse matrix

$$M^{-1} = \text{diag}(D_1^{-1}, D_2^{-1}, \ldots, D_m^{-1}) = \text{diag}(E_1, E_2, \ldots, E_m) \tag{14.9}$$

is explicitly computed, before the iteration process of the KSM commences, the preconditioner can be efficiently applied within the KSM iteration in terms of a dense matrix-vector multiplication (GeMV) per inverse block E_i. Thus, the iteration for the preconditioned KSM remains a memory-bound process, as so is the GeMV kernel, independently of the block size m_i.

14.2.5.2 AC in Block-Jacobi Preconditioning

The use of customized precisions in the block-Jacobi preconditioner has to take into account its numerical effects in order to ensure that the quality of the preconditioner is not diminished and convergence rate of the iteration is maintained. Exploiting the independence of the Jacobi blocks, this implies that there exist certain limits to relaxing the memory precision format: (1) the regularity of the block must be preserved; and (2) the data range of the values must be covered by the precision format. The first requirement has implications on the length of the significand and the exponent, while the second requirement primarily has implications on the exponent only.

Precisely, in order to preserve the regularity of the block-Jacobi preconditioner, each block E_i can be stored in a certain precision, HP, SP, or DP, depending on its condition number, as follows [26]:

$$\begin{cases} \text{HP} & \text{if } \tau_h^L < \kappa_1(D_i) \leq \tau_h^U, \\ \text{SP} & \text{if } \tau_s^L < \kappa_1(D_i) \leq \tau_s^U, \text{ and} \\ \text{DP} & \text{otherwise.} \end{cases} \tag{14.10}$$

Here, $\kappa_1(D_i) = \kappa_1(E_i) = \|D_i\|_1 \|D_i^{-1}\|_1 = \|D_i\|_1 \|E_i\|_1$, and the thresholds τ are set as $\tau_h^L = 0$ and $\tau_h^U = \tau_s^L$. Note, however, that this scheme refers to the storage precision. During the iteration, every time the block E_i (with entries stored in the corresponding format as determined by the thresholds in (14.10)) is recovered from memory, its contents are transformed into DP values prior to the application of the block as part of a GeMV; see property P3.

The practical choice of τ is related to the accuracy level that the preconditioner should preserve. In particular, in order to ensure an accuracy a for the preconditioner a storage format with round-off error u can be considered valid for a block D if $\kappa(D) \leq a/u$. Furthermore, the value τ for this format is computed as $\tau = a/u$.

14.2.5.3 Practical Implementation: Flexible Format and CPEN

As was already introduced for the Jacobi relaxation method, the number formats that can be used to maintain the floating point values in memory does not need to be restricted to those defined by IEEE [10]. Instead, we can be rather flexible when storing the data, utilizing any number of bits to store a floating point number, and dividing these bits between significand and exponent in almost any manner. For performance reasons though, the number of bits per floating-number should be an integer multiple of a byte. The split of this resource between significand and exponent bits should balance representation granularity (bits in the significand) versus range (bits in the exponent), while ensuring that enough bits are dedicated to the later in order to avoid overflows.

When these ideas are applied blockwise to a block-Jacobi preconditioner, it is possible to customize the storage format even further. In particular, the "customized precision format with exponent normalization (CPEN)" in [24] advocates for normalizing the exponents of all the numbers in a block with respect to a certain baseline or reference value (for example, that of the smallest exponent in the block), and then minimize the number of bits dedicated to the exponents in this block by representing only the difference with respect to the baseline.

The exploitation of CPEN requires an initial clustering step that splits the original dataset into subsets (clusters) containing values of similar magnitude. Interestingly, many real linear systems problems inherently consist of clusters that accumulate values of similar magnitude. The consequence is that, in the context of block-Jacobi preconditioning, the exponents of the numbers contained in the preconditioner blocks exhibit a compact distribution of the range.

The bits that are saved with CPEN can then be dedicated to obtain a more compact representation of the block contents (just compression) or, alternatively, to enlarge the bits dedicated to the significand (AC). The latter option presents the appealing property of allowing to fit more significand bits into a 16-bit or 32-bit block. In consequence, more Jacobi blocks can use 16-bit or 32-bit storage without impacting the blocks' regularity. This technique exploits P1–P3 and P5.

14.2.5.4 Experimental Evaluation

In this subsection we assess the benefits attained by a realization of the BiCGSTAB solver [11] enhanced with the adaptive precision block-Jacobi preconditioner relative to the BiCGSTAB solver with the full precision variant of block-Jacobi. For this evaluation, we consider a subset of nonsymmetric matrices with at least 10^6 nonzeros from the SuiteSparse matrix collection for which a BiCGSTAB solver needs at least 100 iterations to converge. The experiment was run using one node of the Summit supercomputer at Oak Ridge National Laboratory, containing two 22-core IBM POWER9 CPUs with 256 GB of RAM and 6 NVIDIA TESLA V100 (SXM2 form factor) GPUs with 16 GB of HBM2 memory. For our experiments, we use only a single NVIDIA V100 GPU with a peak double precision performance of 7.8 TFlop/s and a theoretical peak memory bandwidth of 900 GB/s. The BiCGSTAB solver available in the Ginkgo library was employed.

The outcome of this experiment is reported in Fig. 14.2. The black and gray dots on top of the two plots in that figure represent whether BiCGSTAB (without a preconditioner), BiCGSTAB+(full precision) block-Jacobi and block-Jacobi and BiCGSTAB+adaptive block-Jacobi converged for that matrix. The absence of a dot means that the method not converge. The red dots represent the relative number of iterations, while the green dots the relative time of adaptive precision block-Jacobi compared with the full precision variant. A value greater than 1 thus means that the adaptive precision variant outperforms the full precision block-Jacobi for that specific problem. The adaptive precision preserves 1 digit (top) or two digits (bottom) of the full precision block-Jacobi preconditioner. In more detail, we ran

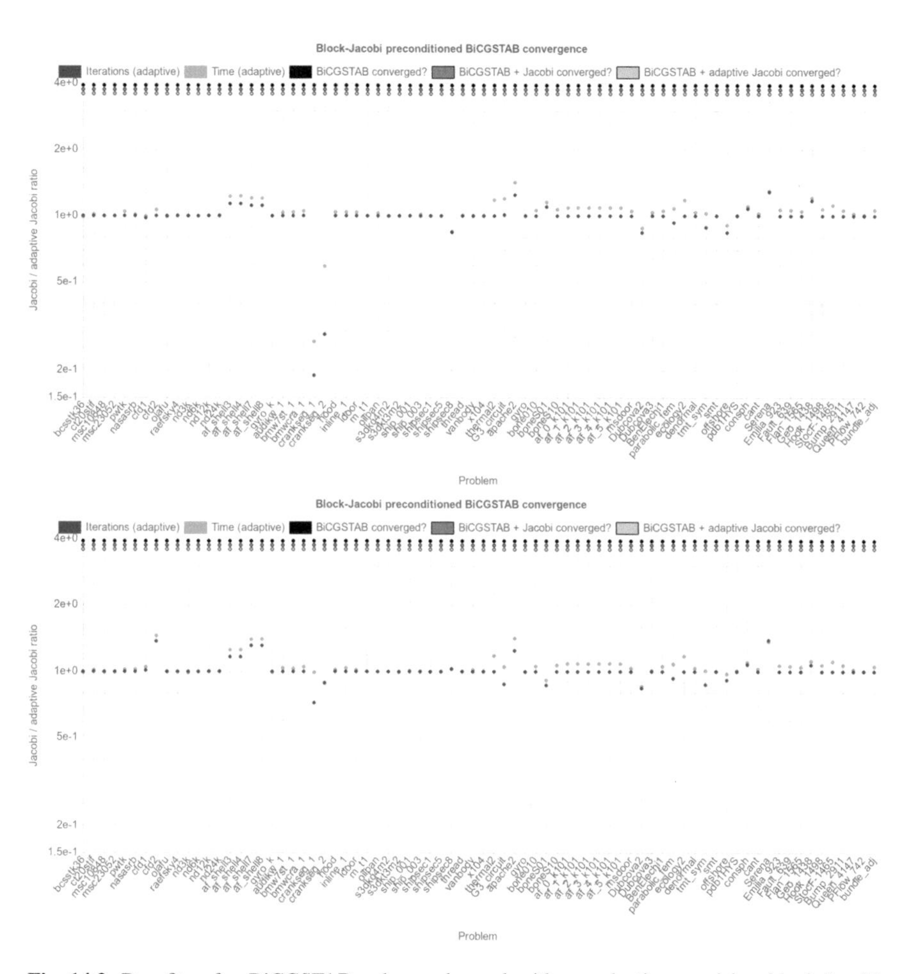

Fig. 14.2 Benefits of a BiCGSTAB solver enhanced with an adaptive precision block-Jacobi preconditioner relative to a realization that includes a conventional full precision preconditioner in the same method

two parameter settings where the automatic precision detection procedure was adapted to assign precisions such that either 1 or 2 decimal digits are preserved when applying the preconditioner. This reflects the assumption that the preconditioner provides 1 and 2 digits of accuracy, respectively.

The results from this experiment show that the BiCGSTAB solver enhanced with any of the precision-variants converged for all problems (black and gray dots on top of the plot). Furthermore, the benefit of adaptive precision is highly dependent on the problem case, in particular, on the conditioning of the diagonal blocks of the preconditioner. This is natural as when most of the blocks are well-conditioned, the majority of the preconditioner can be stored in lower precision, reducing the memory access cost, while for problems with ill-conditioned blocks, there is no difference between the two variants, since all blocks need to be stored in full precision in order to preserve the quality of the preconditioner. On average, for those cases with well-conditioned blocks, we observe performance gains that vary between 10 and 30%.

14.2.5.5 Discussion

There exist several efforts to introduce mixed precision and/or AC in the computation of preconditioners. For example, the Jacobi relaxation method presented earlier can be leveraged to realize a sparse triangular solver that computes an incomplete factorization preconditioner [21].

Alternatively, there are some previous efforts that combine a preconditioner stored in SP in memory with a solver that operates in DP [27, 28]. (Note that the adaptive-precision preconditioner discussed in this section generalizes these ideas to include any reduced precision customized format for the preconditioner, including HP, SP as particular cases.)

Related to this, Carson et al. [29] suggest the use of an incomplete factorization preconditioner computed in lower precision inside an iterative F-GMRES framework and the authors even extend this approach by cascading multiple formats of decreasing precision [30].

14.2.6 Summary

This section provided practical and strong evidence of the benefits that software AC techniques yield in the iterative solution of sparse linear systems arising in scientific applications. Concretely, (1) we re-visited MIPR as a general technique for operating with mixed precision formats; (2) we discussed the theory and practical aspects on how to gradually increase precision as a stationary solver converges to the solution of the linear system; and (3) we presented the adoption of customized formats to accommodate the different accuracy levels required by a block-Jacobi preconditioner that is leveraged in the solution of a linear system via a KSM.

Finally, we remark that some of the software AC techniques that were exposed in this section are not exclusive of linear algebra. Instead, we expect that, with some effort, they can be extended to other domains. An straight-forward example appears in graphs algorithms and from there big data applications, by taking advantage of the connection between graphs and basic sparse linear algebra kernels such as SpMV.

14.3 Deep Learning

The use of Deep Neural Networks (DNNs) is becoming ubiquitous in areas like computer vision (e.g., image recognition and object detection) [31, 32], speech recognition [33], language translation [34], and many more [35]. DNNs provide very competitive pattern detection capabilities and, more specifically, Convolutional Neural Networks (CNNs) classify very large image sets with remarkable accuracy [36]. Indeed, DNNs already play a very significant role in the large production systems of major IT companies and research centers, which has in turn driven the development of advanced software frameworks for the deep learning area [37] as well as DNN-specific hardware accelerators [38, 39]. As an example, deep learning solutions are being coupled with physical computational models for solving pattern classification problems in the context of large-scale climate simulations [40]. Despite all these accomplishments, deep learning models still suffer from several fundamental problems: the neural network topology is determined through a long and iterative empirical process, the training procedure has a huge cost in terms of time and computational resources, and the inference process of large network models incurs considerable latency to produce an output, which is not acceptable in domains requiring real-time responses like autonomous driving.

The DNN training process typically relies on the backpropagation procedure [41], which requires solving an optimization problem aimed at discovering the values of network weights that better fit the training data. A possible way to carry out the backpropagation process is the Gradient Descent (GD) method [42], which aims at fitting the weights to the training data by considering, at each iteration, the steepest descent direction in terms of an error function. A popular variant of the GD procedure is the Stochastic Gradient Descent (SGD) method [43], which computes the gradient against several randomly chosen samples at each iteration. Today's common practice to train DNNs is to split the data set into several subsets, called batches, and let each iteration of SGD to compute a descent direction or gradient that contains contributions of all the samples belonging to the same batch. SGD converges faster than GD since it updates network parameters at the end of each batch once all samples are processed.

To tackle the large amount of Floating Point computations required to train a DNN, GPUs are usually employed [44]. They exploit the large amount of data-level parallelism of deep learning workloads. Although GPUs and other hardware accelerators have been successfully employed to boost the training process, data exchanges involving different accelerators may incur significant performance penalties.

Section 14.3.1 focuses on the use of adaptive precision for accelerating training of classical CNN architectures, including a method to reduce cost of host-to-device memory transfer operations. Section 14.3.2 discusses model optimization and architecture search in presence of constraints for edge devices applications.

14.3.1 Multiprecision and Approximate Computing in Deep Learning

We describe and evaluate a method to accelerate the training of DNNs by reducing the cost of data transfers across heterogeneous high-end architectures integrating multiple GPUs. By relying on DNNs tolerance to data representation formats smaller than the commonly used 32-bit Floating Point (FP) standard [45, 46], this section describes how to dynamically adapt the size of data sent to GPU devices without hurting the quality of the training process. Our solution is designed to efficiently use the incoming bandwidth of the GPU accelerators. It relies on an adaptive scheme that dynamically adapts the data representation format required by each DNN layer and compresses network parameters before sending them over the parallel system. This scheme enables DNNs training to progress in a similar rate as if the 32-bit FP format was used. This section makes the following contributions:

- *Adaptive Weight Precision (AWP)* algorithm, which dynamically adapts the numerical representation of DNN weights during training. AWP relies on DNNs' tolerance for reduced data representation formats. It defines the appropriated data representation format per each network layer during training without hurting network accuracy.
- *Approximate Data Transfer (ADT)* procedure to compress DNN's weights according to the decisions made by the AWP algorithm. ADT relies on both thread- and SIMD-level parallelism and is compatible with architectures like IBM's POWER or x86. ADT is able to compress large sets of weights with minimal overhead, which enables the large performance benefits of our approach.
- It evaluates the combination of ADT and AWP, which call A^2DTWP, on two high-end systems: The first is composed of two x86 Haswell multicore devices plus four Tesla GK210 GPU accelerators and the second system integrates two POWER9 chips and four NVIDIA Volta V100 GPUs. Our evaluation considers the Alexnet [31], the VGG [47] and the Resnet [48] network models applied to the ImageNet ILSVRC-2012 dataset [49]. Our experiments report average performance benefits of 6.18 and 11.91% on the x86 and the POWER systems, respectively. Our solution does not reduce the quality of the training process since networks final accuracy is the same as if they had been trained with the 32-bit Floating Point format.

Many proposals describe how data representation formats smaller than the 32-bit Floating Point IEEE standard can be applied to deep learning workloads without harming their accuracy [45, 50, 51]. This section presents the first approach that uses reduced data formats to minimize data movement during DNN training. This section is particularly relevant from the high-performance computing perspective since it proposes a methodology to accelerate DNNs training in heterogeneous high-end systems, which are extensively used to run deep learning workloads [44].

This section is structured as follows: Sect. 14.3.1.1 motivates our proposals. Section 14.3.1.3 describes the Adaptive Weight Precision algorithm (AWP).

Section 14.3.1.4 details the Approximate Data Transfer (ADT) procedure. Section 14.3.1.8 explains the experimental setup of this section. Section 14.3.1.9 describes the experiments we conduct to evaluate AWP and ADT on three state-of-the-art neural networks. Section 14.3.1.2 describes the most relevant related work. Finally, Sect. 14.3.3 summarizes the main conclusions of this section.

14.3.1.1 Training Deep Neural Networks on Multi-GPU Environments

DNNs training process typically requires applying backpropagation [41], which involves solving a large optimization problem. In this context, the Stochastic Gradient Descent (SGD) algorithm [43], which computes the gradient of the cost function of an artificial neural network with respect to its weights, is commonly applied. Each iteration of SGD processes a set of tens or hundreds of samples called batch. When applying the SGD algorithm, the value of the gradient is updated by combining the contributions of all samples contained within a single batch. By organizing the training samples into batches and just updating the gradient values at the end of each batch, a largely parallel procedure is obtained since all samples in each batch can be processed in an independent way.

The use of heterogeneous nodes composed of multicore CPU and GPU devices is becoming prominent to train DNNs due to the large amount of data-level parallelism that such process involves [44]. At the beginning of each iteration, network weights are updated in the multicore CPU device and sent to the GPUs. The different samples of the batch that correspond to the current iteration are distributed across the GPUs and processed in a parallel way. Processing each sample requires running several times highly parallel numerical kernels like the GEneral Matrix-Matrix (GEMM) multiplication, which fit very well with GPUs architecture. Once all samples are processed, their respective contributions to the gradient are sent back to the multicore CPU device, which uses them to update the gradient and readjust the values of the weights [52]. A new set of weights is then sent to the GPUs and a new batch of samples is processed. This process is repeated until the neural network provides satisfactory results with respect to some test data.

This sequence of data exchanges involving different GPUs requires large bandwidth capacity and may constitute a fundamental performance bottleneck. Section 14.3.1.9 provides a performance profile of the training process and shows how data transfers to the GPU require a very significant amount of time. In this context, our approach consists in reducing as much as possible the amount of data sent to the GPUs every time a batch of samples is processed. This is achieved by reducing the number of bits required to represent each network weight. While previous approaches exploit reduced data formats to speed up arithmetic and reduce memory requirements [51], we improve performance by compressing data before sending them across multi-GPU systems. The ADT procedure carries out the compression process. To drive data compression we use the AWP algorithm, which defines data representation requirements per each network layer. Combining both ADT and AWP produces a significant performance gain, which is reported by Sect. 14.3.1.9.

14.3.1.2 Existing Approaches

A rich body of literature exists describing the effects of using data representation formats smaller than the 32-bit Floating Point standard while training neural networks. Previous work provides theoretical analysis on the ability to learn under limited-precision scenarios of simple networks [53]. In recent years, researchers have shown that low precision arithmetic is well suited for deep neural networks training [54–56], particularly when combined with stochastic rounding [45]. New data representation formats targeting dynamic and low accuracy opportunities for deep learning have been proposed [46]. While these approaches have a very large potential for reducing DNNs training costs, they do not target the data movement problem and, as such, they are orthogonal to the approach presented by this section.

There is a methodology for training deep neural models using 16-bit FP numbers without modifying hyperparameters or losing network accuracy [51]. This previous approach avoids losing accuracy by keeping a 32-bit copy of weights, scaling the loss function to preserve small gradient updates, and using 16-bit arithmetic that accumulates into single-precision registers. This previous approach exploits the tolerance of DNN to data representation formats smaller than the 32-bit FP standard, as our proposal does. However, our goal is fundamentally different since we reduce data motion in the context of heterogeneous high-end architectures while this previous approach aims at reducing the computing and storage costs of DNN training. This approach can be combined with A^2DTWP by decompressing network weights to half-precision to reduce GPU computing time. This reduction would increase the impact of data motion in the overall performance, which implies that the benefits of A^2DTWP could be even larger.

There are several proposals aimed at improving the Stochastic Gradient Descent (SGD) method like the Asynchronous SGD [52] and its variants [57, 58]. Other approaches [59, 60] exploit model parallelism instead of data-level parallelism to orchestrate large-scale parallel executions of deep learning workloads. If the different parallel instances of this model-level parallel scheme had different precision requirements, it would be possible to apply approaches close to the ones we present in this section.

Some previous approaches reduce networks storage and energy requirements to run inference on mobile devices [61]. While these approaches achieve very large storage reductions, they target inference on embedded systems with limited hardware resources. Other approaches achieve large gradient compression ratios in the context of distributed training in mobile devices [62]. These approaches achieve very remarkable speed ups in low network bandwidth environments and are applied to scenarios that require frequent and costly allreduce communications. While they are very valuable in the mobile computing arena, the scope of these approaches is not high-performance computing.

Several approaches target synchronization costs of SGD gradient updates in the context of parallel executions. They either quantize gradients to ternary levels $\{-1, 0, 1\}$ to reduce the overhead of gradient synchronization [63], or they propose a family of algorithms allowing for lossy compression of gradients called Quantized

SGD (QSGD) [64]. Techniques based on sparsifying gradient updates by removing the smallest gradients by absolute value [65] can also reduce SGD synchronization costs. While these approaches apply techniques based on small data representation formats to reduce the synchronization costs of SGD gradient updates, A^2DTWP targets the cost of sending DNNs weights to the GPU accelerators. Therefore, these approaches are orthogonal to A^2DTWP and can be combined with it to reduce as much as possible training communication cost. In particular, techniques targeting synchronization costs of SGD gradient updates can be used to reduce GPU to CPU data transfer overhead while A^2DTWP targets CPU to GPU communication cost. Therefore, the combination of A^2DTWP with techniques targeting synchronization costs of SGD gradient updates would reduce both CPU to GPU and GPU to CPU communication overhead.

To the best of our knowledge, A^2DTWP is the first approach able to accelerate the training of deep neural networks in multi-GPU high-end systems by reducing data motion. A^2DTWP combines the use of an algorithm to dynamically change DNNs weights data representation format during training with a highly tuned data compression and decompression procedure. Our solution successfully reduces data motion and achieves a significant performance improvement on cutting-edge high-end systems. While there are previous proposals exploiting mixed precision scenarios to accelerate training, they are orthogonal to our approach as they speed up arithmetic and reduce memory footprint. Importantly, our proposal can be combined with these previous approaches to obtain a highly optimized training method that minimizes data transfers and accelerates arithmetic in the context of multi-GPU systems.

14.3.1.3 The Adaptive Weight Precision (AWP) Algorithm

The Adaptive Weight Precision (AWP) algorithm relies on the tolerance of DNNs to data representation formats smaller than the 32-bit Floating Point standard. Indeed, previous work indicates that, unlike scientific codes focused on solving partial differential equations or large linear systems, neural networks do not always require 32-bit representation during training [45, 50]. Even more, adding stochastic noise to certain variables during the learning phase improves DNNs accuracy [66–68]. Nevertheless, when facing unknown scenarios in terms of new workloads or parameter settings, the data representation requirements of DNNs are non-trivial to be determined and, to make things more complicated, they may change as the training phase progresses.

The AWP algorithm dynamically determines data representation requirements per each network layer by monitoring the evolution of the l^2-norm of the weights. AWP identifies the number of bits that are needed to represent DNNs weights and guarantees the progress of the training process. AWP assigns the same data representation format to all weights belonging to a certain network layer. The training starts with a relatively small data representation that is independently increased for each layer.

Algorithm 3 Adaptive weight precision (AWP) algorithm

1: BitsPerLayer := $[B_0, B_1, \ldots, B_{NumLayers}]$ ▷ List storing the number of bits corresponding to the data representation of each layer
2: IntervalCounter := [0, 0, ..., 0] ▷ List storing the number of times the relative change rate fails to meet the threshold per layer
3: **for** batch := 0 ... NumBatches **do**
4: Apply backpropagation to batch
5: **for** layer := 0 ... NumLayers **do**
6: $\delta := \frac{(|W_{batch,layer}|-|W_{batch-1,layer}|)}{|W_{batch-1,layer}|}$
7: **if** δ < T **then**
8: IntervalCounter$_{layer}$ += 1
9: **end if**
10: **if** IntervalCounter$_{layer}$ == INTERVAL **then**
11: BitsPerLayer$_{layer}$ += N
12: IntervalCounter$_{layer}$:= 0
13: **end if**
14: **end for**
15: **end for**

Algorithm 3 displays a pseudo-code description of AWP. Once the backpropagation process has been applied to a given batch, AWP iterates over all network layers. The algorithm computes per each batch and network layer the l^2-norm of all its weights' values and derives the relative change rate δ of the l^2-norm with regard to the previously processed batch. For the batch i, the change rate is defined as $\delta_i = (|W_i| - |W_{i-1}|)/|W_{i-1}|$, where W_i is the vector of weights of a certain layer while batch i is processed. Every time the change rate is below a given threshold T for a certain layer, the algorithm accounts for it by increasing the $IntervalCounter$ parameter. The algorithm increases N bits of precision if the change rate is below T during a certain number of batches defined by the parameter *INTERVAL* and sets the $IntervalCounter$ parameter of the corresponding layer to zero. Section 14.3.1.9 describes how we determine the values of parameters *T*, *INTERVAL*, and *N*.

14.3.1.4 The Approximate Data Transfer (ADT) Procedure

The Approximate Data Transfer (ADT) procedure compresses network's weights before they are transferred to the GPUs. In the context of DNNs training on heterogeneous multi-GPU nodes, CPU multicore devices are typically responsible for orchestrating the parallel run and updating DNN parameters. Once the process of a batch starts, the updated parameters including the weights W are sent to each GPU. If the set of parameters does not fit in GPUs' main memory, they are sent on several phases as the different GPUs need them. The different samples of each batch are evenly distributed across all GPUs. Therefore, each GPU computes its contribution to the gradient ΔW by processing its corresponding set of samples. The CPU multicore device subsequently gathers all contributions to the gradient and combines them to update the weights $W \leftarrow W - \mu(\frac{1}{n}\sum_i^n \Delta W_i)$, where μ is the learning rate.

Data movement involving different GPU devices increases as the network topology becomes more complex or the number of training samples grows, which can saturate the system bandwidth and become a major performance bottleneck. The proposed technique mitigates this issue by compressing network weights before they are sent to the GPU devices. The AWP algorithm described in Sect. 14.3.1.3 determines, for all weights belonging to a particular network layer, the number of bits to send. In this context, to efficiently compress and decompress network weights, ADT uses of two procedures that constitute its fundamental building blocks. These procedures are complementary and applied either before or after data transfers to GPU devices.

- **Bitpack** compresses the weights discarding the less significant bits on the CPU side;
- **Bitunpack** converts the weights back to the IEEE-754 32-bit Floating Point format on the GPUs.

Figure 14.3 provides an example including a multicore CPU and two GPU devices to describe the way both Bitpack and Bitunpack procedures operate. All neural network parameters (weights and biases) are updated at the CPU level, which is where the Bitpack procedure takes place. We do not apply the Bitpack procedure to the network biases since we do not observe any significant performance benefit from compressing them. Since each output neuron requires just one bias parameter, the total number of them is significantly smaller than the total number of weights. At the beginning of each SGD iteration the compressed weights are sent to each GPU together with the biases and the corresponding training samples. Each GPU

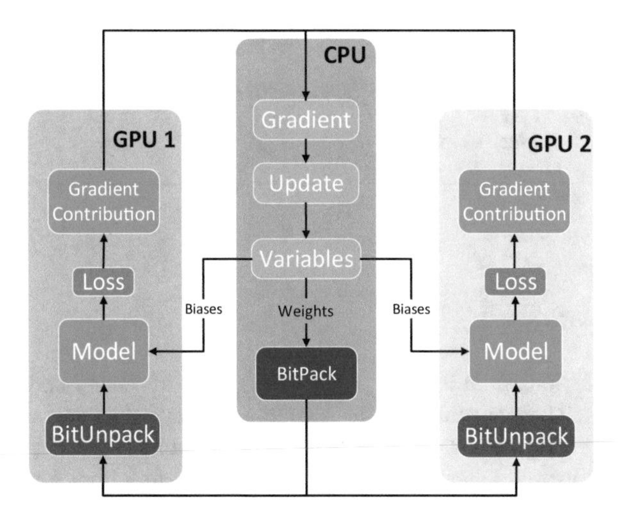

Fig. 14.3 The ADt on a 2-GPU system. Variables include: weights which go through the ADt procedure and biases which are sent directly to the GPUs to build the network model together with the unpacked weights

uncompresses the weights, builds the neural network model, and, finally, computes its specific contribution to the gradient. These contributions are sent to the CPU, which gathers all of them, computes the gradient and updates network parameters.

The Bitpack operation runs on CPU multicore devices. To boost Bitpack we use OpenMP [69] and Single Instruction Multiple Data (SIMD) intrinsics. OpenMP is used to run Bitpack on several threads. The use of SIMD instructions allows Bitpack to operate at the SIMD register level, which avoids incurring large performance penalties in the process of producing the reduced-size weights. We implement two versions of Bitpack. One version uses Intel's AVX2 [70] instruction set and the other one relies on AltiVec [71]. Bitpack can be implemented on top of any SIMD instruction set architecture supporting simple byte shuffling instructions at the register level. The Bitunpack procedure runs on the GPUs.

It can be trivially parallelized since each weight is mapped to a single 32-bit FP variable, which means that GPUs can process a large amount of weights simultaneously and efficiently build the DNN model. In fact, Bitunpack incurs negligible overhead as Sect. 14.3.1.9 shows.

ADT manipulates the internal representation of network weights by discarding some bits. We use the standard 32-bit IEEE-754 single-precision Floating Point format [72] (1 bit sign, 8 bits exponent and 23 bits mantissa) for all the computation routines. The Bitpack method considers network weights as 32-bit words where rounding to N bits means discarding the lowest $32 - N$ bits.

Algorithm 4 High-level pseudo-code version of bitpack

```
1: W                                  ▷ Array of 32-bit Floating Point values containing weights
2: Pw                                 ▷ Array containing the reduced precision weights
3: RoundTo                            ▷ Number of bytes to keep per weight
4: POffset := 0                       ▷ Indicates the current size (in bytes) of Pw
5: for weight in W do
6:     Pw[POffset : POffset+RoundTo] := weight[0 : RoundTo]    ▷ Copy most significant
   RoundTo bytes to Pw
7:     POffset := POffset + RoundTo
8: end for
```

14.3.1.5 Bitpack

A high-level version of the Bitpack procedure in terms of pseudo-code is illustrated by Algorithm 4. The algorithm requires a couple of arrays: the input array W, which contains all the weights of a certain network layer, and an output array Pw, which stores the compressed versions of these weights. The algorithm goes through the entire W input array, per each weight, copies the most significant $RoundTo$ bytes to the output array Pw. Our Bitpack implementation manipulates data at the byte granularity. We do not observe significant performance benefits when operating at finer granularity in the experiments we run. The AWP algorithm described in

Sect. 14.3.1.3 determines the data representation format per each network layer. The number of bits of the chosen format is rounded to the nearest number of bytes that retains all of its information (e.g., if AWP provides the value 14, $RoundTo$ will be set to 2 bytes). The Pw array is sent to the GPUs once the Bitpack procedure finishes compressing network weights.

Deep networks usually contain tens or even hundreds of millions of weights [31, 47, 73], which makes any trivial implementation of Algorithm 4 not applicable in practice. We mitigate compression costs by observing that Algorithm 4 is trivially parallel since processing one weight just requires the $RoundTo$ parameter. Algorithm 5 shows how to parallelize the Bitpack procedure by using OpenMP threads. Each thread takes care of a certain portion of the array storing network weights.

Algorithm 5 Bitpack with OpenMP

```
1: W                                    ▷ Array of 32-bit Floating Point values containing weights
2: Pw                                   ▷ Array containing the reduced precision weights
3: RoundTo                              ▷ Number of bytes to keep per weight
4: NumThreads                           ▷ Number of OpenMP threads
5: #pragma omp parallel for
6: for weight in W do
7:     POffset := Corresponding position in Pw
8:     Pw[POffset : POffset+RoundTo] := weight[0 : RoundTo]    ▷ Copy the most
   significant RoundTo bytes to Pw
9: end for
```

14.3.1.6 Single Instruction Multiple Data Bitpack

Since all weights within one layer are processed in the same way by the Bitpack procedure, we can leverage Single Instruction Multiple Data (SIMD) instructions to vectorize it. Most state-of-the-art architectures implement SIMD instruction set: IBM's AltiVec [71], Intel's Advanced Vector Extensions (AVX) [70], and ARM's Neon [74]. In our experiments we use Intel's AVX2 [70], which implements a set of SIMD instructions operating over 256-bit registers, and IBM's AltiVec instruction set [71], which has SIMD instructions operating over 128-bit registers. Section 14.3.1.8 describes the specific details of our evaluation considering both x86 and POWER architectures.

Figure 14.4 shows the byte-level operations of SIMD-based Bitpack applied to eight 32-bit weights and implemented with AVX2. The *RoundTo* parameter is set to 3, which implies discarding the last 8 bits of each weight since the target data representation is 24-bit long. First, eight 32-bit Floating Point weights are loaded to a 256-bit register. In the next step, we use *_mm256_shuffle_epi8* to shuffle the least significant eight bits of each weight to the least significant bits of their respective 128-bit lane (see the grey area of Fig. 14.4 Step 2) and pack the rest of the bits together. Afterwards we use *_mm256_permutevar8x32_epi32* to do the same

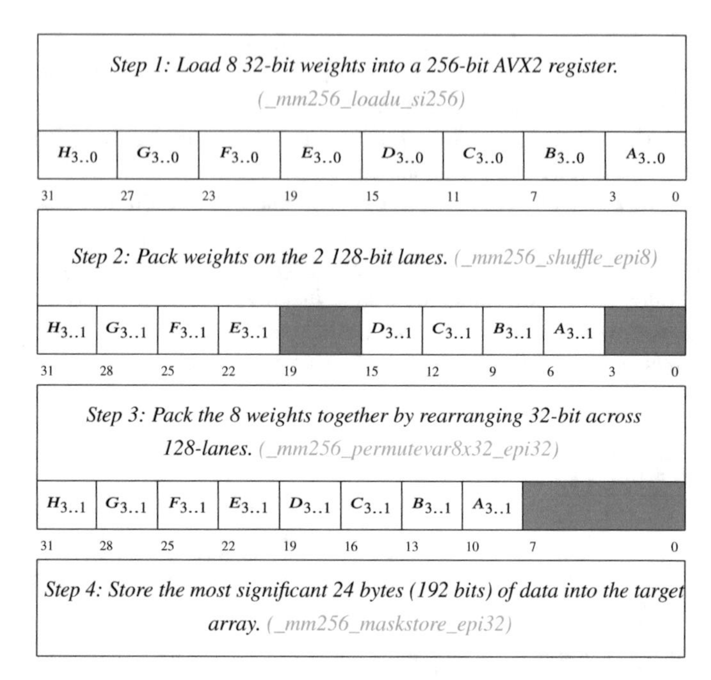

Fig. 14.4 Bitpack implemented with AVX2, RoundTo=3

operation across the two 128-bit lanes. Finally, we use *_mm256_maskstore_epi32* to just store the resulting 192 bits to the target array. Not all AVX2 instructions operate over the entire 256-bit register. Instead, many of them conceive the register as two 128-bit lanes and operate on them separately. This is the reason why we cannot carry out Steps 2 and 3 by using a single AVX2 instruction.

Algorithm 6 Bitpack with OpenMP + AVX2

```
 1: W                              ▷ Array of 32-bit Floating Point values containing weights
 2: Pw                             ▷ Array containing the reduced precision weights
 3: RoundTo                        ▷ Number of bytes to keep per weight
 4: #pragma omp parallel for
 5: for weights in W do
 6:     _mm256_loadu_si256                          ▷ Load 8 32-bit weights
 7:     _mm256_shuffle_epi8                         ▷ Compress at each 128-bit lane
 8:     _mm256_permutevar8x32_epi32    ▷ Shuffle the compressed weights into the most
    significant bits
 9:     _mm256_maskstore_epi32         ▷ Store compressed weights to the target array
10: end for
```

Algorithm 6 summarizes our implementation of the Bitpack procedure with AVX2. It exploits two-level parallelism: first, the input array of weights is distributed across several threads. Second, within each thread, the compression of each eight 32-bit weights subset is performed at the register level by means of byte shuffling

Algorithm 7 Bitunpack on GPU

```
 1: Pw                                        ▷ Array containing compressed weights
 2: W                  ▷ Array of 32-bit Floating Point values containing weights
 3: RoundTo                      ▷ The number of bytes that are going to be kept
 4: for UnitId := 0 ... NumUnit do
 5:     Distribute W and Pw across all the computation units in the GPU
 6:     POffset := 0
 7:     for weight in W do
 8:         weight := Pw[POffset : POffset+RoundTo] ≪ (4 - RoundTo) * 8
 9:         POffset := POffset + RoundTo
10:     end for
11: end for
```

instructions. This sophisticated procedure exploiting parallelism at both thread and SIMD register levels uses all the available hardware and avoids costly memory accesses.

14.3.1.7 Bitunpack

Once data in reduced-size format reaches the target GPU, the Bitunpack procedure immediately restores them into their original IEEE-754 32-bit Floating Point format. We display pseudo-code describing this process in Algorithm 7. Bitunpack reads the reduced-sized weights from array Pw and assigns additional bits to them. Bitunpack gives zero values to these additional bits. We distribute the Bitunpack process across the whole GPU, which enables an extremely parallel scheme exploiting GPUs manycore architecture.

The Bitunpack routine is developed using CUDA [75]. Our code runs in parallel on N CUDA threads and the CUDA runtime handles the dynamic mapping between threads and the underlying GPU compute units. Since each thread involved in the parallel run targets a different portion of the Pw array, our Bitunpack procedure exposes a large amount of parallelism able to exploit the large number of compute units integrated into high-end GPU devices.

14.3.1.8 Experimental Setup

The experimental setup considers a large image dataset, three state-of-the-art neural network models, and two high-end platforms. The following sections describe all these elements in detail.

Image Dataset We consider the ImageNet ILSVRC-2012 dataset [49]. The original ImageNet dataset includes three sets of images of 1000 classes each: training set (1.3 million images), validation set (50,000 images) and testing set (100,000 images). Considering 1000 classes makes the training process around 170 hours long, which is prohibitively expensive for large experimental campaign considering different network models, batch sizes and hardware platforms. To reduce the execution time of our experiments we consider a subset of 200 classes for both the training and the validation dataset, which keeps the training time under manageable margins. We refer to the 200 and 1000 classes datasets as ImageNet200 and ImageNet1000, respectively. Since it is a common practice [47], we evaluate the ability of a certain network in properly dealing with the ImageNet ILSVRC-2012 dataset in terms of the top-5 validation error computed over the validation set.

DNN Models and Training Parameters We apply the AWP algorithm along with the ADT procedure on three state-of-the-art DNN models: a modified version of Alexnet [31] with an extra fully connected layer of size 4096, the configuration A of the VGG model [47] and the Resnet network [48]. All hidden layers are equipped with a Rectified Linear Units (ReLU) [31]. The exact configurations of the three neural networks are shown in Table 14.1. The Alexnet model is composed of 5

Table 14.1 Neural network configurations: the convolutional layer parameters are denoted as "conv<receptive field size>-<number of channels>". The ReLU activation function is not shown for brevity. The building blocks of Resnet and the number of times they are applied are shown in a single cell

<table>
<tr><th>Alexnet</th><th>VGG</th><th>Resnet-34</th></tr>
<tr><td colspan="3">input(224x224 RGB image)</td></tr>
<tr><td>conv11-64</td><td>conv3-64</td><td>conv7-64</td></tr>
<tr><td colspan="3">maxpool</td></tr>
<tr><td>conv5-192</td><td>conv3-128</td><td>conv3-64
conv3-64
x3</td></tr>
<tr><td colspan="2">maxpool</td><td></td></tr>
<tr><td>conv3-384</td><td>conv3-256
conv3-256</td><td>conv3-128
conv3-128
x4</td></tr>
<tr><td colspan="2">maxpool</td><td></td></tr>
<tr><td>conv3-384</td><td>conv3-512
conv3-512</td><td>conv3-256
conv3-256
x6</td></tr>
<tr><td colspan="2">maxpool</td><td></td></tr>
<tr><td>conv3-256</td><td>conv3-512
conv3-512</td><td>conv3-512
conv3-512
x3</td></tr>
<tr><td colspan="2">maxpool</td><td>avgpool</td></tr>
<tr><td colspan="2">FC-4096</td><td rowspan="2"></td></tr>
<tr><td>FC-4096
FC-4096</td><td>FC-4096</td></tr>
<tr><td colspan="3">FC-200</td></tr>
<tr><td colspan="3">softmax</td></tr>
</table>

convolutional layers and 4 fully connected ones, VGG contains 8 convolutional layers and 3 fully connected ones, and Resnet is composed of 33 convolutional layers and a single fully connected one.

We use momentum SGD [76] to guide the training process with momentum set to 0.9. The training process is regularized by weight decay and the L_2 penalty multiplier is set to 5×10^{-4}. We apply a dropout regularization value of 0.5 to fully connected layers. We initialize the weights using a zero-mean normal distribution with variance 10^{-2}. The biases are initialized to 0.1 for Alexnet and 0 for both VGG and Resnet networks. For the Alexnet and VGG models we consider training batch sizes of 64, 32, and 16. To train the largest network we consider, Resnet, we consider batch sizes of 128, 64 and 32. The 16 batch size incurs in a prohibitively expensive training process for Resnet and, therefore, we do not use it in our experimental campaign.

For Alexnet we set the initial learning rate to 10^{-2} for the 64 batch size and decrease it by factors of 2 and 4 for the 32 and 16 batch sizes, respectively. In the case of VGG we set the initial learning rate to 10^{-2} for the 64, 32, and 16 batch sizes, as in the state-of-the-art [47]. In the case of Resnet the learning rate is 10^{-2} for the batch size of 32 and 0.1 for the rest. For all network models we apply exponential decay to the learning rate throughout the whole training process in a way the learning rate decays every 30 batches by a factor of 0.16, as previous work suggests [73]. For Resnet we obtain better results by adapting precision at the Resnet building blocks level [48] instead of doing so in a per layer basis.

Implementation Our code is written in Python on top of Google Tensorflow [37]. Tensorflow is a data-flow and graph-based numerical library where the actual computation is carried out according to a computational graph constructed beforehand. The computational graph defines the order and the type of computations that are going to take place. It supports NVIDIA's NCCL library.

To enable the use of both Bitpack and Bitunpack routines, we integrate them into Tensorflow using its C++ API. Tensorflow executes the two routines before sending the weights from the CPU to the GPU and right after receiving the weights on the GPU side, respectively. The Bitpack routine is implemented using the OpenMP 4.0 programming model. There are two versions of this routine using either Intel's AVX2 or AltiVec instructions, as explained in Sect. 14.3.1.4. Bitunpack is implemented using CUDA 8.0 and CUDA 10.0, respectively, on the two platforms [75].

Hardware Platforms We conduct our experiments on two clusters featuring the x86 and POWER architectures. The x86 machine is composed of two 8-core Intel Xeon ®E5-2630 v3 (Haswell) at 2.4 GHz and a 20 MB L3 shared cache memory each. It is also equipped with two Nvidia Tesla K80 accelerators, each of which hosts two Tesla GK210 GPUs. It has 128 GB of main memory, distributed in 8 DIMMs of 16 GB DDR4 @ 2133 MHz. The 16-core CPU and the four GPUs are connected via a PCIe 3.0 x8 8GT/s. The operating system is RedHat Linux 6.7. Overall, the peak performance of the two 8-core sockets plus the four Tesla GK210 GPUs is 6.44 TFlop/s.

The POWER machine is composed of two 20-core IBM POWER9 8335-GTG at 3.00 GHz. It contains four NVIDIA Volta V100 GPUs. Each node has 512 GB of main memory, distributed in 16 DIMMS of 32 GB @ 2666 MHz. The GPUs are connected to the CPU devices via a NVIDIA NVLink 2.0 interconnection [77]. The operating system is RedHat Linux 7.4. The peak performance of the two 20-core sockets plus the four V100 GPUs is 28.85 TFlop/s.

14.3.1.9 Evaluation

In this section we evaluate the capacity of the AWP algorithm and the ADT procedure to accelerate DNNs training. We show how our proposals are able to accelerate the training phase of relevant DNN models without reducing the accuracy of the network.

Methodology Our experimental campaign considers batch sizes of 64, 32, and 16 for the Alexnet and VGG models and 128, 64 and 32 for the Resnet network. For each model and batch size, the *baseline* run uses the 32-bit Floating Point precision for the whole training. The data representation formats we consider to transfer weights from the CPU to the GPU are: 8-bit (1 bit for sign, 7 bits for exponent), 16-bit (1 bit for sign, 8 for exponent, 7 for mantissa), 24-bit (1 bit for sign, 8-bits for exponent and 15 bits for mantissa) and 32-bits (1 bit for sign, 8 bits for exponent and 23 bits for mantissa). We train the network models with dynamic data representation by applying the AWP algorithm along with the ADT procedure. We denote this approach combining ADT and AWP as A^2DTWP. For each DNN and batch size, we select the data representation format that first reaches the 35, 25, and 15% accuracy thresholds for Resnet, Alexnet, and VGG, respectively, and we denote this approach as *oracle*. For the case of the *oracle* approach, data compression is done via ADT. The closer A^2DTWP is to *oracle*, the better is the AWP algorithm in identifying the best data representation format.

During training we sample data in terms of elapse time and validation error every 4000 batches. The total number of training batches corresponding to the whole ImageNet200 dataset are 16020, 8010, 4005, and 2002 for batch sizes 16, 32, 64, and 128, respectively. The values of AWP parameters T, $INTERVAL$, and N are determined in the following way: In the case of T we monitor the execution of several epochs until we observe a drop in the validation error. We then measure the average change, considering all layers, of weights' l^2-norm during this short monitoring period. The obtained values of T are -5×10^{-2}, -2×10^{-3} and -2×10^{-5} for Alexnet, VGG and Resnet, respectively. We set the $INTERVAL$ parameter to 4000 for both AlexNet and VGG and 2000 for Resnet. These values correspond to a single batch (for the ImageNet200 dataset and batch sizes 64 and 128) and avoid premature precision switching due to numerical fluctuations. We set N to 8 since the smallest granularity of our approach is 1 byte. AWP initially applies 8-bit precision to all layers. We use ImageNet200 to generate Figs. 14.5, 14.6, and 14.7. Figure 14.8 uses ImageNet1000.

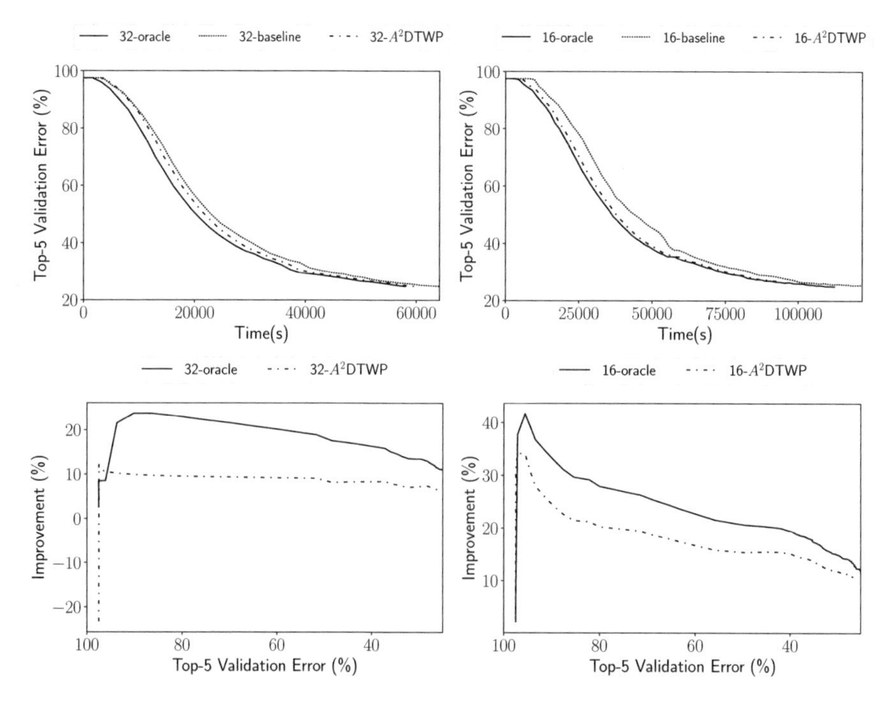

Fig. 14.5 Alexnet training considering 32 and 16 batch sizes. The two upper plots show the top-5 validation error evolution of *baseline*, *oracle* and A^2DTWP. The two bottom plots provide information on the performance improvement of *oracle* and A^2DTWP against *baseline* during the training process. Experiments run on the x86 system

Evaluation on Alexnet The evaluation considering the Alexnet model on the x86 system is shown in Fig. 14.5, which plots detailed results considering batch sizes of 32 and 16, and Fig. 14.7, which shows the total execution time of the *oracle* and A^2DTWP policies normalized to the *baseline* for the 64, 32, and 16 batch sizes on both the x86 and the POWER systems. The two top plots of Fig. 14.5 depict how the validation error of the *baseline*, *oracle*, and A^2DTWP policies evolves over time for the 32 and the 16 batch sizes until the 25% accuracy is reached. The two bottom plots provide information regarding the performance improvement of both *oracle* and A^2DTWP over the 32-bit *baseline* with regard to a certain validation error. Such performance improvement is computed by looking at the time required by the *oracle* and A^2DTWP techniques to reach a certain validation error with respect to the *baseline*.

It can be observed in the upper left-hand side plot of Fig. 14.5 how the *oracle* and the A^2DTWP approaches are 10.82 and 6.61% faster than the baseline, respectively, to reach the 25% top-5 validation error when using a 32 batch size. The upper right-hand side plot shows results considering a 16 batch size. The improvements achieved by the *oracle* and A^2DTWP approaches are 11.52 and 10.66%, respectively. This demonstrates the efficiency of the ADT procedure in

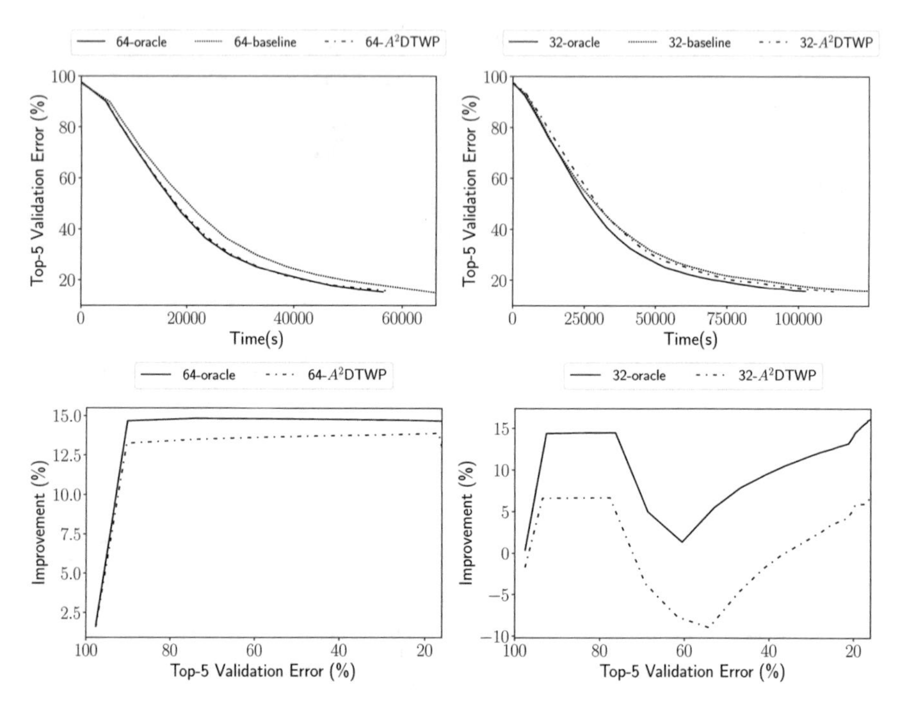

Fig. 14.6 VGG training considering 64 and 32 batch sizes. The two upper plots show the top-5 validation error evolution of *baseline*, *oracle* and *A^2DTWP*. The two bottom figures provide information on the performance improvement of *oracle* and *A^2DTWP* against *baseline* during the training process. Experiments run on the x86 system

compressing and decompressing the network weights without undermining the performance benefits obtained from sending less data from the CPU device to the GPU. It also demonstrates the capacity of AWP to quickly identify the best data representation format per layer.

The two bottom plots of Fig. 14.5 provide information on performance improvement of *oracle* and *A^2DTWP* over the *baseline* during the training process. For the 32 batch size, *oracle* reaches a peak improvement of 24.11% when the 90% validation error is reached and steadily declines from that point although it keeps a significant improvement of 10.82% over the *baseline* once the 25% top-5 validation error is reached. *A^2DTWP* falls in-between the *baseline* and the *oracle* and keeps its improvement above 7.03% until it reaches the 27% top-5 validation error. Once it reaches the 25% validation error *A^2DTWP* is 6.51% faster than the *baseline*. In conclusion, the *A^2DTWP* policy is able to provide performance improvements that are close to the ones achieved by the best possible accuracy. For the 16 batch size, the performance benefits of the *oracle* policy reach a 41.64% peak at the 94% validation error point. The *A^2DTWP* policy reaches its maximum performance benefit, 34.21%, when the validation error is 97%. At the 25% validation error point, the *oracle* and the *A^2DTWP* policies reach 13.00 and 10.75% performance

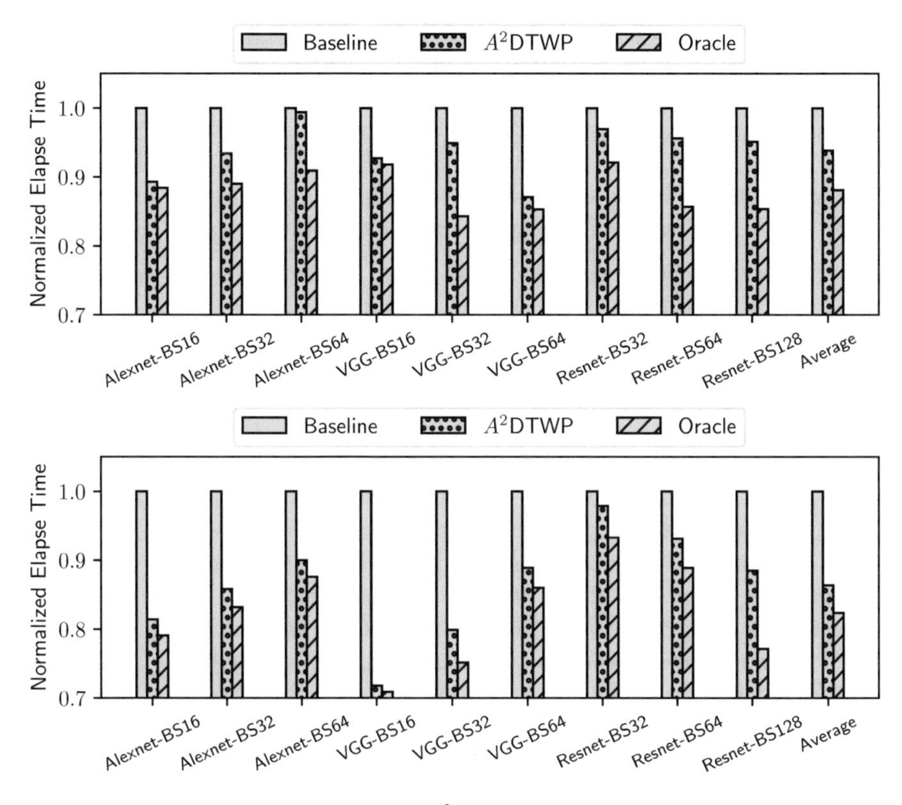

Fig. 14.7 Normalized execution times of the A^2DTWP and the *oracle* policies with respect to the baseline. Results obtained on the x86 system appear in the upper plot while the evaluation on the POWER system appears at the bottom

improvement, respectively. Overall, results considering the Alexnet network for batch sizes 32 and 16 confirm that A^2DTWP, which combines both the AWP algorithm and the ADT procedure, successfully delivers very similar performance benefits to the best possible accuracy.

Figure 14.7 shows the normalized execution time of the *oracle* and A^2DTWP policies with respect to the 32-bit FP *baseline* on the x86 and the POWER systems. The top chart reports performance improvements of 10.75, 6.51, and 0.59% for batch sizes 16, 32, and 64 in the case of Alexnet running on the x86 system. For the 64 batch size, the marginal gains of A^2DTWP over the *baseline* are due the poor performance of the 8-bits format employed by A^2DTWP at the beginning of the training process. This format does not contribute to reduce the validation error for the 64 batch case, which makes the A^2DTWP policy to fall behind the *baseline* at the very beginning of the training process. Although A^2DTWP eventually increases its accuracy and surpasses the *baseline*, it does not provide the same significant performance gains for Alexnet as the ones observed for batch sizes 16 and 32.

A^2DTWP performance improvements on the POWER system in the case of Alexnet are 18.61, 14.25, and 10.01% with respect to the *baseline* for batch sizes 16, 32, and 64, respectively. The POWER system achieves larger performance improvements than x86 since the Bitpack procedure can be further parallelized over the 40 cores of the POWER9 multicore chips than the 16 cores available in the Haswell multicore devices of the x86 system. This mitigates the costs of weights' compression and thus provides larger performance improvements.

Evaluation on VGG Figure 14.6 shows results for batch sizes 64 and 32 when using the VGG architecture on the x86 system. The upper figures display the temporal evolution of the validation error until the 15% top-5 validation error is reached. Like in Alexnet, both the A^2DTWP and the *oracle* policies outperform the *baseline*. In the case of batch size 64, both *oracle* and A^2DTWP display a similar evolution in terms of validation error, which translates to very close performance improvement over the baseline. They maintain an overall improvement of over 13.00% against the *baseline* during most of their training. The A^2DTWP technique outperforms the baseline by 12.88% when reaches 15% of top-5 validation error while the *oracle* policy achieves the same improvement. For batch size 32 the final improvement achieved by A^2DTWP over the baseline is 5.02%. This improvement is not as large as the one achieved for the 64 batch size since the AWP algorithm does not identify a numerical precision able to beat the *baseline* until the 57% validation error is reached, as it can be seen in the bottom right-hand side plot of Fig. 14.6.

Figure 14.7 shows the normalized execution time of A^2DTWP and *oracle* with respect to the *baseline* for VGG considering batch sizes of 16, 32 and 64 on the x86 and POWER systems. When applied to the VGG model on the x86 system, A^2DTWP outperforms the 32-bit Floating Point *baseline* by 12.88, 5.02, and 7.31% for batch sizes 64, 32, and 16, respectively. Despite the already described issues suffered by the A^2DTWP technique when applied to the 32 batch size, this approach achieves very remarkable performance improvements over the baseline in all considered scenarios.

The performance improvements observed when trying VGG on the POWER system are even higher. A^2DTWP outperforms the *baseline* by 28.21, 20.19, and 11.13% when using the 16, 32, and 64 batch sizes, respectively. The performance improvement achieved on the POWER system are larger than the ones observed for x86 since the Bitpack procedure can be parallelized over 40 cores when running on the POWER system. We observe the same behavior for Alexnet.

Evaluation on Resnet We display the normalized execution time of the A^2DTWP and the *oracle* policies when applied to the Resnet model using batch sizes of 128, 64 and 32 in Fig. 14.7. In the case of Resnet we do not show detailed plots describing the evolution of the validation error during training because its behavior is very close to some previously displayed scenarios like VGG. On the x86 system, A^2DTWP beats the 32-bit Floating Point baseline by 4.94, 4.39, and 3.11% for batch sizes of 128, 64, and 32, respectively, once a top-5 validation error of 30% is reached. The relatively low performance improvement achieved in the case of 32 batch size is due to a late identification of a competitive numerical precision, as it happens in the case of VGG and batch size 32.

The performance gains on the POWER system display a similar trend as the ones achieved on x86. While they show the same low improvement for the 32 batch size, 2.12%, A^2DTWP achieves 6.92 and 11.54% performance gains for batch sizes 64 and 128, respectively. A^2DTWP achieves the largest performance improvement with respect to the 32-bit *baseline* when run on the POWER system due to the same reasons as Alexnet and VGG.

Average Performance Improvement The average performance improvement of A^2DTWP over the *baseline* considering the Alexnet, VGG and Resnet models reach 6.18 and 11.91% on the x86 and the POWER systems, respectively. As we explain in previous sections, A^2DTWP obtains larger improvements on the POWER system than on x86 since the ADT procedure can be further parallelized over the 40 cores of the POWER9 multicore devices. In contrast, the two Haswell devices of the x86 system offer just 16 cores for ADT.

The combination of the AWP algorithm and the ADT procedure properly adapts the precision of each network layer and compresses the corresponding weights with a minimal overhead. The large performance improvement obtained while training deep networks on two high-end systems demonstrate the effectiveness of A^2DTWP.

Performance Profile of A^2DTWP This section provides a detailed performance profile describing the effects of applying A^2DTWP when training the VGG network model with batch size 64 on the x86 and POWER systems described in Sect. 14.3.1.8. To highlight these effects we also show a performance profile of applying 32-bit Floating Point format during training. The main kernels involved in the training process and their corresponding average execution time in milliseconds are shown in Tables 14.2 and 14.3. Each kernel can be invoked multiple times by different network layers and it can be overlapped with other operations while processing a batch. Tables 14.2 and 14.3 display for all kernels the average execution time of their occurrences within a batch when run on the x96 and the POWER systems, respectively (Table 14.4).

Results appearing in Table 14.2 show how time spent transferring data from the CPU to the GPU accelerators when applying A^2DTWP on the x86 system, 52.27 ms, is significantly smaller than the cost of performing the same operation when using the 32-bit configuration, 153.93 ms. This constitutes a 2.94x execution time reduction that compensates the cost of the operations involved in the ADT routine, Bitpack and Bitunpack, and in the AWP algorithm, the l^2-norm computation. On POWER we observe a similar reduction of 3.20x in the time spent transferring data from the CPU to the GPUs when applying A^2DTWP. These reductions in terms of CPU to GPU data transfer time are due to a close to 3x reduction in terms of weights size enabled by A^2DTWP. The average execution time of operations where the A^2DTWP technique plays no role remains very similar for the 32-bit Floating Point baseline and A^2DTWP in both systems, as expected. Tables 14.2 and 14.3 indicate

Table 14.2 Performance profiles of both the A^2DTWP and the 32-bit floating point approaches expressed in milliseconds on the x86 system. We consider the VGG network model with batch size 64

	32-bit FP	A^2DTWP
Data transfer CPU→GPU	153.93	52.27
Data transfer GPU→CPU	68.51	73.55
Convolution	128.72	126.13
Fully connected	33.51	34.17
Gradient update	54.39	52.86
AWP (l^2-norm)	N/A	3.88
ADT (Bitpack)	N/A	19.71
ADT (Bitunpack)	N/A	4.51

Table 14.3 Performance profiles of both the A^2DTWP and the 32-bit Floating Point approaches expressed in milliseconds on the POWER system. We consider the VGG network model with batch size 64

	32-bit FP	A^2DTWP
Data transfer CPU→GPU	39.12	12.21
Data transfer GPU→CPU	17.34	17.87
Convolution	69.78	71.21
Fully connected	12.66	13.51
Gradient update	41.29	42.98
AWP (l^2-norm)	N/A	0.93
ADT (Bitpack)	N/A	10.51
ADT (Bitunpack)	N/A	1.11

that performance gains achieved by A^2DTWP are due to data motion reductions, which validates the usefulness of A^2DTWP.

Tables 14.2 and 14.3 also display the overhead associated with AWP and ADT in terms of milliseconds. The AWP algorithm spends most of its runtime computing the l^2-norm of the weights, which takes a total of 3.88 ms within a batch on the x86 system. On POWER, the cost of computing the l^2-norm of the weights is 0.93 ms. The other operations carried out by AWP have a negligible overhead. The two fundamental procedures of ADT are the Bitpack and Bitunpack routines, which take 19.71 and 4.51 ms to run within a single batch on the x86 system. For the case of POWER, Bitpack and Bitunpack take 10.51 and 1.11 ms, respectively. Overall, measurements displayed at Table 14.2 indicate that AWP and ADT constitute 1.05 and 6.60% of the total batch execution time, respectively, on x86. On the POWER system, AWP and ADT constitute 0.54 and 6.82% of the total batch execution time according to Table 14.3. Figures 14.5, 14.6, and 14.7 account for this overhead in the results they display.

Experiments with ImageNet1000 We run experiments considering ImageNet1000 to confirm they display the same trends as executions with ImageNet200. Network parameters are the same as the ones described in Sect. 14.3.1.8. AWP parameters are the ones described in Sect. 14.3.1.9. The experimental setup of the evaluation considering ImageNet1000 is the same as the one we use for ImageNet200. We consider batch sizes that produce the fastest 32-bit FP training for each one of the network models: 64, 64, and 128 for Alexnet, VGG, and Resnet, respectively.

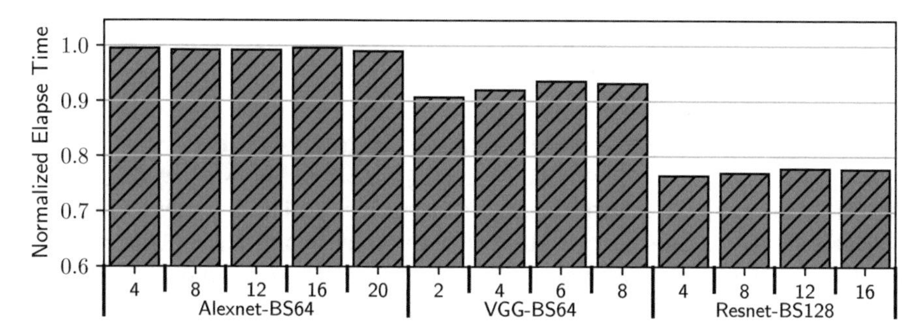

Fig. 14.8 Normalized execution time of A^2DTWP with respect to *baseline* considering the Imagenet1000 data set. Training for Alexnet, VGG, and Resnet considers up to 20, 8, and 16 epochs, respectively

Figure 14.8 displays results corresponding to the experimental campaign with ImageNet1000 on the x86 system. In the x-axis we display different epoch counts for each one of the three models: 4, 8, 12, 16, and 20 epochs for Alexnet; 2, 4, 6, and 8 for VGG; and 4, 8, 12, and 16 epochs for Resnet. The y-axis displays the normalized elapsed time of A^2DTWP with respect to the 32-bit Floating Point *baseline* per each model and epoch count. For the case of Alexnet with batch size 64, A^2DTWP is slightly faster than the *baseline* as it displays a normalized execution time of 0.995, 0.992, 0.992, 0.996, and 0.990 after 4, 8, 12, 16, and 20 epochs, respectively. Figure 14.7 also reports small gains for the case of Alexnet with batch size 64, which confirms that experiments with ImageNet1000 show very similar trends as the evaluation with ImageNet200. When applying A^2DTWP to VGG with 64 batch size, it displays a normalized execution time of 0.907, 0.920, 0.936, and 0.932 with respect to the *baseline* after running 2, 4, 6, and 8 training epochs, respectively. For the Resnet example, we observe normalized execution times of 0.765, 0.770, 0.778, and 0.777 for A^2DTWP after 4, 8, 12, and 16 training epochs, respectively, which constitutes a significant performance improvement.

In terms of validation error, both A^2DTWP and *baseline* display very similar top-5 values at the end of each epoch. For example, for the case of VGG, the Floating Point 32-bit *baseline* approach displays a validation error of 88.04% after 2 training epochs while A^2DTWP achieves a validation error of 89.97% for the same epoch count, that is, an absolute difference of 1.93%. After 4, 6, and 8 training epochs absolute distances of top-5 validation errors between A^2DTWP and *baseline* are 3.09, 0.47, and 0.71%, respectively. Top-5 validation error keeps decreasing in an analogous way for both *baseline* and A^2DTWP as training goes over more epochs, although A^2DTWP is significantly faster. Our evaluation indicates that A^2DTWP can effectively accelerate training while achieving the same validation error as the 32-bit FP *baseline* when considering ImageNet1000.

14.3.2 Constrained Architecture Search on HPC Systems for IoT Deployment

Designing an economically viable artificial intelligence system has become a formidable challenge in view of the increasing number of published methods, data, models, newly available deep learning frameworks as well as the hype surrounding special-purpose hardware accelerators as they become commercially available.

The availability of large-scale datasets with known ground truths [78–89] and the widespread commercial availability of computational performance—usually achieved with graphic-processing units (GPUs)—has driven the current growth of interest in deep learning and the emergence of related new businesses. Smart homes [90], smart grids [91], and smart cities [92] trigger a natural demand for the Internet of Things (IoT), which are products designed to be low in cost and feature low energy consumption and fast reaction times due to the inherent constraints given by final applications that typically demand autonomy with long battery lifetimes or fast real-time operation. Experts estimate that there will be some 30 billion IoT devices in use by 2020 [93], many of which serve applications that benefit from artificial intelligence deployment.

In this context, we propose an automatic way to design deep learning models that satisfy user-defined constraints specifically tailored to match typical IoT requirements, such as inference latency bounds. Additionally, our approach is designed in a modular manner that allows future adaptations and specialization for novel network topology extensions to different IoT devices and lower precision contexts.

Section 14.3.2.1 presents low-level transprecision implementations and how they are modularly integrated into PyTorch [94], a commonly used deep learning framework. Section 14.3.2.2 demonstrates how transprecision computing delivers trade-offs that outperform traditional models. Those models do not only outperform a fixed baseline model, but they offer substantially improved quality over an individual synthesized regular model which meets the same constraints in terms of memory footprint.

14.3.2.1 Transprecision Emulation Framework with PyTorch

We base our transprecision studies on floatx [95], an efficient C++ header-only library. As opposed to other software packages that implement variable precision, such as the GNU MPFR [96] library,[1] floatx is designed to focus on reduced data types. Relying on back-end implementations that have a fixed upper precision turns the library performant due to native support of arithmetic operations. We denote with $T_{w,t}$ the IEEE 754 [97] conform numeric representation of storage width

[1] Available at http://www.mpfr.org/ (version 4.0.1, February 2018).

$1 + w + t$. Floatx implements reduced precision floating point formats $T_{w,t}$ that are coherent with the IEEE 754 standard [97]. It defines the storage encoding, special cases (Nan, Inf), and rounding behavior of arithmetic operations. A number $v = (-1)^s * 2^e * m$ is represented by a sign s, an exponent e and the significand m. The exponent field width w limits the dynamic range, and the trailing significant field width t determines the precision. Standard formats [97] *half*, *float*, and *double* correspond to $T_{5,10}$, $T_{8,23}$, and $T_{11,52}$ respectively. We used round to nearest with a ties to even policy throughout. Casts are implicitly well-defined through IEEE 754 representation and rounding rules.

We integrate floatx modularly into PyTorch [94] to enable large-scale numerical studies. We define a model as directed acyclic graph (DAG) as $M = (V, E)$ where V is the set of nodes defining the kernels and E is the set of edges that defines the flow of tensor data between kernels. Each vertex $v \in V$ defines a kernel operation that is characterized by a list of inputs, internal trainable weights, and an output. Typical kernels are element-wise activation functions, elementary operations (such as addition, subtraction, element-wise multiplication, ...), dense layers (resulting in matrix multiply between input and weights), and convolutional layers, reshaping, shuffling, indexing, cropping, merging, and stacking operations among more specialized kernels used in recent models. Next, we define the quantization operator $Q_{w,t}(.)$ as element-wise casting values to a reduced precision type $T_{w,t}$ with w exponent bits and t mantissa bits implemented by floatx. The quantization operator is applied to data in the natively stored format and the output is returned in the native format, namely in the floating point 32-bit standard format. This modular approach allows injecting quantization at arbitrary points in the model to perform numerical studies.

Similar to reference work [98], we define the extrinsic approach of applying quantization to full tensors at input and output level of operations. The extrinsic approach covers studying transprecision to compress model weights and activations. To that end, our experiments are based on utility source code that traverses any computational graph and introduces a parameterized input quantization step for each input of all kernels of the model. We decide to simultaneously study weight and activation compression.

We use a predefined list of 30 well-established network architectures as presented in Table 14.4. We study the accuracy of all models on CIFAR10 [82] depending on the reduced number formats. We explore the global effect of the number format where one single data type is applied to the entire model. We evaluated a full grid search over 184 floatx configurations for all model topologies. Each configuration corresponds to the reduced precision data type of format $T_{w,t}$ consisting of $w \in [1, 8]$ exponent bits and $t \in [1, 23]$ mantissa bits. Globally applying the same precision configuration enables us to explore the full solution space with a brute-force approach. The knowledge of the full behavior allows to directly answer optimization problems optimally. We evaluated all transprecision configurations, on all models, on all 10,000 validation samples of the CIFAR10 [82] image classification dataset.

Table 14.4 Established reference network architectures

Family	Variant	Max batch size	Instances
ResNet	18, 34, 50, 101, 152	1024–128	5
PreActResNet	18, 34, 50, 101, 152	1024–128	5
ResNeXt29	2x64d, 4x64d, 32x4d	256–64	3
DenseNet	121, 161, 169, 201	256–128	4
LeNet	–	1024	1
GoogLeNet	–	256	1
MobileNet	–	1024	1
MobileNetv2	–	512	1
PNASNet	Type A, Type B	1024, 512	2
DPN	26, 92	512, 128	2
SENet18	–	1024	1
VGG	11, 13, 16, 19	1024	4
Total			30

Optimal configurations are selected from the full grid search based on satisfying the quality constraint $q(T_{w,t}) \geq Q$. We define the requested quality Q relative to the obtained accuracy of the full precision model operating with IEEE 754 32-bit formats as $Q := Q_{float} - \Delta Q$. We measure quality as top1 accuracy which corresponds to the total amount of correctly classified images out the 10,000 validation samples. Figure 14.9 shows the configurations of the optimal number formats for the different quality constraints. All results are achieved with a medium-sized exponent field in the range of 4 to 6-bit. The mantissa width ranges between 5 and 10-bit to achieve equivalent results as in the reference. That range is reduced down to 1 to 6-bits if quality degenerations up to 5% points are allowed. The strictest constraint of zero-quality-loss—that requires to classifying all out of the 10,000 samples as in the full precision case—is satisfied by reduced transprecision formats. On average 12.4-bits are enough to obtain fully accurate results. The weaker the constraints, the more aggressive reductions are achievable. The bit-width can be further reduced to 8.8 and 8.1-bit if one, and up to 5% points of quality reductions are allowed. The results demonstrate that findings generalize well among the variation of network topologies.

14.3.2.2 Constrained Search and Performance Characterization for IoT

Compression, quantization, and pruning techniques reduce heavy computational needs based on the inherent error resilience of deep neural networks [99]. Mobile nets [100] or low-rank expansions [101] change the topology into layers that require fewer weights and reduce workloads. Quantization studies the effect of using reduced precision floating point or fixed-point formats [98, 102], whereas compression attempts to reduce the binary footprint of activation and weight maps

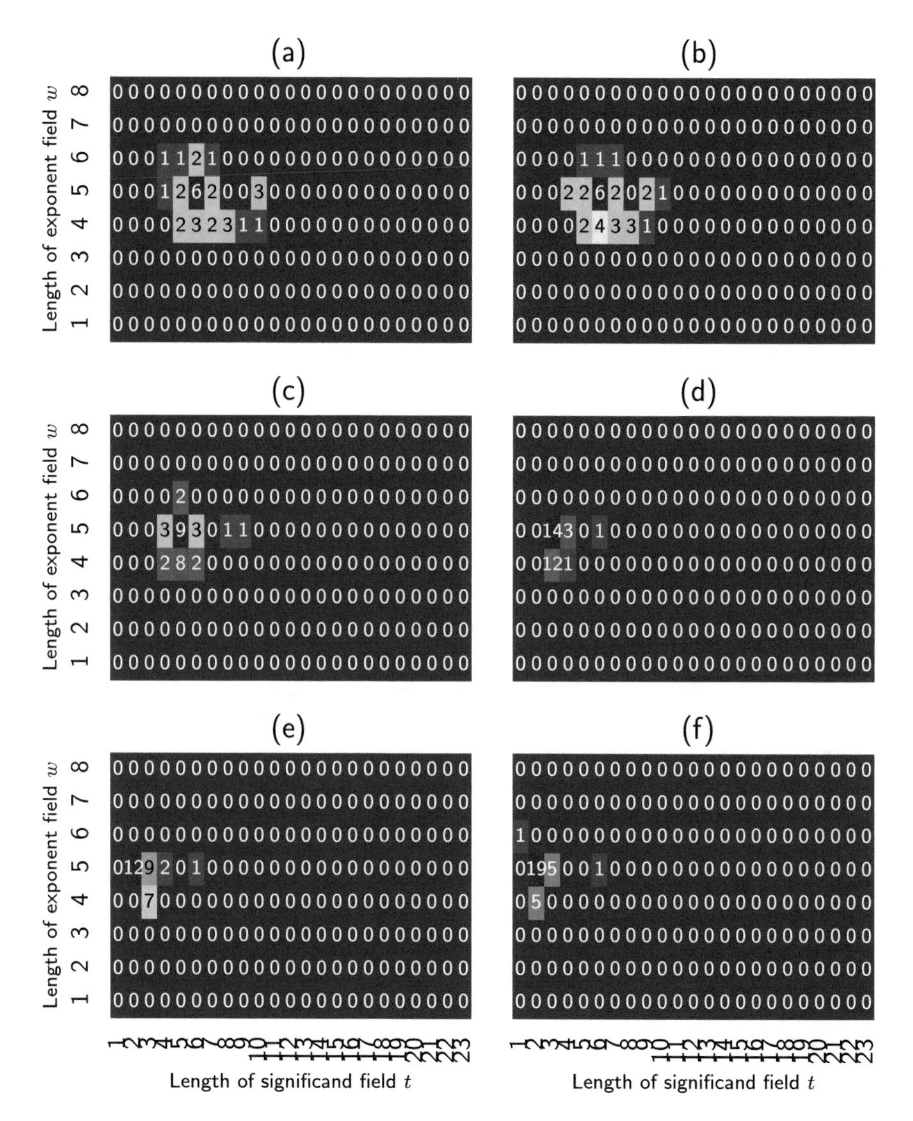

Fig. 14.9 Data type configurations for different quality constraints of the considered reference models. Too low and too high exponent field widths are not required. Allowing for larger quality margins against the reference allows to further reduce the mantissa field. (**a**) $\Delta Q = 0.000\%$. (**b**) $\Delta Q < 0.01\%$. (**c**) $\Delta Q < 0.1\%$. (**d**) $\Delta Q < 1\%$. (**e**) $\Delta Q < 2\%$. (**f**) $\Delta Q < 5\%$

[103]. Pruning approaches avoid computation by enforcing sparsity [104]. We extensively use the developed integration of floatx into PyTorch as detailed in Sect. 14.3.2.1, to assess data format-specific aspects of networks. The novelty of our work is that we jointly evaluate network topologies in combination with reduced precision.

Narrow-Space NAS as Alternative Automated architecture search has the potential to discover better models [105–112]. We recently proposed a narrow-space search that takes constraints into account [113]. In contrast to solving a joint optimization problem in one step, our proposed union of narrow-space searches takes a modular approach that separates the search process of finding architectures that strictly satisfy constraints from the training of candidate networks. That way, we can analyze 10,000 architectures with no training cost and select only a small subset of suitable candidates for training.

Our constraint neural network search (NAS) uses configurable random laws to bias the synthesized model to focus around a property of interest. We skew the probability density function of the number of parameters of generated networks to be concentrated around a given constraint. We manually design random laws by defining lower and upper bounds of uniform distributions to generate models of interest. However, since this manual process requires a human-in-the-loop to adjust structural design choices, we used the tournament selection variant of genetic algorithms [114] that automatically adjusts the random laws of valid configurations. Calibration data maps design goals for inference time on a low-cost IoT device to analytical properties of the model, i.e., the number of parameters.

We used the constraint NAS to generate over 3000 regular models covering the design space over multiple orders of magnitude. For each regular model, we applied a global full grid search over all 184 configurations number formats $T_{w,t}$ with $w \in [1, 8]$ and $t \in [1, 23]$. Quantization is applied to all model parameters and all activation maps. Henceforth, the memory footprint is proportionally improved relative to obtainable gains stemming from modifications of the number formats. Our study allows for evaluating reduced precision accounting for opportunity costs. Quantization effects caused by reduced number formats lower the accuracy simultaneously with the memory footprint. Henceforth, for fairness, we take the NAS generated, regular model with reduced memory footprint to compare regular against transprecision models that meet the same strict constraint. That way, we conclude that all reduced precision models outperform, with a wide margin, regular models that fit into the same memory footprint.

Results and Trade-Offs To study transprecision computing, we ran full design-space explorations on the well-established CIFAR-10 [82] classification task and compared our results with those obtained with established reference models. Figure 14.10 shows the tradeoff between model size and accuracy, including manually and automatically generated results of the aggregate search spaces. The Pareto optimal front follows a smooth curve that saturates towards the best accuracy obtainable for large models. The number of parameters is logarithmic and the

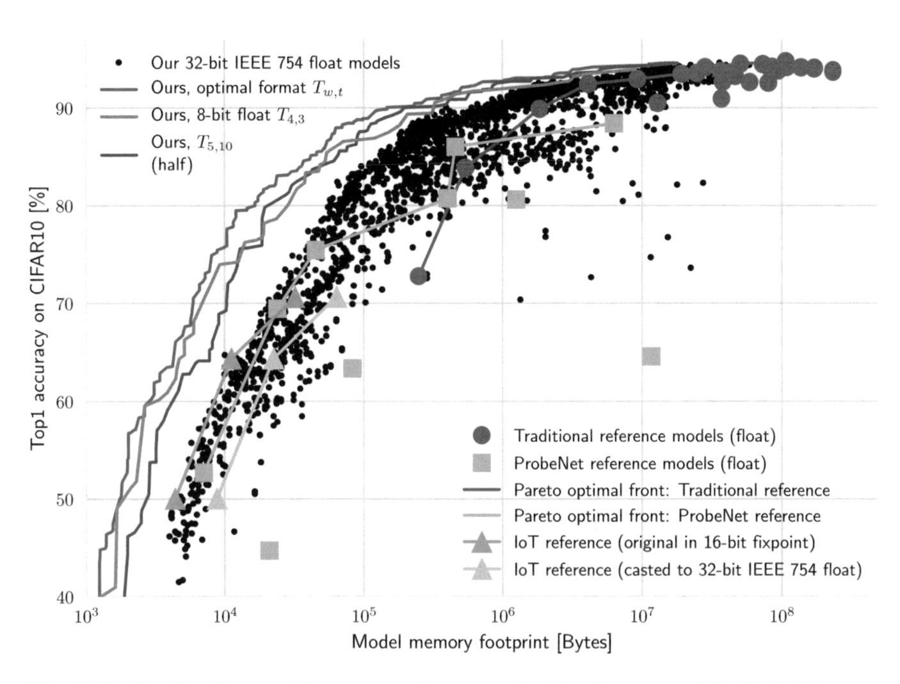

Fig. 14.10 Results of our architecture search compared with reference models. Each dot represents a model according to its size and the obtained accuracy on the CIFAR-10 validation set. Our search finds results over five orders of magnitude and, in particular, finds various models that are much smaller than out-of-the box models. In the restricted IoT domain, our search delivers models that outperform the reference with a wide margin for fixed constraints

accuracy scales linearly. Even very small models with fewer than 1000 parameters can achieve accuracies of greater than 45%. The accuracy increase per decade of added parameters is on the order of 30, 15, 3, and $<$ 2% points and then decreases very quickly. This effect allows us to construct models having several orders of magnitude fewer parameters. It also provides economically interesting solutions for IoT devices that are powerful enough to process data in real-time. We compare our results with three sources of reference models: (a) traditional reference models, (b) ProbeNets [115] that are designed to be small and fast and (c) models designed to run on the parallel ultra-low power (PULP) platform [116]. Traditional models include 30 reference topologies including variants of VGG [117], ResNets [118], GoogleNet [119], MobileNets [120] dual-path nets (DPNs) [121], and DenseNets [122], where most of them (28/30) exceed 1 M parameters. ProbeNets were originally introduced to characterize the classification difficulty and are considerably smaller by design [115]. They act as reference points for manually designed networks that cover the relevant lower tail in terms of parameters. In the IoT-relevant domain (<10 M parameters), our search outperforms all the listed reference models.

The top three fronts in Fig. 14.10 show the results of our precision analysis. For each trained model, we evaluated the effect of running models with all configurations of type $T_{w,t}$ and plot the Pareto optimal front. We considered three cases: (1) running all models with half-precision, (2) running all models with the type T_{4_3}, which is the best choice for types that are 8 bits long, and (3) running each model with its individual best tradeoff type $T_{w,t}$. We demonstrate empirically that reduced precision pushes the Pareto optimal front. Under a given memory constraint, accuracy improves by more than 7% points for half and by another 1% points or more for the model individual format.

14.3.3 Conclusion

This section proposes A^2DTWP, which reduces data movement across heterogeneous environments composed of several GPUs and multicore CPU devices in the context of deep learning workloads. The A^2DTWP framework is composed of the AWP algorithm and the ADT procedure. AWP is able to dynamically define the weights data representation format during training. AWP is effective without any deterioration on the learning capacity of the neural network. To transform AWP decisions into real performance gains, we introduce the ADT procedure, which efficiently compresses network's weights before sending them to the GPUs. This procedure exploits both thread- and SIMD-level parallelism. By combining AWP with ADT we are able to achieve a significant performance gain when training network models such as Alexnet, VGG, or Resnet. Our experimental campaign considers different batch sizes and two different multi-GPU high-end systems.

This section is the first in proposing a solution that relies on reduced numeric data formats to mitigate the cost of sending DNNs weights to different hardware devices during training. While our evaluation targets heterogeneous high-end systems composed of several GPUs and CPU multicore devices, techniques presented by this section are easily generalizable to any context involving several hardware accelerators exchanging large amounts of data. Taking into account the prevalence of deep learning-specific accelerators in large production systems [39], the techniques described by this section are applicable to a wide range of scenarios involving different kinds of accelerators.

Trademarks

IBM is trademark of International Business Machines Corporation, registered in many jurisdictions worldwide. Other product and service names might be trademarks of IBM or other companies.

References

1. Golub, G., & Loan, C. V. (1996). *Matrix computations*, 3rd edn. Baltimore: The Johns Hopkins University Press.
2. Demmel, J. W. (1997). *Applied numerical linear algebra*. Philadelphia: SIAM.
3. Dongarra, J. J., Duff, I. S., Sorensen, D. C., & van der Vorst, H. A. (1998). *Numerical linear algebra for high-performance computers*. Philadelphia, PA: Society for Industrial and Applied Mathematics.
4. Anderson, E., Bai, Z., Dongarra, J., Greenbaum, A., McKenney, A., Du Croz, J., Hammarling, S., Demmel, J., Bischof, C., & Sorensen, D. (1990). Lapack: a portable linear algebra library for high-performance computers. In *Proceedings of the 1990 ACM/IEEE Conference on Supercomputing, Supercomputing'90, (Los Alamitos, CA, USA)* (pp. 2–11). Piscataway: IEEE Computer Society Press.
5. Blackford, L. S., Demmel, J., Dongarra, J., Duff, I., Hammarling, S., Henry, G., Heroux, M., Kaufman, L., Lumsdaine, A., Petitet, A., Pozo, R., Remington, K., & Whaley, R. C. (2002). An updated set of basic linear algebra subprograms (BLAS). *ACM Transactions on Mathematical Software, 28*, 135–151 (2002)
6. Horowitz, M. (2014). Computing's energy problem (and what we can do about it). In *2014 IEEE International Solid-State Circuits Conference Digest of Technical Papers (ISSCC)* (pp. 10–14).
7. Ginkgo. (2019). https://ginkgo-project.github.io
8. Buluç, A., Williams, S., Oliker, L., & Demmel, J. (2011). Reduced-bandwidth multithreaded algorithms for sparse matrix-vector multiplication. In *36th IEEE International Parallel & Distributed Processing Symposium IPDPS* (pp. 721–733).
9. Bell, N., & Garland, M. (2008). Efficient sparse matrix-vector multiplication on CUDA. NVIDIA Technical Report NVR-2008-004.
10. I. S. Commitee. (2000). IEEE standard for modeling and simulation (m&s) high level architecture (HLA) - framework and rules. IEEE Std. 1516–2000 (pp. i–22).
11. Saad, Y. (2003). *Iterative methods for sparse linear systems*, 2nd edn. Philadelphia: SIAM.
12. Wulf, W. A., & McKee, S. A. (1995). Hitting the memory wall: Implications of the obvious. *SIGARCH Computer Architecture News, 23*, 20–24.
13. Molka, D., Hackenberg, D., Schöne, R., & Müller, M. S. (2010). Characterizing the energy consumption of data transfers and arithmetic operations on x86–64 processors. In *International Green Computing Conference 2010, Chicago, IL, USA, 15–18 August 2010* (pp. 123–133).
14. Higham, N. J. (2002). *Accuracy and stability of numerical algorithms*, 2nd edn. Philadelphia: SIAM.
15. Buttari, A., Dongarra, J. J., Langou, J., Langou, J., Luszczek, P., & Kurzak, J. (2007). Mixed precision iterative refinement techniques for the solution of dense linear systems. *International Journal of High Performance Computing Applications, 21*(4), 457–486.
16. Baboulin, M., Buttari, A., Dongarra, J. J., Langou, J., Langou, J., Luszczek, P., Kurzak, J., & Tomov, S. (2009). Accelerating scientific computations with mixed precision algorithms. *Computer Physics Communications, 180*(12), 2526–2533.
17. Barrachina, S., Castillo, M., Igual, F. D., Mayo, R., & Quintana-Ortí, E. S. (2008). Solving dense linear systems on graphics processors. In E. Luque, T. Margalef, & D. Benítez (Eds.), *Euro-Par 2008 – Parallel Processing* (pp. 739–748). Berlin: Springer.
18. Strzodka, R., & Göddeke, D. (2006). Pipelined mixed precision algorithms on FPGAs for fast and accurate PDE solvers from low precision components. In *IEEE Proceedings on Field–Programmable Custom Computing Machines (FCCM 2006)*. Piscataway: IEEE Computer Society Press.
19. Anzt, H., Heuveline, V., & Rocker, B. (2010). Mixed precision error correction methods for linear systems Convergence analysis based on Krylov subspace methods. In K. Jonasson (Ed.) *PARA 2010, Part II, LNCS 7134* (pp. 237–248). Heidelberg: Springer.

20. Haidar, A., Tomov, S., Dongarra, J., & Higham, N. J. (2018). Harnessing GPU tensor cores for fast FP16 arithmetic to speed up mixed-precision iterative refinement solvers. In *Proceedings of the International Conference for High Performance Computing, Networking, Storage, and Analysis, SC'18, (Piscataway, NJ, USA)* (pp. 47:1–47:11). Piscataway: IEEE Press.
21. Anzt, H., Dongarra, J., & Quintana-Ortí, E. S. (2015). Adaptive precision solvers for sparse linear systems. In *Proceedings of the 3rd International Workshop on Energy Efficient Supercomputing, E2SC'15, (New York, NY, USA)* (pp. 2:1–2:10). New York: ACM.
22. Grützmacher, T., & Anzt, H. (2019). A modular precision format for decoupling arithmetic format and storage format. In G. Mencagli, D. B. Heras, V. Cardellini, E. Casalicchio, E. Jeannot, F. Wolf, A. Salis, C. Schifanella, R. R. Manumachu, L. Ricci, M. Beccuti, L. Antonelli, J. D. Garcia Sanchez, & S. L. Scott (Eds.), *Euro-Par 2018: Parallel Processing Workshops* (pp. 434–443). Cham: Springer.
23. Grützmacher, T., Cojean, T., Flegar, G., Göbel, F., & Anzt, H. (2019). A customized precision format based on mantissa segmentation for accelerating sparse linear algebra. *Concurrency and Computation: Practice and Experience, 32*(2), e5418. e5418 cpe.5418.
24. Anzt, H., Flegar, G., Grützmacher, T., & Quintana-Ortí, E. S. (2019). Toward a modular precision ecosystem for high-performance computing. *The International Journal of High Performance Computing Applications, 33*(6), 1069–1078.
25. Grützmacher, T., Anzt, H., Scheidegger, F., & Quintana-Ortí, E. S. (2018). High-performance GPU implementation of PageRank with reduced precision based on mantissa segmentation. In *2018 IEEE/ACM 8th Workshop on Irregular Applications: Architectures and Algorithms (IA3)* (pp. 61–68).
26. Anzt, H., Dongarra, J., Flegar, G., Higham, N. J., & Quintana-Ortí, E. S. (2019). Adaptive precision in block-Jacobi preconditioning for iterative sparse linear system solvers. *Concurrency and Computation: Practice and Experience, 31*(6), 1–12.
27. Tadano, H., & Sakurai, T. (2008). On single precision preconditioners for krylov subspace iterative methods. In I. Lirkov, S. Margenov, & J. Waśniewski (Eds.), *Large-Scale Scientific Computing* (pp. 721–728). Berlin: Springer.
28. Gropp, W. D., Kaushik, D. K., Keyes, D. E., & Smith, B. F. (2000). Latency, bandwidth, and concurrent issue limitations in high-performance CFD. In *Proceedings of the First MIT Conference on Computational Fluid and Solid Mechanics*.
29. Carson, E., & Higham, N. J. (2017). A new analysis of iterative refinement and its application to accurate solution of ill-conditioned sparse linear systems. *SIAM Journal on Scientific Computing, 39*(6), A2834–A2856.
30. Carson, E., & Higham, N. J. (2018). Accelerating the solution of linear systems by iterative refinement in three precisions. *SIAM Journal on Scientific Computing, 40*(2), A817–A847.
31. Krizhevsky, A., Sutskever, I., & Hinton, G. E. (2012). Imagenet classification with deep convolutional neural networks, In F. Pereira, C. J. C. Burges, L. Bottou, & K. Q. Weinberger (Eds.), *Advances in Neural Information Processing Systems* (vol. 25, pp. 1097–1105). New York: Curran Associates.
32. Szegedy, C., Liu, W., Jia, Y., Sermanet, P., Reed, S., Anguelov, D., Erhan, D., Vanhoucke, V., & Rabinovich, A. (2015). Going deeper with convolutions. In *Proceedings of the IEEE Conference on Computer Vision and Pattern Recognition* (pp. 1–9).
33. Hinton, G., Deng, L., Yu, D., Dahl, G. E., Mohamed, A.-R., Jaitly, N., Senior, A., Vanhoucke, V., Nguyen, P., Sainath, T. N., & Kingsbury, B. (2012). Deep neural networks for acoustic modeling in speech recognition: The shared views of four research groups. *IEEE Signal Processing Magazine, 29*(6), 82–97.
34. Wu, Y., Schuster, M., Chen, Z., Le, Q. V., Norouzi, M., Macherey, W., Krikun, M., Cao, Y., Gao, Q., Macherey, K., Klingner, J., Shah, A., Johnson, M., Liu, X., Kaiser, Ł., Gouws, S., Kato, Y., Kudo, T., Kazawa, H., et al. (2016). Google's neural machine translation system: Bridging the gap between human and machine translation. arXiv:1609.08144.
35. Ciregan, D., Meier, U., & Schmidhuber, J. (2012). Multi-column deep neural networks for image classification. In *2012 IEEE Conference on Computer Vision and Pattern Recognition* (pp. 3642–3649).

36. Krizhevsky, A., Sutskever, I., & Hinton, G. E. (2012). ImageNet classification with deep convolutional neural networks. In *Proceedings of the 25th International Conference on Neural Information Processing Systems, NIPS'12* (pp. 1097–1105). New York: Curran Associates.
37. Abadi, M., Barham, P., Chen, J., Chen, Z., Davis, A., Dean, J., Devin, M., Ghemawat, S., Irving, G., Isard, M., Kudlur, M., Levenberg, J., Monga, R., Moore, S., Murray, D. G., Steiner, B., Tucker, P., Vasudevan, V., Warden, P., et al. (2016). Tensorflow: A system for large-scale machine learning. In *Proceedings of the 12th USENIX Conference on Operating Systems Design and Implementation, OSDI'16* (pp. 265–283). Berkeley: USENIX Association.
38. Merolla, P. A., Arthur, J. V., Alvarez-Icaza, R., Cassidy, A. S., Sawada, J., Akopyan, F., Jackson, B. L., Imam, N., Guo, C., Nakamura, Y., Brezzo, B., Vo, I., Esser, S. K., Appuswamy, R., Taba, B., Amir, A., Flickner, M., Risk, W., Manohar, R., et al. (2014). A million spiking-neuron integrated circuit with a scalable communication network and interface. *Science, 345*(6197), 668–673.
39. Jouppi, N. P., Young, C., Patil, N., Patterson, D., Agrawal, G., Bajwa, R., Bates, S., Bhatia, S., Boden, N., Borchers, A., Boyle, R., Cantin, P.-L., Chao, C., Clark, C., Coriell, J., Daley, M., Dau, M., Dean, J., Gelb, B., et al. (2017). In-datacenter performance analysis of a tensor processing unit. In *Proceedings of the 44th Annual International Symposium on Computer Architecture, ISCA'17* (pp. 1–12). New York: ACM.
40. Kurth, T., Zhang, J., Satish, N., Mitliagkas, I., Racah, E., Patwary, M. A., Malas, T., Sundaram, N., Bhimji, W., Smorkalov, M., Deslippe, J., Shiryaev, M., Sridharan, S., Prabhat, P. D. (2017). Deep learning at 15pf: Supervised and semi-supervised classification for scientific data. In *Proceedings of the International Conference for High Performance Computing, Networking, Storage and Analysis, SC'17* (pp. 7:1–7:11). New York: ACM.
41. Werbos, P. J. (1974). *Beyond Regression: New Tools for Prediction and Analysis in the Behavioral Sciences*. Ph.D. Thesis, Harvard University.
42. Press, W. H., Flannery, B. P., Teukolsky, S. A., & Vetterling, W. T. (1988). *Numerical recipes in C: The art of scientific computing*. New York: Cambridge University Press.
43. Kiefer, J., & Wolfowitz, J. (1952). Stochastic estimation of the maximum of a regression function. *The Annals of Mathematical Statistics, 23*, 462–466.
44. You, Y., Buluc, A., & Demmel, J. (2017). Scaling deep learning on GPU and knights landing clusters. In *Proceedings of the International Conference for High Performance Computing, Networking, Storage and Analysis, SC'17*, pp. 9:1–9:12. New York: ACM.
45. Gupta, S., Agrawal, A., Gopalakrishnan, K., & Narayanan, P. (2015). Deep learning with limited numerical precision. In *Proceedings of the 32nd International Conference on Machine Learning, ICML 2015, Lille, France, 6–11 July 2015* (pp. 1737–1746).
46. Köster, U., Webb, T., Wang, X., Nassar, M., Bansal, A. K., Constable, W., Elibol, O., Gray, S., Hall, S., Hornof, L., Khosrowshahi, A., Kloss, C., Pai, R. J., & Rao, N. (2017). Flexpoint: An adaptive numerical format for efficient training of deep neural networks. In I. Guyon, U. V. Luxburg, S. Bengio, H. Wallach, R. Fergus, S. Vishwanathan, & R. Garnett (Eds.), *Advances in Neural Information Processing Systems*(vol. 30, pp. 1742–1752). New York: Curran Associates.
47. Simonyan, K., & Zisserman, A. (2014). Very deep convolutional networks for large-scale image recognition. CoRR, vol. abs/1409.1556.
48. He, K., Zhang, X., Ren, S., & Sun, J. (2016). Deep residual learning for image recognition. In *2016 IEEE Conference on Computer Vision and Pattern Recognition (CVPR)* (pp. 770–778).
49. Deng, J., Dong, W., Socher, R., Li, L.-J., Li, K., & Fei-Fei, L. (2009). ImageNet: A large-scale hierarchical image database. In *IEEE Conference on Computer Vision and Pattern Recognition (CVPR09)*.
50. Bottou, L., & Bousquet, O. (2008). The tradeoffs of large scale learning. In *Advances in Neural Information Processing Systems* (pp. 161–168).
51. Micikevicius, P., Narang, S., Alben, J., Diamos, G. F., Elsen, E., García, D., Ginsburg, B., Houston, M., Kuchaiev, O., Venkatesh, G., & Wu, H. (2018). Mixed precision training. In *Seventh International Conference on Learning Representations (ICLR)*.

52. Dean, J., Corrado, G. S., Monga, R., Chen, K., Devin, M., Le, Q. V., Mao, M. Z., Ranzato, M., Senior, A., Tucker, P., Yang, K., & Ng, A. Y. (2012). Large scale distributed deep networks. In *NIPS'12: Proceedings of the 25th International Conference on Neural Information Processing Systems*.
53. Holi, J. L., & Hwang, J. N. (1993). Finite precision error analysis of neural network hardware implementations. *IEEE Transactions on Computers, 42*, 281–290.
54. Courbariaux, M., Bengio, Y. & David, J. (2014). Low precision arithmetic for deep learning. CoRR, vol. abs/1412.7024.
55. RÃŋos, J. O., Armejach, A., Khattak, G., Petit, E., Vallecorsa, S., & Casas, M. (2020). Evaluating mixed-precision arithmetic for 3d generative adversarial networks to simulate high energy physics detectors. In *2020 19th IEEE International Conference on Machine Learning and Applications (ICMLA)* (pp. 49–56).
56. RÃŋos, J. O., Armejach, A., Petit, E., Henry, G., & Casas, M. (2021). Dynamically adapting floating-point precision to accelerate deep neural network training. In *2021 20th IEEE International Conference on Machine Learning and Applications (ICMLA)*.
57. Niu, F., Recht, B., Re, C., & Wright, S. J. (2011). Hogwild! A lock-free approach to parallelizing stochastic gradient descent. In *Proceedings of the 24th International Conference on Neural Information Processing Systems, NIPS'11* (pp. 693–701). New York: Curran Associates.
58. Zhang, S., Choromanska, A., & LeCun, Y. (2014). Deep learning with elastic averaging SGD. CoRR, vol. abs/1412.6651.
59. Coates, A., Huval, B., Wang, T., Wu, D. J., Ng, A. Y., & Catanzaro, B. (2013). Deep learning with cots HPC systems. In *Proceedings of the 30th International Conference on International Conference on Machine Learning - Volume 28, ICM'13* (pp. III–1337–III–1345), JMLR.org.
60. Le, Q. V., Monga, R., Devin, M., Corrado, G., Chen, K., Ranzato, M., Dean, J., & Ng, A. Y. (2011). Building high-level features using large scale unsupervised learning. CoRR, vol. abs/1112.6209.
61. Han, S., Liu, X., Mao, H., Pu, J., Pedram, A., Horowitz, M. A., & Dally, W. J. (2016). EIE: Efficient inference engine on compressed deep neural network. CoRR, vol. abs/1602.01528.
62. Lin, Y., Han, S., Mao, H., Wang, Y., & Dally, W. J. (2017). Deep gradient compression: Reducing the communication bandwidth for distributed training. CoRR, vol. abs/1712.01887.
63. Wen, W., Xu, C., Yan, F., Wu, C., Wang, Y., Chen, Y., & Li, H. (2017). Terngrad: Ternary gradients to reduce communication in distributed deep learning. CoRR, vol. abs/1705.07878.
64. Alistarh, D., Li, J., Tomioka, R., & Vojnovic, M. (2016). QSGD: Randomized quantization for communication-optimal stochastic gradient descent. CoRR, vol. abs/1610.02132.
65. Aji, A. F., & Heafield, K. (2017). Sparse communication for distributed gradient descent. CoRR, vol. abs/1704.05021.
66. Murray, A. F., & Edwards, P. J. (1994). Enhanced MLP performance and fault tolerance resulting from synaptic weight noise during training. *IEEE Transactions on Neural Networks, 5*, 792–802.
67. Bishop, C. M. (1995). Training with noise is equivalent to Tikhonov regularization. *Neural Computation, 7*, 108–116.
68. Audhkhasi, K., Osoba, O., & Kosko, B. (2013). Noise benefits in backpropagation and deep bidirectional pre-training. In *The 2013 International Joint Conference on Neural Networks (IJCNN)* (pp. 1–8).
69. Dagum, L., & Menon, R. (1998). Openmp: An industry-standard API for shared-memory programming. *IEEE Computing in Science & Engineering, 5*, 46–55.
70. Lomont, C. (2011). Introduction to intel advanced vector extensions. Intel white paper.
71. Gwennap, L. (1998). AltiVec vectorizes PowerPC. *Microprocessors Report* (vol. 12, pp. 1–5).
72. IEEE standard for floating point arithmetic (2008). IEEE Std 754–2008 (pp. 1–70).
73. Krizhevsky, A. (2014). One weird trick for parallelizing convolutional neural networks. CoRR, vol. abs/1404.5997.
74. Seo, H., Liu, Z., Großschädl, J., & Kim, H. (2015). Efficient arithmetic on arm-neon and its application for high-speed RSA implementation. *IACR Cryptology ePrint Archive, 2015*, 465.

75. NVIDIA Corporation. (2018). CUDA toolkit documentation. v10.0.130 ed.
76. Qian, N. (1999). On the momentum term in gradient descent learning algorithms. *Neural Networks, 12*(1), 145–151.
77. NVIDIA Corporation. (2016). Nvlink fabric
78. Deng, J., Dong, W., Socher, R., Li, L.-J., Li, K., & Fei-Fei, L. (2009). ImageNet: A large-scale hierarchical image database. In *IEEE Conference on Computer Vision and Pattern Recognition (CVPR)* (pp. 248–255).
79. Stallkamp, J., Schlipsing, M., Salmen, J., & Igel, C. (2011). The German traffic sign recognition benchmark: A multi-class classification competition. In *The 2011 International Joint Conference on Neural Networks* (pp. 1453–1460).
80. Deng, L. (2012). The MNIST database of handwritten digit images for machine learning research [best of the web]. *IEEE Signal Processing Magazine, 29*(6), 141–142.
81. Xiao, H., Rasul, K., & Vollgraf, R. (2017). Fashion-MNIST: A novel image dataset for benchmarking machine learning algorithms. https://www.bibsonomy.org/bibtex/2de51af2f6c7d8b0f4cd84a428bb17967/andolab and https://arxiv.org/abs/1708.07747
82. Krizhevsky, A., & Hinton, G. (2009). Learning multiple layers of features from tiny images. http://citeseerx.ist.psu.edu/viewdoc/summary?doi=10.1.1.222.9220
83. Coates, A., Ng, A., & Lee, H. (2011). An analysis of single-layer networks in unsupervised feature learning. In G. Gordon, D. Dunson, & M. DudÃŋk (Eds.) *Proceedings of the Fourteenth International Conference on Artificial Intelligence and Statistics*. Proceedings of Machine Learning Research (vol. 15, pp. 215–223), Fort Lauderdale: PMLR.
84. Netzer, Y., Wang, T., Coates, A., Bissacco, A., Wu, B., & Ng, A. Y. (2011). Reading digits in natural images with unsupervised feature learning. In *NIPS Workshop on Deep Learning and Unsupervised Feature Learning* (vol. 2011, p. 5).
85. Zhou, B., Lapedriza, A., Khosla, A., Oliva, A., & Torralba, A. (2018). Places: A 10 million image database for scene recognition. *IEEE Transactions on Pattern Analysis and Machine Intelligence, 40*, 1452–1464.
86. Cimpoi, M., Maji, S., Kokkinos, I., Mohamed, S., & Vedaldi, A. (2014). Describing textures in the wild. In *Proceedings of the 2014 IEEE Conference on Computer Vision and Pattern Recognition, CVPR'14* (pp. 3606–3613). Washington: IEEE Computer Society.
87. Bossard, L., Guillaumin, M., & Van Gool, L. (2014). Food-101 – mining discriminative components with random forests. In D. Fleet, T. Pajdla, B. Schiele, & T. Tuytelaars (Eds.) *Computer Vision – ECCV 2014* (pp. 446–461). Cham: Springer.
88. Nilsback, M. E., & Zisserman, A. (2008). Automated flower classification over a large number of classes. In *2008 Sixth Indian Conference on Computer Vision, Graphics Image Processing* (pp. 722–729).
89. Quattoni, A., & Torralba, A. (2009). Recognizing indoor scenes. In *2009 IEEE Conference on Computer Vision and Pattern Recognition* (pp. 413–420).
90. Li, W., Logenthiran, T., Phan, V.-T., & Woo, W. L. (2019). A novel smart energy theft system (sets) for IoT based smart home. *IEEE Internet of Things Journal, 6*, 5531–5539.
91. Fenza, G., Gallo, M., & Loia, V. (2019). Drift-aware methodology for anomaly detection in smart grid. *IEEE Access, 7*, 9645–9657.
92. Gaber, M. M., Aneiba, A., Basurra, S., Batty, O., Elmisery, A. M., Kovalchuk, Y., & Rehman, M. H. U. (2019). Internet of things and data mining: From applications to techniques and systems. *Wiley Interdisciplinary Reviews: Data Mining and Knowledge Discovery, 9*(3), e1292.
93. Nordrum, A. (2016). The internet of fewer things [news]. *IEEE Spectrum, 53*, 12–13.
94. Pytorch. Retrieved 22 May, 2019, from https://pytorch.org/
95. Flegar, G., Scheidegger, F., Novakovic, V., Mariani, G., Tomas, A., Malossi, C., & Quintana-Ortí, E. (2019). Float x: A c++library for customized floating-point arithmetic. *ACM Trans. Math. Softw.* (to appear)
96. Fousse, L., Hanrot, G., Lefèvre, V., Pélissier, P., & Zimmermann, P. (2007). MPFR: A multiple-precision binary floating-point library with correct rounding. *ACM Transactions on Mathematical Software (TOMS), 33*(2), 13.

97. Zuras, D., Cowlishaw, M., Aiken, A., Applegate, M., Bailey, D., Bass, S., Bhandarkar, D., Bhat, M., Bindel, D., Boldo, S., et al. (2008). IEEE standard for floating-point arithmetic. IEEE Std 754-2008, pp. 1–70. http://www.dsc.ufcg.edu.br/cnum/modulos/Modulo2/IEEE754_2008.pdf
98. Loroch, D. M., Pfreundt, F.-J., Wehn, N., & Keuper, J. (2017). Tensorquant: A simulation toolbox for deep neural network quantization. In *Proceedings of the Machine Learning on HPC Environments,MLHPC'17* (pp. 1:1–1:8). New York: ACM.
99. Rybalkin, V., Wehn, N., Yousefi, M. R., & Stricker, D. (2017). Hardware architecture of bidirectional long short-term memory neural network for optical character recognition. In *Proceedings of the Conference on Design, Automation & Test in Europe* (pp. 1394–1399). European Design and Automation Association.
100. Howard, A. G., Zhu, M., Chen, B., Kalenichenko, D., Wang, W., Weyand, T., Andreetto, M., & Adam, H. (2017). Mobilenets: Efficient convolutional neural networks for mobile vision applications. CoRR, vol. abs/1704.04861.
101. Jaderberg, M., Vedaldi, A., & Zisserman, A. (2014). Speeding up convolutional neural networks with low rank expansions. CoRR, vol. abs/1405.3866.
102. Hill, P., Zamirai, B., Lu, S., Chao, Y., Laurenzano, M., Samadi, M., Papaefthymiou, M. C., Mahlke, S. A., Wenisch, T. F., Deng, J., Tang, L., & Mars, J. (2018). Rethinking numerical representations for deep neural networks. CoRR, vol. abs/1808.02513.
103. Cavigelli, L., & Benini, L. (2018). Extended bit-plane compression for convolutional neural network accelerators. CoRR, vol. abs/1810.03979.
104. Ashiquzzaman, A., Ma, L. V., Kim, S., Lee, D., Um, T., & Kim, J. (2019). Compacting deep neural networks for light weight IoT SCADA based applications with node pruning. In *2019 International Conference on Artificial Intelligence in Information and Communication (ICAIIC)* (pp. 082–085).
105. Miikkulainen, R., Liang, J., Meyerson, E., Rawal, A., Fink, D., Francon, O., Raju, B., Shahrzad, H., Navruzyan, A., Duffy, N., & Hodjat, B. (2019). Chapter 15 - evolving deep neural networks. In R. Kozma, C. Alippi, Y. Choe, & F. C. Morabito (Eds.), *Artificial Intelligence in the Age of Neural Networks and Brain Computing* (pp. 293–312). Cambridge: Academic Press.
106. Xie, L., & Yuille, A. (2017). Genetic CNN. In *Proceedings of the IEEE International Conference on Computer Vision* (pp. 1379–1388).
107. Zhong, Z., Yan, J., & Liu, C. (207). Practical network blocks design with q-learning. CoRR, vol. abs/1708.05552.
108. Zoph, B., & Le, Q. V. (2016). Neural architecture search with reinforcement learning. CoRR, vol. abs/1611.01578.
109. Zoph, B., Vasudevan, V., Shlens, J., & Le, Q. V. (2018). Learning transferable architectures for scalable image recognition. In *The IEEE Conference on Computer Vision and Pattern Recognition (CVPR)*.
110. Cai, H., Chen, T., Zhang, W., Yu, Y., & Wang, J. (2018). Efficient architecture search by network transformation. In *Thirty-Second AAAI Conference on Artificial Intelligence*.
111. Baker, B., Gupta, O., Naik, N., & Raskar, R. (2016). Designing neural network architectures using reinforcement learning. CoRR, vol. abs/1611.02167.
112. Wistuba, M., Rawat, A., & Pedapati, T. (2019). A survey on neural architecture search. arXiv:1905.01392.
113. Scheidegger, F., Benini, L., Bekas, C., & Malossi, C. (2019). Constrained deep neural network architecture search for IoT devices accounting for hardware calibration. In *Advances in Neural Information Processing Systems*.
114. Goldberg, D. E., & Deb, K. (1991). A comparative analysis of selection schemes used in genetic algorithms. In *Foundations of genetic algorithms* (vol. 1, pp. 69–93). Amsterdam: Elsevier.
115. Scheidegger, F., Istrate, R., Mariani, G., Benini, L., Bekas, C., & Malossi, C. (2021). Efficient image dataset classification difficulty estimation for predicting deep-learning accuracy. *The Visual Computer volume 37*, 1593–1610. https://link.springer.com/article/10.1007/s00371-020-01922-5

116. Conti, F., Rossi, D., Pullini, A., Loi, I., & Benini, L. (2016). Pulp: A ultra-low power parallel accelerator for energy-efficient and flexible embedded vision. *Journal of Signal Processing Systems, 84*, 339–354.
117. Simonyan, K., & Zisserman, A. (2014). Very deep convolutional networks for large-scale image recognition. arXiv:1409.1556.
118. He, K., Zhang, X., Ren, S., & Sun, J. (2016). Deep residual learning for image recognition. In *Proceedings of the IEEE Conference on Computer Vision and Pattern Recognition* (pp. 770–778).
119. Szegedy, C., Vanhoucke, V., Ioffe, S., Shlens, J., & Wojna, Z. (2016). Rethinking the inception architecture for computer vision. In *The IEEE Conference on Computer Vision and Pattern Recognition (CVPR)*.
120. Howard, A. G., Zhu, M., Chen, B., Kalenichenko, D., Wang, W., Weyand, T., Andreetto, M., & Adam, H. (2017). Mobilenets: Efficient convolutional neural networks for mobile vision applications. CoRR, vol. abs/1704.04861.
121. Chen, Y., Li, J., Xiao, H., Jin, X., Yan, S., & Feng, J. (2017). Dual path networks. In I. Guyon, U. V. Luxburg, S. Bengio, H. Wallach, R. Fergus, S. Vishwanathan, & R. Garnett (Eds.), *Advances in Neural Information Processing Systems 30* (pp. 4467–4475). New York: Curran Associates.
122. Huang, G., Liu, Z., van der Maaten, L., & Weinberger, K. Q. (2017). Densely connected convolutional networks. In *The IEEE Conference on Computer Vision and Pattern Recognition (CVPR)*.

Chapter 15
Approximations in Deep Learning

Etienne Dupuis, Silviu Filip, Olivier Sentieys, David Novo, Ian O'Connor, and Alberto Bosio

15.1 Introduction

Deep Neural Networks (DNNs) [1], and in particular, Convolutional Neural Networks (CNNs), are currently one of the most intensively and widely used predictive models in the field of machine learning. CNNs have been shown to give very good results for many complex tasks such as object recognition in images/videos, drug discovery, natural language processing, autonomous driving, and playing complex games [2–5].

Despite these benefits, the computational workload involved in CNNs is often out of reach for low-power embedded devices and/or is still very costly when ran on datacenter-style Component-Off-The-Shelf (COTS) hardware platforms. To give an example, the amazing performance of AlphaGo [5] required 4 to 6 weeks of training executed on 2000 CPUs and 250 GPUs for a total of about 600 kW of power consumption (while the human brain of a Go player requires about 20 W), which translates to over 2 TJ of energy consumption. Thus, a lot of research effort from both industrials and academics has been concentrated on defining/designing custom hardware platforms supporting these types of algorithms, to improve performance and/or energy efficiency [6–8].

E. Dupuis · I. O'Connor · A. Bosio
University of Lyon, ECL, INSA Lyon, CNRS, UCBL, CPE Lyon, INL, UMR5270, Ecully, France
e-mail: etienne.dupuis@ec-lyon.fr; ian.oConnor@ec-lyon.fr; alberto.bosio@ec-lyon.fr

D. Novo
LIRMM, Université de Montpellier, CNRS, Montpellier, France
e-mail: david.novo@lirmm.fr

S. Filip · O. Sentieys (✉)
University of Rennes, INRIA/IRISA, Rennes, France
e-mail: silviu.filip@inria.fr; Olivier.Sentieys@inria.fr

A. Bosio et al. (eds.), *Approximate Computing Techniques*,
https://doi.org/10.1007/978-3-030-94705-7_15

CNNs show inherent resilience to insignificant errors due to their iterative nature and the underlying learning process. Therefore, an intrinsic tolerance to inexact computation is clear, and using the AxC paradigm to improve power and speed characteristics is, therefore, relevant [9]. Indeed, CNNs mesh well with AxC techniques, especially with fixed-point arithmetic or low-precision floating-point implementations (it has been shown that even binary or ternary weights and arithmetic can be used), which moreover expose large fine-grain parallelism. They are therefore ideally suited for hardware acceleration using Field Programmable Gate Arrays (FPGAs) and/or Application-Specific Integrated Circuit (ASIC) implementations, as acknowledged by the large body of work on this topic. Although accelerators have demonstrated significant performance/energy gains compared to GPU/CPU implementations, they still require further efficiency to address future performance requirements [10].

The goal of this chapter is to present an up-to-date view of the state-of-the-art solutions applying AxC techniques to CNNs for both inference and training phases. It is structured as follows: Sect. 15.2 presents the background & context of using DNNs, the main focus of the chapter. Section 15.3 overviews AxC methods found in the literature that improve deep neural network inference performance. Approximation techniques for improving the training part of neural network design, which accounts for the majority of computing time and resources, are presented in Sect. 15.4. Section 15.5 discusses DNN accelerator research and the dedicated approximation methods, whereas Sect. 15.6 presents incumbent directions for AxC research in DL. Section 15.7 concludes the chapter.

15.2 Background

Artificial intelligence (AI) is a broad field of study focused on replicating or simulating the intelligence of living beings (human or not). It encompasses various methods and techniques. These range from design space exploration methods like ant colony optimization that focuses on finding increasingly efficient paths through simple random exploration and reward-based reinforcement, to more complex approaches such as genetic algorithms that evolve a population towards a hopefully optimized solution by iteratively picking the best candidates and mutating them. In the last couple of decades, Machine Learning (ML) algorithms have gained the most traction, producing effective predictions/answers based on some trained behavior/model.

15.2.1 Context: From AI to DNNs

The ML subset of AI is focused on algorithms able to improve themselves through seeing already labeled input-output sample pairs and constructing models that attempt to match the expected outputs to this given data. An example is email filtering, deciding whether or not an email is spam based on its provenance, recipients, object, and other (meta)data. Generally, a model for this task is trained (i.e., it *learns*) on a set of already labeled set of spam email data (the *training data set*) until it reaches the desired behavior (the *expected response*) with sufficient accuracy. It is then used with unseen data in the hope that it will still prove to be accurate (i.e., generalize well). Evaluation of this generalization ability is frequently done on a so-called *test* or *validation* data set, different from the training data.

While email filtering can seem like a simple task, there are a plethora of use cases of varying difficulty where ML modeling is used, ranging from security (e.g., in fraud detection) and business data analysis (e.g., churn rate measurement) to computer vision, self-driving technologies, and other complex tasks. The model inputs can be both raw data or high-level features (for instance, statistical aggregates of multiple input data samples) or other complex features that are task-dependent (e.g., the presence of a horizontal line in an image). It is the task of the model to interpret this data and construct useful responses. Among the many tasks suitable for ML one can mention classification, regression, and semantic segmentation.

Traditionally, high-level features needed by a model were derived following a feature extraction step that was often performed by a human, requiring expert knowledge of relevant information. More recently, however, through the rise of DNNs in the ML ecosystem of approaches, this step can be performed automatically, the model is trained to discover relevant features, thus avoiding both the need for human expertise and the induced biases that might result from this.

Artificial neural networks are based on the notion that the computation performed by a neuron is centered around a weighted sum of its input values. This is shown in Fig. 15.1a, where multiple inputs $\{x_k\}_{k=1}^n$ are summed (scaled with *weights* $\{w_{ki}\}_{k=1}^n$) together with an optional *bias* term b_i. The neuron *output* y_i is determined by the application of a *nonlinear activation function* f to this weighted sum. There are many activation functions used in practice, but among the most common are $f(x) = \text{ReLU}(x) := \max\{0, x\}$ and $f(x) = \tanh(x) := (e^x - e^{-x})/(e^x + e^{-x})$.

Such neurons are grouped together to form *layers*. The present chapter is focused on feedforward networks, where the outputs of a layer are then used as inputs for subsequent layers.[1] This is exemplified in Fig. 15.1b. The inputs and outputs of a layer are also known as *input* and *output activations*, respectively. When discussing visual data, they are also known as input and output *feature maps*. The first and last layer in the network are generally known as the *input* and *output layer*, respectively. In between them, there is a number of intermediate layers, called *hidden layers*. The main characteristic of DL and DNNs is that the number of hidden layers can grow quite large, from two layers up to even one thousand.

[1] There are classes of *recurrent* neural networks that allow outputs of a layer to be connected to inputs of previous layers. While they are not discussed here any further, they are frequently used to process sequential data (e.g., speech, text).

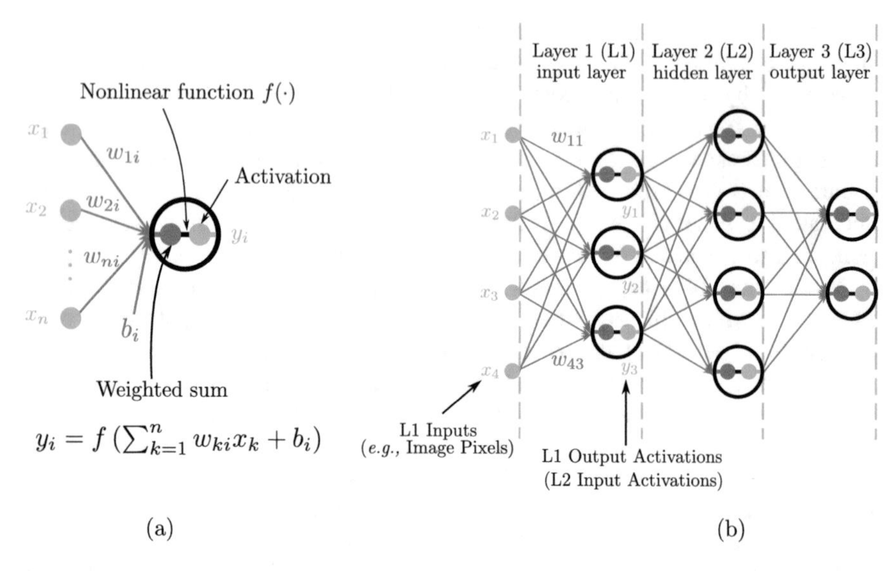

Fig. 15.1 A basic DNN example and the associated terminology (adapted from [11, Figure 1.3]). (**a**) Artificial neuron. (**b**) Simple neural network example

The process of using an artificial neural network with a set of given parameters (e.g., weights and bias terms) is called *inference*. For the neural network to be useful, its inference output has to match as closely as possible an expected/ideal output. This is measured through a *loss* function ℓ that compares how far the resulting output on (subsets of) the training and test data sets is to the expected output. Thus, the goal of *training* a neural network is to find/learn a set of parameters that minimizes the average loss over a large training set.

To train a network, its weights (w_{ij}) are usually updated using a form of Stochastic Gradient Descent (SGD) iterative optimization process. This means that weight is updated by a scaled version of the partial derivative of the loss function ℓ with respect to the weight. In the most basic form, at iteration t, the weight update formula is given by:

$$w_{ij}^{t} = w_{ij}^{t-1} - \alpha \frac{\partial \ell}{\partial w_{ij}^{t-1}}, \tag{15.1}$$

where α is called the learning rate.[2] The partial derivatives of ℓ can be computed efficiently through a process called *backpropagation* [12]. It is effectively an application of the *chain rule* from calculus, and it works by passing values backward through the network to compute how ℓ is affected by each weight. At each layer, the

[2] The deep learning optimization literature describes many ways how to perform the parameter updates and how to choose the learning rate.

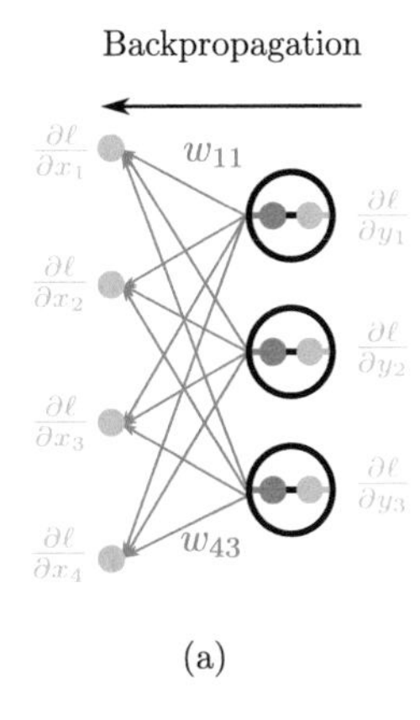

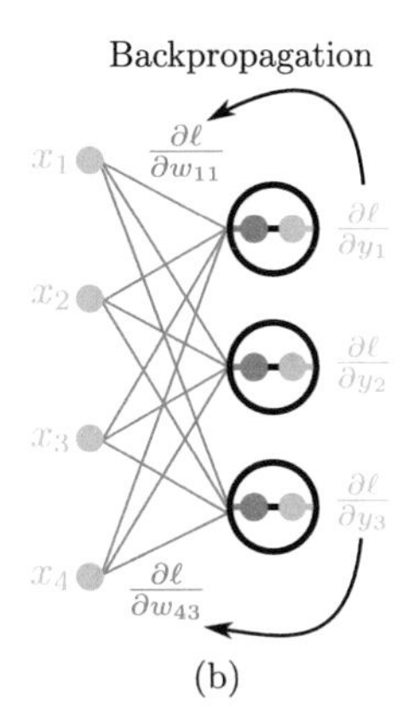

Fig. 15.2 A backpropagation example through a neural network (adapted from [11, Figure 1.6]). (**a**) Compute the gradient of the loss relative to the layer inputs ($\frac{\partial \ell}{\partial x_i} = \sum_j w_{ij} \frac{\partial \ell}{\partial y_j}$). (**b**) Compute the gradient of the loss relative to the weights ($\frac{\partial \ell}{\partial w_{ij}} = \frac{\partial \ell}{\partial y_j} x_i$)

procedure is twofold and is exemplified in Fig. 15.2. To backpropagate through a layer: (a) compute the gradient of the loss with respect to the weights, $\partial \ell / \partial w_{ij}$, from the layer inputs (i.e., the forward activations x_i) and the gradients of the loss relative to the layer outputs, $\partial \ell / \partial y_j$; and (b) compute the gradient of the loss relative to the layer inputs, $\partial \ell / \partial x_i$, from the layer weights, w_{ij}, and the gradients of the loss relative to the layer outputs, $\partial \ell / \partial y_j$.

Computing the gradients of the loss function ℓ over the entire dataset is generally much too complicated in practice, which is why the loss is usually taken only on a (small) subset, called a *mini-batch*, of the training data. The use of batches allows taking advantage of single instruction multiple data (SIMD)-like parallelism on modern GPUs while keeping the complexity of gradient computation manageable. A complete iteration of the training process is called an *epoch* and requires passing through all of the mini-batches, applying (15.1) for each one of the corresponding average losses ℓ. Training is carried out for several epochs until convergence to an appropriate solution is reached.

Both inference and training amount in most part to the same type of computations (i.e., matrix/vector additions and multiplications). There are important differences, however. For one, as the previous paragraph suggests, training is much more expensive, since apart from passing through the entire training data multiple times, it also requires that intermediate outputs and partial derivatives be stored when performing backpropagation. Secondly, due to the gradient update rule, the precision requirements for training are generally higher than for inference, thus also affecting performance. The effect is that the inference quantization techniques that will be discussed in this chapter are not usually directly applicable to training as well.

15.2.2 Deep Learning Landscape

While artificial neural networks have a long history dating as far back as the 1940s, practical applications using digital neurons did not arrive until the late 1980s, when the LeNet-5 [13, 14] network architecture was used for hand-written digit recognition. It is only in the early 2010s, however, with the synergy of three major factors, that artificial neural network models have started to take off, under the names *deep learning* and *deep neural networks*. These factors are: (1) the availability of large and labeled datasets that are needed to train complex models; (2) the advance in computational power of units such as GPUs that allow DNN training to be executed in reasonable time (days or weeks instead of years); (3) development of new algorithmic techniques (e.g., the Adam gradient descent optimization algorithm [15]) that enable improved accuracy at a larger scale.

The importance of large and comprehensive datasets cannot be overstated. If not careful, a small training dataset used in conjunction with a complex DNN can easily lead to *overfitting* (i.e., the model matches the training data extremely well but does not generalize to unseen data accurately). For computer vision, arguably the most popular dataset in recent years has been ImageNet [16], a collection of one million high-resolution images that are generally associated with the ILSVRC [17] image recognition contest that uses 1000 labeled categories. Smaller datasets such as MNIST [18] and CIFAR [19] have also been used extensively in DNN research for inference and training acceleration.

Apart from the data, the choice of model (network architecture and associated parameters) is also crucial in the success of a DL approach. In what follows (Sect. 15.2.3), our focus is on CNN models suited to process visual data.

The current surge of interest in DL is also facilitated by the availability of tools and frameworks that allow for the easy prototyping and design of DNN models. Prominent examples include Tensorflow [20] and Pytorch [21]. The open-source nature of these alternatives offers the possibility to design extensions that can be leveraged throughout a model's life-cycle (from initial prototype to deployment).

Depending on the intended use of DNN models, they can be found in different environments with various computing power and energy consumption characteristics. At one end of the spectrum there are edge devices characterized by low-power and limited computational capabilities, while at the other end power-hungry cloud devices with a high-performance computing profile are dominant.

15.2.3 Convolutional Neural Networks

The most basic layer in a feedforward network is the Fully Connected (FC) or dense layer. It is characterized by the fact that each neuron in the layer is connected to all the neurons in the previous layer. FC layers are parameterized by the number of neurons they contain. An example is shown in Fig. 15.1, which only has FC layers.

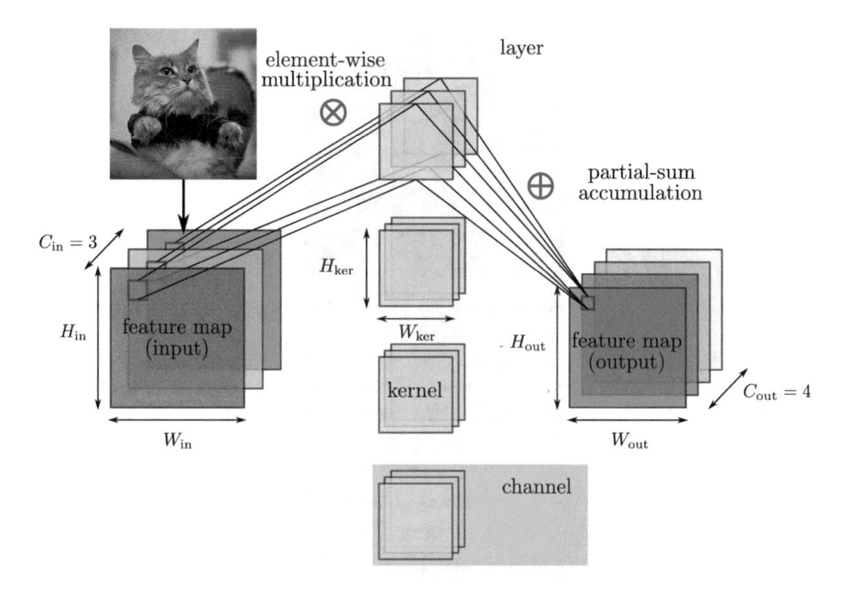

Fig. 15.3 Expanded view of a typical 2D CONV layer inside a CNN

While the expressive power of networks using only FC layers is impressive, it comes at the cost of a very large number of connections (and hence network parameters), making them hard to train and easily prone to overfitting. This is why other types of structured layers, with fewer parameters, but which are more efficient for certain tasks, have been explored. In the case of visual data, this has led to the development of CNNs, a staple of DL today.

The main elements that have led to the introduction of CNNs are Convolutional (CONV) layers, composed of high-dimensional convolutions that allow extraction of shift-invariant features from the input. An example is Fig. 15.3, showing a traditional 2D CONV layer. In this context, the input activation is structured as a 3D set of input feature maps, with input width (W_{in}), input height (H_{in}) and input channel (C_{in}) dimensions. The weights of the layer are structured as a 3D filter, with kernel width (W_{ker}), kernel height (H_{ker}) and input channel (C_{in}) dimensions. For each input channel, the corresponding input feature map is transformed through a 2D convolution with the appropriate kernel in the filter. The convolution results at each point are summed across all the input channels to generate the output partial sums. The results of these partial sums comprise one output feature map with output width (W_{out}) and output height (H_{out}) dimensions. Several 2D filters can be stacked together to generate additional output channels, denoted with C_{out} in this case.

Depending on the size of 2D kernels and their count, the output feature maps can be large and deep, motivating the use of pooling (i.e., subsampling) layers that reduce the scale of feature maps. Pooling is similar to convolution, with a kernel sliding over the input matrix, but instead of performing matrix multiplication, an

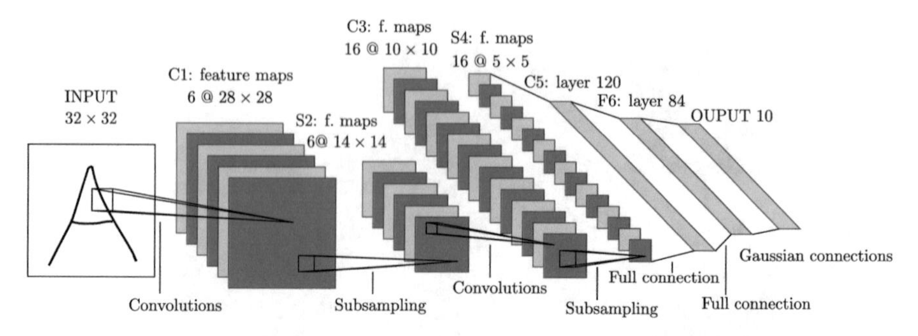

Fig. 15.4 A visual representation of LeNet-5 (adapted from [14, Fig. 2]), an early example of a CNN that promoted the subsequent development of Deep Learning. It contains the main layers that are usually found in CNNs: convolutional, pooling, and fully connected

aggregation operation function is applied. The most common such operations are taking the maximum element or the average. A visual example of a simple CNN mixing in all these layers is given in Fig. 15.4.

Another frequently used layer is Batch Normalization [22] (BN). It contains two trainable parameters that are used to re-center and re-scale the distribution of the values of a feature map, to improve training performance. While there are also more recent and complicated layers, such as depth-wise convolutions [23] or Inception modules [24], their specifics are not important for the rest of this chapter.

15.2.4 Performance and Energy Profiles of Recent Models

To gauge the complexity of current DNN models, there is a need for a set of metrics that allow for a fair comparison between models. In this study, the metrics used are (1) the model accuracy over a validation dataset, (2) the total number of weights in the model, and (3) the number of FLOating-Point operations (FLOPs) necessary to carry out one complete inference. Accuracy is measured in terms of the frequently used top-1 and top-5 percentages (i.e., the proportion of correct predictions on the labeled validation dataset and the probability that the correct result is among the top five predictions). The number of weights allows estimating the total memory storage requirements for the model, whereas the FLOP count hints at the required computing power needed to execute the model at a certain frequency.

Table 15.1 shows a comparison using these metrics on some popular DNNs for image classification on the ImageNet dataset (adapted from [29]). For a long time, the only metric of interest was the network accuracy, resulting in models that were costly to train and operate. The cost of training and inference became so large at

Table 15.1 Recent evolution of DNNs for image classification on the ImageNet dataset

Model name	AlexNet [25]	GoogLeNet [24]	ResNet-50 [26]	MobileNet V2 [27]	EfficientNet B1 [28]
Year	2012	2014	2016	2018	2019
Top-1 accuracy	57.2%	69.8%	76.2%	72.0%	79.1%
Top-5 accuracy	84.7%	93.3%	92.97%	90.6%	94.4%
Number of weights	62M	6.4M	26M	3.5M	7.8M
FLOPs	1.5B	2B	4.1B	0.3B	0.7B

Table 15.2 Estimated cost of training recent NLP models in terms of power, time, and CO_2 emissions.

Model	Hardware	Power (W)	Hours	CO_2e (lbs)
$\text{Transformer}_{\text{base}}$ [31]	P100×8	1415.78	12	26
$\text{Transformer}_{\text{big}}$ [31]	P100×8	1515.43	84	192
ELMo [32]	P100×3	517.66	336	262
$\text{BERT}_{\text{base}}$ [33]	V100×64	12041.51	79	1438
$\text{BERT}_{\text{base}}$ [33]	TPUv2×64	–	96	–
NAS [34]	P100×8	1515.43	274.12	626.155
NAS [34]	TPUv2×1	–	32.623	–
GTP-2 [35]	TPUv2×32	–	168	–

one point that there is now an open engineering consortium called MLCommons[3] that benchmarks DL models and fosters innovation in the field. Thus, there is an increasing interest for faster, lighter, and overall more efficient models that are compatible with edge device resource constraints and operate more efficiently in the cloud. The last two columns in Table 15.1 reflect this, with newer network models achieving competitive accuracy with less memory and a smaller FLOP count.

Some examples of the scale at which modern DNN training costs stand for recent NLP models are given in Table 15.2 (adapted from [30, Table 3]) and showcase the significant resources needed for training the state-of-the-art models.

The need for efficient DL computations coupled with the resilience of DNNs to approximation (due to the stochastic nature of training methods and a high level of inner redundancy [36]) has paved the way for the development of a large number of approximation methods, a part of which are described in the rest of this chapter.

15.3 Approximation for Inference

DNN inference is a very computation-intensive task, having large memory and computation power requirements. For example, inference on a single image using the original ResNet34 [37] model requires 3.6 billion FLOPs and storing 22

[3] https://mlcommons.org/en/.

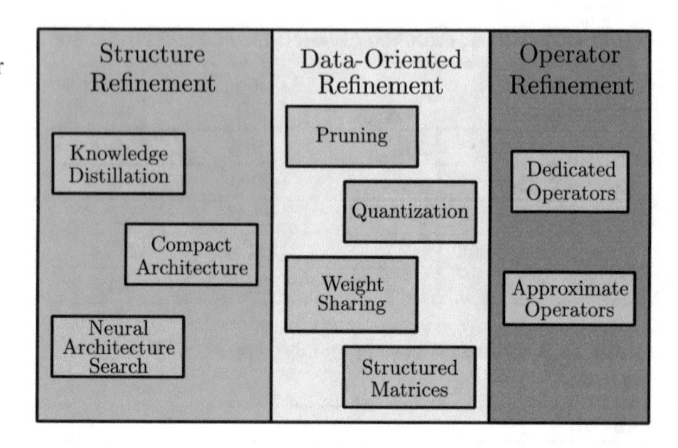

Fig. 15.5 Different types of approximation techniques for DNN inference

million weights plus temporary feature maps. While the execution of such tasks has moved from traditional CPUs having a latency-oriented design to more parallel hardware like GPUs or even custom ASICs/FPGAs, inference is still a costly task, and thus susceptible to benefit from performance improvements when using approximate computing. Consequently, this section describes AxC methods found in the literature that improve deep neural network inference performance.

One can distinguish three different classes of methods (see Fig. 15.5), usable in isolation or combined, to approximate DNN inference. The first one, *structure refinement transformations*, includes the methods that modify the computational structure (i.e., the network layers and their parameters) of the input model. Some notable examples include knowledge distillation [38, 39] which uses the model as a teacher to help train smaller students models or compact architectures [23, 40] where layers are transformed into more hardware friendly ones. The second class, *data-oriented refinement transformations*, focuses on optimizing the finite precision data representation(s) of the model while maintaining the initial computational structure intact. Notable examples are *pruning* [41, 42] (i.e., setting less important parameters to zero to increase sparsity) and *quantization* [43, 44] (i.e., changing the types of the parameters and intermediate results to more efficient representations). While network structure refinement substantially changes the network structure, data type refinement does not, giving the possibility to emulate the inference of the approximated network in the original structure to measure its accuracy loss.

The third class of approaches relies on *operator refinement transformations* which modify the arithmetic operators used inside the CNN implementation (e.g., addition and multiplication) to further improve energy efficiency. Such methods are not discussed any further here since they mainly depend on the hardware implementation of the CNN. More details on the approximate operators that fall in this class can be found in Chaps. 3 and 4.

In the rest of this section, data-oriented refinement methods are covered. Specifically, Sect. 15.3.1 gives an overview of various quantization methods, Sect. 15.3.2 discusses weight sharing approaches, whereas pruning is analyzed in Sect. 15.3.3.

15.3.1 Quantization

Full precision DNNs usually rely on 32-bit floating-point values for representing parameters. For standard backpropagation-based training, using high precision weights makes sense since the gradient update rule generally modifies these weights by a small factor of the corresponding gradient terms. While full precision `float32` DNNs offer excellent result quality, they can generally be compressed and accelerated using lower precision arithmetic with minimal or no loss in the accuracy. Methods for addressing data quantization in DNNs are varied, ranging from simple binary and ternary networks to larger fixed-point and custom floating-point formats. This section gives an overview of the main ones.

Analysis of existing approaches relies on various aspects, such as (1) what parts of the network are being quantized, (2) homogeneity/heterogeneity of the number formats used inside the layers, (3) the type of representations being used, and (4) how and when is quantization performed (during or after the network has been trained).

What to Quantize The most obvious quantization targets are the *network parameters* (e.g., weights and biases). Reducing the number of bits used to represent them primarily brings a memory footprint reduction for on-device storage of the network. Latency improvements are potentially achievable with binary, ternary, and bit-shift (i.e., power of two values) quantized parameters [45–47]. More generally, if faster execution times are to be obtained, *activation function inputs and outputs* also need to be quantized. An example is [48], which proposes an efficient 8-bit integer quantization scheme for both weights and activations. Additionally, one can quantize the weight and activation *gradients* used during backpropagation (see, for instance, [43, 49]) to accelerate training, an aspect discussed in Sect. 15.4.

When and How to Perform Quantization There are two established ways quantization can be performed for efficient inference and a third, emerging method.

The first among the established approaches is *Quantization-Aware Training* (QAT). The idea is to use a network parameter update procedure for several epochs (starting from scratch or after a baseline `float32` training method is run) to adjust parameters in the quantization format(s) such that generalization accuracy is hopefully kept the same or is at worst minimally degraded. Much research has focused on such fine-tuning methods (see, for instance, [43, 44, 47, 48, 50, 51]), mainly because they achieve good results, especially for extremely low-precision formats (i.e., binary and ternary encodings).

While training is a powerful approach to compensate for a model's accuracy drop due to quantization, it is not always applicable in real-world scenarios (e.g., for online learning) since it is costly, time-consuming and generally requires a full-size training dataset. This can be a problem when the data is proprietary, privacy, and regulatory issues are in effect (e.g., medical data that cannot be uploaded to the cloud for remote processing), or when using pre-trained off-the-shelf models for which data is no longer available. As such, there has been a push for faster *Tost-*

Training Quantization (PTQ) methods without any fine-tuning. It has been observed that for down to 8-bit word lengths, PTQ results are close to full precision ones for several models [52] (e.g., AlexNet, VGG, and ResNet), but it becomes significantly more difficult to maintain accuracy when targeting lower precision formats. Work focused on PQT includes [52–56].

A possible issue with QAT and PQT methods is that both generate networks that are *sensitive* to how quantization is carried out (e.g., the target word length). As such, there has been recent work [57, 58] on methods for *robust quantization* that provide intrinsic tolerance of the model to a large family of quantization formats and policies by directly specifying it in the training loss function. Such approaches are interesting for battery-powered edge devices, where depending on the state of charge, a network model capable of operating effectively at various quantization levels would be highly beneficial.

Granularity of Applying a Quantization Format Initially, quantization approaches were homogeneous, with one word length being used for the entire network. This is the case for early works on binary [59] and ternary [46] weight networks, for instance. Such approaches can suffer from significant accuracy loss since different layers tend to have different sensitivities to quantization levels/noise. Subsequent work has focused more on a heterogeneous, layer-wise optimization of the quantization format [53, 60–64].

There have been various metrics proposed to estimate the overall effect of a fixed-point quantization format inside a layer on the overall accuracy of the network. One example is [65], which uses a Signal to Quantization Noise Ratio (SQNR) to empirically measure how suitable a fixed-point format is. The approach in [60] generalizes the work from [65] using an adversarial noise to formulate the quantization error. Another adaptive quantization method is [66], which uses the loss function gradient to determine an error margin for each parameter such as to not degrade accuracy and assign a precision accordingly. Recent work [63, 64, 67] also proposes using second-order information (Hessian-based) to gauge the sensitivity of each layer. From an Information Theory perspective, [68] uses the entropy of weights and activations as a saliency indicator to set fixed-point quantization levels at each layer. Another popular statistical sensitivity measure is based on the Kullback-Leibler divergence, which is used to measure layer sensitivity in [53, 62] and is a core component for fine-tuning low-precision integer weights in NVIDIA's TensorRT inference acceleration library.

On a different granularity level, [69] proposes looking at the distribution of weight values over the entire network to aggressively quantize weights in dense regions and more gently those in sparse ones. Compared to `float32` baselines, such an approach can achieve under 1% accuracy loss for large networks (ResNet-152 & DenseNet-101) with a 4-bit format in the dense areas and a 16 bit one for the sparse regions ($< 1\%$ of parameters).

Quantization Formats There have been various representations used to quantize deep neural networks. At the extreme, there are *Binary Neural Networks* (BNNs), where weights and activations are stored with one of two possible values. If a $\{0, 1\}$ (or equivalently a $\{-1, +1\}$) encoding is used, then multiplications can be implemented efficiently using XNOR gates, making BNNs compelling on FPGA and ASIC targets, but also for emerging computing paradigms such as neuromorphic [70] or in-memory computing [71].

Among the first investigations of binary networks is BinaryConnect [59], which maintains a full precision copy of the weights to be updated during backpropagation, but are binarized for inference. Activations are kept in full precision, meaning full precision accumulations are still required during the forward propagation. The effect of binary activations is considered in [45, 47, 72]. These early papers are the basis for most subsequent research on BNNs.

The XNOR-Net approach [47] expands on the initial BNN ideas by proposing a model where a gain term is added to the network at the level of each dot product in the convolutional layers. Computed from statistics of weights and activations before binarization, the gain was a way to improve the accuracy of BNNs on the ImageNet dataset. Such gain terms are nevertheless costly to compute in practice, which is why later work modified their use. For instance, [43] proposes gain terms that are only based on the non-binarized weights of the network, meaning that they never need to be recomputed after training. Additionally, [73] also advocates binarizing fully connected layers by adding neuron-specific scaling factors, further improving compression without a drastic decrease in the accuracy. A generalization of the BNN concept to multiple binary bases used for quantizing weights and activations is presented in [74], further reducing the accuracy gap between full precision and binary architectures, at the expense of a higher computational cost (compared to previous BNN methods). Changes to the backpropagation process in BNN training [75] can also be effective for limiting accuracy loss.

Ternary neural networks offer a better representation of the (pseudo) normal distribution of weights that is frequently observed after training. For instance, [76] achieved good results on small networks with weights quantized to $\{-1, 0, +1\}$ and 3-bit fixed-point activations. For greater flexibility, [46] proposes using a threshold α for picking the ternary weights (-1 if $w < -\alpha$, 0 if $|w| < \alpha$ and $+1$ if $w \geqslant \alpha$), while keeping activations in full precision. This is further expanded in [77], which uses ternary weights from a set $\{-\alpha^n, 0, +\alpha^p\}$, where α^n and α^p are learnable parameters. By also quantizing activations to 8-bits and adding residual edges to branches in the architecture that are sensitive to quantization, [78] offers comparable accuracy results to `float32` for a ResNet-101 model on the ImageNet dataset, with no additional low-precision (re)training. In a more aggressive compression strategy, [79] proposes the use of ternary activations $\{-1, 0, +1\}$ and binary scalable weights $\{-\alpha, +\alpha\}$.

Extremely low-bit-width networks like the ones just presented are susceptible to non-negligible accuracy loss, which is why there has been work focusing on non-binary *integer* and *fixed-point*-based quantization. Among the early proponents of integer quantization, there is [43], which extends the idea of BNNs to arbitrary word lengths for weights, activations, and gradients. For fixed-point arithmetic, [65] explored the use of various bit-width combinations (4, 8, and 16 bits) of weights and activations. Notable results with integer arithmetic are presented in [48], which showcases how 8-bit integer quantization on ARM CPUs can achieve near-identical accuracy compared to baseline `float32` models based on MobileNet architectures for classification and detection tasks, but with improved on-device latency. Good quantization results with 4-bit weights and activations are presented in [52] by combining three complementary methods for minimizing quantization error at the tensor level. Heterogeneous/mixed-precision quantization approaches also heavily focus on integer/fixed-point formats [53, 61–64].

One problem with low-precision integer/fixed-point formats is that they have limited dynamic range, which might make them inappropriate, especially for networks used in Natural Language Processing (NLP) tasks, where weights tend to have values that are more than $10\times$ larger than the largest magnitude values found in popular CNNs [80, Fig. 1]. While not that widespread, there has been some work looking into low-precision *floating-point* quantization for CNN inference. For instance, [81] explores the use of up to 8-bit (scaled) floating-point formats for weight and activation quantization in classification networks such as GoogLeNet, ResNet, and MobileNet, without any accuracy degradation. More recently, [82, 83] show how an 8-bit floating-point quantization format (4-bit mantissa and 3-bit exponent) can be used in FPGA-based accelerators for deep CNN inference, without any retraining. Another approach [80] consists of an *adaptive* floating-point quantization method, where the exponent range of quantized values is dynamically shifted at each network layer (through changing the bias term of the exponent), yielding competitive results on NLP networks and tasks.

At a coarser level, it is also possible to improve dynamic range by sharing the exponent between parameters, storing only the mantissa and one copy of the exponent. This is the so-called *Block Floating-Point* (BFP) format. For instance, [84] propose using BFP with an 8-bit mantissa for weight storage, showing negligible to no accuracy loss on CNN workloads (VGG16, ResNet-18, ResNet-50, and GoogLeNet-based networks). On the FPGA side of things, [85] showcases a BFP-based CNN accelerator design that uses 16-bit activations and 8-bit weights, reducing memory requirements compared to a `float32` baseline without any retraining/fine-tuning. Another way to increase the dynamic range is to employ a *logarithmic representation*, which also allows multiplications to be replaced with simple binary shift operations. For instance, [86] shows that a log representation can achieve higher classification accuracy than fixed-point formats operating at the same word length. 8-bit log floating-point quantization was also shown [87] to perform close to baseline `float32` values with several CNN classification networks.

A summary of these aforementioned formats (minus the binary and ternary encoding that generally require just 1 or 2 bits to represent) is given in Fig. 15.6.

Looking at the *value distribution* of the data (weights and activations) is a good way to explore what number formats and/or encodings are better suited for a particular network model. Uniform precision was the go-to alternative for a long time, but more recent work is concentrated around non-uniform quantization. This is because the actual distributions of trained weights tend to follow bell-shaped

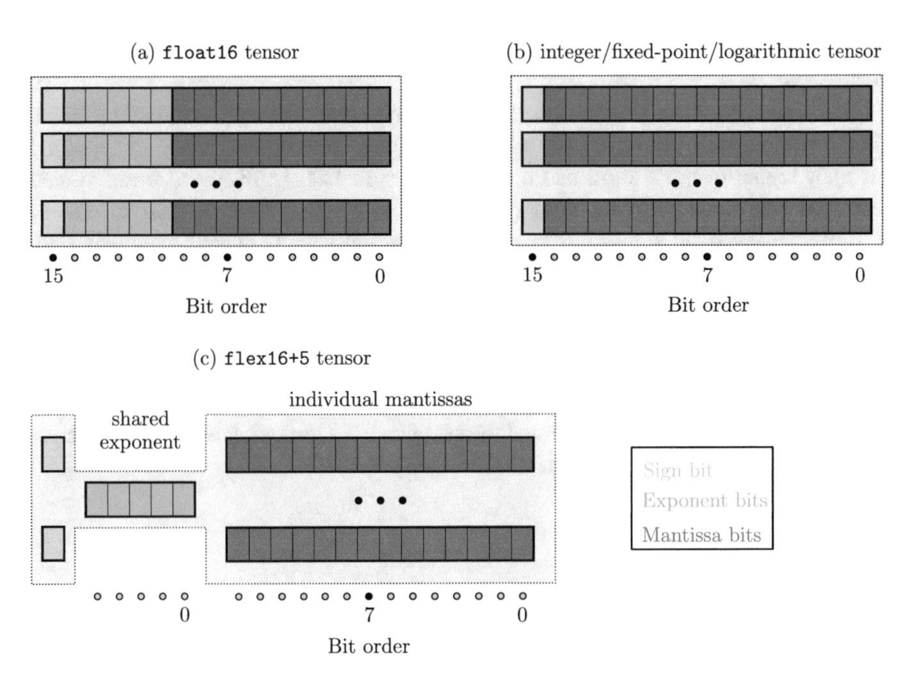

Fig. 15.6 Diagrams for bit representations of various numerical formats discussed in the context of DL quantization in this chapter. Red, green, and blue shading are used to represent mantissa (M), exponent (E), and sign (S) bits, respectively. In (a), the 16-bit IEEE 754 `float16` *floating-point* format is shown (corresponding to $(-1)^S \times 2^{E-15} \times 1.M_2$ for normalized values), with 1 sign bit, 5 exponent bits, and 10 mantissa bits. (b) illustrates a 16-bit *signed integer* format. By choosing a *fixed* splitting point for integer (I) and fractional (F) parts in the mantissa ($M := I.F$), it can also serve as a representation for a *fixed-point* format (namely to $(-1)^S \times I_2.F_2$). Additionally, (b) can represent a form of *logarithmic number system* (see, for instance, [88]), with the encoded value being $(-1)^S \times 2^M = (-1)^S \times 2^{I.F}$. Part (c) exemplifies a *block floating-point* format, namely the `flex16+5` format [89] with a 15-bit mantissa and 5-bit shared exponent

curves. In this direction, [90] focuses on balancing the quantization values based on the distribution of the data. The quantizer can also be trained alongside the model [51, 91] and it is also possible to use reinforcement learning [62] and meta learning [92] approaches to determine good choices for the quantizer.

Choosing Quantized Values There are various methods for quantizing data, ranging from simple heuristics like those used to convert network weights into binary values depending on their sign [59] or projecting real-valued parameters to (one of) the closest discrete points [48], to loss functions that regularize the network and force parameters into quantized states upon the convergence of the training algorithm [93].

One notable approach is [44], which incrementally quantizes network weights to power of two terms. The set of non-quantized weights is progressively shrunk during retraining, with their values being updated to counter any accuracy loss induced by quantization. Knowledge distillation can also be a valid way to pick quantization values [94, 95].

It is also possible to cast this task as a mathematical optimization problem. For instance, [65] converts pre-trained weights to fixed-point values by looking at their signal-to-noise ratio as an optimization metric. In [82], the mean square error of the quantized data with respect to the original data is used to choose the precise 8-bit floating-point quantization format (mantissa and exponent size) and corresponding values. In more involved approaches, the Alternating Direction Method of Multipliers (ADMM) can be used to optimize the quantized values with low-precision formats [96, 97]. Regularization terms and parameters that emphasize quantized solutions are also available. The work of [93] looks at using mean squared quantization error regularization to drive weights to quantized values and how ℓ_2 regularization can lead to sparse weight designs. Regularization also is an effective approach for doing robust quantization [57, 58].

15.3.2 Weight Sharing

Weight sharing compresses the network by assigning shared values to parameters. This transforms plain weight data storage into a reduced number of shared values in a dedicated memory, together with the indices of these values in the weight matrix.

Figure 15.7 shows an example. The first matrix corresponds to a 5×5 convolutional kernel (filter) with values computed during training. The matrix contains $N = 25$ values ranging from 0 to 20. Each value can be represented using $B = 5$ bits, resulting in a total size of $N \cdot B = 25 \cdot 5 = 125$ bits. There are 5 shared values, namely "a," "b," "c," "d," and "e," replacing the 25 original values, as shown in the second matrix.

Accordingly, the size of an element in the weight matrix can be reduced from B to $\log_2(K)$ bits, with K being the number of different shared values. The size of the stored data then becomes $N \cdot \log_2(K) + K \cdot B$, instead of $N \cdot B$.

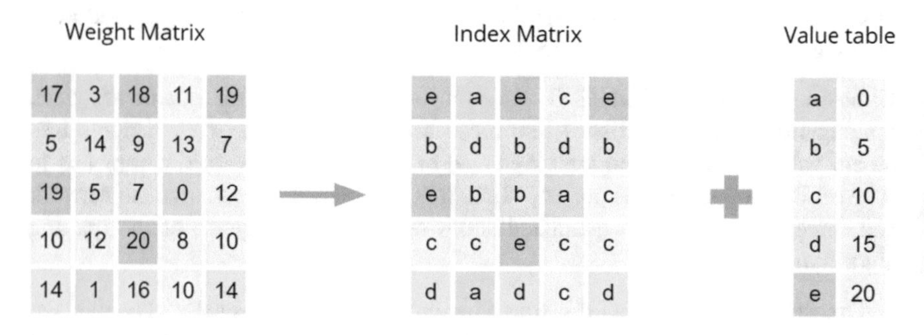

Fig. 15.7 Weight sharing techniques allow network compression by storing indices instead of values

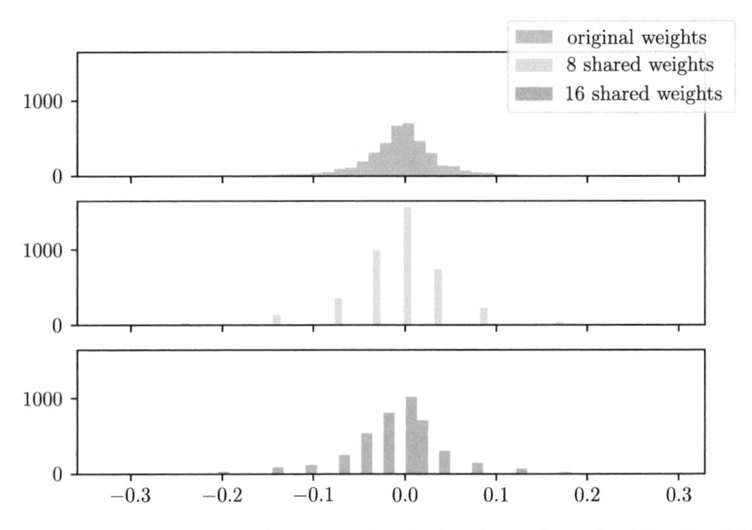

Fig. 15.8 Distribution of the weights composing the first layer of a trained ResNet50V2 [26], original (top), with only 8 (middle) and 16 (bottom) shared values

Depending on the number of shared values used, the distribution of the weights inside a layer will change. An example of this before and after weight sharing (with 8 and 16 shared values) can be found in Fig. 15.8.

Weight sharing approaches can be classified by the method used to group weights together and by the granularity level it is applied at. Each of these aspects will be explained in some detail in the following paragraphs.

Grouping Methods One of the first approaches involving weight sharing that showed it can be a viable option for compressing neural networks is HashedNets [98]. The weights of the network in this setting are randomly grouped into hash buckets sharing the same value. These shared values are then trained and updated using backpropagation. The authors test their approach on the MNIST dataset with two custom fully connected networks with 3 and 5 layers.

However, instead of applying random grouping before the network even sees any data, it is also possible to approximate an already-trained network by determining groups based on weight values. In this vein, DeepCompression [99] uses the K-means algorithm to iteratively group the weights in a network in a global 3-step compression approach involving network pruning, weight sharing, and parameter encoding. The K-means algorithm is used to cluster similar values together, followed by an iterative retraining phase. Different initialization options for the shared values are considered, with experiments showing that uniform initialization over the entire range of weight values works best. Applied to the AlexNet and VGG architectures on the ImageNet dataset, the compression algorithm achieves 35$\times$ and 49$\times$ compression, respectively, with negligible accuracy loss.

The most common way of doing K-means clustering is through the Lloyd algorithm [100], which uses mean square error minimization to solve the clustering. However, this clustering approach does not imply performance loss minimization when taking into account quantization as well. The use of mean square error minimization does not necessarily lead to high accuracy during inference, even when uniform initialization of the clusters is used, as suggested with DeepCompression [99]. Because of this, [101] proposes to use Hessian-weighted K-means clustering to minimize accuracy loss. The approach consists of replacing the mean square error with the distortion of the Hessian matrix (second-order derivative) of the loss function. With this change, it can achieve a higher compression rate than DeepCompression, but with similar accuracy loss.

It is also possible to consider weight distribution when performing clustering. For instance, [102] proposes a clustering method based on weight entropy, using importance (magnitude) and frequency of the weights to group them. Thus, frequent non-zero (low importance) values are grouped, as well as rarer, but higher magnitude (high importance) values.

During the iterative process of training weights, clustering them, and training them again, previously clustered weights will sometimes diverge from the shared values at retraining time, making convergence to a good network model difficult. This is why, rather than applying iterative clustering and retraining, [103] proposes the Deep-K-means approach that adds a regularization term in the training objective function, enforcing weights to stay clustered during training. After training is finished, the K-means algorithm is used to group the obtained weight values.

Other clustering algorithms can also be used. One main issue with using the K-means algorithm in this context is that it targets multi-dimensional data, whereas weights clustering is a 1-D problem. One example of approach using another clustering algorithm is DP-Net [104], which is based on a dynamic programming clustering algorithm that enables weight sharing in constant time, reducing the clustering complexity compared to the K-means algorithm.

Weight Sharing Granularity The weights of a network can be shared at different levels of granularity, as shown in Fig. 15.9. While this can be done for the entire network, as initially proposed in HashedNets [98], each layer has a different weight distribution, covering a different range. Hence, sharing values for the whole network usually does not offer good enough representation power to limit accuracy loss.

On the other hand, sharing the values at the layer level offers a better representation of the original network, as shown with Deep Compression [99]. Such an approach also allows different levels of compression to be used for each layer. The first and last layers are generally more sensitive to compression and require a higher number of shared values to keep accuracy loss acceptable. It is even possible to target a smaller scope, like sharing values at the level of a (convolutional) kernel—but then of course the compression rate will be much lower.

While reducing the scope allows a better representation of the initial weight distribution, thus keeping accuracy loss low, it is possible to improve compression performance. For example, Deep-K-means [103] shares values at a level that is

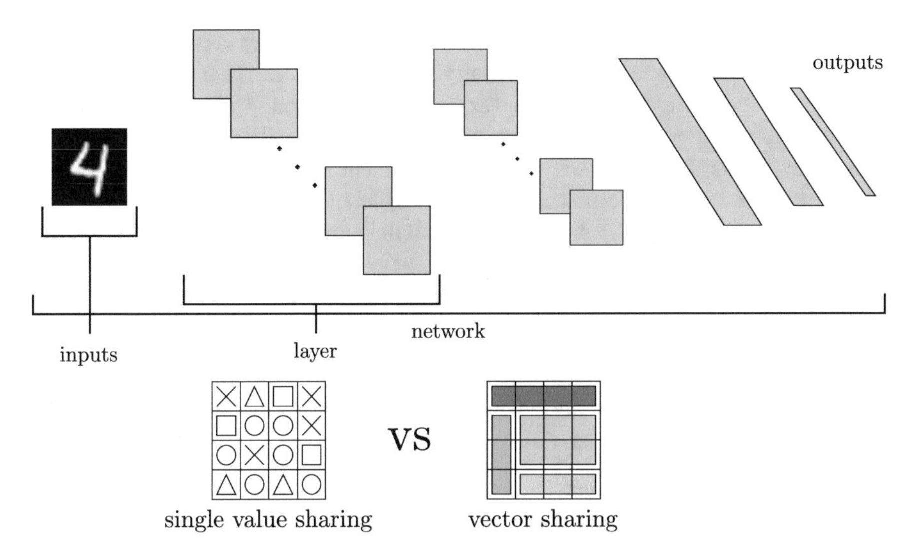

Fig. 15.9 The various scopes of applying weight sharing

optimal for the very efficient row-stationary dataflow used in DNN hardware accelerators.

Even if weight sharing leads to good compression rates, it does not enable inference acceleration by itself. This can be achieved if inputs are also discretized, reducing the number of combination operations and allowing the use of a pre-computed look-up table multiplier. This approach is used in LookNN [105], which applies K-means to the input feature map to achieve a nonlinear quantization whereas the remaining feature maps are quantized in the traditional linear way.

Values can also be shared at a smaller level, as in Q-CNN [106]. Here, layers are decomposed into sub-vectors, which are then clustered using the K-means algorithm. Sharing vectors like this reduces the number of possibilities when performing products. This enables the layer response to be approximated using product pre-computation with a look-up table.

15.3.3 Network Sparsification (Pruning)

DNNs tend to be more complex as their accuracy rate improves and this complexity usually carries with it the fact that the network is over-parameterized. On the other hand, it has been argued for a long time [41] that structure is more important than density in neural networks, with sparse models having the ability to generalize up to as well as their dense counterparts. Removing model parameters has the direct effect of reducing the size of the model, but it can also be used for speeding up the inference process by reducing the number of computations. Depending on the objective, different parts of the network can be more interesting to prune than others. For instance, fully connected layers usually concentrate most of the network weights in a CNN and should be targeted for high compression. Convolutional layers, however, contain fewer model parameters but account for most of the computations. Since they generate the majority of data movement in the model, they should be targeted when model performance and energy efficiency are important.

Pruning methods can be classified by how they are applied to the network, the granularity of the pruning, and finally the saliency determination approach. All these criteria are discussed in the following paragraphs.

Target Regions The loss in accuracy incurred by removing parameters can be recovered by retraining the remaining parameters using the initial training dataset if it is still available. This pruning process can be performed at different steps of the network life-cycle, either prior, during, or after training the model.

It has been shown that some parts of DNNs are more resilient to approximation than others. As such, pruning each layer at the same rate is not very efficient for accuracy. But at the same time, choosing the optimal sparsity level for the whole network is a complicated task. For example, [107] proposes to heuristically optimize the pruning ratio of each layer using reinforcement learning.

Similar to pruning weights, feature maps can also be pruned during the forward pass of the network. This process is called *dynamic sparsity* and is used in many accelerators to avoid zero or near-zero computations [108, 109]. Such approaches require dedicated architectures, but since the focus is only on data type refinement methods for this survey, they will not be discussed further.

Pruning Granularity Depending on the pruning objective (compression or performance), one can choose to focus on weight removal at various sparsity levels (Fig. 15.10). For instance, even though removing an entire structure (e.g., a convolution kernel) allows reducing the computational complexity of the model, and thus, improving performance, it also has the effect of inducing a higher accuracy loss.

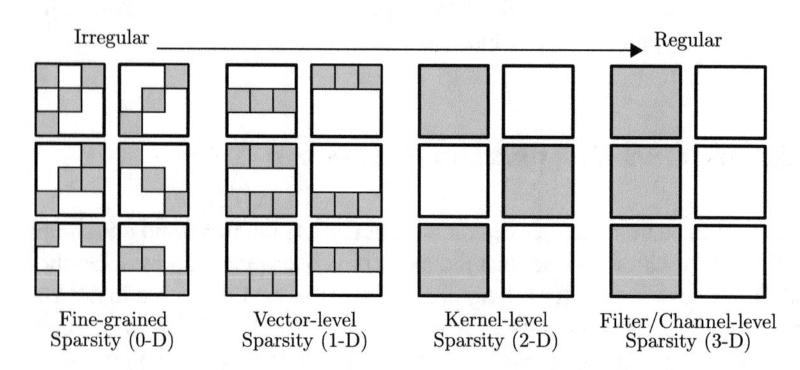

Fig. 15.10 Different granularities of pruning in a 4-dimensional weight tensor for DNN inference (adapted from [110, Figure 1])

The lowest pruning level is at the weight level, the goal being that of removing the individual parameters with the lowest saliency [41, 99]. Although this generally results in the lowest accuracy loss, it does not systematically offer latency or energy improvements because sparse tensor computations are quite difficult to accelerate. Its main purpose is therefore to compress the network in memory.

To accelerate computations, a regular sparsity pattern is usually required. This is called *structured pruning* and aims at removing (spatially close) groups of weights so that network inference can be simplified. To achieve this, [111] iteratively reorders pruned weights to prune larger structures, whereas [112] uses different pruning strategies depending on the hardware, optimizing for the full utilization of available SIMD units.

As previously hinted, it is also possible to remove convolution kernels, thus simplifying the processing of pruned convolutional kernels. An example is [113], which progressively removes convolutional kernels through greedy-based fine-tuning. The method is applied to transfer learning applications, resulting in a $2\times$ speedup on ImageNet-class CNNs.

Another interesting structure amenable for removal is a channel. Once channels are removed, one can remove the corresponding filters that take these channels as input. The filters producing these channels in the previous layer can also be removed [114]. A representative approach is [115] which removes channels based on importance, resulting in a $2{-}5\times$ speedup on multiple ImageNet-class CNNs with under 1% accuracy loss. In subsequent work, [107] proposes to pick the pruning ratio of each layer using reinforcement learning.

Weight Saliency Determination Removing part(s) of the network usually requires knowing which regions are least important for ensuring network accuracy. This is called *saliency determination* and it can be conducted using different methods, as described next. A simple way is to use heuristics like weight magnitude or examining the ℓ_1/ℓ_2 norm of a group of weights, whereas more recent work employs optimization algorithms to address the trade-offs between accuracy loss and compression/acceleration.

The earliest methods removed small magnitude weights because they tend to have the least impact on accuracy [41, 116]. They work iteratively by fine-tuning unpruned weights to recover lost accuracy [99]. It has been shown recently that one can also remove redundant connections in FC layers since for weights having the same value, only one needs to be kept [117]. If accuracy is degraded too much during the pruning process, some methods can be used post-pruning to restore certain weights and improve accuracy [118, 119]. For convolutional layer filter removal, it is possible to rank filters based on their ℓ_1 norm and prune the lowest ranking filters of each layer [120]. Instead of ranking filters at the layer level, one can also do it at a global, network-wide level by first doing a layer-wise filter ordering using ℓ_2 norms and then computing affine mappings that enable inter-layer filter rankings [121, 122]. Such global approaches lead to a Pareto set of approximated networks that offer various trade-offs between performance and accuracy.

In [113], the authors also consider a Taylor expansion criterion that approximates accuracy degradation due to feature map removal. This is done using activation and gradient values already computed during a regular training iteration. Other approaches use weight gradients to compute saliency. For instance, [123] proposes a sequential two-step process where (1) gradient-based information is used to grow the network (adding "dormant" connections and neurons that are deemed important for accuracy) and (2) regular magnitude-based pruning of weights and connections.

Another method to identify representative structures inside a network is [115], which uses a two-step process involving Least Absolute Shrinkage and Selection Operator (LASSO) regression for channel selection and then a least squares-based reconstruction approach of subsequent feature maps in the network.

It is also possible to state the problem of selecting which parts of the network to remove as an optimization problem. One example is [114], which relies on the correlation between feature maps of the current layer and the next one to determine the importance of filters. In another approach [124], the optimization problem features the model's energy efficiency as an objective. It is based on an energy estimation methodology capable of approximating both the power of MAC operations and data access (which is more complicated to compute, depending on the data reuse technique). The resulting iterative process involves local fine-tuning to recover accuracy loss in a layer before moving on to subsequent layers.

By formulating weight pruning as a non-convex optimization problem, it is possible to address it using an ADMM approach [125]. Using the desired sparsity level as a constraint to be satisfied and the loss of the network as the objective to minimize, ADMM can be used in a two-step process. Since convergence can be quite slow, the target error is increased to accelerate convergence and the resulting accuracy loss is compensated by network retraining. The method can also be extended to address high sparsity target problems, by introducing a more progressive algorithm using partial weight pruning with a moderate pruning rate [126].

Another idea is to encourage weights to group around zero using regularization. The closer weights are to zero, the less accuracy loss will be induced by removing them. For example, [127, 128] used group LASSO [129] regularization to obtain structured sparsity, with the same factor being applied to all the weight groups. In [130], ℓ_1 regularization is applied to the scaling factor of batch normalization layers to identify important channels. Different regularization factors can be assigned to different groups, such as in [131], where ℓ_2 regularization is used to transfer the model's representational capacity to a fraction of its filters. An incremental approach for choosing these factors can also be used [132]. In [133], feature map channels are gradually zeroed during training using a dynamic regularization factor (whose value depends on the current compression ratio in the network), allowing safe removal of corresponding filters without a significant drop in the accuracy.

Another recent approach to optimize pruning is through architecture search. Usually, pruning methods target a fully trained network and recover any accuracy loss using fine-tuning because it is hard to train a sparse network. Recently, however, the idea that a classic network contains sub-networks that, trained from scratch, can perform as well as the original network but with fewer parameters and computation, was introduced [42]. This idea was also explored in [134], which claims that directly training (using some form of random initialization) a model found at the end of a classic three-step pruning process (training, pruning, and fine-tuning) can perform as well, if not better, in fewer training steps. The issue is that, in the beginning, none of these studies provided a method for finding an efficient smaller architecture without doing full model training beforehand. This is starting to change, with [135] proposing to use a bee colony exploration algorithm to find an appropriate DNN pruning scheme. It is also possible to reduce the fine-tuning cost by using an external network trained to predict weights of a certain network structure, facilitating a fast exploration of various possible architectures [136].

15.4 Approximation for Training

The state-of-the-art models used in deep learning applications require a considerable hardware infrastructure to be designed properly. There are various challenges related to computing, storage, network/communication, as well as memory capacity and bandwidth that can potentially hinder the scalability of current solutions to future models and applications. This is most visible during the training part of neural network design, which accounts for the majority of the computing time and resources.

Accelerating training at the arithmetic level has thus become a hot research topic, but early work in this direction did not necessarily translate to a wide adoption and availability of low/mixed-precision training hardware. For example, BinaryConnect [59] introduced a CNN training methodology with binary ($+1$ and -1) weights, with all other operations and data structures (e.g., tensors) in full `float32` precision. This binarization was soon extended to include activations [72], followed by experiments with quantization levels of $2, 4$ and 6 bits for weights and activations [137], but with backpropagation gradients still computed and stored in full precision. Binarization for all tensor operations, including gradient computations, is considered in XNOR-Net [47]. While ensuring impressive efficiency gains, these approaches lead to non-trivial accuracy loss for larger CNN models that have since been introduced and adopted in practice.

To manage accuracy loss, DoReFaNet [43] uses different quantization bit-widths for weights, activations, and gradients, but still incurs some accuracy loss and requires exploring different bit-width configurations on a per-network basis, which can be impractical for large models. The approach introduced in [138] improves on previous accuracy results by doubling or tripling the number of inputs and outputs of layers in popular CNN models, but again requires that gradients be computed and stored in full precision and does not achieve the same accuracy as the baseline non-quantized trained model.

Studies with fixed-point arithmetic on DNNs have also been conducted since the early 1990s [139–142] and more recently [143] has shown that a 16-bit fixed-point representation coupled with stochastic rounding can be used to train CNNs on the

MNIST and CIFAR-10 datasets without accuracy loss. Nevertheless, it is unlikely that this approach would work on larger CNNs trained on larger datasets.

There have also been several proposals for quantizing recurrent neural network (RNN) training. For instance, in [144], training for quantized versions of gated recurrent units and long short-term memory cells with few bits for weights and activations are investigated, with a slight loss in accuracy with respect to base full precision models. A different approach [145] evaluates binary, ternary and exponential quantization for weights used in various RNN models trained for speech recognition and language modeling. Similar to the CNN-centered methods evoked so far, however, all these approaches use full precision gradients, and therefore do not improve computation cost during backpropagation.

15.4.1 *Mixed-Precision Training Approaches*

The most widespread approach to increase performance and efficiency of DNN training at the arithmetic level is through the use of *mixed precision*.

On the commercial side, NVIDIA has offered the possibility to do low-precision training since the Pascal architecture in 2016 and mixed-precision training (combining `float16` and `float32` arithmetic) has really taken off with the subsequent introduction of TensorCore units in their Volta and Turing architectures in 2017–2018. TensorCores are, in essence, programmable $4 \times 4 \times 4$ matrix-multiply-and-accumulate units (performing the operation $D = A \times B + C$, where A, B, C, and D are 4×4 matrices, with A and B stored using `float16` and C and D being either `float16` or `float32` matrices). An execution of a large number of such units provides a huge performance boost (several times when compared to NVIDIA's previous Pascal hardware) to convolution and matrix operations with mixed-precision operands and results. Over at Google, their newer (from version V2 onward) Tensor Processing Units (TPUs) offer similar support for mixed-precision training with the introduction of `bfloat16`, a 16-bit floating-point format that, when compared to `float16`, trades in mantissa bits for exponent bits (a 5-bit exponent and 10-bit mantissa for `float16` versus an 8-bit exponent and 7-bit mantissa for `bfloat16`). Intel and ARM are also adopting `bfloat16` in their push to offer AI-enhanced hardware, while AMD has introduced software support for `bfloat16` in recent versions of their ROCm platform. As of May 2020, the Ampere architecture from NVIDIA also introduces `bfloat16` operator support in their third version of TensorCore units.

15.4.1.1 Mixed 16-32-Bit Precision Training

An important remark about backpropagation training that should guide the choice of number formats is how the values contained within various quantities (activations, gradients, and parameters) vary during successive training iterations. It is noted

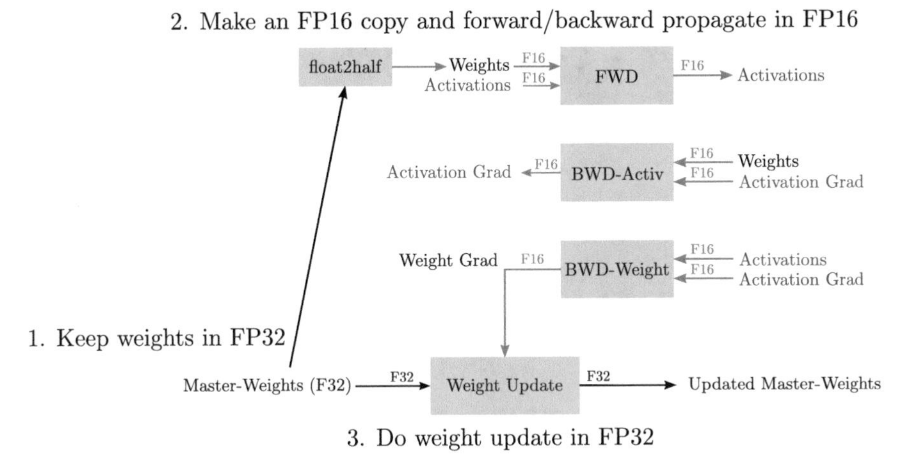

Fig. 15.11 Mixed-precision training iteration for a network layer (adapted from [147, Fig. 1])

in [146] that "activations, gradients and parameters have very different ranges," whereas "gradient ranges slowly diminish during training." There is also the idea that a higher numerical precision should be used when updating the parameters than when using them during the back and forward propagation operations [146, Sec. 6]. Recent accelerated training approaches (at the arithmetic level) follow these observations.

An Approach for `float16`-Based Training Acceleration In [147], NVIDIA TensorCores are used to perform mixed `float16` and `float32` operations during each training iteration. The process is illustrated in Fig. 15.11: a full precision copy of the weights is always stored and updated at each iteration, whereas the gradient computations of the weights and activations are done using `float16` quantizations of the weights. The dot product and reduction (i.e., sums of elements across a vector) operations are performed with a `float32` accumulator (as is enabled by TensorCores), which, according to [147], is needed in some cases to maintain the same model accuracy as with a baseline `float32` approach.

The main reason for using 32-bit values for the weight updates is that during later iterations of training, the update gradients become too small to be used with `float16` addition, which will result in them getting clipped when $\mathbf{w}^t \gg \varepsilon \frac{\partial \ell}{\partial \mathbf{w}^t}$ and adversely affect the final model accuracy. For `float16`, this happens when the ratio between weight and update is larger than 2048.

A related issue when gradients become too small is that they might not be accurately representable in `float16`, even though the dynamic range of the weight/activation gradients at each layer is much smaller than the 2^{40} range associated with `float16`. This means that a scaling approach might be applicable. This is indeed what is advocated in [147], where gradient values can be shifted to `float16`-representable ranges by scaling the loss value computed during the

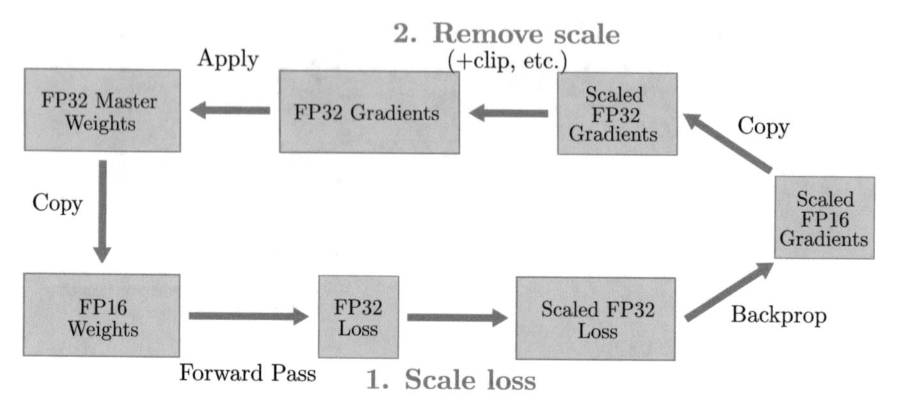

Fig. 15.12 The loss scaling procedure for updating the master weights in mixed-precision training

forward pass, before performing backpropagation. By chain rule calculus during backpropagation, all gradient values will then be scaled by the same amount. Weight gradients will have to be unscaled back before weight update to ensure the same update process as with `float32` training. The entire procedure is summarized in Fig. 15.12. Although not explored in [147], the scaling factor can be chosen automatically: start with a very large scaling factor (e.g., 2^{24}), if gradient overflows (with `Inf` or `NaN`) decrease the scale by a factor of 2 and skip the current update, whereas if no overflow has occurred for some time (e.g., 2000 iterations), increase the scale by a factor of 2.

The results presented in [147, Sec. 4] show that mixed-precision training is a viable alternative (in the sense that it gives comparable results to baseline `float32` training) for various tasks such as image classification (with tests on AlexNet, VGG-D, GoogLeNet (Inception v1), Inception v2 & v3, and ResNet50), object detection, speech recognition, machine translation, language modeling and Generative Adversarial Networks (GAN) generation.

In addition to the speed benefit that such a mixed-precision training approach brings (which varies from $2\times$ to $6\times$ with respect to baseline training on the experiments carried out in [147] on a Volta GPU), the memory consumption for training is roughly halved, since the dominating quantities are the activations (due to larger batch sizes and the fact that they need to be stored for reuse during backpropagation), which are stored in `float16`.

Enabling `bfloat16`-based training methods It seems that the need for loss scaling can be avoided if the `float16` format and associated operations are replaced with `bfloat16` (this is shown in [148], where experiments with various state-of-the-art networks in image classification, speech recognition, language modeling, generative networks, and industrial recommendation systems show the versatility of `bfloat16`-based training). This is due to the fact that `bfloat16` has the same exponent range as `float32` and the lower mantissa width does not adversely impact the final model accuracy. There are also additional hardware-related benefits that come with the combination of `bfloat16` and `float32`. Core computational primitives such as FMA units can be built using 8-bit multipliers, leading to a significant area and power savings while preserving the full dynamic range of `float32`.

The appeal of using `bfloat16` is that it also does not require any changes to the training model (as designed for a baseline `float32` approach). The increasing (planned) hardware support from several vendors seems to suggest it will soon be the *de facto* choice for performing DNN training, replacing the aforementioned `float16` approach. This statement is strengthened by the added support of `bfloat16` on NVIDIA's Ampere GPU architecture.

Fixed-Point-Based Training Mixed-precision training approaches that are based mostly on integer/fixed-point arithmetic has also been proposed recently. These methods [89, 146, 149, 150] use during computation integer tensors with tensor-wide shared exponents. The format explored in [146] has an 11-bit mantissa and a 5-bit shared exponent, tested on custom maxout [151] networks for the MNIST, CIFAR-10, and SVHN datasets. At each layer, every weight, bias, activation input & output, gradient vectors, and matrices have different exponent values. These exponents are updated based on a *passive* over/underflow detection policy which is run periodically during training. Because it is just reacting to the presence of overflows in the networks, it can potentially impede convergence of the training process.

To address this problem, [89] proposes widening the dynamic fixed-point format to a 16-bit mantissa and a 5-bit shared exponent, a format which they call `flexpoint` (`flex16+5`). They also introduce a new algorithm (Autoflex) for adjusting the shared exponents in an *adaptive* the way each time a tensor is written to, using tensor-wide statistics gathered at previous iterations. This essentially eliminates the appearance of overflow errors, leading to results on par with baseline `float32` training on AlexNet, ResNet-110 and Wasserstein GAN models. Choosing the bit-widths that resulted in the `flex16+5` format was done such that the mantissa can encode most of the variability of values inside a tensor during one training epoch and that for weight update operations there will be sufficient mantissa overlap between tensors to ensure accurate computation (which seems to eliminate the need for 32-bit master copies of the weights during the update process).

The Flexpoint approach would require the presence of dedicated hardware for it to truly show its effectiveness. That is why in [149] another dynamic fixed-point representation that can leverage already existing general-purpose hardware (through the use of existing integer operations) is presented. The mantissa is again 16-bit, while the shared exponent is stored as an 8-bit integer. The matrix multiply and dot product operations needed for the training procedure are done using 16-bit input 32-bit output integer FMAs, with some intermediate accumulations converted to `float32` in order to avoid overflows in long addition chains. Similar to [147], a float32 master copy of the weights is kept at each iteration for the update process. Tests are carried out on Intel XeonPhi Knights-Mill hardware for several CNN models (ResNet-50, GoogLeNet-v1, VGG-16, and AlexNet) on ImageNet, showing an 1.8× speedup over baseline float32 training on the same platform.

While using tensors with shared exponents can lead to performance and efficiency gains in the just discussed methods, [150] identifies three potential roadblocks in their use for training acceleration: (1) whereas dot product operations can be area-efficient with such formats, other operations might be less efficient; (2) exponent sharing can lead to data loss if magnitudes are too large or too small, making exponent selection critical; (3) data loss can happen if the tensor value distributions are too wide to be captured by the allotted number of mantissa bits. To address them, [150] proposes a hybrid approach, where all dot product operations are performed with shared exponent formats, while other operations are kept in floating-point. Since training operations are dominated by dot products, there will be little overhead to using floating-point for the remaining operations.

By using tiling for matrix multiplications (with shared exponent at tile level) and wider weight storage for the weight update process (similar to other approaches), [150] can limit data loss when compared to baseline float32 training on a large range of tasks, with little silicon density penalty. Investigating the design space, they find that the hybrid approach is most convenient for 24 × 24 tile sizes, 8 to 12-bit mantissa and 16-bit size for weight storage.

15.4.1.2 Mixed 8-16-Bit Precision Training

While combined 16-32-bit training seems to be the most widespread approach currently, for accelerating DNN training, there has also been work recently to push the envelope further with 8-bit tensor datatypes and multiplication operators coupled with 16-bit accumulators and weight updates [152, 153] (instead of the 16-32-bit mix advocated in Sect. 15.4.1.1).

According to [152, Sec. 1], there are three main elements that can significantly impact model test accuracy when using extremely low precision formats during training: (a) all operands in a tensor matrix multiply operations (GEMMs and convolutions) are in 8-bit formats (2% degradation over a baseline float32 training loop on ResNet18 with the ImageNet dataset), (b) GEMM accumulation results reduced from 32 to 16 bits (while critical to reducing the area and power of 8-bit hardware, such a move also leads to significant degradation—1% with respect to the same ResNet18 baseline) and (c) reducing weight updates from 32 to 16-bits (high precision weight updates and gradients require expensive parameter copies to be kept in memory, whereas reducing their precision can also lead to significant degradation—1.7% with respect to the ResNet18 baseline).

To cope with these problems, [152] advocates the choice of a 5-bit exponent and 2-bit mantissa floating-point format to represent weights, activations, errors, and gradients in matrix multiply operations (forward, backward, and gradient), coupled with a 6-bit exponent 9-bit mantissa format for all the accumulation results.

These format choices are motivated by how data is distributed inside networks in practice, with a focus on striking a balance between representation accuracy and dynamic range. To optimize the accuracy of the accumulation, a blocked approach (which is standard in high performance basic linear algebra routines) is used. The multiplications are done in the 8-bit format, whereas the accumulation is done in 16-bits to more accurately model the result (i.e., try to avoid *stagnation/swamping* from appearing: small $x_k y_k$ terms cannot contribute to $\sum_{k=1}^{n} x_k y_k$ in the floating-point computation path).

Another way to improve on the overall accuracy of summation results is to use stochastic rounding, which shows similar results to block accumulation (see [152, Fig. 3]). In the context of deep learning, it seems that using stochastic rounding is more natural for the weight update process (in the dot product AXPY operations) since the weight gradient is accumulated into the weight over mini-batches during several epochs (so not at once in a complete dot product operation!).

The precision settings for all the operations done during training are summarized in Fig. 15.13. In terms of results, a large spectrum of neural networks for both image classification and object recognition are used (AlexNet and ResNet 18 and 50 versions for the ImageNet and CIFAR10 datasets) with both SGD and ADAM-based optimizers. A loss scaling approach similar to [147] is used to preserve the dynamic range of back-propagated errors with small magnitude.

In both [147, 152], the hardware complexity of the floating-point computation pipeline is dominated by the accumulator bandwidth (32 and 16 bit, respectively), and in many cases, this size seems much too conservative. The follow-up work [154] introduces an analytical method for predicting the precision requirements for partial sum accumulation in the three GEMM accumulation units from Fig. 15.13. It studies in what (precision/format) scenarios the variance of the accumulator units is maintained when doing dot product computations in reduced precision.

One downside of all these aforementioned methods is that they require certain knobs to be finely tuned (such as appropriate chunk-based accumulator design, stochastic rounding techniques, loss scaling, and maintaining some layers of the

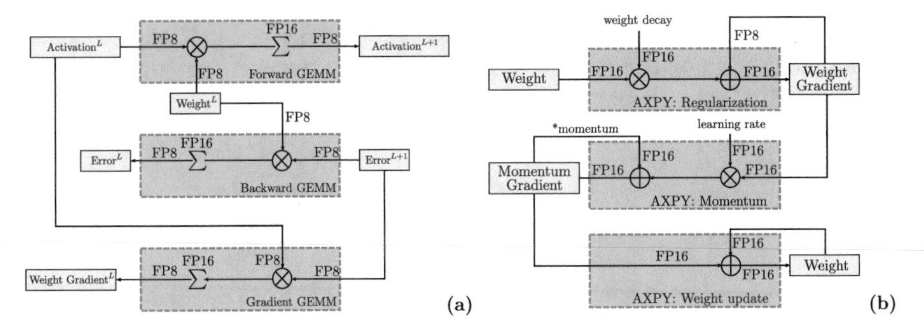

Fig. 15.13 Summary of the precision settings for (**a**) the GEMM operations during the forward and backward passes in backpropagation and (**b**) the AXPY operations during a standard SGD weight update process (adapted from [152, Fig. 2])

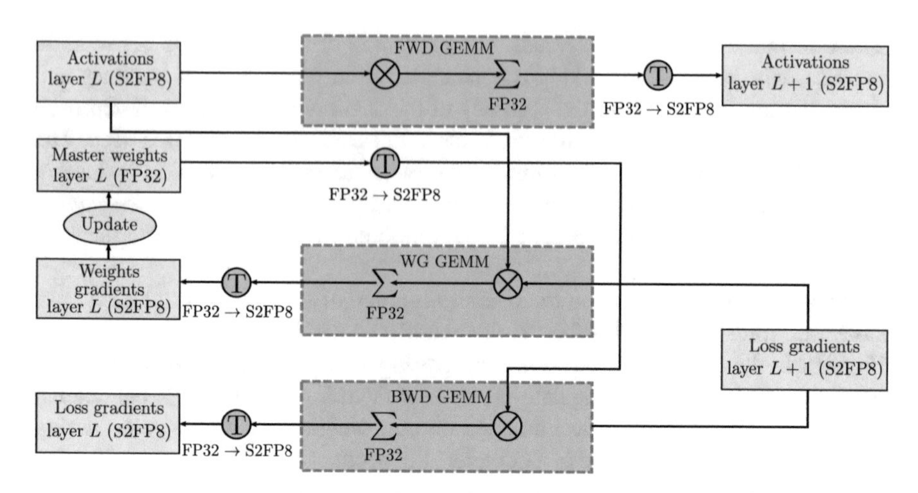

Fig. 15.14 The low-precision training flow with the S2FP8 format, where the truncation function T corresponds to $T(X) = \left[2^{-\beta}\left\{\text{round}_{\text{FP8}}(2^{\beta}|X|^{\alpha})\right\}\right]^{1/\alpha}$. The forward and backward GEMM operations use only S2FP8 values, whereas the weight update step uses FP32 master weights (adapted from [155, Fig. 4])

network in higher precision—in particular the first and last ones), necessitating experimentation on a network-by-network basis. To eliminate the need for such fine-tuning, [155] proposes a new, tensor level 8-bit floating-point format. Given an N-element tensor $X = \{X_i\}_{i=1}^{N}$, instead of encoding each element directly in an 8-bit floating-point format, X is stored using N 8-bit floating-point values $\{Y_i\}_{i=1}^{N}$ and two extra factors α and β that account for statistical information about X and capture its dynamic range. This tensor format is called S2FP8 and its use in the training procedure (for the forward and backward passes and the gradient update computations) is summarized in Fig. 15.14.

Tests on the effectiveness of this approach (FP32 vs S2FP8) are performed on residual networks of varying depths on the CIFAR10 and ImageNet datasets, the Transformer network on an English-Vietnamese translation dataset and neural collaborative filtering network architecture. The authors of [155] state that the extra hardware complexity required to handle the conversion operations and the management of the α, β parameters at each layer is small.

15.4.2 Low-Precision Training Algorithm Design

Section 15.4.1 reviewed how mixed-precision computation can be used to speed up neural network training algorithm execution, with minimal or no loss to the final test accuracy for the resulting model. Such methods are attractive because they do not require any changes to the problem's hyper-parameters (such as learning rate scheduling), making them potentially easy to use (for instance, the use of mixed-precision training with NVIDIA GPUs is straightforward with the use of their Automatic Mixed Precision (AMP) support for major deep learning frameworks).

An orthogonal and complementary direction is the development of learning algorithms tailored for low-precision computation. One such approach is MuPPET [156], which advocates for an automatic *intra* epoch numerical precision switch of training quantization and computation levels. It proposes a metric that estimates how much information each new training step obtains for a given quantization level, by quantifying the diversity of computed gradients across epochs. This allows for a heuristic runtime policy that progressively increases the working precision/format such that the final test accuracy is comparable to that of baseline `float32` training. The approach is designed to take advantage of the myriad of numerical precisions that have started to appear in modern hardware (e.g., 4 and 8-bit integer computations and 16-bit floating-point formats). For each iteration/mini-batch and a working fixed-point precision q, a block floating-point training scheme (similar to [89, 146, 149, 150]) with both values and scale factors stored as q-bit integers and stochastic rounding for quantization is used. Similar to most other mixed-precision approaches, a `float32` master copy of the weights is always kept in memory and updated at each iteration with the low-precision loss function gradients computed most recently. To test this approach, five levels of precision (8-, 12-, 14-, and 16-bit fixed-point formats and ultimately `float32`) were used in [156] for training AlexNet, ResNet18/20, and GoogLeNet networks with the CIFAR-10/100 and ImageNet datasets on an NVIDIA RTX 2080 Ti GPU. A comparison with baseline `float32` training shows a $1.25 - 1.32\times$ speedup for MuPPET, whereas with respect to [147], it achieves a $1.23\times$ speedup for AlexNet and comparable performance for ResNet18 and GoogLeNet.

Following [157] "there is always a tradeoff with standard training algorithms: as the number of bits is decreased, noise that limits statistical accuracy is increased." To limit the loss in statistical accuracy when doing low-precision training, they propose HALP (High Accuracy Low Precision), a low-precision variant of stochastic gradient descent which uses low precision for most of the time in its innermost loop, while infrequently recentering the weight parameters with higher precision in an outer loop to counteract the noise effect of low-precision quantization. The idea of the algorithm is based on the Stochastic Variance Reduced Gradient (SVRG) approach, introduced in [158], and a *bit centering* representation, where each number is represented as the sum of a high precision offset term, modified only infrequently, and a low-precision *delta* term, which is modified at each inner iteration.

For strongly convex problems, the authors show that the HALP approach can produce arbitrarily accurate solutions retaining the same linear asymptotic convergence rate as SVRG in full precision. On non-convex problems (namely CNN and LSTM neural network training), HALP (with a 16-bit low-precision format and 32-bit high precision one) is empirically shown to improve on low-precision variants of SGD and SVRG and equals or outperforms full precision SVRG and SGD. It can also be used to effectively fine-tune low-precision trained results as well, as the authors show on a ResNet18 model, closely matching the result obtained from de facto SGD training in full precision. On ImageNet, such variance-reduced mixed-precision training algorithms can obtain the state-of-the-art timing results [159].

A simpler approach for a low-precision training algorithm is SWALP (Stochastic Weight Averaging in Low-Precision Training) [160]. It is based on the recent *Stochastic Weight Averaging* (SWA) method [161]. SWA was introduced as an SGD variant that shows improved generality in deep learning training. Low-precision training on the other hand produces extra quantization noise and generally tends to underperform when the learning rate is low. Averaging weights that have been rounded both down and up during quantization can potentially reduce quantization effects and is the reason why the authors of [160] propose that SWA can be beneficial for low-precision training. The SWALP approach consists of quantizing in low precision all numbers during training, including the gradient accumulator (and potentially the velocity vector for momentum-based approaches). On a theoretical level, the authors can show that SWALP can converge to an optimal solution for quadratic objectives and a smaller noise ball than low-precision SGD for strongly convex objective functions. Empirically, for non-convex objectives, an 8-bit SWALP approach (with an 8-bit block floating-point format with 8-bit shared exponents) can match full precision SGD baselines in DNN training tasks such as for VGG-16 and Preactivation ResNet-164 on CIFAR-10/100 datasets.

15.5 Support for Approximation in DNN Accelerators

DNN models can be executed in different environments, ranging from high-power data center servers to low-power edge devices. Within this large space, there is an even wider one representing the different backends that can be used. Backends are differentiated in terms of both software and hardware. The solutions vary from general-purpose frameworks and computing units to application-specific frameworks and computing units.

Like many applications, DNNs were initially executed on latency-oriented CPUs, but ever since the start of the 2010s, there has been a major shift towards parallel hardware. Examples include GPUs for performance-oriented scenarios and microcontrollers for low-power devices. Still, due to their static and general-purpose data path, General-Purpose Processors (GPPs) are not able to efficiently process DNNs in all application scenarios, motivating the need for dedicated hardware accelerators.

The first proposed hardware accelerators were ASICs [162–164], and they achieved orders of magnitude improvements in energy efficiency compared to GPPs. This gain nevertheless comes at the expense of flexibility, with the design cost being very high. FPGAs, on the other hand, provide a good balance between flexibility, design cost, and performance [165, 166].

Independently of the target (ASIC or FPGA), hardware accelerators adopt the same strategy of maximizing data reuse, an element that has been extensively

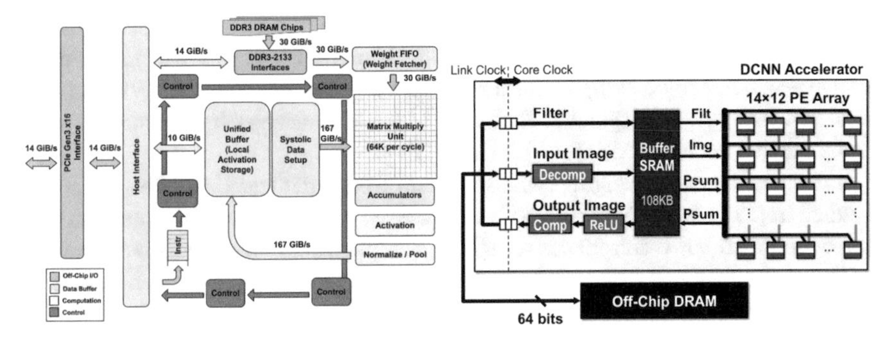

Fig. 15.15 Comparing the systolic arrays based architecture of the ASIC Google TPU [164] (top) and the Processing Element (PE) of the FPGA grid-based Eyeriss [163] architecture (bottom)

studied by Chen et al. in [163]. The main architectures adopted by re-configurable accelerators such as FPGAs is a dedicated grid of Processing Element (PE) [163], while the main architecture adopted by ASICs are based on more generic systolic arrays [164]. This is mainly because a systolic array is more flexible once designed and can efficiently process matrix products, while a PE array requires tuning some parameters for efficiently executing a DNN (like the number of PEs and the size of the memory bus), making them more suitable for re-configurable accelerators. A schematic view of the two approaches is given in Fig. 15.15.

Since DeepCompression [99] proved that approximation techniques can significantly improve DNN processing efficiency with very small accuracy loss, approximation for DNN acceleration has become quite popular, at the same time posing new challenges for efficient processing. For example, accelerating a sparse DNN (after application of pruning methods like those presented in Sect. 15.3.3) requires adapting the dataflow to take advantage of the available sparsity, whereas accelerating a reduced precision DNN (after application of quantization methods as described in Sect. 15.3.1) requires implementing dedicated operators.

15.5.1 Architectures for Accelerating Inference

Almost, if not all, dedicated DNN hardware accelerators rely on reduced precision computations. This is mainly because a 32-bit floating-point is not mandatory to achieve high accuracy, and has a prohibitive computing cost. Most accelerators use 16-bit or 8-bit representations, like [163]. Some accelerators are dedicated to specific quantization formats, such as [167] that targets acceleration of fully-binarized DNNs, or [168] for accelerating logarithmic representations.

Whereas reduced precision acceleration-based solutions mainly require changes to the arithmetic operators, accelerating pruned DNNs with a sparse representation requires changes to the dataflow. Lu et al. proposed to use the combination of two structures representing the COOrdinates (COO) of the values and the values as Compressed Sparse Rows (CSR) [169], and developed an accelerator to take advantage of these representations. It is also possible to take advantage of Feature Map (FM) sparsity. Due to the use of ReLU activations, FMs contain a large number of zeros which can be skipped during the next layer computation. CNVlutin [170] explores this dynamic sparsity. It is also possible to accelerate structured sparse DNNs with a dedicated dataflow like in [171].

DNNs with shared weights can also benefit from a dedicated dataflow. This was studied in [172], which targets DNNs compressed using the DeepCompression [99] three-step method. It introduces an efficient implementation of the sparse matrix-vector multiplications with weight sharing that are central to the approach from [99].

15.5.2 Architectures for Accelerating Training

Accelerating DNN algorithms on hardware targets such as FPGAs faces many challenges, including limited on-chip memory, external memory bandwidth, and computational resources. Compared to the design of inference accelerators, on-chip training is a less studied topic, but it is feasible [173].

An example is [174], which targets training acceleration for embedded Xilinx Zynq All Programmable System on Chip (APSoC) devices. It essentially implements a version of the method introduced in [160] with predominantly 8-bit integer arithmetic. The Arm-based processor on the device is used for 32-bit floating-point weight updates, whereas the FPGA logic evaluates all the 8-bit integer matrix multiplications needed during the backpropagation computation path. The overall hardware platform is configured using a software-based High-Level Synthesis (HLS) flow with Xilinx tools. On the Intel side of things, [175] has proposed a Register-Transfer Level (RTL) compiler that performs SGD-based training on Intel FPGAs for various CNNs with 16-bit fixed-point arithmetic.

15.6 Perspectives

Due to the rapid evolution of the field of deep learning, it is difficult to give an accurate prediction of how to approximate computing techniques that will impact DL acceleration in the future. This section presents an overview of three different research directions that figure to grow in importance in the years to come.

15.6.1 Approximation for Attention-Based Architectures

While the focus of the previous sections is mostly directed at CNN-based models, in recent years alternative structures such as Transformer attention architectures [31] have led to the state-of-the-art accuracy results in NLP-based tasks (e.g., language modeling). Subsequent models, like BERT [33], RoBERTa [176] and GPT [177], although impressive, have a large memory footprint, increased latency, and power consumption that are prohibitive for efficient deployment on embedded edge devices and even on data centers. Due to their expressive power, Transformer-based models are also beginning to be adapted for other tasks, such as computer vision applications [178, 179].

Their increasing usage is driving interest for efficient approximation methods that specifically target Transformer models. While work in this direction is still in its early stages, there are already some approaches based on quantization [180, 181], knowledge distillation [182, 183], and pruning [184, 185].

15.6.2 Edge AI

One area where training acceleration with reduced precision and increased energy efficiency is becoming important is incremental/lifelong learning scenarios on edge devices (e.g., in autonomous driving, IoT, and robotics). Compared to a cloud-based scenario, training locally avoids transferring data back and forth between data centers and IoT devices, helping reduce communication and latency and improve privacy.

Such on-chip training is feasible [173], but extremely challenging. The training acceleration methods described in Sect. 15.4 usually cannot be applied directly to this context and alternatives need to be considered.

A training framework specifically designed for such scenarios is E^2-Train [186], which proposes three complementary strategies: (a) stochastic mini-batch dropping to eliminate what can be considered “unnecessary costs,” (b) input-dependent selective layer update where a different subset of CNN layers are updated for every mini-batch, and (c) predictive sign gradient descent, a variation of an extremely low-precision SGD algorithm, signSGD [187]. Besides this approach, other algorithmic & arithmetic-level methods have started to appear [188, 189]. It is expected that this area of research will grow in importance in the years to come, with on-site learning becoming paramount in certain application domains.

15.6.3 Analog In-Memory Computing

The recent explosive growth in highly data-centric applications related to DL has motivated the appearance of analog in-memory computing solutions [190–194] as alternatives to traditional von Neumann computing systems. Hereby important computational tasks, such as vector-matrix multiplications, are performed in place in the memory itself by exploiting the physical attributes of the memory devices (e.g., Kirchhoff’s current summation law). Besides alleviating the costs in latency and energy associated with data movement, in-memory computing also has the potential to significantly improve the computational time complexity by using large crossbar memory arrays [195]. However, this comes at the expense of imprecision in the mixed-signal computations and becomes a form of approximate computing. For instance, the mapping of synaptic weights onto some of those memory devices suffers from non-ideal analog storage in the form of stochastic distribution of conductance values and temporal drifting. Accordingly, Joshi et al. [196] have proposed a custom noise-injection training method to increase the robustness of the resulting network to such non-idealities and achieve a software equivalent accuracy. Given the game-changing advantages in computing efficiency of analog in-memory computing, more work is expected on this nascent field in the future.

15.7 Conclusion

In this chapter, a comprehensive survey of approximation techniques applied to Deep Learning is provided. These techniques target various improvements, some geared towards the training of DNN models, others that focus on DNN inference. Depending on the objective, various methods can be applied, whether for the improvement (reduction) of memory requirements by using compression techniques or for the reduction of the computational workload by using acceleration techniques.

Such a wide range of approximation techniques involves various implementation changes, ideally resulting in a backend adaptation that maximizes the expected performance improvement. These adaptations can be implemented at the software level using dedicated frameworks and/or at the hardware level in dedicated accelerators.

To compare the various methods available, it is desirable to use the same input DNN and workload, but this is not always feasible. This is mostly due to the large range of DNN topologies that have appeared over the years: while some methods can be applied almost automatically to multiple topologies, some require manual tuning as the size of the search space increases exponentially with the DNN size. There is also a wide range of workloads, from small "toy" datasets to more recent and challenging large-scale datasets. Some methods cannot perform equally well in both contexts. The difference in backend compatibility with the various approximation methods also regularly involves manual tuning steps, which are hard to reproduce and compare to other backends.

Most of the recent methods give very promising results and pave the way for further research, by proving that approximations can be applied at various levels, from the topology of the DNN, to the data value and type, and including the backends-DNN codesign (hardware or software).

Acknowledgments This work has been funded by the French National Research Agency (ANR) through the AdequatedDL research project (ANR-18-CE23-0012).

References

1. LeCun, Y., Bengio, Y., & Hinton, G. (2015). Deep learning. *Nature, 521*(7553), 436–444.
2. Deng, L., Li, J., Huang, J.-T., Yao, K., Yu, D., Seide, F., Seltzer, M., Zweig, G., He, X., Williams, J., et al. (2013). Recent advances in deep learning for speech research at Microsoft. In *2013 IEEE International Conference on Acoustics, Speech and Signal Processing* (pp. 8604–8608). Piscataway: IEEE.
3. Krizhevsky, A., Sutskever, I., & Hinton, G. E. (2017). ImageNet classification with deep convolutional neural networks. *Communications of the ACM, 60*(6), 84–90.
4. Chen, C., Seff, A., Kornhauser, A., & Xiao, J. (2015). Deepdriving: learning affordance for direct perception in autonomous driving. In *Proceedings of the IEEE International Conference on Computer Vision* (pp. 2722–2730).
5. Silver, D., Huang, A., Maddison, C. J., Guez, A., Sifre, L., van den Driessche, G., Schrittwieser, J., Antonoglou, I., Panneershelvam, V., Lanctot, M., et al. (2016). Mastering the game of go with deep neural networks and tree search. *Nature, 529*, 484–489.
6. Wang, C., Gong, L., Yu, Q., Li, X., Xie, Y., & Zhou, X. (2016). DLAU: A scalable deep learning accelerator unit on FPGA. *IEEE Transactions on Computer-Aided Design of Integrated Circuits and Systems, 36*(3), 513–517.
7. Chen, Y.-H., Krishna, T., Emer, J. S., & Sze, V. (2016). Eyeriss: An energy-efficient reconfigurable accelerator for deep convolutional neural networks. *IEEE Journal of Solid-State Circuits, 52*(1), 127–138.
8. Liu, Z., Dou, Y., Jiang, J., Xu, J., Li, S., Zhou, Y., & Xu, Y. (2017). Throughput-optimized FPGA accelerator for deep convolutional neural networks. *ACM Transactions on Reconfigurable Technology and Systems (TRETS), 10*(3), 1–23.
9. Sung, W., Shin, S., & Hwang, K. (2015). Resiliency of deep neural networks under quantization. arXiv:1511.06488.
10. Tann, H., Hashemi, S., Bahar, R. I., & Reda, S. (2017). Hardware-software codesign of accurate, multiplier-free deep neural networks. In *2017 54th ACM/EDAC/IEEE Design Automation Conference (DAC)* (pp. 1–6). Piscataway: IEEE.
11. Sze, V., Chen, Y.-H., Yang, T.-J., & Emer, J. S. (2020). Efficient processing of deep neural networks. *Synthesis Lectures on Computer Architecture, 15*(2), 1–341.
12. Rumelhart, D. E., Hinton, G. E., & Williams, R. J. (1986). Learning representations by back-propagating errors. *Nature, 323*(6088), 533–536.
13. Le Cun, Y., Jackel, L. D., Boser, B., Denker, J. S., Graf, H. P., Guyon, I., Henderson, D., Howard, R. E., & Hubbard, W. (1989). Handwritten digit recognition: Applications of neural network chips and automatic learning. *IEEE Communications Magazine, 27*(11), 41–46.
14. Lecun, Y., Bottou, L., Bengio, Y., & Haffner, P. (1998). Gradient-based learning applied to document recognition. *Proceedings of the IEEE, 86*, 2278–2324.
15. Kingma, D. P., & Ba, J. (2014). Adam: A method for stochastic optimization. arXiv:1412.6980.
16. Deng, J., Dong, W., Socher, R., Li, L.-J., Li, K., & Fei-Fei, L. (2009). ImageNet: A large-scale hierarchical image database. In *IEEE Conference on Computer Vision and Pattern Recognition CVPR09* (pp. 248–255).
17. Russakovsky, O., Deng, J., Su, H., Krause, J., Satheesh, S., Ma, S., Huang, Z., Karpathy, A., Khosla, A., Bernstein, M., Berg, A. C., & Fei-Fei, L. (2015). ImageNet large scale visual recognition challenge. *International Journal of Computer Vision (IJCV), 115*(3), 211–252.
18. LeCun, Y., & Cortes, C. (2010). MNIST handwritten digit database. http://yann.lecun.com/exdb/mnist/
19. Krizhevsky, A. (2009). *Learning Multiple Layers of Features from Tiny Images*. Masters Thesis, University of Toronto.
20. Abadi, M., Agarwal, A., Barham, P., Brevdo, E., Chen, Z., Citro, C., Corrado, G. S., Davis, A., Dean, J., Devin, M., Ghemawat, S., Goodfellow, I., Harp, A., Irving, G., Isard, M., Jia, Y., Jozefowicz, R., Kaiser, L., Kudlur, M., et al. (2015). TensorFlow: Large-scale machine learning on heterogeneous systems. Software available from tensorflow.org.

21. Paszke, A., Gross, S., Massa, F., Lerer, A., Bradbury, J., Chanan, G., Killeen, T., Lin, Z., Gimelshein, N., Antiga, L., Desmaison, A., Kopf, A., Yang, E., DeVito, Z., Raison, M., Tejani, A., Chilamkurthy, S., Steiner, B., Fang, L., Bai, J., & Chintala, S. (2019). Pytorch: An imperative style, high-performance deep learning library. In H. Wallach, H. Larochelle, A. Beygelzimer, F. d'Alché-Buc, E. Fox, & R. Garnett (Eds.), *Advances in Neural Information Processing Systems* (vol. 32, pp. 8024–8035). New York: Curran Associates.
22. Ioffe, S., & Szegedy, C. (2015). Batch normalization: Accelerating deep network training by reducing internal covariate shift. arXiv:1502.03167.
23. Howard, A. G., Zhu, M., Chen, B., Kalenichenko, D., Wang, W., Weyand, T., Andreetto, M., & Adam, H. (2017). Mobilenets: Efficient convolutional neural networks for mobile vision applications. CoRR, ArXiv, vol. abs/1704.04861.
24. Szegedy, C., Liu, W., Jia, Y., Sermanet, P., Reed, S. E., Anguelov, D., Erhan, D., Vanhoucke, V., & Rabinovich, A. (2014). Going deeper with convolutions. CoRR, vol. abs/1409.4842.
25. Krizhevsky, A., Sutskever, I., & Hinton, G. E. (2012). ImageNet Classification with Deep Convolutional Neural Networks. In F. Pereira, C. J. C. Burges, L. Bottou, & K. Q. Weinberger, (Eds.), *Advances in Neural Information Processing Systems, 25* (pp. 1097–1105). New York: Curran Associates.
26. He, K., Zhang, X., Ren, S., & Sun, J. (2015). Deep residual learning for image recognition. CoRR, vol. abs/1512.03385.
27. Sandler, M., Howard, A., Zhu, M., Zhmoginov, A., & Chen, L.-C. (2018). Mobilenetv2: Inverted residuals and linear bottlenecks.
28. Tan, M., & Le, Q. V. (2019). Efficientnet: Rethinking model scaling for convolutional neural networks. CoRR, vol. abs/1905.11946.
29. Stojnic, R., Taylor, R., Kerkez, V., & Viaud, L. (2020). Papers with code, State of the Art models on the ImageNet dataset. Retrieved Nov. 19, 2020
30. Strubell, E., Ganesh, A., & McCallum, A. (2019). Energy and policy considerations for deep learning in NLP. arXiv:1906.02243.
31. Vaswani, A., Shazeer, N., Parmar, N., Uszkoreit, J., Jones, L., Gomez, A. N., Kaiser, Ł. & Polosukhin, I. (2017). Attention is all you need. *Advances in Neural Information Processing Systems, 30*, 5998–6008.
32. Peters, M. E., Neumann, M., Iyyer, M., Gardner, M., Clark, C., Lee, K., & Zettlemoyer, L. (2018). Deep contextualized word representations. arXiv:1802.05365.
33. Devlin, J., Chang, M.-W., Lee, K., & Toutanova, K. (2018). BERT: Pre-training of deep bidirectional transformers for language understanding. arXiv:1810.04805.
34. So, D. R., Liang, C., & Le, Q. V. (2019). The evolved transformer. arXiv:1901.11117.
35. Radford, A., Wu, J., Child, R., Luan, D., Amodei, D., & Sutskever, I. (2019). Language models are unsupervised multitask learners. *OpenAI Blog, 1*(8), 9.
36. Chippa, V. K., Chakradhar, S., Roy, K., & Raghunathan, A. (2013). Analysis and characterization of inherent application resilience for approximate computing. In *2013 50th ACM/EDAC/IEEE Design Automation Conference (DAC)*, pp. 1–9.
37. He, K., Zhang, X., Ren, S., & Sun, J. (2016). Deep residual learning for image recognition. In *Proceedings of the IEEE Conference on Computer Vision and Pattern Recognition* (pp. 770–778).
38. Hinton, G., Vinyals, O., & Dean, J. (2015). Distilling the knowledge in a neural network. arXiv:1503.02531.
39. Tang, J., Shivanna, R., Zhao, Z., Lin, D., Singh, A., Chi, E. H., & Jain, S. (2020). Understanding and improving knowledge distillation. arXiv:2002.03532.
40. Iandola, F. N., Moskewicz, M. W., Ashraf, K., Han, S., Dally, W. J., & Keutzer, K. (2016). Squeezenet: Alexnet-level accuracy with 50x fewer parameters and <1mb model size. CoRR, vol. abs/1602.07360.
41. Cun, Y. L., Denker, J. S., & Solla, S. A. (1990). *Optimal brain damage* (pp. 598–605). San Francisco: Morgan Kaufmann.
42. Frankle, J., & Carbin, M. (2018). The lottery ticket hypothesis: Training pruned neural networks. CoRR, vol. abs/1803.03635.

43. Zhou, S., Wu, Y., Ni, Z., Zhou, X., Wen, H., & Zou, Y. (2016). DoReFa-net: Training low bitwidth convolutional neural networks with low bitwidth gradients. arXiv:1606.06160.
44. Zhou, A., Yao, A., Guo, Y., Xu, L., & Chen, Y. (2017). Incremental network quantization: Towards lossless CNNs with low-precision weights. arXiv:1702.03044.
45. Courbariaux, M., Hubara, I., Soudry, D., El-Yaniv, R., & Bengio, Y. (2016). Binarized neural networks: Training deep neural networks with weights and activations constrained to $+1$ or -1. arXiv:1602.02830.
46. Li, F., Zhang, B., & Liu, B. (2016). Ternary weight networks. arXiv:1605.04711.
47. Rastegari, M., Ordonez, V., Redmon, J., & Farhadi, A. (2016). XNOR-Net: ImageNet classification using binary convolutional neural networks. In *European Conference on Computer Vision* (pp. 525–542). Berlin: Springer.
48. Jacob, B., Kligys, S., Chen, B., Zhu, M., Tang, M., Howard, A., Adam, H., & Kalenichenko, D. (2018). Quantization and training of neural networks for efficient integer-arithmetic-only inference. In *Proceedings of the IEEE Conference on Computer Vision and Pattern Recognition* (pp. 2704–2713).
49. Wu, S., Li, G., Chen, F., & Shi, L. (2018). Training and inference with integers in deep neural networks. arXiv:1802.04680.
50. Choi, J., Wang, Z., Venkataramani, S., Chuang, P. I.-J., Srinivasan, V., & Gopalakrishnan, K. (2018). Pact: Parameterized clipping activation for quantized neural networks. arXiv:1805.06085.
51. Zhang, D., Yang, J., Ye, D., & Hua, G. (2018). LQ-nets: Learned quantization for highly accurate and compact deep neural networks. In *Proceedings of the European Conference on Computer Vision (ECCV)* (pp. 365–382).
52. Banner, R., Nahshan, Y., & Soudry, D. (2019). Post training 4-bit quantization of convolutional networks for rapid-deployment. In *Advances in Neural Information Processing Systems* (pp. 7950–7958).
53. Cai, Y., Yao, Z., Dong, Z., Gholami, A., Mahoney, M. W., & Keutzer, K. (2020). ZeroQ: A novel zero shot quantization framework. In *Proceedings of the IEEE/CVF Conference on Computer Vision and Pattern Recognition* (pp. 13169–13178).
54. Choukroun, Y., Kravchik, E., Yang, F., & Kisilev, P. (2019). Low-bit quantization of neural networks for efficient inference. In *2019 IEEE/CVF International Conference on Computer Vision Workshop (ICCVW)* (pp. 3009–3018). Piscataway: IEEE.
55. Nagel, M., Baalen, M. V., Blankevoort, T., & Welling, M. (2019). Data-free quantization through weight equalization and bias correction. In *Proceedings of the IEEE International Conference on Computer Vision* (pp. 1325–1334).
56. Zhao, R., Hu, Y., Dotzel, J., De Sa, C., & Zhang, Z. (2019). Improving neural network quantization without retraining using outlier channel splitting. arXiv:1901.09504.
57. Alizadeh, M., Behboodi, A., van Baalen, M., Louizos, C., Blankevoort, T., & Welling, M. (2020). Gradient ℓ_1 regularization for quantization robustness. arXiv:2002.07520.
58. Shkolnik, M., Chmiel, B., Banner, R., Shomron, G., Nahshan, Y., Bronstein, A., & Weiser, U. (2020). Robust quantization: One model to rule them all. arXiv:2002.07686.
59. Courbariaux, M., Bengio, Y., & David, J.-P. (2015). BinaryConnect: Training deep neural networks with binary weights during propagations. In *Advances in Neural Information Processing Systems* (pp. 3123–3131).
60. Zhou, Y., Moosavi-Dezfooli, S.-M., Cheung, N.-M., & Frossard, P. (2017). Adaptive quantization for deep neural network. arXiv:1712.01048.
61. Wu, B., Wang, Y., Zhang, P., Tian, Y., Vajda, P., & Keutzer, K. (2018). Mixed precision quantization of convnets via differentiable neural architecture search. arXiv:1812.00090.
62. Wang, K., Liu, Z., Lin, Y., Lin, J., & Han, S. (2019). HAQ: Hardware-aware automated quantization with mixed precision. In *Proceedings of the IEEE Conference on Computer Vision and Pattern Recognition* (pp. 8612–8620).
63. Dong, Z., Yao, Z., Gholami, A., Mahoney, M. W., & Keutzer, K. (2019). HAWQ: Hessian aware quantization of neural networks with mixed-precision. In *Proceedings of the IEEE International Conference on Computer Vision* (pp. 293–302).

64. Dong, Z., Yao, Z., Cai, Y., Arfeen, D., Gholami, A., Mahoney, M. W., & Keutzer, K. (2019). HAWQ-V2: Hessian aware trace-weighted quantization of neural networks. arXiv:1911.03852.
65. Lin, D., Talathi, S., & Annapureddy, S. (2016). Fixed point quantization of deep convolutional networks. In *International Conference on Machine Learning* (pp. 2849–2858).
66. Khoram, S., & Li, J. (2018). Adaptive quantization of neural networks. In *International Conference on Learning Representations*.
67. Shen, S., Dong, Z., Ye, J., Ma, L., Yao, Z., Gholami, A., Mahoney, M. W., & Keutzer, K. (2020). Q-bert: Hessian based ultra low precision quantization of bert. In *Association for the Advancement of Artificial Intelligence (AAAI)* (pp. 8815–8821).
68. Zhu, X., Zhou, W., & Li, H. (2018). Adaptive layerwise quantization for deep neural network compression. In *2018 IEEE International Conference on Multimedia and Expo (ICME)* (pp. 1–6). Piscataway: IEEE.
69. Park, E., Yoo, S., & Vajda, P. (2018). Value-aware quantization for training and inference of neural networks. In *Proceedings of the European Conference on Computer Vision (ECCV)* (pp. 580–595).
70. Esser, S. K., Merolla, P. A., Arthur, J. V., Cassidy, A. S., Appuswamy, R., Andreopoulos, A., Berg, D. J., McKinstry, J. L., Melano, T., Barch, D. R., et al. (2016). From the cover: Convolutional networks for fast, energy-efficient neuromorphic computing. *Proceedings of the National Academy of Sciences of the United States of America, 113*(41), 11441.
71. Sun, X., Liu, R., Peng, X., & Yu, S. (2018). Computing-in-memory with SRAM and RRAM for binary neural networks. In *2018 14th IEEE International Conference on Solid-State and Integrated Circuit Technology (ICSICT)* (pp. 1–4). Piscataway: IEEE.
72. Hubara, I., Courbariaux, M., Soudry, D., El-Yaniv, R., & Bengio, Y. (2016). Binarized neural networks. In D. D. Lee, M. Sugiyama, U. V. Luxburg, I. Guyon, & R. Garnett (Eds.), *Advances in Neural Information Processing Systems* (pp. 4107–4115, vol. 29). New York: Curran Associates.
73. Tang, W., Hua, G., & Wang, L. (2017). How to train a compact binary neural network with high accuracy? In *Association for the Advancement of Artificial Intelligence (AAAI)* (pp. 2625–2631).
74. Lin, X., Zhao, C., & Pan, W. (2017). Towards accurate binary convolutional neural network. In *Advances in Neural Information Processing Systems* (pp. 345–353).
75. Darabi, S., Belbahri, M., Courbariaux, M., & Nia, V. P. (2018). Bnn+: Improved binary network training. arXiv:1812.11800.
76. Hwang, K., & Sung, W. (2014). Fixed-point feedforward deep neural network design using weights +1, 0, and -1. In *2014 IEEE Workshop on Signal Processing Systems (SiPS)* (pp. 1–6). Piscataway: IEEE.
77. Zhu, C., Han, S., Mao, H., & Dally, W. J. (2016). Trained ternary quantization. arXiv:1612.01064.
78. Kundu, A., Banerjee, K., Mellempudi, N., Mudigere, D., Das, D., Kaul, B., & Dubey, P. (2017). Ternary residual networks. arXiv:1707.04679.
79. Wan, D., Shen, F., Liu, L., Zhu, F., Qin, J., Shao, L., & Tao Shen, H. (2018). TBN: Convolutional neural network with ternary inputs and binary weights. In *Proceedings of the European Conference on Computer Vision (ECCV)* (pp. 315–332).
80. Tambe, T., Yang, E.-Y., Wan, Z., Deng, Y., Reddi, V. J., Rush, A., Brooks, D., & Wei, G.-Y. (2020). Algorithm-hardware co-design of adaptive floating-point encodings for resilient deep learning inference. In *2020 57th ACM/IEEE Design Automation Conference (DAC)* (pp. 1–6). Piscataway: IEEE.
81. Settle, S. O., Bollavaram, M., D'Alberto, P., Delaye, E., Fernandez, O., Fraser, N., Ng, A., Sirasao, A., & Wu, M. (2018). Quantizing convolutional neural networks for low-power high-throughput inference engines. arXiv:1805.07941.
82. Wu, C., Wang, M., Chu, X., Wang, K., & He, L. (2020). Low precision floating-point arithmetic for high performance FPGA-based CNN acceleration. arXiv:2003.03852.

83. Wu, C., Wang, M., Li, X., Lu, J., Wang, K., & He, L. (2020). Phoenix: A low-precision floating-point quantization oriented architecture for convolutional neural networks. arXiv:2003.02628.
84. Song, Z., Liu, Z., & Wang, D. (2017). Computation error analysis of block floating point arithmetic oriented convolution neural network accelerator design. arXiv:1709.07776.
85. Lian, X., Liu, Z., Song, Z., Dai, J., Zhou, W., & Ji, X. (2019). High-performance FPGA-based CNN accelerator with block-floating-point arithmetic. *IEEE Transactions on Very Large Scale Integration (VLSI) Systems, 27*(8), 1874–1885.
86. Miyashita, D., Lee, E. H., & Murmann, B. (2016). Convolutional neural networks using logarithmic data representation. arXiv:1603.01025.
87. Johnson, J. (2018). Rethinking floating point for deep learning. arXiv:1811.01721 [cs].
88. Fu, H., Mencer, O., & Luk, W. (2010). FPGA designs with optimized logarithmic arithmetic. *IEEE Transactions on Computers, 59*(7), 1000–1006.
89. Köster, U., Webb, T., Wang, X., Nassar, M., Bansal, A. K., Constable, W., Elibol, O., Gray, S., Hall, S., Hornof, L., Khosrowshahi, A., Kloss, C., Pai, R. J., Rao, N., et al. (2017). Flexpoint: An adaptive numerical format for efficient training of deep neural networks. In *Advances in Neural Information Processing Systems* (pp. 1742–1752).
90. Zhou, S.-C., Wang, Y.-Z., Wen, H., He, Q.-Y., & Zou, Y.-H. (2017). Balanced quantization: An effective and efficient approach to quantized neural networks. *Journal of Computer Science and Technology, 32*(4), 667–682.
91. Jung, S., Son, C., Lee, S., Son, J., Han, J.-J., Kwak, Y., Hwang, S. J., & Choi, C. (2019). Learning to quantize deep networks by optimizing quantization intervals with task loss. In *Proceedings of the IEEE Conference on Computer Vision and Pattern Recognition* (pp. 4350–4359).
92. Wang, T., Wang, J., Xu, C., & Xue, C. (2020). Automatic low-bit hybrid quantization of neural networks through meta learning. arXiv:2004.11506.
93. Choi, Y., El-Khamy, M., & Lee, J. (2020). Learning sparse low-precision neural networks with learnable regularization. *IEEE Access, 8*, 96963–96974.
94. Bai, H., Wu, J., King, I., & Lyu, M. (2019). Few shot network compression via cross distillation. arXiv:1911.09450.
95. Polino, A., Pascanu, R., & Alistarh, D. (2018). Model compression via distillation and quantization. arXiv:1802.05668.
96. Chen, S., Wang, W., & Pan, S. J. (2019). Deep neural network quantization via layer-wise optimization using limited training data. In *Proceedings of the AAAI Conference on Artificial Intelligence* (vol. 33, pp. 3329–3336).
97. Leng, C., Li, H., Zhu, S., & Jin, R. (2017). Extremely low bit neural network: Squeeze the last bit out with admm. arXiv:1707.09870.
98. Chen, W., Wilson, J. T., Tyree, S., Weinberger, K. Q., & Chen, Y. (2015). Compressing neural networks with the hashing trick. CoRR, vol. abs/1504.04788.
99. Song Han, W. J. D., & Mao, H. (2016). Deep compression: Compressing deep neural networks with pruning, trained quantization and Huffman coding. arXiv:1510.00149 [cs.CV].
100. Lloyd, S. P. (1982). Least squares quantization in PCM. *IEEE Transactions on Information Theory, 28*, 129–137.
101. Choi, Y., El-Khamy, M., & Lee, J. (2016). Towards the limit of network quantization. CoRR, vol. abs/1612.01543.
102. Park, E., Ahn, J., & Yoo, S. (2017). Weighted-entropy-based quantization for deep neural networks. In *2017 IEEE Conference on Computer Vision and Pattern Recognition (CVPR)* (pp. 7197–7205).
103. Wu, J., Wang, Y., Wu, Z., Wang, Z., Veeraraghavan, A., & Lin, Y. (2018). Deep k-means: Re-training and parameter sharing with harder cluster assignments for compressing deep convolutions. CoRR, vol. abs/1806.09228.
104. Yang, D., Yu, W., Zhou, A., Mu, H., Yao, G., & Wang, X. (2020). DP-net: Dynamic programming guided deep neural network compression. arXiv:2003.09615 [cs.LG]

105. Razlighi, M. S., Imani, M., Koushanfar, F., & Rosing, T. (2017). LookNN: Neural network with no multiplication. In *Design, Automation & Test in Europe Conference & Exhibition (DATE), 2017* (pp. 1775–1780). Piscataway: IEEE.
106. Wu, J., Leng, C., Wang, Y., Hu, Q., & Cheng, J. (2015). Quantized convolutional neural networks for mobile devices. CoRR, vol. abs/1512.06473.
107. He, Y., & Han, S. (2018). ADC: Automated deep compression and acceleration with reinforcement learning. CoRR, vol. abs/1802.03494.
108. Huan, Y., Qin, Y., You, Y., Zheng, L., & Zou, Z. (2016). A multiplication reduction technique with near-zero approximation for embedded learning in IoT devices. In *2016 29th IEEE International System-on-Chip Conference (SOCC)* (pp. 102–107). Piscataway: IEEE.
109. Huan, Y., Qin, Y., You, Y., Zheng, L., & Zou, Z. (2017). A low-power accelerator for deep neural networks with enlarged near-zero sparsity. arXiv:1705.08009.
110. Mao, H., Han, S., Pool, J., Li, W., Liu, X., Wang, Y., & Dally, W. J. (2017). Exploring the regularity of sparse structure in convolutional neural networks. arXiv:1705.08922.
111. Ji, Y., Liang, L., Deng, L., Zhang, Y., Zhang, Y., & Xie, Y. (2018). Tetris: TilE-matching the tremendous irregular sparsity. In *32nd Conference on Neural Information Processing Systems (NeurIPS)*.
112. Yu, J., Lukefahr, A., Palframan, D., Dasika, G. S., Das, R., & Mahlke, S. (2017). Scalpel: Customizing DNN pruning to the underlying hardware parallelism. In *2017 ACM/IEEE 44th Annual International Symposium on Computer Architecture (ISCA)* (pp. 548–560).
113. Molchanov, P., Tyree, S., Karras, T., Aila, T., & Kautz, J. (2016). Pruning convolutional neural networks for resource efficient transfer learning. CoRR, vol. abs/1611.06440.
114. Luo, J.-H., Wu, J., & Lin, W. (2017). Thinet: A filter level pruning method for deep neural network compression. In *2017 IEEE International Conference on Computer Vision (ICCV)* (pp. 5068–5076).
115. He, Y., Zhang, X., & Sun, J. (2017). Channel pruning for accelerating very deep neural networks. CoRR, vol. abs/1707.06168.
116. Hassibi, B., & Stork, D. (1992). Second order derivatives for network pruning: Optimal brain surgeon. In *Advances in Neural Information Processing Systems (NIPS)*.
117. Srinivas, S., & Babu, R. V. (2015). Data-free parameter pruning for deep neural networks. In *The British Machine Vision Conference (BMVC)*.
118. Guo, Y., Yao, A., & Chen, Y. (2016). Dynamic network surgery for efficient DNNs. In *Advances in Neural Information Processing Systems (NIPS)*.
119. Narang, S., Diamos, G., Sengupta, S., & Elsen, E. (2017). Exploring sparsity in recurrent neural networks. ArXiv, vol. abs/1704.05119.
120. Li, H., Kadav, A., Durdanovic, I., Samet, H., & Graf, H. (2017). Pruning filters for efficient convnets. ArXiv, vol. abs/1608.08710.
121. Chin, T.-W., Ding, R., Zhang, C., & Marculescu, D. (2020). Towards efficient model compression via learned global ranking. In *2020 IEEE/CVF Conference on Computer Vision and Pattern Recognition (CVPR)*.
122. Chin, T., Ding, R., Zhang, C., & Marculescu, D. (2019). LeGR: Filter pruning via learned global ranking. CoRR, vol. abs/1904.12368.
123. Dai, X., Yin, H., & Jha, N. K. (2017). Nest: A neural network synthesis tool based on a grow-and-prune paradigm. CoRR, vol. abs/1711.02017.
124. Yang, T.-J., Chen, Y., & Sze, V. (2017). Designing energy-efficient convolutional neural networks using energy-aware pruning. In *2017 IEEE Conference on Computer Vision and Pattern Recognition (CVPR)* (pp. 6071–6079).
125. Zhang, T., Ye, S., Zhang, K., Tang, J., Wen, W., Fardad, M., & Wang, Y. (2018). A systematic DNN weight pruning framework using alternating direction method of multipliers. ArXiv, vol. abs/1804.03294.
126. Ye, S., Zhang, T., Zhang, K., Li, J., Xu, K., Yang, Y., Yu, F., Tang, J., Fardad, M., Liu, S., Chen, X., Lin, X., & Wang, Y. (2018). Progressive weight pruning of deep neural networks using ADMM. CoRR, vol. abs/1810.07378.

127. Lebedev, V., & Lempitsky, V. (2016). Fast convnets using group-wise brain damage. In *2016 IEEE Conference on Computer Vision and Pattern Recognition (CVPR)* (pp. 2554–2564).
128. Wen, W., Wu, C., Wang, Y., Chen, Y., & Li, H. (2016). Learning structured sparsity in deep neural networks. ArXiv, vol. abs/1608.03665.
129. Yuan, M., & Lin, Y. (2006). Model selection and estimation in regression with grouped variables. *Journal of the Royal Statistical Society, Series B, 68*, 49–67.
130. Liu, Z., Li, J., Shen, Z., Huang, G., Yan, S., & Zhang, C. (2017). Learning efficient convolutional networks through network slimming. CoRR, vol. abs/1708.06519.
131. Ding, X., Ding, G., Han, J., & Tang, S. (2018). Auto-balanced filter pruning for efficient convolutional neural networks. In *Association for the Advancement of Artificial Intelligence (AAAI)*.
132. Wang, H., Zhang, Q., Wang, Y., & Hu, H. (2019). Structured pruning for efficient convnets via incremental regularization. *2019 International Joint Conference on Neural Networks (IJCNN)* (pp. 1–8).
133. Luo, J.-H., & Wu, J. (2020). Autopruner: An end-to-end trainable filter pruning method for efficient deep model inference. *Pattern Recognit., 107*, 107461.
134. Liu, Z., Sun, M., Zhou, T., Huang, G., & Darrell, T. (2019). Rethinking the value of network pruning. ArXiv, vol. abs/1810.05270.
135. Lin, M., Ji, R., Zhang, Y. X., Zhang, B., Wu, Y., & Tian, Y. (2020). Channel pruning via automatic structure search. ArXiv, vol. abs/2001.08565.
136. Liu, Z., Mu, H., Zhang, X., Guo, Z., Yang, X., Cheng, K., & Sun, J. (2019). Metapruning: Meta learning for automatic neural network channel pruning. In *2019 IEEE/CVF International Conference on Computer Vision (ICCV)* (pp. 3295–3304) .
137. Hubara, I., Courbariaux, M., Soudry, D., El-Yaniv, R., & Bengio, Y. (2017). Quantized neural networks: Training neural networks with low precision weights and activations. *The Journal of Machine Learning Research, 1*, 6869–6898.
138. Mishra, A., Nurvitadhi, E., Cook, J. J., & Marr, D. (2017). WRPN: Wide reduced-precision networks. arXiv:1709.01134.
139. Holt, J. L., & Baker, T. E. (1991). Back propagation simulations using limited precision calculations. In *IJCNN-91-Seattle International Joint Conference on Neural Networks* (vol. 2, pp. 121–126). Piscataway: IEEE.
140. Presley, R. K., & Haggard, R. L. (1994). A fixed point implementation of the backpropagation learning algorithm. In *Proceedings of SOUTHEASTCON'94* (pp. 136–138). Piscataway: IEEE.
141. Simard, P. Y., & Graf, H. P. (1994). Backpropagation without multiplication. In *Advances in Neural Information Processing Systems* (pp. 232–239).
142. Savich, A. W., Moussa, M., & Areibi, S. (2007). The impact of arithmetic representation on implementing MLP-BP on FPGAs: A study. *IEEE Transactions on Neural Networks, 18*(1), 240–252.
143. Gupta, S., Agrawal, A., Gopalakrishnan, K., & Narayanan, P. (2015). Deep learning with limited numerical precision. In *International Conference on Machine Learning* (pp. 1737–1746).
144. He, Q., Wen, H., Zhou, S., Wu, Y., Yao, C., Zhou, X., & Zou, Y. (2016). Effective quantization methods for recurrent neural networks. arXiv:1611.10176.
145. Ott, J., Lin, Z., Zhang, Y., Liu, S.-C., & Bengio, Y. (2016). Recurrent neural networks with limited numerical precision. arXiv:1608.06902.
146. Courbariaux, M., Bengio, Y., & David, J.-P. (2014). Training deep neural networks with low precision multiplications. arXiv:1412.7024.
147. Micikevicius, P., Narang, S., Alben, J., Diamos, G., Elsen, E., Garcia, D., Ginsburg, B., Houston, M., Kuchaiev, O., Venkatesh, G., & Wu, H. (2017). Mixed precision training. arXiv:1710.03740.

148. Kalamkar, D., Mudigere, D., Mellempudi, N., Das, D., Banerjee, K., Avancha, S., Vooturi, D. T., Jammalamadaka, N., Huang, J., Yuen, H., Yang, J., Park, J., Heinecke, A., Georganas, E., Srinivasan, S., Kundu, A., Smelyanskiy, M., Kaul, B., & Dubey, P. (2019). A study of bfloat16 for deep learning training. arXiv:1905.12322.
149. Das, D., Mellempudi, N., Mudigere, D., Kalamkar, D., Avancha, S., Banerjee, K., Sridharan, S., Vaidyanathan, K., Kaul, B., Georganas, E., et al.. (2018). Mixed precision training of convolutional neural networks using integer operations. arXiv:1802.00930.
150. Drumond, M., Tao, L., Jaggi, M., & Falsafi, B. (2018). Training DNNs with hybrid block floating point. In *Advances in Neural Information Processing Systems* (pp. 453–463).
151. Goodfellow, I. J., Warde-Farley, D., Mirza, M., Courville, A., & Bengio, Y. (2013). Maxout networks. arXiv:1302.4389.
152. Wang, N., Choi, J., Brand, D., Chen, C.-Y., & Gopalakrishnan, K. (2018). Training deep neural networks with 8-bit floating point numbers. In *Advances in Neural Information Processing Systems* (pp. 7675–7684).
153. Mellempudi, N., Srinivasan, S., Das, D., & Kaul, B. (2019). Mixed precision training with 8-bit floating point. arXiv:1905.12334.
154. Sakr, C., Wang, N., Chen, C.-Y., Choi, J., Agrawal, A., Shanbhag, N., & Gopalakrishnan, K. (2019). Accumulation bit-width scaling for ultra-low precision training of deep networks. arXiv:1901.06588.
155. Cambier, L., Bhiwandiwalla, A., Gong, T., Nekuii, M., Elibol, O. H., & Tang, H. (2020). Shifted and squeezed 8-bit floating point format for low-precision training of deep neural networks. arXiv:2001.05674.
156. Rajagopal, A., Vink, D. A., Venieris, S. I., & Bouganis, C.-S. (2020). Multi-precision policy enforced training (MuPPET): A precision-switching strategy for quantised fixed-point training of CNNs. arXiv:2006.09049.
157. De Sa, C., Leszczynski, M., Zhang, J., Marzoev, A., Aberger, C. R., Olukotun, K., & Ré, C. (2018). High-accuracy low-precision training. arXiv:1803.03383.
158. Johnson, R., & Zhang, T. (2013). Accelerating stochastic gradient descent using predictive variance reduction. In *Advances in Neural Information Processing Systems* (pp. 315–323).
159. Jia, X., Song, S., He, W., Wang, Y., Rong, H., Zhou, F., Xie, L., Guo, Z., Yang, Y., Yu, L., Chen, T., Hu, G., Shi, S., & Chu, X. (2018). Highly scalable deep learning training system with mixed-precision: Training ImageNet in four minutes. arXiv:1807.11205.
160. Yang, G., Zhang, T., Kirichenko, P., Bai, J., Wilson, A. G., & De Sa, C. (2019). Swalp: Stochastic weight averaging in low-precision training. arXiv:1904.11943.
161. Izmailov, P., Podoprikhin, D., Garipov, T., Vetrov, D., & Wilson, A. G. (2018). Averaging weights leads to wider optima and better generalization. arXiv:1803.05407.
162. Chen, Y., Luo, T., Liu, S., Zhang, S., He, L., Wang, J., Li, L., Chen, T., Xu, Z., Sun, N., & Temam, O. (2014). Dadiannao: A machine-learning supercomputer. In *2014 47th Annual IEEE/ACM International Symposium on Microarchitecture* (pp. 609–622).
163. Chen, Y., Emer, J., & Sze, V. (2016). Eyeriss: A spatial architecture for energy-efficient dataflow for convolutional neural networks. In *ACM/IEEE 43rd Annual International Symposium on Computer Architecture (ISCA)*.
164. Jouppi, N. P., Young, C., Patil, N., Patterson, D., Agrawal, G., Bajwa, R., Bates, S., Bhatia, S., Boden, N., Borchers, A., Boyle, R., Cantin, P., Chao, C., Clark, C., Coriell, J., Daley, M., Dau, M., Dean, J., Gelb, B., et al. (2017). In-datacenter performance analysis of a tensor processing unit. In *2017 ACM/IEEE 44th Annual International Symposium on Computer Architecture (ISCA)* (pp. 1–12).
165. Guo, K., Sui, L., Qiu, J., Yao, S., Han, S., Wang, Y., & Yang, H. (2016). Angel-eye: A complete design flow for mapping CNN onto customized hardware. In *2016 IEEE Computer Society Annual Symposium on VLSI (ISVLSI)* (pp. 24–29).
166. Reddy, R., Reddy, B. M., & Reddy, B. (2018). DLAU: A scalable deep learning accelerator unit on FPGA. *International Journal of Research, 5*, 921–928.

167. Guo, P., Ma, H., Chen, R., Li, P., Xie, S., & Wang, D. (2018). FBNA: A fully binarized neural network accelerator. In *2018 28th International Conference on Field Programmable Logic and Applications (FPL)* (pp. 51–513).
168. Kudo, T., Ueyoshi, K., Ando, K., Hirose, K., Uematsu, R., Oba, Y., Ikebe, M., Asai, T., Motomura, M., & Takamaeda-Yamazaki, S. (2018). Area and energy optimization for bit-serial log-quantized DNN accelerator with shared accumulators. In *2018 IEEE 12th International Symposium on Embedded Multicore/Many-Core Systems-on-Chip (MCSoC)* (pp. 237–243).
169. Lu, Y., Wang, C., Gong, L., & Zhou, X. (2017). SparseNN: A performance-efficient accelerator for large-scale sparse neural networks. *International Journal of Parallel Programming, 46*, 648–659.
170. Albericio, J., Judd, P., Hetherington, T. H., Aamodt, T. M., Jerger, N. E., & Moshovos, A. (2016). Cnvlutin: Ineffectual-neuron-free deep neural network computing. In *2016 ACM/IEEE 43rd Annual International Symposium on Computer Architecture (ISCA)* (pp. 1–13).
171. Zhu, C., Huang, K., Yang, S., Zhu, Z., Zhang, H., & Shen, H. (2020). An efficient hardware accelerator for structured sparse convolutional neural networks on FPGAs. In *IEEE Transactions on Very Large Scale Integration (VLSI) Systems, 28*, 1953–1965.
172. Han, S., Liu, X., Mao, H., Pu, J., Pedram, A., Horowitz, M., & Dally, W. (2016). EIE: Efficient inference engine on compressed deep neural network. In *2016 ACM/IEEE 43rd Annual International Symposium on Computer Architecture (ISCA)* (pp. 243–254).
173. Tao, Y., Ma, R., Shyu, M.-L., & Chen, S.-C. (2020). Challenges in energy-efficient deep neural network training with FPGA. In *Proceedings of the IEEE/CVF Conference on Computer Vision and Pattern Recognition Workshops* (pp. 400–401).
174. Fox, S., Faraone, J., Boland, D., Vissers, K., & Leong, P. H. (2019). Training deep neural networks in low-precision with high accuracy using FPGAs. In *2019 International Conference on Field-Programmable Technology (ICFPT)* (pp. 1–9). Piscataway: IEEE.
175. Venkataramanaiah, S. K., Ma, Y., Yin, S., Nurvithadhi, E., Dasu, A., Cao, Y., & Seo, J.-S. (2019). Automatic compiler based FPGA accelerator for CNN training. In *2019 29th International Conference on Field Programmable Logic and Applications (FPL)* (pp. 166–172). Piscataway: IEEE.
176. Liu, Y., Ott, M., Goyal, N., Du, J., Joshi, M., Chen, D., Levy, O., Lewis, M., Zettlemoyer, L., & Stoyanov, V. (2019). RoBERTa: A robustly optimized BERT pretraining approach. arXiv:1907.11692.
177. Brown, T. B., Mann, B., Ryder, N., Subbiah, M., Kaplan, J., Dhariwal, P., Neelakantan, A., Shyam, P., Sastry, G., Askell, A., Agarwal, S., Herbert-Voss, A., Krueger, G., Henighan, T., Child, R., Ramesh, A., Ziegler, D. M., Wu, J., Winter, C., et al. (2020). Language models are few-shot learners. arXiv:2005.14165.
178. Carion, N., Massa, F., Synnaeve, G., Usunier, N., Kirillov, A., & Zagoruyko, S. (2020). End-to-end object detection with transformers. In *European Conference on Computer Vision* (pp. 213–229). Berlin: Springer.
179. Strudel, R., Garcia, R., Laptev, I., & Schmid, C. (2021). Segmenter: Transformer for semantic segmentation. arXiv:2105.05633.
180. Zadeh, A. H., Edo, I., Awad, O. M., & Moshovos, A. (2020). GOBO: Quantizing attention-based NLP models for low latency and energy efficient inference. In *2020 53rd Annual IEEE/ACM International Symposium on Microarchitecture (MICRO)* (pp. 811–824). Piscataway: IEEE.
181. Kim, S., Gholami, A., Yao, Z., Mahoney, M. W., & Keutzer, K. (2021). I-BERT: Integer-only BERT quantization. arXiv:2101.01321.
182. Sanh, V., Debut, L., Chaumond, J., & Wolf, T. (2019). DistilBERT, a distilled version of BERT: Smaller, faster, cheaper and lighter. arXiv:1910.01108.
183. Jin, J., Liang, C., Wu, T., Zou, L., & Gan, Z. (2021). KDLSQ-BERT: A quantized BERT combining knowledge distillation with learned step size quantization. arXiv:2101.05938.

184. Mao, J., Yang, H., Li, A., Li, H., & Chen, Y. (2021). TPrune: Efficient transformer pruning for mobile devices. *ACM Transactions on Cyber-Physical Systems, 5*(3), 1–22.
185. Wang, H., Zhang, Z., & Han, S. (2021). SpAtten: Efficient sparse attention architecture with cascade token and head pruning. In *Proceedings of the International Symposium on High-Performance Computer Architecture (HPCA)*.
186. Wang, Y., Jiang, Z., Chen, X., Xu, P., Zhao, Y., Lin, Y., & Wang, Z. (2019). E2-train: Training state-of-the-art CNNs with over 80% energy savings. In *Advances in Neural Information Processing Systems* (pp. 5138–5150).
187. Bernstein, J., Wang, Y.-X., Azizzadenesheli, K., & Anandkumar, A. (2018). signSGD: Compressed optimisation for non-convex problems. arXiv:1802.04434.
188. Fu, Y., You, H., Zhao, Y., Wang, Y., Li, C., Gopalakrishnan, K., Wang, Z., & Lin, Y. (2020). FracTrain: Fractionally squeezing bit savings both temporally and spatially for efficient DNN training. arXiv:2012.13113.
189. Fu, Y., Guo, H., Li, M., Yang, X., Ding, Y., Chandra, V., & Lin, Y. (2021). Cpt: Efficient deep neural network training via cyclic precision. arXiv:2101.09868.
190. Shafiee, A., Nag, A., Muralimanohar, N., Balasubramonian, R., Strachan, J. P., Hu, M., Williams, R. S., & Srikumar, V. (2016). ISAAC: A convolutional neural network accelerator with in-situ analog arithmetic in crossbars. In *Proceedings of the International Symposium on Computer Architecture (ISCA)* (pp. 14–26).
191. Chi, P., Li, S., Xu, C., Zhang, T., Zhao, J., Liu, Y., Wang, Y., & Xie, Y. (2016). PRIME: A novel processing-in-memory architecture for neural network computation in ReRAM-based main memory. In *Proceedings of the International Symposium on Computer Architecture (ISCA)* (pp. 27–39).
192. Ankit, A., Hajj, I. E., Chalamalasetti, S. R., Ndu, G., Foltin, M., Williams, R. S., Faraboschi, P., Hwu, W.-M. W., Strachan, J. P., Roy, K. et al. (2019). PUMA: A programmable ultra-efficient memristor-based accelerator for machine learning inference. In *Proceedings of the International Conference on Architectural Support for Programming Languages and Operating Systems (ASPLOS)* (pp. 715–731).
193. Sebastian, A., Boybat, I., Dazzi, M., Giannopoulos, I., Jonnalagadda, V., Joshi, V., Karunaratne, G., Kersting, B., Khaddam-Aljameh, R., Nandakumar, S. R., Petropoulos, A., Piveteau, C., Antonakopoulos, T., Rajendran, B., Gallo, M. L., & Eleftheriou, E. (2019). Computational memory-based inference and training of deep neural networks. In *Proceedings of the Symposium on VLSI Technology* (pp. T168–T169).
194. Demler, M. (2018). Mythic multiplies in a flash. Microprocesser Report.
195. Sebastian, A., Le Gallo, M., Khaddam-Aljameh, R., & Eleftheriou, E. (2020). Memory devices and applications for in-memory computing. *Nature Nanotechnology, 15*(7), 529–544.
196. Joshi, V., Le Gallo, M., Haefeli, S., Boybat, I., Nandakumar, S. R., Piveteau, C., Dazzi, M., Rajendran, B., Sebastian, A., & Eleftheriou, E. (2020). Accurate deep neural network inference using computational phase-change memory. *Nature Communications, 11*(1), 1–13.

Index

A
Absolute energy model
ALU, 205
cache memory energy, 206
CPU, 205
memory energy, 206
system, 205
ACCEPT framework, 164
Accuracy analysis, 178
Accuracy and *Cost* tradeoff, 4
Accuracy-aware compilers
accuracy–performance tradeoff space
approximation knob, 180
fixed approximation, 180
runtime system, 180
accuracy specification
accuracy goal, 180
accuracy metric function, 179
output abstraction, 179
approximate kernel, 183–185
approximation, 178
configuration representation, 179
optimization formulation, 179
optimization search space, 182–183
profitable tradeoffs, 178
sensitivity profiling-based techniques, 178
static analysis-based techniques, 178
transformations (*see* Accuracy-aware transformations)
Accuracy-aware optimization techniques
analysis-based compilation, 182–183
sensitivity profiling-based compilation, 182
Accuracy-aware transformations
accuracy-aware software, 181
approximate hardware components, 181
computation find instructions, 181
data representation, 181
flexible framework, 178
program semantics, 178, 180
Accuracy constraint, 189
Accuracy degradation, 4
Accuracy estimation, 35, 193
Accuracy metric (error power), 147
Accuracy metric determination
analytical techniques
IA, 165–168
perturbation theory, 168–171
simulation-based techniques, 171–172
Accuracy metric function, 179
Accuracy/performance tradeoff space, 180, 210
AC source error model, 146
Active attacks, 328
Adaptive precision
in preconditioner
AC in Block-Jacobi preconditioners, 427
block-Jacobi preconditioners, 426
experimental evaluation, 428–430
KSMs, 425–426
practical implementation, 427–428
in stationary solvers
AC in Jacobi relaxation method, 423
experimental evaluation, 424–425
Jacobi relaxation method, 422–423
practical implementation, 423–424
stationary methods, 425
Adaptive Weight Precision (AWP) algorithm, 432, 433, 435–436, 442, 444, 446–450, 458

A. Bosio et al. (eds.), *Approximate Computing Techniques*,
https://doi.org/10.1007/978-3-030-94705-7

Adversarial attacks on neural networks (NNs)
adversarial attack, 336–337
perturbations, 336
types of attacks, 337–338
Affine arithmetic (AA), 165
Algebraic Decision Diagrams (ADDs), 91
Algorithmic data-sketching techniques, 181
Algorithmic level AxC
aim, 109
approximation based (*see* Approximation based approaches)
energy consumption, 109
HEVC (*see* HEVC video codec)
processing complexity, 109
skip-based (*see* Skip-based approaches)
Algorithmic Parrot transformation, 120
Algorithm selection, 118
Almost Correct Adder (ACA), 149
ALU/FPU arithmetic operation, 197
ALU/FPU operations, 199
Analog in-memory computing, 501–502
Analysis-based compilation, 182–183
Analytical expression, 170
Application-based fault tolerance (ABFT) encode, 394
Application domains, 177
Application quality metrics, 151
Application-specific integrated circuits (ASICs), 83
Applied Micro's (APM) X-Gene 2 micro-server, 50
ApproxHPVM, 209
Approximate adders, 81, 95
Approximate application-specific processing cores
artificial neurons, 48
convolutional neural network, 48
DSP, 48
Goya, 48
NVIDIA Tensor Cores, 48
significance driven design methodology, 49
TPU, 48
Approximate Array Multiplier (AAM), 149
Approximate Booth multipliers, 96
Approximate circuits
approximation methods, 84
arithmetic circuits, 82
automated methods, 97–98
design abstraction, 83
error analysis methods, 82
error metrics, 85
EvoApprox library, 99–100
formal error analysis, 87–92
multi-objective optimization problem, 92–94
number representation, 83
problem-specific approaches, 94–97
quality configurable circuits, 85–87
target technology, 83
Approximate circuits using less hardware, 44
Approximate compressor, 47
Approximate computing (AxC/AC)
accuracy level, 3
aim, 3
applications, 3, 6
approximate arithmetic primitives, 325
categories, 44
comprehensive analysis, 7
in deep learning (*see* Deep learning (DL))
design paradigm, 215
design space, 4, 5
digital communication receiver, 146
emerging research field, 4
error analysis, 145
error-resilient properties, 43
intuitive observation, 3
low-end IoT devices, 323
non-determinism, 324
opportunity, 177
Pareto frontier, 4
post-Moore's Law computing, 145
on safety-critical systems, 391 (*see also* Safety-critical systems)
for scientific applications (*see* Scientific applications)
security, 323
and security solutions
cryptography (*see* Cryptography based on approximate primitives)
defending against adversarial attacks on NNs (*see* Adversarial attacks on neural networks (NNs))
and security threats
active attacks, 328
cloning and counterfeiting, 330–331
Hardware Trojans, 329–330
passive side channel attacks, 326–327
reverse engineering, 328–329
stochastic computing, 323
techniques, 324, 416
Approximate Computing Techniques (AxCTs), 216–217
Approximate data transfer (ADT), 432, 433, 436–438

Approximate digital Integrated Circuits (AxICs)
applications, 350
ASLAN, 357
barrier, 358
Bayesian conditional probability, 381
Bayesian inference theory, 379
Bayesian Network (BN), 379
Bayesian Theory, 380
BN model, 382
critical defects, 351
defect modeling, 353–354
design and manufacturing flow, 350
error magnitude, 357, 360
error rate, 357
fault analysis, 360
fault injection, 358, 359, 379
fault selection, 360
fault simulation, 354–355, 358
faulty operators, 382
functional approximation, 350, 357
functional test, 352–353
in-field application, 351
in-field testing, 352
inherent resiliency property, 349
multi-level circuits, 357
operations and functions, 381
over-scaling based approximation, 357
prediction average error, 382
production testing, 351–352
random variable, 381
SASIMI, 357
structural test, 353
test generation, 355–356
testing, 351
testing phase
acceptable behaviors, 361
error thresholds, 363
fault classification, 363–366
metrics, 362
simple arithmetic circuit, 362
test pattern generation, 363, 366–368
test set application, 363, 368–371
two-bit multiplier, 358, 359
user-define threshold, 380
verification, 351
verification phase, 360–361
Approximated Logic Synthesis (ALS), 357
Approximate fault tolerance, 395–396
Approximate homomorphic encryption, 333–335
Approximate kernels, 182
accuracy loss, 183
application's input, 183
elements, 184
empirical observations, 184
end-to-end acceptability, 184
loop, 184
multiple transformations, 184
probabilistic accuracy specifications, 185
profiling-based approaches, 184
sensitivity profiling, 185
structure and functionality, 184
Approximate memories
DRAM, 67–70
pessimistic operating parameters, 65
power/energy consumption, 65
SRAM, 65–67
Approximate multi-bit full adders, 95
Approximate multipliers, 81
Booth's algorithm, 96
classes, 96
combinational n-bit multiplier, 95
divide-and-conquer strategy, 96
DRUM, 96
$2n$-bit, 97
RoBA, 96
specialized accelerators, 96
structure, 96
TOSAM, 96
Approximate storage, 155, 325
Approximate techniques facilitating voltage scaling, 44
Approximate TMR (ATMR)
approximation intensity, 397
and FATMR, 396
hardware ATMR based on data precision approximation
accuracy assessment, 399–400
area usage assessment, 400–401
converters, 398
exhaustive fault injections, 403–404
fault tolerance, 397
pseudo-C code, 397
random accumulated fault injection, 401–403
TMR implementation, 397
software ATMR based on successive approximation
evaluation, 406–411
Newton-Raphson method, 404
proposed ATMR method, 405
"71-37-14" (ATMR configuration), 405
traditional TMR, 405
Approximation-based techniques, 44
algorithm selection, 118
mathematical functions (*see* Mathematical functions approximation)

Approximation-based techniques (*cont.*)
  memoization, 119–120
  neural network, 120
  parameter adjustment, 118–119
Approximation errors
  accuracy metrics, 148–151
  AC source error model, 146
  application quality metric, 145
  categories, 147
  definition, 147
  occurrence probability, 147
  output quality, 146
  Widrow model, 146
Approximation for inference
  classes of methods, 476
  data-oriented refinement transformations, 476
  DNN inference, 475–476
  network sparsification (pruning) (*see* Pruning)
  operator refinement transformations, 476
  pruning, 476
  quantization (*see* Quantization)
  structure refinement transformations, 476
  weight sharing
    granularity, 484–485
    grouping methods, 483–484
    shared values, 482
    techniques, 482
Approximation for training
  accelerating training, 489
  DoReFaNet, 489
  fixed-point arithmetic on DNNs, 489–490
  low-precision training algorithm design, 496–498
  mixed-precision training approaches (*see* Mixed-precision training)
  recurrent neural network (RNN) training, 490
Approximation intensity, 397
Approximation knob, 180
Approximation level control, 127, 128
Approximation strategies
  combination, 64
  dynamic prediction, 56–58
  memory technologies, 54
  NTC, 54
  path redistribution (*see* Path redistribution, voltage down-scaling)
  power savings, 65
  precision scaling and operand truncation, 55–56
  SDC, 54
  special storage modules, 55
  timing errors, 54
  voltage down-scaling, 54
ApproxIt framework, 114
ARC framework, 164
Arithmetic/affine arithmetic, 19
Arithmetic circuits, 46
Arithmetic customization, 11
Arithmetic error metrics
  approximation, 87
  Boolean function, 87
  discrete random variable, 89
  error quantification, 87
  mean relative error, 88
  mean-squared error, 88
  natural binary representation, 88
  proposition, 89
  PSNR, 88
  relative worst-case error, 88
  statistically oriented, 88
  weighted mean error distance, 88
  worst-case arithmetic error, 87, 88
Arithmetic operators, 81
ARM A7-based system, 62
ARMv8-based multicore CPUs, 51
ARMv8-compliant CPUs, 51
Array load/store, 201
Array safety constraints, 201
Artificial intelligence (AI), 11
  backpropagation, 470
  basic DNN example and associated terminology, 470
  edge AI, 501
  email filtering, 469
  high-level features, 469
  inference, 470
  mini-batch, 471
  ML modeling, 469
  neurons, 469
  SGD iterative optimization process, 470
Automated approximation methods, 85
Automated Behavioral Synthesis of Approximate Computing Systems (ABACUS), 224–225
Automated characterization process, 50
Automated circuit optimization, 84
Automated functional approximation methods
  approximate circuits, 98
  basic techniques, 97
  benchmark problems, 98
  candidate designs, 97
  CGP, 97
  common (exact) circuit, 97
  Pareto front, 97

Automated Test Equipment (ATEs), 351–352, 370
Automatic Methodology for Sequential Logic Approximation (ASLAN), 222–224
Automatic synthesis, 47
Automatic Test Pattern Generation (ATPG), 356
Average-case arithmetic error, 88
Average-case error analysis, 92
Average relative error, 63, 64
AxC error characterization
  analytical techniques
    finite precision arithmetic, 156–157
    inexact operators, 157–160
    voltage-overscaled circuits, 160
  simulation-based techniques
    exhaustive simulations, 161–162
    Monte Carlo simulations, 162
    pre-characterization, analytical techniques, 162–163
    voltage-overscaled circuits, 160
AxC filters, 127
AxC impacts analysis metrics
  application quality metrics, 151
  bitwise, 149–150
  interval-based, 150–151
  statistical, 148–149
AxC techniques emulation
  abstraction levels, 155
  error modeling, 155
  finite precision arithmetic, 152, 154
  inexact arithmetic operators, 151–154
  intermediate accuracy metric, 151, 152
  operator overloading, 155
  voltage overscaling, 155

B
Basic algorithmic approximation techniques, 97
Bellerophon, 245–246
Binary neural network (BNNs), 478–479
Bin Packing, 118
Bi-partite methods, 122
2-Bit approximate multipliers, 84
8-Bit approximate multipliers, 100, 103
16-Bit approximate multipliers, 103
Bit-error-rate (BER) simulations, 147, 272
4-Bit ripple carry adder, 45
Bitwise Error Rate (BWER), 149
Bitwise metric, 149–150
Bivariate, 124
Bivariate non-uniform segmentation, 124
Bjøntegaard Delta Bit Rate (BD-BR), 133
BLAS (Basic Linear Algebra Subprograms), 416
Block-based adders, 158
Block class modification, 127
Blockers, 237
Block Floating-Point (BFP) format, 480
Block-Jacobi preconditioners, 426–429
Boolean function, 91
Boolean Matrix Factorization (BMF), 227–228
Bounded loops, 202
Brakerski-Gentry-Vaikuntanathan scheme, 334
Branch-and-bound technique, 117, 266
Built-in self-test (BIST) architectures, 353, 370

C
Candidate designs, 97
Carry-Look-Ahead Adder (CLA), 160
Carry propagation, 55
Carry-select adders, 95, 160
Cartesian genetic programming (CGP), 97, 98
Cheon, Kim, Kim and Song (CKKS), 335
Chisel framework
  approximate hardware specifications, 195
  approximate instructions and data selection, 193
  constraint simplification, 203–205
  energy objective construction, 205–207
  exact program, 193
  final optimization problem statement, 207–208
  intermediate language, 196–197
  optimization algorithm, 196
  optimization constraint construction, 204–205
  reliability and accuracy specifications, 193, 194
  reliability constraint construction, 198–205
  reliability predicates, 197–198
  typical program inputs, 195
Chisel's optimization algorithm
  approximate instructions and variables selection, 196
  decision variables specification, 196
  energy savings objective computation, 196
  ILP, 196
  reliability and accuracy constraints computation, 196
Chromosome, 226
Circuit approximation problem, 81
Circuit level functional approximation
  automated synthesis, 47
  inexact units, 46–47
Circuit models, 99

Circuit simulation, 101, 103
Circuit Under Test (CUT), 229
CKKS (Cheon-Kim-Kim-Song) scheme, 334
Clang-Chimera mutator
  AST pattern matching logic, 248
  configuring and running, 253–254
  defining and registering, 252–253
  definition, 249
  fine-grained matcher function, 250
  fine-grained matching rules, 250
  getStatementMatcher function, 251
  getXXXMatcher functions, 249
  mutate function, 252
  mutation rules methods, 250
  variables, 250
Clang/LLVM Just-In-Time compiler (LLVM-JIT), 246
C++ library Boost, 166
C++ library LibAffa, 166
Clock frequency, 44
Clock-to-output delay, 59
Cloning, 330
CMOS technology, 2, 3
Coarser-grained optimization, 209
Coding Block (CB), 132
Coding Tree Block (CTB), 132
Coding-tree partitioning (CT), 134
Combinatorial optimization problem, 188
Commercial off-the-shelf (COTS), 388
Common (exact) circuit, 84
Compilers, 177
Compile-time static analysis, 177
Compressors, 96
Computational complexity, 117, 127
Computational kernels, 113
Computation skipping, 111
Computer arithmetic, 11, 21
Computing devices, 2, 3
Computing system layers, 5, 6
Conditional Probability Tables (CPTs), 380
Conditionals, 201, 202
Configuration, 197
Conjunctive Normal Form (CNF), 90, 361
Constant, 198
Construction, 269
Continuous-amplitude random variable, 157
Continuous-amplitude signal quantization, 156
Conventional guardband-based techniques, 55
Convolutional (CONV) layers, 473
Convolutional neural networks (CNNs), 431
  Batch Normalization (BN), 474
  computational workload, 467
  CONV layers, 473
  FC layers, 472–473
  intrinsic tolerance, 468
  typical 2D CONV layer, 473
COordinate Rotation DIgital Computer (CORDIC), 114
CORDIC algorithm, 121
Counterfeiting, 330
Critical systems, 18
Cryptographic accelerators, 325
Cryptography based on approximate primitives
  approximate homomorphic encryption, 333–335
  homomorphic encryption, 331–333
  implementations and standardization, 335–336
CT_FLOAT library, 30
CT partitioning OSSE approximation
  approximation management, 135
  coarse solution predictor design, 135
  quality and cost evaluation, 135
Cumulative distribution function (CDF), 314
Customization, 39
Customized fixed-point *vs.* floating-point arithmetic
  application-level, 32–39
  classical fixed-point choice, 30
  number representations, 30
  operator-level, 30–32
Cyclic redundancy check (CRC), 394–395

D

Dadda multiplier, 48
Data-oriented processing, 109
Data-oriented refinement transformations, 476
Data structure, 112
Data-type transformations, 224
Deblocking filter processes, 127
Deep learning (DL)
  analog in-memory computing, 501–502
  ANNs, 472
  approximation
    for attention-based architectures, 500–501
    for inference (*see* Approximation for inference)
    for training (*see* Approximation for training)
  approximation in DNN accelerators
    ASICs, 498
    DeepCompression, 499
    Feature Map (FM) sparsity, 500
    training acceleration, 500
  DNNs (*see* Deep Neural Networks (DNNs))

FC layers, 472
HPC systems for IoT deployment
constrained search and performance characterization for IoT, 454–458
transprecision emulation framework with PyTorch, 452–454
multiprecision and approximate computing
ADT procedure, 432, 436–438
AWP algorithm, 432, 435–436
Bitpack procedure, 438–439
Bitunpack procedure, 441
DNNs training, 432
evaluation, 441–451
existing approaches, 434–435
experimental setup, 441–444
SIMD-based Bitpack, 439–440
training DNNs on multi-GPU environments, 433
performance and energy profiles of models, 474–475
physical computational models, 431
Tensorflow and Pytorch, 472
Deep neural networks (DNNs), 40, 371, 467, 472
hardware accelerators, 431
multiprecision and approximate computing (*see* Deep learning)
training process, 431
use, 431
Delay fault model, 354
Design-centric error mitigation scheme, 55
Design-for-testability (DfT), 352, 356
Design-Space Exploration (DSE), 145, 217
Design space exploration tools
approximate arithmetic circuits, 226–227
Approximate Computing Techniques (AxCTs), 216–217
Automated Behavioral Synthesis of Approximate Computing Systems (ABACUS), 224–225
Automatic Methodology for Sequential Logic Approximation (ASLAN), 222–224
AxC automation tools, 218
AxC leverages, 216
Boolean Matrix Factorization (BMF), 219, 227–228
CIRCA, 229–230
error-resilient applications, 216
fields of application, 216
growth, 215
issues and open challenges, 217
machine learning applications, 215
power-aware and branch-aware word-length optimization, 218, 220
software applications
ACCEPT framework, 236–238
ASAC, 239–240
EnerJ, 230–232
Intel's Approximate Computing Toolkit (iACT), 240–241
precimonious, 234–235
REACT modeling framework, 238–239
Self-tuning Approximation for Graphic Engines (SAGE), 235–236
variable-accuracy algorithms, 232–234
Statistically Certified Approximate Logic Synthesis (SCALS), 225–226
Substitute-And-SIMplIfy (SASIMI), 221–222
Systematic Methodology for Automatic Logic Synthesis of Approximate circuits (SALSA), 220–221
tools, 218, 219
Digital circuits, 82
Digital computing machine, 151
Digital signal processing, 160
Directed acyclic graphs (DAGs), 170, 309
Direction Cosine Matrix update algorithm, 123
Direct quality metric determination, 164
Discrete Cosine Transform (DCT), 220
Discrete optimization techniques, skipping, 116–117
Displacement damage (DD), 390
Distance computation, 35
Distance (error) metric, 90
Distortion, 180
Distortion function, 190
Divide-and-conquer strategy, 96
2D K-means clustering algorithm, 35
DNN accelerators
for machine learning applications, 325
DNN-based classification, 325
DNN inference, 475
DNNs data type approximation
accuracy loss, 372
application-dependent outcome, 373–374
8-bit floating-point variables (FP8), 375
classification, 375
Critical Faults, 374
custom data type, 372
darknet framework, 372
experiments, 374
fault injection, 372, 377
floating-and fixed-point versions, 375
floating-point and fixed-point representations, 372

DNNs data type approximation (*cont.*)
FloatX library, 372
Golden Custom, 373
Golden Standard, 373
LeNet-5 data type, 374, 375
LeNet-5 fault injection outcomes, 375, 376
LeNet-5 fault list, 377, 378
libfixmath library, 372
safety and memory, 377
Domain-specific quantization scheme, 83
DRAM refresh rate, 67, 155
Duplication with comparison (DWC), 389
Dynamic operand truncation, 63
Dynamic Power Management (DPM), 110
Dynamic Random Access Memory (DRAM), 155
alternative approaches, 68
critical data structures, 67
DIMM, 67
energy savings, 70
error behaviour, 67, 68
evaluation results, 68
HaRMony scheme, 69
HRM benefits, 67
inherent application error resiliency, 70
MCUs, 67
memory errors rate, 68
nominal circuit parameters, 67
power consumption, 67
Raspberry Pi 3, 69
reliability, 68
resilient/non-critical data, 67
SLIMpro, 67
virtualization software, 69
WER metric, 68, 69
X-Gene 2, 67, 68
Dynamic range unbiased multiplier (DRUM), 96
Dynamic sensitivity analysis, 210
Dynamic Voltage and Frequency Scaling (DVFS), 49, 109, 155

E

Earliest Deadline First (EDF), 291
Earliest Deadline First with Virtual Deadlines (EDF-VD), 290
Early termination, 114–115
Easy-to-compute function, 96
ECC corrected errors, 51
Electrical characteristics, 82
Electronic Design Automation (EDA), 47
Elementary approximate arithmetic circuits, 81
EmETXe-i87M0 platform, 133
EMEURO, 120
Energy consumption, 12, 48
computing systems, 1, 2
manipulated data, 1
mobile broadband networks, 1
type of operations, 1
Energy cost, 2
Energy efficiency, 1
Energy-efficient gains, 49
Energy model parameters, 195
Energy objective construction
absolute energy model, 205–206
energy consumption, 205
relative energy model, 206–207
relative energy savings, 205
Energy-per-instruction costs, 47
Energy reduction opportunities, 137
Equally Segmented Adder (ESA), 157
Equivalent faults, 354
Error amplitude, 160
Error analysis, 270
Error analysis methods, 85
circuit simulation accuracy, 101–103
computational requirements, 100–101
Error characterization, 146
Error criterion, 122
Error detection and correction, 394–395
Error detection by duplicated instructions (EDDI), 394
Error distance (ED), 148, 149
Error fitness function, 245
Error injection, 164
Error magnitude/error significance, 87
Error metrics, 85
Error model, 270
Error pattern, 158
Error probability, 89
Error-prone floating-point operands, 62
Error-prone instructions, dynamic prediction
approximation-based schemes, 56
bitwidth truncation, 56
carry propagation, 58
dynamic data-dependent sensitization, 56
LLPPU, 58
LLPs, 56
LSBs, 58
precision scaling-induced quality loss, 56
prediction units, 58
quality degradation, 57
SLPs, 58
Error propagation, 146
Error rate (ER), 35, 89, 149
Error resilience, 172
Error-resilient code, 44

Error-resilient GPU applications, 55
Error-Tolerant Adder Type II (ETAII), 157
Error-tolerant applications, 46
Error variance, 159
E^2-Train, 501
EvoApprox8b, 99
EvoApprox8b-Lite, 99
EvoApprox library, 99–100, 102
Evolutionary Algorithms (EAs), 396
Evolutionary computing-based framework, 47
Evolving Objects (EOs) framework, 246
Exhaustive approach, 189
Exhaustive functional simulations, 161–162
Expected Yield Increase (eYI), 363–366
Experimental analysis, 62
Expression and variable-to-constant transformations, 224

F

Fail moderate, 148
Fail rare, 147
Fail small, 147
Fast error estimation, 271
Fast Fourier Transform (FFT), 230
  additions/subtractions, 38
  DIT, 38
  energy peak, 39
  energy per operation (pJ), 38
  error-energy trade-off, 38
  FxP outperforms FlP, 39
  hardware performance estimation, 38
Fault coverage (FC), 366, 367
Fault injection, 356, 359, 373, 376–379
Fault masking, 410
Fault tolerance technique
  approximate fault tolerance, 395–396
  ATMR (*see* Approximate TMR (ATMR))
  CFT-tool, 394
  classification, 393
  CRC, 394–395
  devices, 393
  DWC techniques, 394
  EDDI, 394
  NVP, 394
  safety-critical systems, 393
  SIHFT techniques, 393–394
  TMR, 393
Fault-tolerant mechanisms, 3
FHEW (Fully Homomorphic Encryption library), 335
Field programmable gate array (FPGA), 83, 388
Filtering block f, 111
Final optimization problem statement
  candidate approximate instructions, 208
  decision variables, 208
  individual instruction, 208
  kernel computation, 207
  off-the-shelf ILP solver, 208
  relative energy, 208
  reliability factors, 208
Finite Impulse Response (FIR), 119, 220
Finite precision arithmetic, 171
  C++-based fixed-point data-types, 154
  commercial high-level tools, 154
  custom floating-point data-types, 154
  dynamic fixed-point data-type, 154
  fixed-point simulation, 154
  Matlab/Simulink, 154
  SystemC fixed-point data-types, 154
FlP competitiveness, 36
FlP representation, real numbers
  computations, 21
  decimal arithmetic, 21
  high dynamic range, 21
  IEEE 754-2008, 21, 22
  mantissa m, 21, 22
  scaling factors, 21
  sign bits, 21
First Order Mutator (FOM), 243
Fixed-point arithmetic, 146, 468, 479–480, 500
Fixed-point computation errors, 122
Fixed-point conversion process
  aim, 12
  cost function $C(\cdot)$, 18
  energy consumption, 17
  fractional part word-length determination, 20
  implementation cost reduction, 17
  integer and fractional part, 17
  integer-part word-length determination, 18–20
  optimization algorithm, 18
  quality degradation, 18
  word-length m, 17
  word-length n, 18
  word-length optimization, 17
Fixed-point data types, 40, 171
Fixed-point format conversion, 157
Fixed-point numbers, 12, 31
Fixed-point (FxP) representation
  fixed-point value, 13
  format propagation, 14–15
  FWL, 13
  IWL, 13
  m and n, 13
  overflow modes, 16–17

Fixed-point (FxP) representation (*cont.*)
  Q-format notation, 13
  quantization process, 15–16
  real numbers, 11
  reduced-precision, 12
  specification, 13
  virtual BP, 13
Fixed-priority EDF (fpEDF), 291
FlexJava, 210
Floating-point adder, 30, 31
Floating-point arithmetic, 278, 446
Floating-Point Cores (FLOPOCO), 26
Floating-point error magnitude, 30
Floating-point instructions, 47, 63
Floating-point multipliers, 31
Floating-point operations, 47
Floating-point operators
  32-bit FlP multiplication energy, 24
  classical hardware overheads, 23
  FlP adder implementation, 22
  FlP addition *vs.* integer addition, 23
  FlP multiplication, 23
  high control overhead, 22
  integer multiplier, 23
  Synopsys Design Compiler, 23
Floating-point (FlP) representation
  addition and multiplication principle, 12
  high-performance computing, 12
  low-energy benefits, 12
  low-precision, 24–25
  operators, 22–24
  potentially high dynamic range, 20
  real numbers, 20–22
  reduced-precision, 25–29
  scaling factor, 11
Floating-point simulation, 170
FlP and FxP arithmetic operators, 36
FlP distance computation, 36
Formal analysis methods, 85
Formal error analysis
  approximate arithmetic circuits, 87
  average-case error analysis, 92
  determination, 87
  error metrics, 87–89
  relaxed equivalence checking, 89–91
  worst-case error analysis, 91–92
fpEDF-Virtual Deadline (fpEDF-VD), 291
FPGA architectures, 11
FPU design, 63
FPU re-execution, 55
Fractional motion vector compensation, 126
Fractional part word-length determination, 20
Full Adder (FA), 159
Functional approximation, 82
Functional interruption (FIs), 388
Functional simulation techniques, 152, 161
Function returning constant analysis, 199, 200
Fuzzy memoization, 119

G

Gate-level error characterization, 159
General-purpose low-energy processors, 40
General-purpose processing cores
  abnormal behaviours, 51–52
  behaviour formalization, 50
  comprehensive characterization, 50
  design enhancements, 53
  DVFS, 49
  dynamic variation, 49
  full energy saving potential, 50
  full precision, 49
  fully automated system-level framework, 50
  severity function, 50
  standard IEEE 754, 49
  static variation, 49
  undervolting effects' mitigation, 52–53
  workloads, 49
Generic simulation-based framework, 162
Genes, 226
Genetic algorithm (GA), 124, 231
Global scheduling technique
  active low-criticality job, 298–300
  fpEDF-VD, 295–296
  high-criticality mode, 295
  service preserving interval, 297–298
  service preserving method, 296–297
  service preserving policy, 300–301
GPU-based deep learning, 83
Gradient Descent (GD) method, 431
Granularity, 115
Graphic processing units (GPUs), 83, 270
Greedy algorithm, 165
Greedy Randomized Adaptive Search Procedure (GRASP), 269
Groups of Frames (GOF), 135

H

Hadamard transforms, 191
Hard errors, 388
Hardware, 5
Hardware-based/software-based mitigation approaches, 52
Hardware design, 40

Hardware design guidelines
finer-grained voltage domains, 53
hardware detectors, 53
stronger error protection, 53
Hardware implementation cost, 19
Hardware security, 328, 330
Hardware Trojans, 329–330
HaRMony scheme, 69
Heartwall program, 64
H.264 encoding algorithm, 190
Heterogeneous architectures, 44, 415
Heuristic search strategies, 210
HEVC decoder use-case
algorithm classification, 125
block selection, 126–127
block transformations, 127–129
control-oriented data, 125
DCT-like process, 125
DF and SOA, 125
domain conservation block, 125
energy consumption trade-off, 130
entropy decoder, 125
intra-/inter-frame type, 125
intra/inter mode selection, 125
intra-prediction, 125
power measurements, 129
signal-oriented data, 125
HEVC encoder use-case
CTB, 132
CT partitioning OSSE approximation, 135
CTUs, 130
decoder processing loop, 131
experimental set-up, 133
hybrid video encoder, 130
IM prediction OSSE approximation, 135–137
intra encoding, 131, 132
OSSE algorithm identification and classification, 134
OSSE combination, 137–139
PB, 131
RD, 131
RDO step, 132
RMD, 132
HEVC intra-frame prediction, 131
HEVC legacy interpolation filters, 127
HEVC test Model (HM), 132
HEVC video codec
algorithmic level, 124
decoder use-case (*see* HEVC decoder use-case)
encoder use-case (*see* HEVC encoder use-case)
Hierarchical analysis, 159–161
High Efficiency Video Coding (HEVC), 110
High-Level Synthesis (HLS), 26, 389, 500
High Order Mutator (HOM), 243
High-performance computing (HPC), 5, 12, 416, 434
Homogeneous error magnitude, 30
Homomorphic encryption
approximate computing, 333
bootstrapping procedure, 333
construction, 333
description, 331–332
Gentry's strategy, 333
limitation, 332
partially homomorphic, 332
potential, 332
RSA scheme, 332
Homomorphic Encryption for Arithmetic of Approximate Numbers (HEAAN), 335
Hybrid techniques, 170
Hyperparameters, 120

I

IDEA tool suite
Bellerophon, 245–246
Clang-Chimera, 241–244
code mutation, 244–245
lack of generic automation tools, 241
mutants, 241
space exploration example, 246–247
user-defined approximation methods, 241
walk-through
BAS12 algorithm, 248
Bellerophon, 248, 254–255
Clang-Chimera mutator (*see* Clang-Chimera mutator)
DCT computation algorithm, 247, 248
Image filter implementation, 3
IM OSSE study, 138
Imprecise MC (IMC) system, 292–293
IM prediction OSSE approximation
approximation management, 136–137
coarse solution predictor design, 135–136
quality and cost evaluation, 137
Inexact arithmetic operators
BALL simulation, 152
bit-accurate simulations, 152
comparison, 153
floating-point simulation, 152
functional simulation, 152
ratio *r*, 153, 154
required time, 153

In-field applications
collateral resiliency effect, 371
DNNs data type approximation (*see* DNNs data type approximation)
Infinite Impulse Response (IIR), 166
Initial population, 247
In-loop filters, 126, 127
Instruction-level approximation granularity, 209
Instruction pipelining, 59
Instruction set architecture (ISA), 49
Instrumental Variable approach, 123
Integer Linear Programming (ILP), 367
Integer-part word-length determination
dynamic range, 18–19
IWL determination, 19–20
scaling operations, 18–20
Intel Running Average Power Limit (RAPL), 133
Intel's Approximate Computing Toolkit (iACT), 240–241
Interleaving perforation, 113, 186
Intermediate accuracy metric, 146, 171, 172
Intermediate language, 196–197
Internet-of-Things (IoT), 1, 349, 452
constrained search and performance characterization, 454–458
Interval arithmetic (IA), 150, 165
Interval-based arithmetic
IA and AA, 165–167
MAA, 167–168
Interval-based metric, 150–151
Intra-mode prediction (IM), 134
Iterative refinement, 112, 114, 115, 216, 421

J

Jacobi preconditioner, 419
Jacobi relaxation method, 422–425
Jacobi solver, 415
JCT-VC H.264/AVC standard, 126
Joint reliability factor, 198

K

Kernel-level accuracy specifications, 183
K-means clustering algorithm
accuracy estimation, 35
accuracy target, 34
bidimensional sets, 33
distance computation function, 35
double-precision floating-point computations, 33
energy estimation, 35
estimation-maximization process, 33
experimental results, 35–37
Gaussian distributions, 35
iterative distance computation, 35
multidimensional space, 32
set of clusters, 33
steps, 34
stopping conditions, 34, 36, 114
vector quantization, 32
Knapsack problem, 189
Kolmogorov–Smirnov hypothesis test, 239
Krylov subspace methods, 415, 417, 419

L

Labeled reliability, 198
LAPACK (Linear Algebra PACKage), 416
LAS (Basic Linear Algebra Subprograms), 416
Latin Hypercube Sampling technique, 239
Least significant bits (LSB), 44, 119, 158, 159, 262
LeNet-5, 371
LibTooling interface, 244
Lightweight technique, 118
Linear algebra
AC for iterative linear system solvers, 420
adaptive precision in stationary solvers (*see* Adaptive precision)
fundamental linear algebra problems, 416
LAPACK and LAS, 416
large-scale linear algebra problems, 416
mixed precision iterative refinement (MPIR), 421–422
sparse linear systems and iterative solvers
iterative solvers and preconditioners, 419
representation of data, 417–419
Linear energy transfer (LET), 390
Linear time-invariant (LTI) systems, 169, 262
LLP prediction unit (LLPPU), 58, 59
LLVM compiler framework, 186
Load Value Approximation (LVA), 217
Locking, 330–331
Long latency paths (LLPs), 56
Look-Up Tables (LUTs), 119, 163, 225
Loop computation, 113
Loop Perforation (LP), 112–113, 192, 216
Loop perforation transformation, 186
Loop performation, 324
Loop-simplify, 186
Low energy, 12, 24, 40, 67, 390
Low power, 388, 389, 467, 472, 498
Low-precision computations, 11, 12

Low-precision floating-point arithmetic, 12
accuracy, 24, 25
artificial intelligence, 24
custom bias, 25
exponent, 24, 25
implicit bias, 25
low-energy benefits, 24
mantissa, 24
Low-precision training, 496–498
LSBs truncation, 56

M

Machine learning-based models, 221010
MapReduce algorithm, 115
Markov Chain Monte Carlo method, 226
Mathematical function approximation
iterative approaches, 121–122
mathematical libraries, 121
multivariate function, 123–124
polynomial approximation, 122–123
scientific computation, 121
table-based techniques, 122
Matrix-based mean ED determination, 159
Matrix-based method, 159
Maximal approximation error criterion, 124
Maximum Error Distance (maximum ED), 149
MC block approximation, 127
Mean Error Distance (mean ED), 157, 158, 160
Mean Square Error of the resulting cluster Centroids (CMSE), 35
Mean Square Minimization method, 124
Mean work to failure (MWTF), 392
Memoization, 119–120
Memoization technique, 217
Memory Controller Units (MCUs), 67
Metering, 330–331
Microprocessors, 270
MIMO decoding, 116
Minifloat representation, 83
Minimum Mean Square Error (MMSE), 124
Minimum uniform wordlength, 265
Minimum wordlength set, 265
Miter, 89
Mitigation approaches
application and system crashes, 53
corrected errors first, 52
nothing abnormal, 52
SDCs alone, 52–53
Mixed-criticality (MC) system
IMC/VPMC systems
global scheduling technique (*see* Global scheduling technique)
MC-DP-fair scheduling, 301–302
partitioned scheduling, 294
Mixed Integer Linear Programming (MILP), 265
Mixed precision iterative refinement (MPIR), 421–422
Mixed-precision training
mixed 8-16-bit precision training, 494–495
mixed 16-32-bit precision training, 490–494
Mobile devices, 349
Modified Interval/Affine Arithmetic (MIA/MAA), 162
asymmetric distributions, 167
error magnitude, 167
error propagation method, 167
inexact operator, 168
range explosion, 168
storage explosion, 168
total error probability, 167
Modified Interval Arithmetic (MIA), 151
Monte-Carlo simulation, 112, 160, 161, 163
Most Probable Mode (MPM), 136
Most significant bit (MSB), 56, 119, 122, 159, 262
Motion compensation (MC) filters, 118, 126
Motion estimation, 191
Multi-kernels system, 164
Multi Layer Perceptron (MLP), 120
Multi-objective optimization problem
constraints, 94
error metrics, 92
feasible solutions, 93
non-dominated solutions, 93, 94
objective functions, 93
optimization problem, 92
Pareto front, 93
suboptimal solutions, 93
worst-case error, 93
Multiple bit upset (MBU), 391
Multiple cell upset (MCU), 391
Multiple fault, 354
Multiple wordlength (MWL), 263
Multipliers, 46, 84
"Multiply and accumulate" (MAC), 81
Multivariate functions, 123–124
Mutants, 244

N

n-bit approximate additions, 46
$2n$-bit approximate multiplier, 97
N-bit floating-point, 30
Nearest Neighbor classifier, 116

Near-threshold computing (NTC), 54
Negative-feedback control system, 192
Network architecture search, 453, 454
Network On Chips (NOCs), 236
Neural network approximation, 120
Neural processing unit (NPU), 120
Neuromorphic computing, 325
Newton-Raphson method, 404
Noise aggregation, 169
Noise gains, 267
Noise generation, 169
Noise propagation, 169
Non-determinism, 324
Non-deterministic AxC techniques, 325
Non-dominated solutions, 93, 94
Normalized energy reduction, 139
Number representations, 83
Numerical accuracy, 145
N-version programming (NVP), 394

O

Observability Don't Care (ODC), 221
On-chip power consumption, 43
ON-set, 90
Operands, 45
Operand truncation, 45, 55–56
Operation and memory accuracy/reliability, 195
Operation transformations, 224
Operator characterization, 30
Operator refinement transformations, 476
Optimal non-uniform segmentation, 123
Optimization algorithms, 20
Optimization based on Search Space Exploration (OSSE), 116, 117
Optimization constraint construction
  configuration, 204
  final constraint, 204
  reliability expression, 204
  validity checking, 204
Original (exact) circuit, 83
OSSE Classification Step, 134
OSSE combination, 138
Out-of-order (OoO), 62
Output abstraction, 179
Output quality estimation metric, 63
Overflow occurrence, 18
Overscaled supply voltage, 160

P

Parallel ultra-low power (PULP) platform, 457
Parameter adjustment, 118–119
Parents, 236
Pareto dominance relation, 85
Pareto front, 93, 94, 139
Pareto-optimal perforations, 190
Pareto optimal solutions, 93, 254
Pareto sets construction, 190
Partial products, 96
Partial product tree, 96
Passive side channel attacks
  AxC, 327
  in cryptography, 327
  power analysis attacks, 326
  security evaluation, 327
  security threat, 326
  self-adaptation circuits, 327
  statistical tests and metrics, 327
  timing attacks, 326
Path delays $D(P)$, 60
Path redistribution, voltage down-scaling
  approximate-based strategies, 65
  comparison between strategies, 62–65
  LLPPU, 58
  path shaping, 61–62
  performance-centric design implementation, 59
  timing errors, 58
  timing properties of pipelined designs, 59–60
  timing wall phenomenon, 60–61
Path shaping, 61–62
Peak signal-to-noise ratio (PSNR), 88, 163, 190
Penalty fitness function, 245
Perceptual limitation, 4
Perforation space exploration, 188–190
Perturbation theory
  fixed-point format conversion, 168
  fixed-point systems, 168
  noise aggregation, 169
  noise propagation, 169
  power expression, 169
PetaBricks framework, 118, 232
Physical fault injection, 391–392
Pipeline stage processes, 59
Point of failure (PoF), 60
Polynomial approximation, 122–123
Polynomial coefficients, 123
Post-silicon technique, 55
Power *vs.* e_{mae} trade-offs, 99
Pre-characterization phase, 162–163
Precision-scaling, 55–56, 216, 324
Precondition, analysis constructs, 203
Precondition generator analysis, 196
Primary Input (PI), 221

Primary Output (PO), 221
Probabilistic analysis, 157–159
Probability density function (PDF), 19, 149
Probability error analysis methods, 85
Probability mass function (PMF), 146, 149
Problem-specific approximation methods
  approximate adders, 95
  approximate implementations, 94
  approximate multiplier, 95–97
  truncation, 94
Processing units, 44
Processor architecture, 2
Process variations, 51
Product of Reliability Factors, 198
Profiling-based optimization, 183
  accuracy specification, 185
  approximation knobs, 185
  loop perforation transformation, 186–187
  sensitivity profiling algorithm, 187–190
  SpeedPress framework, 185
  video encoder perforating, 190–192
Propagation/path sensitization, 356
Propagation process, 19
Pruning
  granularity, 486–487
  over-parameterized, 476, 485
  target regions, 486
  weight saliency determination, 487–489
PyTorch, 452, 453, 456, 472

Q
QEMUKVM hyper-visor, 69
QUAES (QUality assurance, Approximation, Estimation, and search space exploration), 229
Quality configurable circuits
  8-bit and 16-bit, 87
  2-bit multiplier, 86
  definition, 85
  error compensation support, 86
  properties, 87
  QCM, 86
  Synopsys DC, 86
Quality configurable multiplier (QCM), 86
Quality degradation, 63, 115
Quality Evaluation Circuit (QEC), 222
Quality/latency-aware task graph scheduling
  budgeting, 318
  budgeting formulation, 317–318
  Darknet deep learning framework, 318
  deployment results, 319–320
  distributed neural network applications, 319
  intersection over union (IoU), 318
  mapping heuristics, 316–317
  mean average precision (mAP), 318
  Network Delay Distribution (NDD), 313
  quality model, 314–316
  Raspberry Pi 3 devices, 318
  scheduling formulation, 313–314
  simulation results, 319
  YOLOv2 object detection network, 319
Quality metric determination
  analytical techniques, 163
  simulation-based techniques
    direct, 164
    error injection, 164–165
Quality of service (QoS), 151, 239
Quantization
  activation function inputs and outputs, 477
  BFP format, 480
  granularity level, 478
  network parameters, 477
  PTQ methods, 478
  QAT, 477
  quantized values, 481–482
  ternary neural networks, 479
  value distribution of the data, 481–482
  XNOR-Net approach, 479
Quantization-Aware Training (QAT), 477
Quantization error, 157
Quantization error power estimation, 171

R
Random perforation, 187
Rate-Distortion (RD), 131
Rate-Distortion Optimization (RDO), 116, 134, 135
Rate-Energy space, 135
REACT framework, 164
Real-time scheduling
  acceptance ratio *vs.* normalized utilization, 306
  audio/video applications, 288
  communication times, 287
  evaluation, acceptance ratio, 306–307
  evaluation of errors, 307–308
  high-criticality task, 305–306
  inherent error tolerance, 288
  Internet of Things (IoT), 288
  low-criticality task, 305–306
  mixed-criticality settings, 287
  mixed-criticality (MC) system, 289
    DP (Deadline Partition)-Fair, 292
    EDF-VD method, 293
    global scheduling, 291

Real-time scheduling (*cont.*)
high-criticality tasks, 289–290, 292, 293
low-criticality mode, 289–290, 292, 293
multiprocessor system, 289
partitioned scheduling, 290
Quality of Service (QoS), 292
uniprocessor scheduling, 290
network protocols, 288
optimization kernel, 302–303
single multiprocessor device, 288
software simulations, 305
system execution, 287
utilization slack estimation and customization, 303–305
Recognition, Mining, and Synthesis (RMS) applications, 349
Reduced Ordered Binary Decision Diagrams (ROBDDs), 89, 90, 361
Reduced-precision arithmetic, 11
Reduced-precision floating-point arithmetic
AC_FLOAT, 26
16-bit addition/subtraction, 28, 29
32-bit FlP addition/subtraction, 29
CT_FLOAT, 26, 27
exponent, 26
FLOPOCO, 26–28
half-precision, 28
hardware performance comparison process, 27, 28
library, 26, 27
mantissa, 25, 26
properties, 27, 28
rounding modes, 26, 27
single-precision, 28
subnormals, 27
synthesizable C++ libraries AC Datatypes, 25
VFLOAT, 26
Redundancy methods, 394
Redundant fault, 354
Reference design, 4
Reference implementation, 3
Register mapping, 200
Register transfer (RT), 83
Register-transfer level (RTL), 220, 353–354
Relative difference (RD), 101, 102
Relative energy model
ALU, 207
CPU, 207
cross-design parameters, 206
memory and cache, 207
multiple inputs, 207
system, 206
Relaxations, 237
Relaxed equivalence checking
analysis, 100
Boolean function, 90
CNF, 90
ROBDDs, 89
SAT, 90
worst-case error, 90
Reliability, 387, 389, 391, 401–404
Reliability constraint construction
constraint simplification
joint reliability factors ordering, 203
labeled reliabilities ordering, 203
subsumption property, 204
final precondition, 201–203
function, 198
initial postcondition, 199
precondition generator, 198
ALU/FPU operations, 199
array load/store, 201
bounded loops, 201
conditionals, 201
initialization and sequence, 198
scalar load/store, 199
Reliability factors, 198
Reliability predicates, 197–198
Reliability transformer relation, 198
Rely's simplification procedure, 203
Representative approach, 55
Representative inputs, 190
Residual error symbol, 166
Return instruction, 199
Reverse engineering, 328–329
Reward fitness function, 245
Ripple carry adder (RCA), 58, 160
(Ring) Learning With Errors (RLWE) problem, 334
RoBA multiplier, 46
Rosa, 210
Rough Mode Decision (RMD), 132
Rounding-based approximate (RoBA), 96

S
Safety, 192, 387
Safety analysis, 178, 210
Safety constraints, 209
Safety-critical systems
as aerospace and avionics applications, 387–388
approximate computing, 388
error analysis, 391–392
external perturbation, 390–392

fault injections, 389
fault tolerance technique, 393–396
Saliency determination, 487
SAO filter, 126
Satisfactory complexity-accuracy trade-off, 118
Satisfiability (SAT), 89, 90, 361, 365
Scalar store analysis, 200
Scaling operations, 19–20
Scheduling dependent tasks
approximations, task graph scheduling, 310–312
directed acyclic graphs (DAGs), 309
task graph system model, 309–310
Scientific applications
deep learning, 417 (*see also* Deep learning)
A^2DTWP framework, 458
linear algebra, 417 (*see also* Linear algebra)
Search/autotuning algorithm, 189
Search space enumeration, 111
SECDEC ECC protection, 53
Segmentation, 122, 123
Segmented adders, 95
Selective dynamic loop perforation, 113
Self-tuning Approximation for Graphic Engines (SAGE), 235–236
Sensitivity analysis, 210
Sensitivity profiling, 185
Sensitivity profiling algorithm
individual loops, 187–188
Pareto sets construction, 190
perforation space exploration, 188–190
Sensitivity profiling-based compilation, 182
Separate light-weight intelligent processor (SLIMpro), 67
Sequence of operations, 19
Severity function, 50
Shift-and-add BKM algorithm, 121
Short latency paths (SLPs), 46
Signal Flow Graph (SFG), 169
Signal-oriented blocks, 128
Signal-to-noise ratio (SNR), 151
Signal-to-Quantization-Noise Ratio (SQNR), 167
Significance driven design methodology, 49
Silent data corruption (SDCs), 51, 54, 388
Simulated annealing (SA), 268
Simulation-based fault injection, 392
Simulation-based techniques, 146
Single event effects (SEE), 390, 391
Single event transient (SET), 391
Single event upset (SEU), 391
Skip-based approaches
discrete optimization algorithm, 116–117
early termination, 114–115
granularity levels, 111
limitation, 111
loop perforation, 112–114
scope, 111–112
task skipping, 115–116
Skip control configurations, 128
Skip control parameter, 128
"Skipping" instructions, 3
Smart Search Space Reduction (SSSR), 117
Smooth operators, 270
"Smooth" severity, 50
Soft error rate (SER), 390, 391
Soft errors, 388
Software, 5
Software-centric compilation, 209
Software layer, 6
Software-only techniques, 324
Software-oriented techniques, 97
Soundness, 208
Special storage modules, 55
Speculative adders, 95
Speculative Carry Selection Adder (SCSA), 157
SRAM
adaptive technique, 66
approximations, 66
architectural schemes, 66
3D raytracer application, 66
error-correction capabilities, 66
error-pattern transformation scheme, 66
high-order bits, 66
microprocessors' energy efficiency, 66
output quality degradation, 66
power consumption, 66
supply voltage, 65
voltage noise elimination, 67
worst-case behaviour, 67
SRAM array failures, 51
SRAM supply voltage, 45
SSSR techniques, 117
Standard deviation, 161
Standard deviation of the error, 149
Static timing analysis (STA), 59
Static truncation, 63
Statistical error model, 156
Statistically Certified Approximate Logic Synthesis (SCALS), 225–226
Statistically oriented error metrics, 88
Statistical metrics, 148–149
Statistical parameters, 170
Stochastic computing (SC), 323, 325

Stochastic Gradient Descent (SGD) iterative optimization process, 470
Stochastic Gradient Descent (SGD) method, 431
Stopping conditions, 36, 114
Strat 1, 63
Structural Similarity Index Measure (SSIM), 151
Structure refinement transformations, 476
Stuck-at fault model (SaF), 354
Substitute-And-SIMplIfy (SASIMI), 221–222
Substitute Signal (SS), 221
Sum/argmin patterns, 111
Supercomputer, 5
Support for approximation in DNN accelerators
Switching activity, 44
Switching power, 43
Symbolic expressions, 188
Synopsis Design Compiler tool, 227
Systematic Methodology for Automatic Logic Synthesis of Approximate circuits (SALSA), 220–221
System quality degradation, 19
Systems-on-a-chip (SoC), 388
System unresponsiveness, 53

T
Table-based techniques, 122
Target approximation strategies, 63
Target Signal (TS), 221
Task processing, 115
Task skipping, 115–116
Ternary neural networks, 479
Testing-based sensitivity analysis, 210
Threshold, 101
Time-varying impulse response, 170
Timing error-induced quality degradation, 62
Timing errors, 54
Timing wall phenomenon, 60–61
Tost-Training Quantization (PTQ) methods, 478
Total energy, 36
Total ionizing dose (TID), 390
Total power consumption, 44
Tradeoff curve, 180
Tradeoff points, 187, 190
Traditional optimizations, 178
Traditional program optimization, 177
Transfer function, 169
Triple modular redundancy (TMR), 393, 394, 396
  ATMR (*see* Approximate TMR (ATMR))
  traditional TMR, 396
True-value simulator, 355
Truncation, 94
Truncation-and rounding-based scalable approximate multiplier (TOSAM), 96
Truncation-induced quality loss, 62
Truncation perforation, 187

U
Uncertainty, 170
Undervolting characterization studies, 53
Unified Modeling Language (UML) class diagram, 249
Uniform wordlength (UWL) approach, 263
Uninitialized memory, 187

V
Variable-Precision MC (VPMC) system, 293
Verilog Hardware Description Language (HDL), 224
Very large-scale integration (VLSI) testing, 351
Vmin characterization, 51, 52
Voltage down-scaling and hardware approximations
  application-specific processing cores, 48–49
  general-purpose processing cores, 49–53
Voltage-overscaled circuits, 160
Voltage over-scaling, 82
Voltage-scaled designs, 65
Voltage scaling
  delay uncertainty, 45
  reduced energy consumption, 45

W
Wallace tree, 96
WCEGT procedure, 100, 101
Weighted mean error distance, 88
Weight sharing
  granularity, 484–485
  grouping methods, 483–484
  shared values, 482
  techniques, 482
Widrow model, 146, 157
Wordlength optimization (WLO)
  accurate error models, 278
  architectural synthesis, 273–274
  comparing techniques, 274–275

complexity
fast error estimation, 270–271
multi-objective optimization, 271–272
non-integer wordlengths, 271
parallel hardware, 269–270
quantization, 269
signal grouping, 271
system partitioning, 272
cost estimation, 263
cost function, 278
cost of systems, 260
fixed-point refinement, 260–262
floating-point arithmetic implementations, 278
hierarchical techniques, 277
optimization function, 272–273
optimization techniques
categories, 264
classification, 264
local search, 266–267
non-integer wordlengths, 267–268
optimal approaches, 265–266
stochastic optimization, 268–269
techniques, 264
optimizer control block, 263
process, 263
quantization, 278
TABU search, 278
Wordlength propagation task, 270
Wordlength variations, 270
Worst-case absolute error computation, 91
Worst-case arithmetic error, 87
Worst-case error analysis, 91–92
Worst-case error checking, 101
Worst-Case Execution Time (WCET), 289

X
X-Gene 2, 51, 53

Y
Yield Increase Loss (YIL), 369

Z
Zurich Urban Micro Aerial Vehicle Dataset, 318

Printed in the United States
by Baker & Taylor Publisher Services